A BIOGRAPHICAL DICTIONARY OF PEOPLE IN ENGINEERING

From Earliest Records until 2000

A Biographical Dictionary of People in Engineering

From Earliest Records until 2000

CARL W. HALL

Purdue University Press
West Lafayette, Indiana

ISBN 978-1-55753-459-0

Library of Congress Cataloging-in-Publication Data

Hall, Carl W.
   A biographical dictionary of people in engineering from earliest
records until 2000 / Carl W. Hall.
     p.   cm.
Includes bibliographical references and index.
ISBN 978-1-55753-459-0 (alk. paper)
1. Engineers—Biography—Dictionaries. 2. Engineers—History.   I. Title.
TA139.H275 2007
620.0092'2—dc22
[B]
                                                    2007005257

CONTENTS

ACKNOWLEDGMENTS

Thank you to the following people whose helpful suggestions and valuable input helped make this book possible: Arthur E. Bergles, Wesley F. Buchele, Napoleon Curtis, George E. Ferguson, Donna Hull, James J. Pottmyer, William C. Salmon, George A. Schnebli I, Donald Swygert, and William A. Wulf.

Thank you also to the following libraries and their personnel: Arlington County Library, Arlington, Virginia; George Washington University, Gelman Library, Washington, DC; and the U.S. Library of Congress.

Thank you also for those people who put the manuscript in final form for publication at Purdue University Press: Margaret Hunt and Becki Corbin, and at BookMasters, Inc., Sharon Anderson.

FOREWORD

Most engineers, scientists, and technologists have seen and have had access to the well-known publications *Who's Who in America*, *Who's Who in Science and Technology*, *Who's Who in the World*, and other variations of these publications. They look at these references when they need to find concise information about people they are interested in referencing or citing, or they want preliminary and concise information about people they are interested in recruiting or approaching for whatever reasons. However, there is no publication of a similar nature that could be called "who-was-who" in engineering and invention. This void has been filled with Carl W. Hall's book.

This dictionary contains the necrologies up to year 2000 of people whose names appear in engineering and invention literature with names, dates, and a brief biography. The author has compiled the names and entries in this dictionary from histories, biographies, literature, and handbooks. The dictionary includes only people who, as the author indicates in the preface, can be considered as "distinguished, honored, leaders, outstanding, pioneers, prominent, recognized, renowned, and respected."

The necrologies cite the major employers and dates of employment and specialties listed and references to the sources of information. The entries include dates of events, employment, and birth/death of people, and field of expertise. Very useful to the reader is the inclusion of dates when a particular development occurred and when a person was employed. The reader will find references in upper-case letters at the end of each entry, which are given in full at the end of the book.

Owing to the diversity of the many individual people in the different entries, inventors are identified by a number of criteria including their patents. Engineers in turn are recognized by a number of criteria including scholarly study, professional societies, title granted by an employer and academic appointments.

The reader will find especially interesting the author's observations in the preface of the book about the meaning and origin of the words 'engineer' and 'engineering', as well as the words 'scientist' and 'inventor'. The listing will help the searcher recognize those people who can be considered notables, those who have made outstanding contributions, those who have had eminent careers, and those who are giants in engineering, science, and technology. The great inventors can be easily recognized and be given due tribute.

Who will benefit from this dictionary? Historians seeking references; curious, meticulous researchers seeking names of engineers, scientists, and inventors of yesteryear; authors seeking valuable references; and people who would like to have at arm's length a valuable dictionary-like, reference-like volume of all the many people who have made science, engineering, and technology what is today by referring to the time in history.

The reader will find useful lists of people included in the dictionary who have large numbers of patents and those who are Nobel Prize winners. The reader will be made aware of house names, frequently used today without attribution. Finally the dictionary contains an invaluable list of abbreviations and acronyms so common in many publications today without their meaning. This book is a welcome addition to the literature.

—Angel G. Jordan, University Professor Emeritus,
Provost Emeritus, Carnegie Mellon University

PREFACE

Advancements in a field of endeavor are the result of utilizing the information of the past, imagination, and creativity. Thus, the words attributed to Isaac Newton are appropriate: "If I have seen further than others, it is because I have stood on the shoulders of giants."

This is an apt phrase for the field of engineering and invention. The work and contributions of thousands of people from many countries, representing numerous fields of endeavor, over many centuries provide the basis of this text. The ancient cultures of people in China, Arabia, India, and Japan; the renaissance culture of the Greeks, Egyptians, Romans; and in addition the European cultures of the Russians, Germans, French and others, all have contributed to what is now called engineering. The terms 'engineer' and 'engineering' are rather recent even though in present-day literature we apply those terms to those in previous centuries as engineers who at that time were builders, architects, instrument makers, inventors, discoverers, and shop workers.

The precise source of the word 'engineer' is unknown except that it seems to come from the French word 'ingenieur.' Early in the seventeenth century the word appeared as 'enginier', then later as 'engineer.' In the sixteenth century the word was familiar in the Roman literature in the military—an engineer was one who contrives, designs, or invents. Writers document such use of the word as early as the fourteenth century (1380, 1387, 1420, 1592, 1602, 1605, 1611) according to the *Oxford English Dictionary* (1989). In the early years 'engineer' was primarily applied to military applications, and it was not until the eighteenth century that civil engineers, people who applied their skills to civilian applications, became distinguished from those applying their skills to military applications. Then followed the application to numerous fields of engineering as we know them today.

Many people have contributed to the field of engineering who are not engineers or who have not been educated as engineers. Many of the laws of nature have been developed by physicists, mathematicians, chemists and biologists as well as engineers. Many of the practitioners of engineering have as the basis of their expertise one of these 'disciplines' and/or their creativity in developing a device, design, product or process that became commercially viable. The creativity may have resulted in an innovation that often became a patent.

In the eighteenth century some people were identified as engineers who had no academic education or training, a rarity today, yet they contributed much to the field of engineering. Blacksmiths and machinists led the way in the mechanical arts, particularly in the nineteenth century (1800s) with academic programs developed to prepare people to work in those areas. One evidence of a contribution is a written paper, a patent, a product, or a process to meet specific needs. Often entirely new industries developed based on a patent, such as the sewing machine, harvesting equipment, wagons, buggies, automobiles, and airplanes from simple items based on one patent to a device or mechanism or machine based on several patents. The names of people who developed patentable products, processes, or designs often became the name of a company. Others, seeing the opportunity to develop a profitable business, although not engineers, also became associated with an industry. Their names became a part of engineering and are cited in the engineering literature. The dates associated with an invention or development are not precise—the literature varies in dates identified with a development, whether the development date is the exposure of an idea, a public disclosure, a patent, or date of manufacture is often not clear.

Theories, often developed by scientists and mathematicians, became the basis of engineering works, and their names follow them into the literature of engineers and inventors. Names such as Angstrom, Bessel, Boltzmann, Boyle, Charles, Descartes, Galileo, Joule, Laplace, Newton—are a few of these people who would not claim to be engineers but whose names have continued in the literature of engineering for many years. Likewise, numerous inventions by people other than engineers have become a part of much engineering literature. Edison, Ford, Morse, and Wright appear in engineering literature but who would probably not claim to be engineers.

This book contains the necrologies (names, dates, and a brief biography) up to the year 2000 of people whose names appear in engineering and invention literature. The names were culled from histories, biographies, literature, and handbooks, particularly from those reference titles including words such as: distinguished, honored, leaders, outstanding, pioneers, prominent, recognized, renowned, and respected. The necrologies are necessarily abbreviated with the major employers and dates of employment and specialties listed and references to the sources of information.

An explanation of the methods used by the author is in order. A special effort has been made to include dates of events, employment and birth/death of people involved. An entry consists of a person's name, dates of birth and death, nationality and field of expertise. Many people are born in one country and then move to another country for their work. Their nationality is also designated. A Black American inventor is designated as such rather than as being an African American unless born in Africa.

A parenthesis enclosing dates follows the time at which a development occurred and when a person was employed. Brackets enclose dates of birth and death of other people related to the entry, embedded in the principal entry. These people are listed in the index so their contribution is recognized. References vary with regard to the date of a finding. The date may refer to first disclosure in a meeting, paper presented, article published or patent issued. Such dates are often not well prescribed.

References in upper case letters are given in full at the end of each entry. Complete references are given at the back of the book. The number following the reference initials is the volume and in some cases the year of publication. The parenthesis in the text gives the date of the event or activity.

Inventors are usually, but not always, identified by the patents issued or financial returns on a creative contribution. Engineers are recognized by scholarly study, professional societies, academic appointments, literature written by or about a person, title granted by an employer, registration by the state or country in which they practice, and even by self-identification. The variety of means by which a person may be an engineer is represented by the perceptive statement:

"A poet knows he is a poet,
A farmer knows he is a farmer;
But in the years following the Republic (USA)
There were no words in the American vocabulary to describe
What later came to be called the engineer, the scientist, the inventor."

—Mitchell Wilson, *American Science and Invention,* p. 38

ABBREVIATIONS AND ACRONYMS

a.	age
AAAS	American Association for the Advancement of Science
ABC	American Broadcasting Company
ABM	antiballistic missile
AC	alternating current (electricity)
AD	after death of Christ (anno Domini)
AEC	Atomic Energy Commission
AEF	American Expeditionary Forces
AEG	Allgemeine Elektrizitats Gesellschaft
AIChE	American Institute of Chemical Engineers
AIME	American Institute of Mining, Metallurgical, and Petroleum Engineers
AK	Alaska
aka	also known as
AL	Alabama
Alcoa, ALCOA	Aluminum Company of America
ALGOL	algorithm-oriented language (computer)
AM	amplitude modulation
ANSI	American National Standards Institute
AR	Arkansas
ARPA	Advanced Research Project Agency (of DOD)
ASAE	American Society of Agricultural Engineers
ASCE	American Society of Civil Engineers
ASEE	American Society for Engineering Education
ASHRACE	American Society of Heating, Refrigerating, and Air Conditioning Engineers
ASME	American Society of Mechanical Engineers
Ass., Assoc.	Association, Associates
AST	atomized suspension technique
ASTM	American Society for Testing Materials
AT&T	American Telephone and Telegraph Company
ATM	automatic teller machine
AZ	Arizona
b. or dob	born, date of birth
BART	Bay Area Rapid Transit (system)

BBC	British Broadcasting Corporation
BC	before Christ, British Columbia (Canada)
BG	brigadier general
BTL	Bell Telephone Laboratory
C.	Celsius or centigrade temperature
c.	circa (approximately)
ca.	circa (about)
CA	California
CAA	Civilian Air Authority
CAD	computer-aided design
CAT	computerized axial tomography (CAT scan)
CATV	community antenna television
CBS	Columbia Broadcasting System, Inc. (US)
CCC	Civilian Conservation Corps
CEO	chief executive officer
cf.	compare
CIA	Counter Intelligence Agency (US)
CIGR	International Commission of Agricultural Engineering
CNRS	Centre National de la Recherche Scientifique (France)
CO	Colorado
Co. or co.	company
COBOL	common business oriented language (computer)
COMSAT	Communications Satellite Corporation (DC) (US)
Corp.	Corporation
CSA	Confederate States of America
CSIRO	Commonwealth Scientific and Industrial Research Organization (Australia)
CT	Connecticut
d. or dod	d. or date of death
DC	direct current (electricity)
DC	District of Columbia
DE	Delaware
Delco	Dayton Electric Company
Dept.	Department
DOD	Department of Defense
DOE	Department of Energy
DSIR	Department of Scientific and Industrial Research (UK)
ed.	editor, edition
ele.	elevation (as above sea level)
eng.	engineering

engr	engineer
ENIAC	electronic numerical integrator and compiler
EPA	Environmental Protection Agency (US)
ERDA	energy research and development administration (US)
est.	established (such as date or name of)
etc.	et cetera (and so forth)
F.	Fahrenheit
FDA	Food and Drug Administration
ff	(and) following
fl.	flourished (in time period)
FL	Florida
FM	frequency modulated
fNAE	foreign member of NAE
fNAS	foreign member of NAS
fRS	foreign member of Royal Society of London
FRS	fellow of Royal Society of London
GA	Georgia
Gen.	general
GMC	General Motors Corporation
HEW	(Department of) Health, Education, and Welfare (US)
HI	Hawaii
hp, ihp	horsepower, indicated horsepower
HVAC	heating, ventilating, and air conditioning
HVDC	high voltage direct current
IA	Iowa
IANA	Internet Assigned Number Authority
IBM	International Business Machines Corporation
ICAN	Internet Corporation for Assigned Names and Numbers (was IANA)
ICBM	intercontinental ballistic missile
ICI	Imperial Chemical Industries
ID	Idaho
IEEE	Institute of Electrical and Electronics Engineers
IHC	International Harvester Company
IL	Illinois
IN	Indiana
Inst.	Institute, institution
ISBN	International Standard Book Number
ISO	International Standards Organization

ISSN	International Standard Serial Number
IT&T	International Telephone and Telegraph Company
JPL	Jet Propulsion Laboratory
KGB	Committee of State Security (formerly USSR)
KS	Kansas
KY	Kentucky
LA	Louisiana
lab., labs.	laboratory, laboratories
LASER	light amplification by stimulated emission of radiation
LC	Library of Congress
LCI	landing craft infantry
LG	lieutenant general
LSI	large scale integration
LST	landing ship (for) tanks (and personnel)
MA	Massachusetts
MANIAC	mathematical and numerical integrated computer
MASER	microwave amplification by stimulated emission of radiation
MD	Maryland
ME	Maine
MG	major general
MI	Michigan
MIT	Massachusetts Institute of Technology
MN	Minnesota
MO	Missouri
MS	Mississippi
MT	Montana
NACA	National Advisory Committee on Aeronautics (founded 1915)
NAE	National Academy of Engineering (est. 1964)
NAS	National Academy of Sciences (chartered by Congress [1863])
NASA	National Aeronautics and Space Administration (1958)
NATO	North American Treaty Organization
NBC	National Broadcasting Corporation
NBS	National Bureau of Standards
NC	North Carolina
ND	North Dakota
NDRC	National Defense Research Council
NDT	nondestructive testing

NE	Nebraska
NELA	National Electric Lamp Association
NH	New Hampshire
NIH	National Institutes of Health
NIST	National Institute of Standards and Technology
NJ	New Jersey
NM	New Mexico
NOAA	National Oceanographic and Atmospheric Administration
NOL	National Ordnance Laboratory
NPL	National Physical Laboratory (UK)
NRC	National Research Council of National Academies (organized [1916])
NRC	Nuclear Regulatory Commission
NSA	National Security Agency (US)
NSF	National Science Foundation
NUC	National Union Catalogue
NV	Nevada
NY, NYC	New York, New York City
OH	Ohio
OK	Oklahoma
ONR	Office of Naval Research
OPA	Office of Price Administration
OR	Oregon
OSHA	Occupational Safety and Health Administration
OSRD	Office of Scientific Research and Development
OSS	Office of Strategic Services (preceded formation of CIA)
OTA	Office of Technology Assessment
PA	Pennsylvania
pass.	for passim, used throughout text
pat.	patents, patented
PBS	Public Broadcasting Service (Company)
PCB	polychlorinated biphenyl
PET	positron emission tomography
PET	poly(ethylene terephthalate)
Ph. D.	Doctor of Philosophy
PR	Puerto Rico
pres.	president
PRR	Pennsylvania Railroad
PT boat	Patrol torpedo boat
pub.	published

radar	radio direction and ranging
R&D	research and development
RCA	Radio Corporation of America
REA	Rural Electrification Administration
RFC	Reconstruction Finance Corporation
RI	Rhode Island
RR	Railroad, Railway
RS	Royal Society of London
SAM	surface-to-air missile
SC	South Carolina
SCS	Soil Conservation Service
SD	South Dakota
sonar	sound navigation and ranging
SRI	Stanford Research Institute
TAMS	Tibbetts-Abbett-McCarthy-Stratton
TBPi	Tau Beta Pi, an engineering scholastic honorary
TN	Tennessee
TNT	trinitrotoluene
TRW	Thompson, Ramo, Woodridge
TVA	Tennessee Valley Authority
TX	Texas
UCLA	University of California–Los Angeles
UK	United Kingdom
UN	United Nations
univ.	university
U S, US, USA	United States of America
USA	United States Army
USAAF	United States Army Air Force
USAF	United States Air Force
USBM	United States Bureau of Mines
USBR	United States Bureau of Reclamation (or BOR)
USCG	United States Coast Guard
USCS	United States Coast Survey
USDA	United States Department of Agriculture
USDC	United States Department of Commerce
USDI	United States Department of the Interior
USGS	United States Geological Survey
USMA	United States Military Academy
USMC	United States Marine Corps

USN, USNR	United States Navy, United States Navy Reserve
USNA	United States Naval Academy
USPH, USPHS	United States Public Health Service
USPS	United States Postal Service
USSR	Union of Soviet Socialist Republics
UT	Utah
VA	Virginia
VDI	Verein Deutscher Ingenieur
VHF	very high frequency
vp, VP	vice president
VT	Vermont
VTO	vertical take off
VTOL	vertical take off and landing (airplane)
WA	Washington
WHO	World Health Organization
WI	Wisconsin
WPA	Work Projects Administration
WTO	World Trade Organization
WV	West Virginia
WWI	World War I
WWII	World War II
WY	Wyoming

BIOGRAPHIES A–Z

A

ABBE, Ernst Karl, 1840–1905, German physicist, inventor and industrialist; professor at Univ. of Jena (1870), became director of university's astronomical observatory (1871ff), research director for Carl Zeiss [1816–1888] (1866) optical company, partner (1875ff), became president of Zeiss upon the death of Carl Zeiss, made improvements in optics and instruments, called "the first optics engineer" (CSOSB GMI IAI MWBD SAI WWWIS)

ABBOT, Arthur Vaughn, 1854–1906, American mechanical engineer; bridge cable construction, rod-coupling (1877), testing machine, central station heating system, telephone construction (1892), electric railroad (MEIA)

ABBOT, Henry Larcom, 1831–1927, American soldier and civil engineer; railway surveys, water control and flood protection, channel improvements, levees, hydraulics, US Corps of Engineers, NAS (1872) (ANB BDOACE BDOAS BM13 DAB MWBD PS)

ABEL, Frederick Augustus, 1827–1902, British chemist and inventor; explosives, gun-cotton, cordite (with James Dewar [1842–1923]), Abel instrument for determining the flash point of petroleum (DNB RHW WWWIS)

ABELL, Sir Westcott Stile, 1877–1961, English naval architect; ship surveyor, designed ferries, author (DNB MWBD)

ABERT, John James, 1788–1863, American civil and topographical engineer; geodesic survey of Atlantic coast, harbor and river improvements, maps and defenses, commander of Corps of Topographical Engineers (1838–1861) (ANB BDOACE BDOAS WWWIA)

ABNEY, Sir William de Wiveleslier, 1843–1920, English chemist and inventor; photographic emulsion used to map solar spectrum, photography with Army Royal Engineers (ante 1877), School of Military Education, innovations in photography including photochemistry and spectral analysis, invented dry photographic emulsion (1874) (BDOTHOT CDOSB DNB EB RS WWWIS)

ACHESON, Edward Goodrich, 1856–1931, American inventor; discovered silicon carbide (carborundum) (1891), graphite, developed carbon products (ANB BEST MWBD NC23 WWWIS)

ACKER, Charles Ernest, 1868–1920, American electrical engineer, inventor and manufacturer; electrochemistry producing caustic soda and chlorine from molten salt (1896), Acker process, electro-metallurgy, production of alkali metals (1899), halogen gas process (1901), nitrides (1909), obtained over 40 patents (DAB NC13 WWWIA)

ACKERMANN, Rudolph, 1764–1834, German inventor and art publisher; print shop, art lithography, invented method for waterproofing paper and cloth (1801) (MWBD)

ACKERMANN, William Carl, 1913–1988, American civil engineer; water survey, river forecaster, field engineer, hydraulics, hydrology, sedimentation, meteorology (AMOS MT4 WWWIS)

ACKERT, Jakob, 1898–1981, Swiss mechanical and aeronautical engineer; aerodynamics, fluid mechanics, high speed and supersonic flow, at Univ. of Göttingen (1921–1927), Max Planck Institute for Fluid Mechanics research and first director (1925–1928), chief engineer at Escher Wyss Ltd. in Switzerland (1928–1932), developed Ackert-Keller turbine, professor at Federal Inst. of Technology in Zurich (1932–1967), where he founded and was director at Inst. of Aerodynamics (1934ff), fNAE (1976) (MT8)

ACKROYD-STUART—see STUART

ADAIR, John, c.1655–c.1772, Scottish surveyor and cartographer; map making, west country of Scotland, published map books (1680–1686) (EB MWBD)

ADAMS, Comfort Avery, 1868–1958, American mechanical and electrical engineer; studied movement of glaciers (AL), an ice sheet named after him (1890), worked on design of cantilever cranes, large direct current generators, gearless traction motor for industries in Cleveland (OH) area, on faculty at Harvard Univ. (1891ff) for 45 yr. during which he served as dean of engineering (1919), consulting engineer, NAS (1930) (ANB BM38 NCA PS TNAOS WWWIA)

ADAMS, Frederick Upham, 1859–1921, American inventor and mechanical engineer; machine designer, city illumination, railroad roadbeds, aviation, bridge, wrote extensively on relationship of science and society (DAB MEIA NC14 WWWIA)

ADAMS, Isaac, Sr., 1802–1883, American inventor; printing machinery, Adams Power Printing Press (1828–1834), cabinet making, continued development of power printing press (1836ff) (DAB MEIA NC9 WWWIA)

ADAMS, Isaac, Jr., 1836–1911, American industrial chemist; discovered and promoted metal plating, patents on process (1869), formed Boston Nickel Plating Co. (1869), became a glass blower making Geissler tubes (1865–1868) (ACCE IIA)

ADAMS, Julius Walker, 1812–1899, American civil engineer; assistant engineer and engineer of various railroad companies, Brooklyn Navy Yard (1844), editor of *Mechanics' Magazine* (1851), sewer systems and waterworks, consulting engineer for NYC Department of Public Works (1880–1886) (BDOACE NC9 WWWIA)

ADER, Clément, 1841–1926, French engineer and inventor; microphone, stereophonic sound transmission by telephone (1881), built steam-powered airplane called Éole that could take off but not fly successfully (1890) (GI RHW MWBD)

ADLER, Charles, Jr., 1899–1980, American engineer and inventor; invented electric automobile brake (1913), automatic highway-railroad crossing signal (1920), other traffic light signals, including sound-activated safety systems for railroads, highways, airplanes, coordinated transportation systems (1964), 60 patents (ANB WWWIA)

AENEAS THE TACTICIAN (aka TACTICUS), c. 360 BC, Greek general; military science, battlements, incendiary devices, defense against assaults, cryptography—history and methods (EB TEE)

AFFEL, Herman Andrew, 1893–1972, American electrical engineer and inventor; communications engineer (1916–1931), with Lloyd Espenschied [b. 1889] invented co-axial cable system (1929), engineer and executive at Bell Labs. (1931–1958), consulting engineer, numerous community activities (WA WWWIA WWWIS)

AGOSHKOV, Mikhail Ivanich, 1905–??, Soviet mining authority; exploration of ore deposits, mine productivity, USSR Inst. of Mining (1941ff), becoming deputy director (1952), officer of USSR Academy of Sciences (1962) (SMOS)

AGRICOLA, Georgius (aka Georg Bauer), 1494–1555, German mining engineer and metallurgist; initially a physician, underground mining works, assaying, nature of ore deposits, metal extraction, smelting, book *De re metallica* (1556) translated by Herbert Hoover [1874–1964] and his wife (1912) (BDOS CEOS&T EB GI HIW MISAT MMAH PS SAI TAE TBDOS TEE TGS TSOE WWWIS)

AGRIPPA, Marcus Vipsanius, 63–12 BC, Roman engineer and builder; top aide to Caesar Augustus [63 BC–14 AD], collapsible tower for troops, grapnel, anchors, aqueduct (20 BC), public works (ME125 TAE WWWIS)

AIKEN, Howard Hathaway, 1900–1973, American mathematician and computer engineer; communication engineering, at Univ. of Chicago, at Harvard Univ., with others developed Mark I and II computers, applied mathematician, industry with IBM, electromagnetic relays, automatic sequence controlled calculator (1944) (MMAH NC60 NT-CS PS RHW TBDOS WA WWWIA)

AILLERT, Pierre M., d. 1996, age 96, French electrical engineer; electric power generation and distribution, Technical Union of Electricity of France, fNAE (1977) (Bridge)

AIRD, Sir John, 1833–1911, British contractor and builder; reservoirs, canals, docks, water works, worked with Benjamin Baker [1840–1907] (DNB MWBD WWWIS)

AKELEY, Carl Ethan, 1864–1926, American inventor and naturalist; developed taxidermy called Akeley Method, contributed to museums, developed Akeley Camera, concrete and searchlight development during World War I (ANB DAB NC26 WWWIA)

ALBAUGH, Frederic William, 1913–1999, American physical chemist and engineer; application of nuclear energy to civilian and defense, storage of nuclear wastes, petroleum and plutonium, research and development, management, petroleum industry (1935–1943) (1945–1947), during WWII with Manhattan Project (1943–1945), research director at General Electric-Hanford (WA) and Battelle Northwest Labs. (WA) (1947–1976), consultant (1976–1987), NAE (1978) (MT10 WWIE WWWIA)

ALBERTI, Leone Battista, 1404–1472, Roman engineer and architect; camera obscuras, building bridges, cartography, urban design, author (EB TAE WWWIS)

AL-BIRUNI, Abu Rayhan Muhammad Ibn Ahmad, 973–1050, Persian (Iranian) engineer and astronomer; developed gears for astrolabe (orrey), clock (c. 1000) (AITBF GI TAE WWWIS)

ALBONE, Daniel, c. 1860–1906, British engineer and inventor; machines, cycle maker and repairer, axle bearings, funded Ivel Agricultural Motors (1902) and built first prototype of tractor (1902) called Ivel tractor, included power-take-off pulley on tractor, built fire engine, armored vehicles (1904) (BDOTHOT GI TBOF TOIAD)

ALCOCK, Sir John William, 1892–1919, English aviator; flew over Atlantic with Arthur W. Brown (1919), test pilot for Vickers Ltd. (DNB MMAH MWBD)

ALDEN, George I., 1843–1926, American mechanical engineer; steam, hydraulics, abrasives, faculty at Worcester Polytechnic Inst. (1868–1896) as head of department, gave Alden Hydraulic Lab. to WPI, chairman of board of Norton Co. (MEIA NC20 WWWIA)

ALDEN, John Ferris, 1852–1917, American civil engineer; bridge-building, iron and steel construction of buildings and other metal structures (DAB)

ALEKSANDROV, Boris Kapitonovich, 1889–??, Soviet hydraulic engineer; design of Moscow canal, building of hydroelectric plants, chief engineer of Great Volga Directorate for planning hydroelectric plants (1939), institute researcher and teacher (1946ff) (SMOS WWWIS)

ALEKSEEV, Aleksandr Yemelyanovich, 1891–??, Soviet electrical engineer; electric power plant (1908–1919), institute of electromachines (1953), electric traction with AC and DC, rail welding machine, institute professor (PS SMOS WWWIS)

ALEXANDER, Archie Alphonso, 1888–1958, Black American civil engineer; engineering construction, formed engineering firm with partner George F. Higbee, built bridges, viaducts, sewage system in Grand Rapids, a civilian airport at Tuskegee, apartment buildings, several projects in Washington, DC, territorial governor of Virgin Islands (1954) (ANB BCST BIS CBB NC49 PS)

ALEXANDER, John H., 1812–1867, American engineer and geologist; railroad development, coal mining operations, geological survey (MD) (1834ff), coastal survey, standardization of weights and measures, NAS (1863) (BDOAS BM1 DAB NC9 PS WWWIS)

ALEXANDERSON, Ernst Frederik Werner, 1878–1975, Swedish American electrical engineer and inventor; to US (1901), cascade tuning radio receiver (1913), radio transmission triode modulation (1914), high frequency alternator, antennas, tuning, color television, with General Electric Co. (1902–1948), RCA (1952ff), 322 patents (ANB HOCS&T MWBD NC62 PS WA WWWIA)

ALEXANDROV, Boris Kapitonovich, 1889–??, Soviet hydraulic engineer; helped design Moscow canal, planned building of hydroelectric plants, chief engineer and head of Greater Volga directorate for planning hydroelectric plants (1939ff), government and academic work, professor at Moscow Inst. of Energetics (1948) (PS SMOS)

ALFREY, Turner, Jr., 1918–1981, American research scientist, chemical and metallurgical engineer; plastics, mechanical behavior of polymers, polymerization rheology, kinetics, research chemist with Monsanto Chemical Co. and faculty of Polytechnic Inst. of Brooklyn, Dow Chemical Co. (1950ff), retiring as research scientist, 24 US patents, NAE (1977) (MT2 WWWIA)

ALGER, Cyrus, 1781–1856, American industrialist, inventor and metallurgist; iron master, iron foundry (1809), cylinder stove (1822), rifled gun barrel, cannon balls during war (1812), sea wall construction, malleable cast iron plow (1837), cast iron processing (ANB DAB MEIA NC6 MWBD PS WWWIA)

AL-JAZARI, Ismail, 13th century, Persian mechanician; clockmaker, water wheels (HIW TAE WWWIS)

ALLAIRE, James Peter, 1785–1858, American mechanical engineer; brass foundry (NYC) (1813), oldest steam engine works in US as Allaire Works, patented steam boiler (1828) (MEIA NC24 WWWIA)

ALLAIS, Maurice Félix, 1911–??, French economist and engineer; with French government dealing with mining and mining statistics, faculty at Univ. of Paris (1948ff), Allais paradox (1952), game theory, rationality in risky situations (DOT WWWIS)

ALLEN, Anthony Benezet, 1802–1892, American farmer, inventor, manufacturer and dealer; farm machinery, patents related to improvements in mowers (1860), reapers and harvesters (1864), plows (cast iron), arrows, hoes, shovels, corn sheller (1862) (DAB MEIA WWWIA)

ALLEN, Harry Julian, 1910–1977, American aeronautical engineer; contributed to understanding of subsonic to hypersonic flow, ballistic missiles and effect of radiation on space vehicles, reentry vehicles, Langley Memorial Aeronautical Lab. (VA) (1936–1940), Ames Research Lab. of NACA (later NASA) (1940–1969), through various positions becoming director of the Ames Research Center (MT1 WWWIA)

ALLEN, Herbert, 1907–1990, American mechanical engineer and inventor; oil field equipment, forging and extrusion of high-strength metals, technology and equipment for safety and efficiency of oil well drilling, 150 patents, consultant, NAE (1979) (MT5 WWIE WWWIA)

ALLEN, Horatio, 1802–1899, American civil and mechanical engineer and inventor; canal construction, resident engineer, steam locomotive, swiveling truck for locomotive (1831–1832), railway construction, bridges, steam fire engines, consulting, manufacturing, executive (ANB DAB MEIA MWBD NC8)

ALLEN, Jeremiah Mervin, 1833–1903, American consulting engineer; railroad construction and operation, boiler explosions, boiler insurance, active in engineering, business and civic organizations (DAB NC5 WWWIA)

ALLEN, John Ferris, 1852–1917, English American civil and mechanical engineer and inventor; to US (1841), built freight boats, use of Corliss engine, Porter-Allen high-speed engine with governor control, pneumatic riveting, air compressors, with Charles T. Porter [1826–1910] invented water tube boiler (DAB EE MEIA WWWIA)

ALLEN, Zachariah, 1795–1882, American engineer and inventor; controlled hot air house heating, steam engine control (automatic steam engine cut-off using centrifugal ball governor) (1834), city water works, mechanics, fire engine and equipment, textile manufacturing machinery, leather belt power transmission (ANB DAB IIA MEIA MWBD NC8 PS WWWIA)

ALLISON, Robert, 1827–1913, American mechanical engineer and inventor; boring machine (1866), Allison Cataract Steam Pump (1868), rock drill (1877), rotary drill (1882) (MEIA WWWIA)

ALMOND, Thomas Richard, 1846–1905?, English American mechanical engineer and inventor; to US (1866), drill chuck (1876), gas burner for heating (1877), right-angle coupling for shaft (1881), flexible coupling (1895), reaction engine (1905) (MEIA)

ALMY, Darwin, 1848–1916, American farmer, fisherman and mechanical engineer; fishing steamer, steam yachts, steam power (engines and boilers), Almy boiler (1890) (MEIA)

AL-NAWBAKHT, c. 700–800, Persian engineer; city planning, moats, dikes (TAE)

ALTER, David, 1807–1881, American inventor and physicist; spectroscopy, electric clock, Josef von Fraunhofer [1787–1826] lines of solar spectrum (TBDOS WWWIA WWWIS)

ALVAREZ, Luis Walter, 1911–1988, American physicist and inventor; design of radar systems, proton linear accelerator, liquid hydrogen bubble chamber, 25 patents, NAS (1947), Nobel Prize (1968), NAE (1969) (ANB BEST HIW HOCS&T MT5 NYT-A S:TLAW TNAS WWIE WWWIA WWWIS)

AEINOKLES of Corinth, c. 700 BC, Greek industry shipbuilder (TAE)

AMICI, Giovanni Battista, 1786–1863, Italian optician and inventor; invented Amici microscope, designed parabolic mirrors for reflecting telescope (CEOS&T)

AMIRASLANOV, Ali Agamaly Ogly, 1900–1962, Soviet engineer; work dealt with non-ferrous and rare metal deposits, chief engineer (1939–1947) then director (1948–1953) at Inst. of Mineral Raw Materials, taught at Moscow Geological Survey Inst. (1931–1955) (PS SMOS WWWIS)

AMIRIKAN, Arsham, 1899–1990, American structural engineer; improved methods of structural analysis, framing, construction techniques for US Navy, shore facilities, consultant, NAE (1980) (MT5 WWIE WWWIA)

AMMANATI, Bartolommeo (aka Battiforri de Settignano), 1511–1592, Florentine (Italian) architect, engineer and sculptor; bridge over Arno River called Ponte della Trinità (Florence), Neptune fountain, Piazza della Signoria, finished Pitti Palace, begun by Filippo Brunelleschi [1377–1446] (EB MWGD TSOE)

AMMANN, Othmar Hermann, 1879–1965, Swiss American civil engineer; steel bridges, New York Port Authority, chief engineer of George Washington Bridge (NYC-NJ) (1927–1931), consulting engineer, NAE (1965) (AMOS ANB FEL MT1 MWBD NCD,52 NT-CS PS WWWIA WWWIS)

AMMEN, Daniel, 1819–1898, American naval officer and inventor; exploration and mapping, coastal mapping, navigator of several ocean ship missions, commander for several Civil War attacks, signaling and deep sea sounding, US representative on several missions, designed twin-screwed ramming vessel, developed lifeboats, rear admiral (1877), author (ANB NC4 WWWIA)

AMONTONS, Guillaume, 1663–1705, French physicist and inventor; study of friction, instrumentation, thermometry, hygrometer (1687), perfected barometer (1695), gas relationships (1699), air pressure thermometer (1702) (BEST MISAT MWBD NYPL WWWIS)

AMPÈRE, André Marie, 1775–1836, French physicist and inventor; electromagnets, mathematical analysis, electrostatics, experimental physics, invented galvanometer (1824) (BEST EB IIA MISAT NYPL 1000Y SAI SAI-MP TBDOS WWWIS)

AMSLER-LAFFON, Jakob, 1823–1912, Swiss mathematical physicist, inventor and manufacturer; established factory to make scientific instruments and tools, tool for measuring area inside a curve, polar planimeter (1854) (TBDOS WWWIS)

ANDERSON, Arthur R., 1910–1995, American civil engineer; pre-stressed concrete, design and construction of engineering structures, consulting engineer, executive of ABM, NAE (1977) (WWIE WWWIA)

ANDERSON, Carl David, 1905–1991, American physicist and engineer; X-ray photoelectrons, research on gamma rays, cosmic rays (1930ff), faculty and department chairman at California Inst. of Technology (1927ff), Tau Beta Pi, discovered positron (1932), Nobel Prize (1936), NAS (1938) (AMOS BEST BM73 NCI TNAS WWWIA WWWIS)

ANDREWS, George Leonard, 1828–1899, American civil engineer and soldier; graduated from USMA (1851), private practice (1855–1861), fort construction, service in Civil War as brigadier general commanding US volunteers, engineer at Amoskeag Manufacturing Co. (OH), instructor at USMA (1871–1882) (ANB DAB WWWIA)

ANGELINI, Arnoldo M., 1909–1999, Italian electrical and nuclear engineer; nuclear generating stations, hydro and pumped hydro stations, electric transmission and distribution systems, consultant, fNAE (1976) (WWIE)

ANGELL, William Gorham, 1811–1870, American inventor and manufacturer; reed-making, screws, machinery for making screws with American Screw Co. (1858), which was preceded by Eagle Screw Co. (1838), company executive (DAB NC2 WWWIA)

ÅNGSTRÖM, Anders Jonas, 1814–1874, Swedish physicist; studied spectra of flames, one of founders of field of spectroscopy, studied and developed spectral analysis of sun (1861ff), Ångström law (1861), studied heat and magnetism, faculty of Uppsala Univ. (1839ff), Uppsala Observatory (1843ff), Ångström unit of measure (a ten billionth of a m.) named after him (1905) (BEST CDOS L&M MWBD RHW WWWIS)

ANKLITZEN, Konstantin (aka Friar Berchtholdus, aka Berthold Schwarz), 14th century, German monk and inventor; although people in China are often credited with inventing the cannon (1260) and the first cannon was with bamboo (1288), the first cast bronze cannon and gunpowder credited to Berthold Schwarz (1313) (EB GI MWBD S&TF TAE)

ANONYMUS (Lat.), 4th century nameless Roman mechanical inventor; recommended to Emperor Valentinian I (370 AD) that he develop a mechanized army, weapons of war: siege engine, catapult improvement, javelin (TAE)

ANSCHÜTZ-KAEMPFE, Hermann, 1872–1931, German engineer and inventor; worked in German shipyards, invented first practical gyroscope (1908) especially useful for submarines and applied to ships, aircraft, rockets and missiles, first work done by Frenchman Jean-Bernard-Léon Foucault [1819–1868] (1852) and later by American Elmer A. Sperry [1860–1930] (1907) (GSA MWBD)

ANTHEMIOS of Tralles, fl. 500s, d. c. 534, Greek architect, engineer and builder; building construction, hot-air engine, architectural monuments, public works (DOSB GEAPIT TAE WWWIS)

ANTHONY, Gardner Chace, 1856–1937, American mechanical engineer; draftsman, designer of engines, Harris-Corliss Engine Works, boilers, machines for automatic bolt making and nut tapping, threading machinery, faculty and dean of engineering at Tufts College (1898–1927) (MEIA WWWIA)

ANTHONY, William Arnold, 1835–1908, American physicist and electrical engineer; professor of physics (1872–1887), electrical engineer at Mather Electrical Co. (1887–1892), consulting engineer (1892–1894), electrical testing devices, on faculties at Cooper Univ., Columbia Univ. and Cornell Univ. (AMOS BDOAS NC11 WWWIA)

APOLLONIOS of Perga, c. 262–190 BC, Greek mathematician; established properties of conics, introduced names of parabola, ellipse, hyperbola, invented system of tetrads for expressing large numbers (BEST CDOS DOSB EB EOM GI MWBD RHW WOM WWWIS)

APPERT, Nicholas-François, c. 1752–1841, French inventor and food manufacturer; food preservation, thermal stabilization of food, canning pioneer (1804, 1810), bouillon tablet, autoclave, built first successful commercial cannery (1812), at same time King George III of England granted a patent to Englishman Peter Durand [fl. early 1800s] for the same process, who sold the process to Bryan Donkin [1768–1855] and John Hall [fl. 1800s]. (BEST DOTHOS EB GI GLOSB MISAT MWBD NYPL 1000Y PS SAI TBDOS(E&I) WPost2005 WWWIS)

APPIUS CLAUDIUS CRASSAS, 3rd century BC, Roman engineer and builder; artillery, scaling city walls, Appian way (312 BC+), Appian aqueduct (Aqua Appia) (c. 312 BC) (BEST E64 GI TAE TEE TSOE)

APPLEBY, John Francis, 1840–1917, American inventor and mechanical engineer; served in Civil War, cartridge magazine and automatic feed for rifle (1864), wire binder (1869), twine grain binder and knotter (1879) called Appleby Knotter, formed Appleby Reaper Works (1874), harvester rake (1875), numerous patents on grain binders and harvesters (ANB ASAI DAB MEIA MWBD NC11 WWWIS)

APPLETON, Edward Victor, 1892–1965, Scottish physicist and inventor; vacuum tubes (1932), radio signals, reflecting atmospheric layer of ionized gases, Nobel Prize (1947), radar research administrator for Scientific and Industrial Research (DNB TFE&I WWWIS)

APPOLD, J. G., 1800–1865, English inventor; centrifugal pump for land drainage called Appold pump (EE)

APPSTEIN, Maurice, 1910–1987, American electronics engineer; guided missiles, arming and safety systems, radio frequency standards, chief engineer, university teaching and research, NAE (1977) (AMOS MT4)

ARANDA-HEREDIA, Eladio, ??–1983, Spanish agricultural engineer; mechanization of agriculture, professor at Escuela Técnica Superior, leader in international agricultural engineering (CIGR) (AE63)

ARBOGAST, Philip, fl. late 1800s, American inventor; introduced concept of bottle-making machines (1881) which later (1902) was Michael J. Owens [1859–1923] bottle-making machine

ARCHIMEDES (Archimedes of Syracuse), c. 287–212 BC, Greek mathematician, inventor and engineer; some call him the greatest Hellenistic engineer in classical world, mechanics (212 BC), lever, hydrostatics, buoyancy, military arts, engines of war, shipbuilding, specific gravity, "Archimedes screw," compound pulley, author (AIN BEST GI GLOSB IAI MISAT MMAH MWBD NYPL WBEOS SAI TAE TBDOS TEE TGS TS100 WWWIS)

ARCHYTAS of Taras (aka Taranto), c. 350 BC, Greek engineer; mechanics, child's toy rattle, mechanical bird (TAE)

ARCHYTAS of Tarentum, fl. 400–350 BC, Greek (resided in Italy); mathematics of mechanics, inventor of screw (c. 400 BC), use of geometry (EB MMAH TEE WWWIS)

ARÇON, Jean-Claude-Eléonore Le Michaud, 1733–1800, French military engineer; unsinkable floating batteries for artillery (1782), public safety (MWBD)

ARGALL, Philip, 1854–1922, Irish American engineer and metallurgist; copper mining, ore deposits, copper precipitation, gold mining and use of cyanide for gold separation, consulting engineer (DAB WWWIA)

ARGAND, Aimé, 1755–1803, Swiss inventor and physicist; Argand oil lamp (1784), lamps used in lighthouses, used circular wick in lamps (EB HIW MWBD TAE WWWIS)

ARISTOTLE (aka Aristoteles of Stagyra), 384–322 BC, Greek (Hellenistic) natural philosopher; scientist, not generally known as an engineer, rainbows, reflection on water, colors of spectrum, laws of motion, mechanics (wrote first known engineering treatise), gears, levers (BEST DOT DOTHOS GLOSB HOFL MISAT SAI SAI-MP TAE TBDOS TEE TGS WWWIS)

ARKWRIGHT, Sir Richard, 1732–1792, English inventor and manufacturer; water-powered spinning frame (1769), engineering of textile machinery (1771), introduced cotton calico (1773), industry development (BEST CEOS&T GI HIW IIA MISAT MMAH MWBD PS RHW SAI S:TLAW TBDOS TGS WWWIS)

ARMOUR, Philip Danforth, 1831–1901, American industrialist and inventor; formed Armour & Co. for meat packing (1870), of which he became head (1875), introduced refrigeration and invented the refrigerated railroad car, founded Armour Inst. of Technology (1893), with Gustavus F. Swift [1839–1903] developed meat packing business (ASAI MWBD NC7 WWWIA)

ARMSTRONG, Edwin Howard, 1890–1954, American electrical engineer and inventor; radio development, triode feedback circuits (1912), continuous wave transmitter (1912), developed regenerative circuit for radio (1913), superheterodyne circuit (1917), frequency-modulation (FM) 2-path (1933), faculty at Columbia Univ. (1934ff) (AEI ANB BEST DAB EOWB GI IAI I&T16 MISAT MWBD NYPL PS TBDOS TFE&I TGS WA WWWIA WWWIS)

ARMSTRONG, William George, Baron, 1810–1900, English engineer and inventor; hydroelectric machines, hydraulic crane (1846) and machinery (1850), breech loading gun (1855) with rifled barrel made of wrought iron, big guns (1880), shipbuilding, company became a part of Vickers Ltd. (EB FNIE MWBD TBDOS(E&I) WWWIS)

ARNOLD, Aza, 1788–1865, American inventor and machinist; carpenter, wool and cotton manufacturing machinery, engraving machine, self-setting and self-raking saw for sawing machine (1856) (DAB MEIA NC13 WWWIA WWWIS)

ARNOLD, Philip Mills, 1911–1994, American chemical engineer; petroleum fuels, manufacture of aviation fuels, synthetic rubber, polyolefins, fertilizer, with Phillips Petroleum Co. (1937–1976) from research engineer (1937) to manager of research and development (1950) to vice president (1964), 22 US patents, NAE (1970) (AMOS MT8 NC53 WWWIA)

ARRHENIUS, Svante August, 1859–1927, Swedish physicist, chemist and inventor; ionic dissociation (1884), professor at Univ. of Stockholm (1895–1905), Nobel Inst. of Physical Chemistry (1905ff), pioneer in many relationships of physical chemistry and engineering, Nobel Prize (1903), fNAS (1908) (BEST IAI MWBD SAI S:TLAW TGS WWWIS)

D'ARSONVAL, Jacques-Arsène, 1851–1940, French biophysicist and inventor; developed reflecting galvanometer named after him (1880), invented diathermy apparatus, studied electricity in muscle action, directed laboratory at College of France (1882–1910), with Georges Bordas developed freeze drying in Paris (1906) (CDOSB GSA IAI TTOT MWBD WWWIS)

ARTOBOLEVSKII, Ivan Ivanovich, 1905–1977, Soviet mechanical engineer; agricultural engineering institute teaching and research, Moscow Inst. of Chemical Engineering (1932ff) and Moscow Univ., Lab. of Dynamic Machines (1937ff), administration of mechanical engineering, mechanisms, theory of machines (PS SMOS WWWIS)

ARYABHATA I, 476–c. 550, Indian (Hindu) mathematician and astronomer; invented the Indo-Arabic number system (DOSB EB WWWIS)

ASHBURNER, Charles Edward, 1870–1932, British American civil engineer; born in India of British parents, to US (c. 1900), engineer and administrator, city manager of Staunton (VA) (1908–1911) and later several other cities in US (MWBD)

ASHBY, Wallace, 1890–1986, American agricultural engineer; structures for livestock, farm structures research for US Department of Agriculture (1931–1960), served in WWII (AMOS WPost1986)

ASPDIN, Joseph, 1799–1855, English bricklayer and inventor; invented Portland cement (1824), stone mason (EB HIW IG MWBD NYPL TGE:B WA)

ASTARITA, Giovanni, 1933–1997, Italian chemical engineer; kinetics, mass transfer, rheology, mechanics of fluids and solids, educator, faculty at Univ. of Naples, visiting professor at Univ. of Delaware (1965–1995), fNAE (1994) (Bridge MT9 WWWIS)

ATANASOFF, John Vincent, 1903–1995, American physicist, computer engineer and mathematician; designed first semielectronic digital computing device (1939), known as Atanasoff-Berry (Clifford Berry) Computer, which was challenged in court by Mauchly-Eckert patent, acoustics, naval ordnance, fuse, electromagnetism, instrumentation, automation, faculty of Iowa State Univ. (1926–1945), during WWII with Naval Ordnance Lab. (1942–1945) (1951–1952), with various industries, becoming manager of Aerojet General Corp. (1957ff) (AMOS FAL GLOSB HIW NT-CS NYTObit WWWIA)

ATHENAIOS the Mechanic, fl. 100 BC, Greek writer and military engineer; described siege machines, catapults, towers, three-wheeled belfry (GEAPIT TAE)

ATKINS, Jearum, fl. 1840–1880, American inventor; millwright, raking device for reapers known as self-rake or Atkins automaton (1852), developed attachment to reaper sold to and manufactured by J. S. Wright Co. (1853ff) (CDOAB DAB WWWIA)

ATKINSON, Robert D'Escourt, 1898–1982, English astronomer and inventor; design of astronomical clock, instruments, solar energy (TBDOS)

ATWOOD, Donald J., 1924–1994, American electrical engineer; with US Army (1943–1946), instrumentation, inertial guidance systems, automotive, electronics, several positions, advancing to chairman of General Motors Corp. (1954ff), NAE (1980), treasurer of NAE (MT8 WWIE WWWIA)

ATWOOD, John Leland, 1904–1999, American aeronautical engineer; design and development of numerous military aircraft (X-15), design engineer, chief engineer, Army Air Corps at Wright Field (OH) (1928–1929), Douglas Aircraft Co. (1930–1934), North American Aviation (1934–1967), including several executive positions and advancing to CEO (1960–1967), president and CEO of North American Rockwell Corp. (1967–1970), Rockwell International Corp. (1970ff), NAE (1974) (Bridge MT9 NCL WWWIA WWIE)

ATWOOD, Lewis John, 1827–1909, American inventor and manufacturer; brass buckles, buttons, brass manufacturer, scrap metal press (DAB)

AUER, Karl Freiherr von (aka Freiherr Welsbach), 1858–1929, Austrian chemist and inventor; invented incandescent gas mantle, spectroscopy, lighting (Welsbach gas mantle) (1855), metallic filament for incandescent lamps, misch metal and Auer metal (1904) (BEST MWBD TBDOS WWWIS)

AUERBACH, Isaac Levin, 1921–1992, American electrical engineer and inventor; computer scientist, magnetic core, digital communications, radar detection, ICBM guidance, information processing, electronic packaging, industry executive, development and application of digital computer systems, US Navy in WWII (1943–1946), computer designer for what became Sperry-Univac Corp., Burroughs Co. (8 yr.), formed more than a dozen companies (1957ff), the best known of which was Auerbach Associates, 16 patents, NAE (1974) (MT7 NCM WWWIA)

AUSTIN, Herbert, 1st Baron, 1866–1941, English engineer and industrialist; motorcar manufacturer, sheep shearing equipment (1893–1905), three-wheeled automobile (1895), Austin automobile built in his shop (1905ff) (CBD DNB MWBD)

AUSTIN, James Bliss, 1904–1988, American chemical engineer; with US Steel (1928ff), technology of steel, metallurgical thermodynamics, properties of metals, became director of the US Steel Research Lab. (NJ) (1946) and then vice president (PA), retiring (1968), NAE (1967) (MT5 WWIE WWWIS)

AUSTIN, John, fl. 18th century, Scottish inventor; developed improved power loom (1789) built and operated in Glasgow, loom driven by steam engine (1798ff) (BDOTHOT MMAH)

AVOGADRO, Amedeo (aka Conte di Quaregna e Ceretto), 1776–1856, Italian physicist and inventor; behavior of gases, kinetic molecular theory, Avogadro law (1811), electrical properties of various substances, thermal expansion, Avogadro number or constant is 6.022×10^{23} molecules per gram-mole for all gases (BEST EB HIW RHW TFE&I WWWIS)

AYRES, Quincy Claude, 1891–1963, American agricultural and civil engineer; surveying, office engineer, drainage, erosion control flood control, land development, Muscle Shoals (1912–1916), faculty and manager of research and assistant to president at Iowa State Univ. (1920–1957), consulting engineer, author and editor for McGraw-Hill Book Co. (1940ff) (AMOS AE44 WWWIA)

AYRTON, Hertha Marks, 1854–1923, English electrical engineer; first woman member of Inst. Electrical Engineers (British), collaborated with husband William Edward Ayrton [1847–1908] on electric arc, electric searchlights, sand ripples (1906), Ayrton antigas fan in WWI (1915) (DNB EB MWBD WWWIS)

AYRTON, William Edward, 1847–1908, English electrical engineer and inventor; telegraph systems, electrical measuring devices (ammeter, voltmeter), induction, collaborated with Hertha Ayrton [1854–1923] (1885ff) (BDOS CEOS&T DNB MWBD PS TBDOS WWWIS)

B

BABBAGE, Charles, 1792–1871, English mathematician and inventor; developed actuary tables, assisted by Countess of Lovelace [1815–1852], built calculating machine called calculating engine (1822), pioneer in mechanical computation, built the analytical engine that was the forerunner of computers, differential calculating machine called difference engine (1833), invented heliograph and ophthalmoscope, automatic process (c. 1833). See SCHICKHARD(T). (BDOTHOT BEST DNB DOSB E-64 GI GLOSB HIW HOFL IAI MISAT MWBD NYPL 1000Y SAI SAI-MP SS1660 S:TLAW TBDOS TGS WA WWWIS)

BABBITT, Benjamin Talbot, 1809–1889, American inventor and manufacturer; pumps, engines, farm machinery, ordnance, Babbitt's Best Soap, at least 108 patents (DAB MEIA NC8 WWWIA)

BABBITT, Isaac, 1799–1862, American inventor; goldsmith, metalware, Babbitt metal for bearings consisting of 50 parts of tin, 5 parts of antimony and 1 part copper (1839) (DAB MWBD NC13 WA WWWIA WWWIS)

BABCOCK, George Herman, 1832–1893, American mechanical engineer and inventor; inventor of polychromatic printing press (1854ff), formed Babcock and Wilcox (Stephen Wilcox [1830–1893]) firm, produced steam engine, designed and manufactured safety water tube boiler (1856) (1867) (DAB MEIA MWBD NC5 TBDOS(E&I) WW-WIA WWWIS)

BABCOCK, James Francis, 1844–1897, American chemist and inventor; chemical analysis, food investigations, beverages, adulteration of foods, inventor of Babcock fire extinguisher (BDOAS DAB NC10 WWWIA WWWIS)

BABCOCK, Orville Elias, 1835–1884, American engineer and soldier; military engineer in Civil War, chief engineer—defense, bridges, superintendent of buildings and grounds in Washington (DC) (ANB DAB NC16 WWWIA)

BABCOCK, Stephen Moulton, 1843–1931, American agricultural chemist and inventor; faculty at Univ. of Wisconsin (1887–1913), invented Babcock Milk Tester (1890) for determining butterfat content of milk, 10 other inventions (he rejected patenting) for food analysis (EB MISAT MWBD NC22 WWWIA WWWIS)

BACH, Julius Carl von, 1847–1931, German engineer; research in materials, strength of materials, stress, machine design (MWBD WWWIS)

BACHE, Alexander Dallas, 1806–1867, American civil engineer; USMA (1825) to Corps of Engineers (1826–1828), faculty at Univ. of Pennsylvania (1828–1836) (1843–1867), private industry, studied daily variation of magnetic needle, superintendent of US Coast Survey (1843–1867) while at Univ. of Pennsylvania, observation of tide, study of bottom of ocean, NAS (1863), including first president (BDOAS BM1 NC3 WWWIA WWWIA-S&T WWWIS)

BACHE, Hartman, 1797–1872, American civil engineer; surveys, canals, terminal improvements, railroads, Florida defenses, lakes, harbors and western rivers, military roads for Pacific coast (1838ff), retired as Brevet Brigadier General (1865) (BDOACE)

BACHELDER, John, 1817–1906, American inventor and manufacturer; machinist, developed continuous feed sewing machine (1849), horizontal table sewing machine (Elias Howe [1819–1867]) also developed sewing machines), John Bachelder and Allen B. Wilson [1824–1888] patents covered many of the fundamentals of sewing machines, fruit dryer (ANB DAB MEIA NC12 WWWIA)

BACHMAN, Walter C., 1911–1991, American industrial and mechanical engineer; leader in design and engineering of marine—merchant and naval—ships, analysis of stresses, vibration, ship propulsion, performance of moving machinery, chief engineer of Gibbs and Cox, Inc. (1936ff), NAE (1967) (MT7 WWIE WWWIA)

BACON, Francis Thomas, 1561–1626, English natural philosopher; proponent of scientific method, proposed inductive logic, lectured and wrote about scientific method, held several government positions (BEST DOTHOS HOFL IAI IIA MISAT MWBD 1000Y SAIMP TGS WWWIS)

BACON, Nathaniel Terry, 1858–1926, American civil engineer; harbor improvement in New York, railroad surveys in northeast US, expert in Solvay process, engineering manager (BDOACE NC32 WWWIA)

BACON, Roger, c. 1220–1292, English philosopher, scientist and inventor; advocate of experimentation, recorded use of explosives, published formula (1242), contributed to optics, magnifying glasses (1250), spectacles, envisioned use of mechanical innovations such as powered ships, carriages and flying machines (BEST DOSB GI HIW MWBD NYPL 1000Y TGS WWWIS)

BACON, Vinton W., 1916–1997, American civil engineer; sanitary engineering, USPHS (1943–1946), California State Water Control Board (1950–1965), Northwest Pulp and Paper Assoc. (1956–1962), paper and pulp processes, waste treatment, pollution control, NAE (1969), consulting engineer (Bridge WWIE)

BAEKELAND, Leo Hendrik, 1863–1944, Belgian American chemist and inventor; professor at Univ. of Bruges (1887), to US (1889), consultant (1891ff), photographic papers (Velox) (1898), developed Bakelite (1905) made of phenol, artificial shellac, formed Bakelite Corp. in US and England with James Swinburne [1858–1959], invented formaldehyde (1907, 1909), cellophane (1912), acetate (1927), vinyl (1928), plexiglas (1930), acrylic (1936), melmac (1937), styrene (1938), formica (1938), polyester (1940), Bakelite became a unit of Union Carbide and Carbon Corp., over 100 patents, NAS (1936) (AIE ANB BEST BM24 GI GLOSB HFP IAI IIA LOTCS MISAT MWBD NC32 NYPL 1000Y PA SAI TPA WA WWWIA WWWIS)

BAGGALEY, Ralph, 1846–1915, American mechanical engineer; developed railroad air brakes with George Westinghouse [1846–1914], relief valve for air brakes (1872), manufacture of water-chilled rolls, over 20 patents (MEIA)

BAIF—see BAYFIUS

BAILEY, Sir Donald Coleman, 1901–1985, British engineer and inventor; Bailey bridge (1941)—a standardized preformed mobile bridge that could be rapidly erected and adapted to various conditions, first used extensively in WWII (1942), FRS (1946) (CBD DNB EB HIW PS TOIAO)

BAILEY, Frank Harvey, 1851–1921, American engineer and naval officer; marine engineer, chief engineer of Raleigh (1896), chief designer for the Bureau of Steam Engineering, inspector of engineering and ordnance (DAB WWWIA)

BAILEY, Jeremiah, fl. 19th century, American inventor; invented first commercially successful mowing machine (1822), then called a mower-reaper used for forages and later used on reapers (BOI TWABOI WA)

BAILEY, Joseph, 1825–1867, American soldier and civil engineer; construction, army division engineer in Civil War, engineer of defenses of New Orleans (1863), wing dams (ANB DAB WWWIAH-S&T WWWIA)

BAILEY, Richard W., 1885–1957, British mechanical engineer; consulting research engineer, Metropolitan Vickers Electric Co., active in professional organizations and College of Technology (PS RS4 WWW WWWIS)

BAILEY, Stuart L., 1905–1984, American radio engineer; electronic aids for navigation (marine and air), executive of consulting engineers association, NAE (1973) (MT3 WWIE(I))

BAILEY, William Holloway, 1834–1908, American mechanical engineer; brass and copper tube industry, American Tube Works (MA) (1858ff) (MEIA)

BAILEY, William John, 1921–1989, American polymer chemist and inventor; synthesis of biodegradable plastics, faculty at Wayne State Univ. (1947–1951), Univ. of Maryland (1951ff), invented a new class of monomers, developed biodegradable plastics, polymer synthesis, micromolecules, NAE (1990) (MT4 WWIE WWWIA WWWIS)

BAINER, Roy, 1902–1990, American agricultural engineer; mechanization of agriculture, segmentation and decortication of sugar beet seeds, department chairman and dean of college of engineering at Univ. of California-Davis, NAE (1965) (MT5 Res6 WWIE WWWIA)

BAIRD, Jack A., 1921–1986, American electrical engineer; US Navy in WWII (1943–1946), system engineering, research and administration at Bell Labs. and AT&T (1943–1986), serving as vice president of Bell Labs. (1966–1973) and of AT&T (1973–1986), NAE (1971) (MT3 WWIE)

BAIRD, John, 1820–1891, Canadian Scottish mechanical engineer and inventor; mechanics, machine design, iron vessels, valves for steam engines, worked in US for Burden Iron Works (NY) and was general manager of Delmeter Iron Works (NYC) (1850) and several other companies making iron vessels, at least 15 patents (MEIA)

BAIRD, John Logie, 1888–1946, Scottish inventor and electrical engineer; first person to televise an image using mechanical scanning (1923 or 1925), video tape recording (1926), demonstrated color television (1928) based on three primary colors, pioneered fiber optics (CDOS DNB GI MMAH MWBD RHW SAI TBDOS TGE:B WA WWWIS)

BAIRSTOW, Leonard, 1880–1963, English aviation engineer; fatigue and aerodynamics, wind tunnel design, aircraft stability, with government laboratories and Air Board in England, tremendous effect on aeronautical research, validity of Prandtl boundary effects (DNB PS RS11)

BAKER, Arthur Alan, 1897–1996, American mining engineer and geologist; US Geological Survey (USGS) (1921–1973), officially retiring (1967) followed by six years of service, served as associate director of USGS (1956–1967), geology project in Utah (1923–1952) including stratigraphic and structural studies in Utah and surrounding areas (HOWRD WWWIS)

BAKER, Sir Benjamin, 1840–1907, English civil engineer; underground railway, dams, tunnels, bridges (also see Sir John Fowler [1817–1898]), with John Aird [1833–1911] constructed Aswan and Asyut dams (1890–1903) (DNB MWBD TBDOS TBDOS(E&I) TGE:B WWWIS)

BAKER, Edwin Myron, 1893–1943, American chemical engineer; electromechanical engineering, Hooker Electrochemical Co. (1916–1918), faculty and chairman of chemical and metallurgical engineering at Univ. of Michigan (1918–1943), research in gas absorption, inorganic chemical industrial processes, coating of metal surfaces (ACCE NC32)

BAKER, Ira Osborn, 1853–1925, American civil engineer; professor of engineering, books on leveling (1886) and surveying (1891), masonry construction, roads and pavements (1903, 1918), drainage of earth roads (1906), brick pavements, crushed stone (1908) (BDOACE WWWIA)

BALDWIN, Frank Stephen, 1838–1925, American inventor; calculator, self-coupler for railroad cars (1855), corn planter (1860), Baldwin computing engine (1890, 1902), numerous inventions, worked with William S. Burroughs [1855–1898], associated with Jay R. Monroe [1883–1937] with Monroe Calculating Machine Co. (1911) (MEIA MWBD NC16 WWWIA)

BALDWIN, Loammi, 1745–1807, American civil engineer; soldier in Revolutionary War, canals, surveyor, grower of fruit trees, known for development of Baldwin apple (ANB BDOACE DAB MWBD NC10 WWWIA WWWIS)

BALDWIN, Loammi, Jr., 1780–1838, American civil engineer and lawyer; son of Loammi Baldwin [1745–1807], fire engine design, public works, engineer for Union Canal, docks, navy yards, tunnels, dams, artificial lakes, surveying routes for waterworks, canals, drydocks (ANB BDOACE DAB MWBD NC10 PS WWWIA)

BALDWIN, Matthias William, 1795–1866, American inventor and manufacturer; made bookmaker's and engraver's tools, hydraulic presses, noiseless stationary engine (1827), wheels for locomotives and cars (1833), steam generation in locomotives (1836) (MEIA NC9 WWWIA WWWIS)

BALDWIN, Stephen Warner, 1833–1910, American mechanical engineer; single valve automatic engine, farm engines, mowing machines (Clipper), iron works (MEIA)

BALDWIN, William James, 1844–1924, American naval architect and mechanical engineer; construction and repair of iron ships, manager of foundry and machine shops, designed and built large public buildings, consulting engineer (MEIA WWWIA)

BALL, Albert, 1835–1927, American civil and mechanical engineer and inventor; machinist, planers, invented combined repeating and single-loading rifle (1863), firearms (1864), polishing machine (1863–1864), devised and patented diamond drill for channeling through rock (1868), mining and quarrying machinery, rock drill (1871), approximately 135 patents (DAB MEIA)

BALL, Ephraim, 1812–1892, American inventor and manufacturer; manufacture of thresher that was not commercially successful, designed and manufactured Blue Plough (1845), manufactured Hussey Reaper (1845), Ohio or Ball Mower (1854) and New American Harvester (1859–1860) (DAB MEIA NC11 WWWIA WWWIS)

BALL, Frank Harvey, 1847–1920, American mechanical engineer; steam engines for drilling and operating oil wells, valve gear (1875), balanced slide valve (1878), Ball automatic high-speed engine (1890), carburetor (1913, 1919), numerous patents (MEIA WWWIA)

BANCROFT, Edward, 1744–1821, American inventor and writer; discovered dyes used for manufacturing textiles, made first metal planer in US, involved in intrigue among France, England and colonies during revolution (DAB MWBD PS WWWIA WWWIS)

BANCROFT, John Sellers, 1843–1919, American engineer and inventor; developed sleeve-nut and method of making the sleeve-nut (1881), injector (1885), numerous tools for metal operation—drills, grinding, chuck, castings, over 100 patents (MEIA WWWIA WWWIS)

BANDEL, Hannskarl, 1925–1993, German American civil engineer; design and research in German steel industry, cable-stayed bridges, space structures, structural design with Fred N. Severud [1899–1990] in US, becoming partner in firm, then with DRC Consultants, design of unusual and aesthetically pleasing structures, NAE (1978) (MT8)

BANKS, Harvey O., 1910–1996, American civil engineer; water resources, organizing and managing complex resource programs, sanitary engineer for city of Palo Alto (CA) (1934–1935), hydraulic engineer with US Soil Conservation Service (1935–1938), WWII in US Corps of Engineers (1942–1945), California Water Resource Division (1938–1942) (1950–1955), which became the California Department of Water Resources, of which he was first director (1957–1960), with firm of Leeds, Hill and Jewett, Inc. (1961–1969), serving as

president and chairman, which was merged with Camp Dresser and McKee (1979–1982), consulting engineer, NAE (1973) (MT10 WWIE)

BANNEKER, Benjamin, 1731–1806, Black American astronomer, surveyor, civil engineer and inventor; construction of mills (1771), surveyor who helped survey Washington (DC), wooden clock (1753) that worked 40 yr., almanac (1790ff), free black born in Maryland, known for scientific instruments (BDOACE BPOSAI NC5 SAI-MP TBDOS TFM TGS WBEOS WWWIA)

BARBER, John, 1734–1801, English inventor; water turbine (1766), smelting operation (1773, 1792), boiler mounted impulse steam turbine (1776), oil engines of turbine type, first patent for gas turbine (BDOTHOT DNB EE MMAH)

BARDEEN, John, 1908–1991, American electrical engineer, inventor and physicist; co-inventor of transistor (1947) with Walter H. Brattain [1902–1987] and William B. Shockley [1910–1989], superconductivity (1957) with chemical engineer Leon N. Cooper [b. 1924] and John R. Schreiffer [b. 1931], faculty at Univ. of Illinois (1951–1975), Nobel Prizes for developing transistor (1956) and for BCS (named after initials of discoverers) theory of superconductivity (1972), NAS (1954), NAE (1972) (CBY GI GLOSB IAI MMAH MT6 MWBD NYPL 1000Y RHW TBDOS TS WA WWWIA WWWIS)

BARGER, Edgar Lee, 1905–1997, American agricultural engineer; research on tractors and fuels, power units, faculty at Kansas State Univ. (1930–1941), after one year at Univ. of Arkansas became member of faculty at Iowa State Univ. (1942–1951), with Harry George Ferguson [1884–1960] of Ferguson Co. and its successor the Massey-Ferguson Corp. (1951–1971) (AE52 NUC)

BARKAN, Philip, 1925–1996, American mechanical engineer; circuit breaker development, mechanism design, analysis of mechanical systems, General Electric Co. (ante 1977), faculty at Stanford Univ. (1977–1993), over 50 US patents, NAE (1980) (MT9 WWIE WWWIA)

BARKER, George Frederick, 1835–1910, American physicist, chemist and electrician; dynamos and electric machinery, exhibited radium in radioactive materials, faculty at Yale Univ. (1867–1873), Univ. of Pennsylvania (1873–1900), president of AAAS (1879), consultant (BDOAS DAB NC4 WWWIA)

BARKHAUSEN, Heinrich Georg, 1881–1956, German electrical (electronics) engineer and physicist; communications, non-linear switching, electron tubes, electric oscillators, acoustics, magnetism, ferrous materials, Barkhausen effect (1919) faculty at Technische Hochschule Dresden (1911ff) (CEOS&T LDOS MWBD PS TFE&I WWWIS)

BARLOW, Harold E. M., 1899–1989, English electrical engineer and inventor; microwave technology, invented several microwave devices, research and development of radar (1939–1945), superintendent of radio development for Royal Aircraft Establishment (1945–1950), Univ. College of Univ. of London, including head of department and dean of engineering (18 yr.), international cooperation, fNAE (1979) (DNB MT4 WWWIS)

BARLOW, John Whitney, 1838–1914, American military engineer; topographical engineer, with army in Civil War, survey of Yellowstone Park (1871), Muscle Shoals (1890) (DAB WWWIA)

BARLOW, Thomas, 1789–1865, American engineer and inventor; built steamboat (1820), miniature steam locomotive (1827), planetarium (1848), automatic nail and tack Macalpine (1854), rifled cannon (1855) (MEIA)

BARMIN, Vladimir Pavlovich, 1909–??, Soviet mechanical engineer; mechanics applied to compressors, plant designer (1930–1946), becoming chief designer (1940), taught at Moscow Technical College (1931ff) (SMOS)

BARNACK, Oscar, 1879–1936, German engineer; optics, introduced Leica camera (1913, 1924), utilized 35 mm film size, which became a standard (MWBD)

BARNARD, John Goss, 1815–1882, American civil and military engineer; graduate of USMA assigned to engineer corps (1833–1881), retired as brigadier general, worked on coast defenses, improvement of rivers and harbors, fortification, bridges, jetty method of improving mouth of Mississippi River for traffic, NAS (1863) (ANB BDOAS BM5 DAB PS WWWIA)

BARNES, Albert C., 1872–1951, American chemist and entrepreneur; received Doctor of Medicine degree, consulting chemist for H. K. Mulford Co. (ante 1907), with Hermann Hill developed Argyrol (1902), founded the Barnes Co. for manufacturing medicines, supported philanthropic project in art and education, supported Lincoln Univ. (estab. 1854) (PA) (ANB ChemHerit18 WWWIS)

BARNES, James, 1801–1869, American soldier and engineer; chief engineer and railroad superintendent (1836), railroad construction, isostasy, Civil War officer (DAB WWWIA)

BARNES, Kenneth Kirtland, 1921–1977, American agricultural engineer; US Army in WWII (1943–1946), faculty at Purdue Univ. (1950–1952), at Iowa State Univ. (1952–1959), to Univ. of Arizona (1959ff), including head of department (1960–1971) then head of Department of Soils, Water and Engineering, research in tillage machinery, machinery management, vegetable crop mechanization (AE47,58 AMOS WWIE WWWIA)

BARNEY, Nora Stanton Blatch, 1883–1971, English American civil engineer; first woman to receive civil engineering degree at Cornell Univ. (1905), architect, in laboratory of Lee De Forest (1906–1912), married Morgan

Barney (1919), worked for American Bridge Co., NYC Board of Water Supply, into architecture and real estate development (1923ff), activist for women's rights (AWIT)

BARON, Melvin L., 1927–1997, American civil and mechanical engineer; structural mechanics, buried structures, numerical analysis, inelastic behavior of soil and rock, ground shock, wave propagation effect from explosives, partner of Weidlinger Assoc. (45 yr.) and adjunct professor at Columbia Univ., consulting engineer, NAE (1978) (MT10 WWIE WWWIA)

BARON, Thomas, 1921–1985, Hungarian American chemical engineer; to US (1939), theory and practice in fluid dynamics, organic chemical reactions, combustion, catalytic reactions, petroleum extraction, Univ. of Illinois (1948–1951), Shell Development Co. (1951–1981), becoming president (1967–1981), NAE (1977) (MT3)

BARR, Archibald, 1855–1931, Scottish engineer and inventor; military range finders, pressure pumps, gunfire control instruments, chairman of manufacturing company (DNB MWBD)

BARR, Harry F., 1904–1990, American automotive engineer; automotive industry, chief engineer, automotive safety and emission control, various positions with General Motors Corp. (1926–1969), becoming corporate vice president (1963), NAE (1965) (MT6 WWIE)

BARR, Jacob Neff, 1848–1904, American mechanical engineer and inventor; wheel foundry, sand flange and casting car wheels (1878), drill press (1894), brake beam (1901), manager (MEIA)

BARRE, Henry J., 1905–1987, American agricultural engineer; received first PhD in agricultural engineering in US (1938), design of farm structures, handling, drying and aeration of grain, USDA grain storage investigations (1938–1943), faculty and department head at Purdue Univ. (1943–1962), faculty at Ohio State Univ. (1962ff) (AE47 WWIE)

BARRE, Salmon August, 1854–1897, Swedish engineer and explorer; ballooning, North Pole and Arctic exploration by balloon (EB MWBD WWWIS)

BARRELL, Joseph, 1869–1919, American geologist and engineer; mining, metallurgy, survey, popular lecturer, design, drawing, isostasy, geological processes and geological time, NAS (1919) (ANB BM12 DAB WWWIA WWWIS)

BARRERE, Marcel Louis J., 1920–1996, French aeronautical engineer; rocket propulsion, combustion, Office National d'Etudes et de Recherches Aéronautiques (1944–1985), including director of research (1979–1985), consultant (1985ff), fNAE (1984) (MT9)

BARRINGER, Daniel Moreau, 1860–1929, American mining engineer; practiced law (1882–1889), geologist, mining, craters (Great Barringer Meteor Crater in Arizona), consulting mining engineer and geologist

(1889ff), president and director of several mining companies (BEST NC22 WWWIA WWWIS)

BARRUS, George Hale, 1854–1929, American mechanical engineer and inventor; assisted George E. Dixon [1850–1904] in building first steam engine laboratory in US, paper dryer machine, steam and power, calorimeter (1888), steam and water gauges, mechanical stoker (1900) (BDOACE MEIA WWWIA WWWIS)

BARSANTI, Eugenio, 1821–1864, Italian scientist, engineer and inventor; invented free piston atmospheric engine (1854ff) with Felice Matteucci [1803–1887] called Barsanti-Matteucci gas engine, principle used in Otto-Langen engine (1867) (BDOTHOT EE)

BARTH, Carl George Lange, 1860–1939, Norwegian American mechanical engineer; to US (1881), draftsman, designer (1890), chief engineer (1895), work efficiency studies with Frederick W. Taylor [1856–1915] (1890ff), shop management, instructor for International Correspondence Schools (PA) (1847ff) and other educational institutions, consulting engineer (1905ff) (DAB MEIA NC34 WWWIA WWWIS)

BARTHEL, Oliver Edward, 1877–1969, American engineer and inventor; design engineer for Ford Motor Co. (1901–1902), design of valves and racing cars for Ford, Barthel Motor Co. (1901–1904), Olds Motor Works (1905), consulting engineer (1905–1955), working on automotive, marine and aircraft, taper frame for automobile (1919), held 39 patents related to gasoline engine and automotive development (NC55,F PS WWWIA)

BARTON, Sir John, 1771–1834, British inventor; invented machine for making and engraving buttons, machine for making diffraction gratings (EE)

BATCHELDER, John Montgomery, 1811–1892, American civil engineer; in charge of mill, US Coast Survey (ante 1858), Dudley Observatory (NY) (1858ff), compressibility of sea water and other liquids, compressibility of rubber, effects of temperature on compression, deep-sea sounding apparatus, tide meter (BDOAS)

BATCHELDER, John Putnam, 1784–1868, American physician and inventor; inventor and improvement of surgical instruments (DAB NC9 WWWIA)

BATCHELDER, Samuel, 1784–1879, American inventor and manufacturer; developed and managed factories, particularly cotton mills, developed many new devices for cotton manufacture, invented stop-motion drawing frame, invented dynamometer (1837) (DAB NC5 WWWIA)

BATES, David Stanhope, 1777–1839, American civil engineer; surveyor (1810), iron works, practiced law, canal construction, aqueduct masonry (BDOACE NC18)

BATES, Onward, 1850–1936, American civil engineer; machinist, draftsman, inspector, worked on Eads bridge (steel) across Missouri River, worked for several com-

panies as chief engineer and superintendent, mining operations in Mexico, consulting business (1901–1907) (ANB BDOACE NC15 WWWIA)

BATTISTA, Della Porta, 17th century, Italian—see PORTA

BAUDOT, Jean-Maurice-Émile, 1845–1903, French engineer; telegraph code (1874), multiplex system (1894), improved telegraph transmission, one baud is one pulse per second (CES&T MWBD WWWIS)

BAUER, Andreas Friedrich, 1783–1860, German engineer; steam printing presses, co-founder with Friedrich Koenig [1774–1833] of Koenig & Bauer Co. to manufacture printing presses (MWBD)

BAUER, Benjamin Baumzweiger, 1913–1979, Russian American electrical engineer; acoustics, radio engineering, microphones, audio-video, magnetic recording, underwater sound, industry official at Shure Brothers, research and development at Columbia Broadcasting Co. (CBS), NAE (1974) (AMOS MT2 WWWIA)

BAUER, George—see AGRICOLA

BAUER, Sebastian Wilhelm Valentin, 1822–1875, German inventor; submarine, made first underwater escape (1851), made early underwater photographs (1855) (MWBD WWWIS)

BAUMÉ, Antoine, 1728–1804, French chemist and inventor; invented improved hydrometer (1768) graduated to Baumé scale, aerometer (1768), density of fluids (MWBD NYPL WA WWWIS)

BAUSCH, Edward, 1854–1944, American industrialist and inventor; designed and produced microscopes for Bausch and Lomb Optical Co. (NY) (1875), company formed by father John Jacob Bausch [1830–1926] with Henry Lomb [1828–1908] (1853), designed machines to make lens formerly done manually, numerous inventions on application of microscope such as for meat inspection, worked deal with Carl Zeiss [1816–1888], who had a firm in Germany, range finders, president of company (1926) then chairman of the board (1935) (ANB MWBD NC19 WWWIA)

BAXTER, Samuel S., 1905–1982, American civil engineer; public works in Philadelphia, including assistant director and chief engineer and commissioner of water (1923–1972), highways, airports, bridges, water and sewage, NAE (1970) (MT2 WWIE WWWIA)

BAYFIUS, Lazarus de (aka Baif), ??–1547, French naval engineer; wrote extensively on naval art, science and shipbuilding (1536) NUC TAE)

BAYLES, James Copper, 1845–1913, American engineer; in Civil War (1862–1864), founded and edited magazine *Iron Age* (1889), edited magazine *Metal Worker,* electrometallurgy, metals, iron ore, microscopic analysis, sanitation and water supply, consulting engineer for city departments and public utilities (1889ff), lecturer at Cornell Univ. (DAB NC13 WWWIA WWWIAH-S&T)

BAYLEY, William, 1845–1934, American mechanical engineer and inventor; machinist, draftsman, designer, shipbuilder and associated machinery manufacture, patented waterwheel (1871) with diverging discharge buckets, grain conveying apparatus (1881), machinery for paving streets (1884), manufacture of farm machinery (MEIA)

BAZALGETTE, Sir Joseph William, 1819–1891, English civil engineer; drainage (London) (1858–1875), Thames embankment design and construction (1860–1874) (BDOS(E&I) DNB MWBD PS)

BAZIN, Henri-Émile, 1829–1917, French engineer; channel water flow (follow-up on Darcy), wave propagation, fluid flow through orifices, suction dredgers, chief engineer, inspector general (MWBD WWWIS)

BEACH, Alfred Ely, 1826–1890, American mechanical engineer and inventor; son of Moses Yale Beach [1800–1868], typewriter (1856), typewriter for blind (1856), cable railway devices (1864), pneumatic tubes for mail (1865), Beach hydraulic tunneling shield, cofounded magazine *Scientific American* (1845) (ANB MEIA NC8 WWWIA WWWIS)

BEACH, Moses Sperry, 1822–1892, American inventor and mechanical engineer; son of Moses Yale Beach [1800–1868], development of composing and pressroom equipment, invented device to feed paper from a roll to a press (1856), numerous devices for feeding, wetting and cutting paper for printing (MEIA NC13 MWBD WWWIA WWWIS)

BEACH, Moses Yale, 1800–1868, English American journalist and inventor; cabinet maker, mechanics, used gun powder as a fuel for engines, rag cutting machine, managed mechanical operations and then owned newspaper *Sun* (NY), credited for developing syndicated articles (ANB DAB MWBD NC1 WWWIA WWWIS)

BEASLEY, Maria E., c. 1847–1904?, American inventor; prolific inventor in several fields, many inventions related to barrel making, pasting shoe uppers (1882), steam generator (1886) antiderailment device for railroad cars (1898), life rafts (1880) (1882), at least 15 patents (AWIT)

BEATTIE, Horace Smart, 1909–1993, American engineer; inventions, design, business machines, electric typewriter (1957), office products, company executive at International Business Machines (IBM), at IBM for 40 yr., consultant, NAE (1976) (MT7)

BEAU DE ROCHAS, Alphonse-Eugène, 1815–1893, French engineer; four-cycle internal combustion engine patented (1862) but not built, compression and ignition at constant volume in four cycles, called De Rochas engine (MWBD WWWIS)

BEAUFORT, Francis, 1774–1857, English naval officer and hydrographer; survey data, wind velocity, Beaufort Scale for velocity (1805), meteorological measurements (CEOS&T DOT MWBD NYPL TBDOS WWWIS)

BECK, Paul Adams, 1908–1997, Hungarian American metallurgist; patent engineer, to US (1937), American Smelting & Refining Co. (1937–1941), Cleveland Graphite Bronze Co. (1942–1945), deformation and textures of engineering alloys, grain growth, annealing texture, recrystallization, electronic and magnetic characterization of alloy systems, faculties at Univ. of Notre Dame (1945–1951) and Univ. of Illinois (1957ff), NAE (1981) (AMOS Bridge WWIE WWWIA WWWIS)

BECHTEL, Stephen Davison, 1900–1989, American engineer, constructor and industrialist; served in France in WWI, entered construction business with his father (1919), Warren A. Bechtel [1872–1933], W. A. Bechtel Co. formed (1925), of which Stephen D. Bechtel was vice president (1925–1936) and then president (1936–1948), Bechtel Corporation formed (1945), of which W. A. Bechtel Co. was a part, the Bechtel Corporation and its affiliates were involved in numerous construction and building projects, including ships and planes in WWII, petroleum refineries, electric power plants, chemical plants, industrial plants, including Trans Arabian pipeline, worked with several foreign companies, was active in numerous professional and civic organizations, succeeded by son Stephen D. Bechtel, Jr. [b. 1925] (BAC NC24,H WWIA WWWIA)

BECKET, Frederick Mark, 1875–1942, Canadian American metallurgist; to US (1895), silicon reduction (1904), chief metallurgist (1906), production of low-carbon ferroalloys and stainless steels, company officer of Electro Metallurgical Co., later known as Union Carbide Corp. (1906ff) (MWBD WWWIA WWWIS)

BECKMAN, Henry C., 1888–1962, American civil engineer; US Geological Survey (1914–1958), where he served as district engineer, division hydrologist, regional engineer, coordinated stream-gauging programs operated by several districts of the US Corps of Engineers for Mississippi-Missouri River basins (HOWRD)

BECKWITH, Edward Pierrepont, 1877–1966, American chemical engineer; research chemist, instructor in chemical engineering, use of tungsten in filament for electric lamps at General Electric Co. (1904–1908), purification of water, international pilot license (1915), scientific exploration, photographer, consulting engineer, held several patents (NC52 PS)

BECQUEREL, Alexandre-Edmond, 1820–1891, French experimental physicist and inventor; photoelectric effect (relation between light and electricity) first to convert sunlight to electricity later done by English engineer Willoughby Smith [1828–1891] and American Charles Fritts using selenium for photovoltaic cell (1883, 1886), work followed by development of solar cells (1954) and silicon solar cells by Americans Calvin Fuller [1902–1994] and Daryl M. Chapin [b. 1906] at Bell Telephone Labs. (1960), magnetism phosphorescence, spectroscopy, faculty and chair at Conservatoire des Arts et Métiers (1852ff), succeeded his father (Antoine-César Becquerel [1788–1878]) as director of Muséum d'Histoire Naturelle (Paris) (1878) (EB CBD DNB DOSB IAI IG MWBD WOI WWWIS)

BECQUEREL, Antoine-César, 1788–1878, French electrochemist and inventor; voltaic pile, work led to development of batteries, galvanometers, atmospheric electricity, thermoelectric needle to determine internal body temperatures, professor then director of National Museum of Natural History (France) (1837–1878) (DOSB DOTHOS MWBD TGS WOI WWWIS)

BECQUEREL, Antoine-Henri, 1852–1908, French physicist and engineer; son of Alexandre-Edmund Becquerel [1820–1891], originally engineer for bridges and highways, for which he gained an international reputation, Museum of Natural History of France (1892ff) after his father and grandfather held similar post, École Polytechnique (1895ff), radioactivity (1896), Nobel Prize (1903), fNAE (1905) (BEST DOTHOS GLOSB IAI MWBD SAI TBDOS TGS WWWIS)

BEDAUX, Charles Eugene, 1886–1944, French American engineer; efficiency authority, Bedaux system for productivity measurement (MWBD WWWIA)

BEEBE, Charles William, 1877–1962, American naturalist and inventor; invented bathysphere (1930), a sphere for underwater study, with Otis Barton [b. 1899] established depth of 3,028 ft. (1934), early work on a diving bell done by Englishman Edmund (or Edmond) Halley [1656–1742], work on exploration of sea continued by Frenchman Jacques Yves Cousteau [1910–1997] who invented aqua-lung apparatus (1942) for diving (GI GO-FYI MWBD RHW TTIT WWWIS)

BEHN, Carl, 1858–1918, German American mechanical engineer; to US (1878), pioneer in building refrigeration machinery in US, gained interest in Buffalo Refrigerating Machine Co. (1887), which was founded by his father, became head of company (1904) (MEIA)

BEITLER, Samuel R., 1899–1995, American mechanical engineer; faculty member at Ohio State Univ. (1921–1962), thermodynamics, fluid mechanics, flow measurement, national research administration and standards committees of ASME (ASMENews14 WWIE)

BEKKER, Mieczyslaw Gregory, 1905–??, Polish American mechanical engineer; with Polish Ministry of Defense (1931–1939), French Ministry of Armament (1940), Canadian army (1943–1946), chief of Land Locomotion Lab. (US) (1954–1961), with US educational institutions of Stevens Inst. of Technology, Johns Hopkins Univ., Univ. of Michigan, head of general mobility research with General Motors Corp. (1961–1979), specialized in off-road locomotion and all-terrain vehicle systems, consulting engineer (1970ff) (WWIE)

BELGRAND, Marie-François-Eugène, 1810–1878, French engineer; sewage system of Paris, reservoir, hydrographic studies (MWBD WWWIS)

BELIDOR, Bernard Forest de, 1698–1761, French engineer; soldier, artillery, ballistics, inspector, author of significant books on engineering, fortifications, professor of artillery at military school (EWOB MWBD TSOE WWWIS)

BELIN, Edouard, 1876–1963, French engineer; first telephoto transmission (1907), first transatlantic telephoto transmission (1921) (MWBD)

BELL, Alexander Graham, 1847–1922, Scottish American inventor; speech school (1872), deaf instruction, acoustical devices, gramophone, founded magazine *Science* (1883), telephone (1876), photophone (1880), Bell Telephone Co. (1877), cylinder recording (with Charles S. Tainter [1854–1940]) (1887), NAS (1883), credited with developing *National Geographic Magazine* (founded in 1888), US postage stamp (1940) in his honor (AIE ANB BDOAS BEST BM23 CEOS&T DAB EAI (EOET GI GLOSB IAI IIA MMAH MWBD NC6 1000Y SAI-MP S:TLAW TGS WA WWWIA WWWIS)

BELL, Henry, 1767–1830, Scottish engineer; designed steamboat, first commercially successful steamboat operation, ideas led to Fulton's steamship, invented discharge machine (DNB MWBD WWWIA WWWIS)

BELL, Lawrence Dale, 1894–1956, American aircraft designer; with Glenn L. Martin C. (1913), becoming vice president, Consolidated Aircraft Co. (1929–1935), president of Bell Aircraft Co. (1935–1956), aircraft fighters, built Bell X-1, the first US aircraft to fly above Mach 1 (1947), piloted by Gen. Chuck Yeager [b. 1923] (MWBD NCF WWWIA)

BELL, Louis, 1864–1923, American physicist and engineer; absorption spectrum of nitrogen peroxide, ultraviolet spectrum of cadmium, editor of magazine *Electrical World* (1890ff), chief engineer of electric power transmission, consulting engineer (1895ff), illumination (DAB PS WWWIA WWWIS)

BELL, Milo C., 1905–1998, American mechanical engineer; application of engineering to fisheries, particularly salmon, State of Washington Dept. of Fisheries, becoming chief engineer (1930–1933) (1935–1943) (1951–1957), Corps of Engineers (1933–1935), international Pacific Salmon Fisheries Commission (PSFC) first as consultant and then chief engineer (1943–1951), lecturer at Univ. of Washington (1940–1953) and faculty (ante 1975), NAE (1968) (MT10 WWIE)

BELL, Patrick, 1799–1869, Scottish clergyman and inventor; invented reaper with reciprocating knife (1826), one of first successful reaping machines (1826) built (1828–1832), pushed from behind with horses, improved and introduced as Beverly Reaper (1857) (DNB EAI GI MMAH PS)

BELL, Thomas, fl.1770–1785, Scottish inventor; millwright, engraved cylinders to print calico (1783), developed three-color calico printing machine (1784) (BDOTHOT DNB)

BELL, Sir Thomas, 1865–1952, British engineer and shipbuilder; naval architecture, construction of shipyards and shipbuilding, director of shipbuilding works (DNB WWW)

BELLAMY, Charles, 1851–1915, American civil engineer; first registered engineer in US in Wyoming (1907), surveyor, astronomy, mining claims, location of railroads, irrigation ditches, engineering firm of Bellamy & Sons (1913) (ET WWWIA)

BELLIS, G. E., 1838–1909, British inventor; forced lubricated Bellis marine engine, formed G. E. Bellis Co., forced lubrication invented by A. C. Pain [1856–1892] (EE)

BELLMAN, Richard Ernest, 1920–1984, American mathematician and electrical engineer; US Army in WWII, radio and sound, stochastic processes, invented dynamic programming (1953), control theory and systems analysis, known for theory of multistage decision processes, optimization of large-scale systems, faculties at Princeton Univ. (1946–1948), Stanford Univ. (1948–1953), Univ. of Southern California (1965–1984), and Rand Corp. (1953–1965), NAE (1977), NAS (1983) (AMOS MT3)

BELLPAIRE, A., 1820–1893, Belgian engineer; invented long grate, flat-topped firebox for boilers (EE)

BELLPORT, Bernard Philip, 1907–1987, American civil engineer; water resource facilities, chief engineer for US Bureau of Reclamation (USBM), mining companies, consultant, NAE (1970) (MT4 WWIE WWWIA)

BELYAYEV, Anatolii Ivanich, 1906–??, Soviet metallurgist and engineer; electrochemistry of alloy salts, electrometallurgy of light metals, plant engineer then chief engineer of main aluminum plant in Moscow, All-Union Aluminum Inst. (1934–1937), Inst. of Non-Ferrous Metals and Gold (1941ff) (SMOS)

BEMENT, Caleb N., 1790–1868, American agriculturalist and inventor; manufacturer of farm implements, Bement's Expanding Corn Cultivator, Bement's Turnip Drill (1835–1836), Bement's Compound as a yeast substitute (DAB TBDOS WWWIA)

BENDIX, Vincent Hugo, 1882–1945, American inventor and industrialist; automotive and aviation parts, formed Bendix Co. (1907), Bendix self-starter (1912), Bendix Brake Co. (1912), carburetors, Bendix Aviation Corp. (1929), credited with first automatic washing machine (AIE ANB MWBD NC 40,M PS WWWIA)

BENEDICTUS, Édouard, 1879–1930, French artist and inventor; invented laminated glass as used in windshields called shatterproof glass, shatterproof uses (1909) called Triple x (TM) (EB IAI WBE)

BENERITO, Ruth Rogan, 1916–??, American chemist and inventor; taught college chemistry, USDA Southern Regional Research Center (1953–1983), work related to treating cotton, pioneer in wash-and-wear fabrics, more than 50 patents (AWIT NWITPS NWS)

BENEZET, Saint, fl. 1165–1183, French bridge builder; originally a shepherd, thought to have built the Pont d'Avignon under divine guidance, some call him the saint of bridge builders, achieved canonization (FEL)

BENHAM, Henry Washington, 1813–1884, Canadian American engineer and soldier; coastal defense, with Army Corps of Engineers, repair of defenses and harbors of New York City and Boston, pontoon bridges, chief engineer during Civil War (DAB NC4 WWWIA)

BENJAMIN, Charles Henry, 1856–1937, American mechanical engineer; worked with several companies, boiler installation to reduce smoke pollution, centrifugal effects of rapidly rotating equipment, faculties at Univ. of Maine (1880–1887), Case Inst. School of Applied Science (OH) (1889–1907) and Purdue Univ., including dean of engineering (1907–1921) (MEIA WWWIA WWWIS)

BENJAMIN, George Hillard, 1852–1927, American lawyer and engineer; patent authority, expert witness, patents covering underground conduits for electricity, glass melting furnaces, dynamo, pipe couplings, expansion joints, tin plate manufacturing, curing and processing of tobacco, beet sugar and fruits (DAB NC29 WWWIA)

BENTHAM, Sir Samuel, 1757–1831, English naval engineer and architect; boats, barges, shell guns, steam dredges, mass production, pulley block manufacturing (1793), wood working machinery (DNB EE MMAH MWBD)

BENTON, Linn Boyd, 1844–1932, American inventor; part owner of type foundry (1873), which became Benton, Waldo & Co. (1882) and later American Type Founders Co. (1892), invented type foundry (1882), machine for punch engraving (1885), numerous patents relating to type casting, automatic type casting machine (1914) (NCA PS)

BENZ, Karl Friedrich, 1844–1929, German engineer; motor-driven vehicles, engines, horizontal engines, slow-speed engines, induction coils, spark ignition engine (1883), founded Benz and Co. (1883), petrol-driven automobile (1885), differential gears (1885), later merged with Daimler (1926) (E-64 EAI GI HIW MMAH MWBD PS SAI WA WWWIS)

BENZINGER, Theodor Hannes, 1905–1999, German American doctor and inventor; invented first ear thermometer (c. 1959), bio energetics, biothermodynamics, Blanck-Benzinger equation, aerospace medicine, high-altitude physiology, 12 patents (AMOS WPostObit WWWIS)

BERCHTHOLDUS—see Anklitzen Kostanton

BERDAN, Hiram, 1823–1893, American inventor and engineer; reaping machine, mechanical bakery, bread cutter (1859), firearms, rifles, cartridges, Berdan rifle used by US government (1865), torpedoes and torpedo boats (1879ff) numerous patents (MEIA)

BERG, Aksel Ivanovich, 1893–??, Soviet radio engineer; designed and developed tube oscillators, stabilized frequency, amplification and frequency control oscillators, submariner in Civil War in Russia, taught at Naval Engineering School (1925) (BDOS PS SMOS)

BERGEN, William B., 1915–1987, American aerospace engineer; astronautics, structures, aeroelasticity, management of complex systems, executive of several aerospace companies, including Glenn L. Martin Co., Martin Marietta Co., and Rockwell International Corp. (1951–1975), consultant (1975ff) (MT5 WWIE WWWIA)

BERGER, Hans, 1873–1941, German psychiatrist and inventor; discovered electric brain waves (EEG) (1904), invented electroencephalograph (1929, 1931) for measuring the electrical activity of brain, physical aspects of brain function, work confirmed and enhanced by Edgar Douglas Adrian [1889–1977] (CDOS EB NYPL WA WWWIS)

BERGIUS, Friedrich Karl Rudolf, 1884–1948, German industrial chemist and inventor; technical research laboratory, converting coal by catalytic hydrogenation into hydrocarbon to produce gasoline (1912) and lubricating oil, effect of high pressure on chemical reactions, production of sugar from wood, Nobel Prize (1931) (BEST MWBD TBDOS TGS WWWIS)

BERGLUND, Nils H., 1896–1968, Swedish agricultural engineer; test engineer, professor, helped form Swedish Association of Agricultural Engineers and served as its secretary (1927–1966), head of department at Royal Agricultural College of Sweden (AE50)

BERKLEY, James John, 1819–1862, English engineer; railway construction—first in India (1853) (DNB MWBD)

BERKNER, Lloyd Viel, 1905–1967, American physicist and engineer; lighthouses, Byrd Antarctic expedition (1928–1930), study of atmosphere, ionosphere, radio propagation, radar, with Carnegie Institution (DC) working on terrestrial projects (1933–1939), US government research and development for navy in WWII, president of Associated Universities, NAS (1948) (ANB BM61 MWBD WWWIA WWWIS)

BERLIER, Jean-Baptiste, 1843–1911, French engineer; underground railway construction (Paris), city pneumatic system for distribution of letters and telegrams (MWBD)

BERLINER, Emile, 1851–1929, German American inventor; to US (1870), microphone (1877), worked for Bell Telephone Co. (1878), telephone features, duplication of disk record (1887), gramophone done parallel with work of Thomas Edison [1847–1931] (1888), duplication of records, established Berliner Gramophone Co. (PA), which became Victor Talking Machine Co. (NJ), which produced Victrola, vigorous supporter of milk pasteurization, developed airplane engines, over 100 patents, with his son Henry Adler Berliner [1895–1970] devised a working helicopter (1919) (AI AIE ANB ASAI BEST CDOAB GSA IAI MMAH MWBD NC21 NYT-A PA WWWIAH-S&T WWWIA WWWIS)

BERNARDE, Simon, 1779–1839, French military engineer; worked in US then returned to France, developed military defenses for US Corps of Engineers—fortifications, coastal defenses, roads and canals (ANB BDOACE DAB MWBD)

BERNOULLI, Daniel, 1700–1782, Swiss mathematician, natural philosopher and mechanician; son of Johann J. Bernoulli I [1667–1748], acoustics, mechanics, elasticity, heat and thermodynamics, developed Bernoulli principle relating fluid velocity and pressure, hydrostatics, hydrodynamics, technology development CEOST DOTHOS FNIE HOFL ITCOF MWBD TGS WWWIS)

BERTHELOT, Marcellin Pierre, 1827–1907, French chemist and inventor; development of polystyrene (polymer), first synthetic preparation of polystyrene (1866) (IAI WWWIS)

BERTHOUD, Ferdinand, 1727–1807, Swiss clock and instrument maker; improved marine chronometer (previous work by English clockmaker John Harrison [1693–1776]), his nephew Pierre-Louis Berthoud [1754–1813] succeeded him in work on marine clocks and industrialized construction of chronometers in France (EAI MMAH WWWIS)

BERTIN, Louis Émile, 1840–1924, French naval engineer and architect; hydraulics, improved ventilation on ships, artificial waves and application of water swell of ocean, French-built compartmentalized cruiser, director of naval construction (1892), administrative positions in French Navy (DOSB PS WWWIS)

BERTOLA da Novate, fl. 1450s, Italian engineer; hydraulics, designed and built canals and locks (TAE TSOE)

BERTRAND, Henri-Gratien, compte, 1773–1844, French engineer; built military structures especially bridges, bridge across Danube River (1809), became general in military forces (LC MWBD)

BERTSCH, John Charles, 1859–1939, German American mechanical engineer; to US (1892), installed refrigeration machinery, natural draft cooling towers (1898), ice plants, rotary ammonia compressor, developed vacuum oil refining process, worked for Fred W. Wolf Co. (IL), Columbia Brewing Co. (OH), and Frick Co. (GA), consulting engineer (1902ff) (MEIA)

BERZELIUS, Jöns Jacob, 1779–1848, Swedish chemist and inventor; electric forces involved in combining of atoms, atomic weights of elements, introduced desiccator to keep samples free of moisture, gravimetric analysis (HIW WWWIS)

BESSEL, Friedrich Wilhelm, 1784–1846, German astronomer and inventor; prepared star catalogue (1818), professor and director of observatory at Königsberg (1810–1846), method of analysis for astronomical calculations used in many other fields called Bessel functions (1817) (1834), designed heliometer (c. 1844) (BEST LDOS MWBD TBDOS(Po) TBDOS(Wi) TGS WWWIS)

BESSEMER, Sir Henry, 1813–1898, British engineer and inventor; iron and steel processing, cast steel, gunmaking (1854–1856), steel rails, Bessemer process (1855ff) introduced commercially (1878), invented typesetting machine (1838) (BEST CES&T DNB E-64 EAI HIV IAI IIA MMAH MWBD 1000Y SAI S:TLAW TGE:B TGS WWWIS)

BESSON, Jacques, 1540–1576, French engineer and inventor; lathe development (screw cutting) (1560), cams, templates, improved waterwheel (MMAH MWBD NYPL)

BEST, Daniel, 1838–1923, American inventor and manufacturer; invented portable grain elevator (1865), cleaner for grain (1870), conceived and manufactured steam-driven separator (1887), built steam-powered traction engine (1887), steam-driven combined harvester and thresher (1887), formed Daniel Best Agricultural Works Co. and Best Manufacturing Co. (CA) (1887), which was sold to Holt Manufacturing Co. (CA) (1908), which made the crawler type tractor, his son Clarence Leo Best [1878–1951] joined the enterprise with his father and continued with Holt Manufacturing Co. (NC40 WWWIA)

BEST, William Newton, 1860–1922, American mechanical engineer and inventor; machinist, railroad erecting shop, oil-burning locomotive, lighting mines, miners' safety lamps (1895), consulting engineer (MEIA NC21 WWWIA)

BETANCOURT Y MOLINA, Augustin de, 1758–1824, Spanish engineer; shipbuilding, navigation, mechanics, use of steam engine, optical telegraphy (invented by French engineer Claude Chappe [1763–1805]), entered

Russian service (1808), bridge construction, aqueducts, directed civil engineering projects in Russia, studied elasticity of vapors (DOSB PS WWWIS)

BETTENDORF, William Peter, 1857–1910, American inventor, mechanical engineer and manufacturer; machinist, with brother Joseph William Bettendorf [1864–1933] formed Bettendorf Metal Wheel Co. (IA) (1886), Bettendorf Axle Co. (IA) (1895), invented first power-lift sulky plow (1878), Bettendorf metal wheel for wagons and farm implements, steel gear for farm wagons (1892), railroad car parts, Bettendorf brake beam, wheels, frame, bolster, Bettendorf journal box, sold company to International Harvester Co. (1905) (ANB DAB MEIA NC21 WWWIA)

BICH, Marcel, 1914–1994, Italian French inventor and industrialist; began work in Paris on ball point pen (1945), invented Bic disposable pen (1952), Ladislao Bíró [1899–1985] developed improved rolling ball point pen (1938) (1943), the first patent for a rotating ball point pen was issued to John J. Loud (MA) (1888) although not commercially successful (IB TBOF TIY(2) WWWIA)

BICKFORD, William, 1774–1834, English merchant and inventor; invented miner's safety fuse (1831), contributed to safety and productivity of mines and quarries (BDOS(E&I) DNB PS)

BIERIOT, Louis, 1872–1936, French engineer and aviator; invented and built monoplane (WA)

BIGELOW, Erastus Brighton, 1814–1870, American inventor, mechanical engineer and manufacturer; invented first power loom for weaving ingrain carpets, mechanized power looms for lace (1837), weaving loom (1838), looms for many types of fabrics, Bigelow mills and carpets, development borne by The Clinton Co. (MA) established (1843), helped found Massachusetts Inst. of Technology (1861), built loom for weaving Brussels and Wilton carpets (1845–1851), numerous US and English patents (DAB EAI MEIA MWBD MEIA NC3 PS WWWIA)

BILGRIM, Hugo, 1847–1932, Bavarian American mechanical engineer and inventor; to US (1869), draftsman, machinist, instruments, developed accurate outlines for teeth of gears especially bevel wheels, slide valve gears (1877), taught and worked at Franklin Inst. (PA) (1876ff), sole owner (1884) of Brehman Brothers Co., inventions in gear manufacturing (MEIA WWWIA)

BILLINGS, Charles Ethan, 1835–1920, American mechanical engineer and inventor; worked for Colt's Patent Firearms Manufacturing Co. (1855ff), armaments (1866), drop forge (1863–1865), formed Billings & Spencer Co. with Christopher M. Spencer [1833–1922], sewing machines (1869, 1872), tap wrench (1879), ratchet drill (1882) (MEIA NC5 WWWIA)

BIOT, Jean-Baptiste, 1774–1862, French physicist and applied physicist; acoustics, voltaic pile, optical properties of gases, Biot law (1818), balloon ascension (1804), Biot-Savart (Felix Savart [1791–1841]) law (1820) (DOT L&M MWBD WWWIS)

BIOT, Maurice A., 1905–1985, Belgian American mining and electrical engineer; aeronautics, mechanics, served as professor at Louvain Univ., Columbia Univ., Brown Univ., USNR in WWII (1943–1946), consultant with Shell Development Co. (1946–1966) and Cornell Aeronautical Lab. (NY) (1946–1969), NAE (1967) (MT3 WWIE)

BIRDSEYE, Clarence, 1886–1956, American inventor and businessman; quick-freezing food (1923), installed Birds Eye line of frozen food, set up General Seafoods Corp. (1924ff), later called General Foods Co., held approximately 300 patents (AIE ANB GI MWBD NYPL TOIAD WA WWWIA WWWIS)

BIRDSEYE, Claude Hale, 1878–1941, American topographical engineer; mapping of volcano areas, topography, mapping of Grand Canyon, US Geological Survey (USGS) (MWBD NC33 WWWIA)

BIRINGUCCIO, Vannoccio, 1480–c. 1539, Italian metallurgist; armaments, published book (1540), papal arsenal (MWBD TAE WWWIS)

BIRKINBINE, John, 1844–1915, American mechanical engineer; hydraulics, metallurgy, public water supplies, iron ore blast furnaces, iron and steel manufacture, fuels for smelting, magnetic concentration of iron ore, with Thomas A. Edison [1847–1931] (MEIA NC12 WWWIA)

BÍRÓ, Laszlo Joszef (aka Ladislao José Bíró), 1899–1985, Hungarian Argentinean journalist and inventor; to Argentina (1940), with brother Georg Bíró patented first workable rolling ball point pen with quick drying ink (1938) (pat. 1943), merchandised (1945), American John Loud patented a roller-ball marking pen (1888), Bíró sold rights to Schaeffer Co., Marcel Bich [1914–1994] manufactured a ballpoint pen in Argentina (1943), Milton Reynolds redesigned the Biro pen later marketed as Papermate then after WWII manufactured in France as Bic pen (BDOTHOT BEOS GI NYTObit OED PS S&TF TBOF TIY(2) WOI)

BISAT, William Sawney, 1886–1973, British civil engineer and surveyor; construction of large civil engineering works, gravel and sand quarries for commercial use, wrote treatises on fauna of carboniferous rocks, studied drift deposits (PS RS20 WWW)

BISPLINGHOFF, Raymond L., 1917–1985, American aeronautical engineer; aeronautics and astronautics, aeroelasticity, structural dynamics, faculty and administration of department and college at Massachusetts Inst. of Technology (1946–1970), government research administrator (NSF) (NASA) (1970–1974), chancellor at Univ. of Missouri-Rolla (1974–1977), university laboratory director (1977ff), NAE (1965), NAS (1967) (IWAP MT3 WWIE WWWIA)

BISSELL, Melville Reuben, 1843–1889, American inventor and manufacturer; carpet sweeper (1876) preceded by Englishman Joseph Whitworth [1803–1887] (1859), waterproof shoe (1881), seamless felt boot (1882), obtained 11 patents (GI MEIA NC7 MWBD WA)

BJORNSON, Bjorn G., 1898–1969, Icelandic American engineer and inventor; to US (1918), research in electric switching, designed and tested protection of circuits, variable impedance circuit (1928), voice operated electric systems and switches, included echo depressor and noise desensitizer, research supervisor at Bell Telephone Labs., held over 57 patents (NC55 PS)

BLACK, Harold Stephen, 1898–1983, American electrical engineer and inventor; feedback control system, negative feedback amplifier (1927), pulse code modulation, consumer electronics, servomechanisms, biomechanics, consultant (1966ff), 333 patents (AMOS BDOTHOT MHSE NYT-A PS WWIE WWWIA)

BLACK, Joseph, 1728–1799, Scottish chemist, physician and physicist; rediscovered carbon dioxide (1754), specific heats (1760), latent heat (c. 1763), at Univ. of Glasgow (1756–1766), Univ. of Edinburgh (1766ff) (CDOS DNB LOTE MWBD RHW SFLOE)

BLACKBURN, J. Lewis, 1913–1997, American electrical engineer; power system protection, power system relays, Westinghouse Electric Corp. (1936–1978), including engineering supervisor of Relay and Instrument Division (1952) and section manager of all high-speed relaying (1955), engineering consultant (1969ff), NAE (1997) (MT10 WWIE WWWIA)

BLAGONRAVOV, Anatolii Arkadievich, 1894–??, Soviet mechanical engineer; machinery and mechanics of armaments, automatic weapons, Moscow Academy of Artillery (1929–1946) becoming professor (1938) and later president of Academy of Artillery Science (1946–1950), lieutenant general in artillery (SMOS WWWIS)

BLAIR, Henry, 1804–1860, Black American inventor; invented corn seeder (1834), invented cotton seed planter (1836) (BI NBAS)

BLAKE, Eli Whitney, 1795–1886, American inventor and manufacturer; worked for his uncle (Eli Whitney [1765–1825]) in manufacture of firearms, manufacturer of domestic hardware, door lock (1833), fasteners for cupboards (1843), Blake jaw stone crusher (1858) (ANB DAB NC9 PS WWWIA)

BLAKE, Francis, 1850–1913, American physicist and inventor; US Coast and Geodesic Survey, transcontinental longitude surveys, telephone communication (1878), Berliner microphone, 20 patents between

1878 and 1890 (ANB BDOAS DAB NC22 WWWIA WWWIS)

BLAKE, Lyman Reed, 1835–1883, American inventor; shoemaking manufacture, machine to sew uppers to soles of shoes (1858), McKay sewing machine (AIE ANB DAB MEIA WWWIA)

BLAKE, William Phipps, 1825–1910, American geologist and mining engineer; explored various regions in North Carolina, Japan, Alaska and other regions of US for mineral resources, involved in educational and demonstration activities (DBOAS DAB NC25 WWWIA WWWIS)

BLAKEY, William, c. 1712–c. 1792, British steam engineer; made Savery-type (Thomas Savery [1650–1715]) steam engine—a competitor to Newcomen (Thomas Newcomen [1663–1729]) (EE MWBD)

BLANC, Armond, ??–1962, French agricultural engineer; served in WWI in French engineer division (1914–1918), integrated body of agricultural engineers into Génie Rural & Hydraulique Agricola (1919–1955), moving through ranks, becoming general director, engineer of electricity for agriculture for Inst. of Electrotechnique of Grenoble, director and taught at National School of Agricultural Engineering, was president of CIGR (5 yr.), French Court of Law (1955–1962), international award named after him (CIGRNewsletter)

BLANCHARD, Jean-Pierre-François, 1753–1809, French aeronaut and inventor; balloon ascents (1784ff), invented parachute (1785), first flight across English channel (1785) joined by John Jeffries [1745–1819] (ASAI MWBD WBE WWWIS)

BLANCHARD, Thomas, 1788–1864, American inventor and mechanical engineer; device for paring apples (1891), lathe for turning gun barrels, tack-making and counting machines (1806), first US patent on lathe to make irregular shapes (1819), process for binder timber fibers (1836) for ship building, steam carriage, invented machine for cutting paper and folding envelopes, pantograph (1836), over 20 patents (AIE ANB ASAI DAB ID LAI MEIA MAH MWBD NC6 WWWIA WWWIS)

BLATCH, Nora Stanton—see BARNEY, Nora Stanton Blatch

BLEICH, Hans Heinrich, 1909–1985, Austrian American civil engineer; to US (1945), worked in Austria and England (1935–1945), structural design, applied mechanics, aircraft company, airflight structures, vibration analysis, stability of columns, with Chance-Vaught Aircraft (CT) and Hanover Bridge Engineers (NYC), Weidlinger Associates (NYC) (1957–1985), also with Columbia Univ. (1947–1975), consulting engineer, NAE (1978) (AMOS MT3 WWWIA)

BLELOCH, George H., 1835–1891, American machinist and mechanical engineer; with brother organized Bleloch & Co., which was a book and stationary company, manufactured self-threading sewing machine (c. 1871), developed into National Needle Co. (MA) of which he was general manager, part owner of Excelsior Needle Co. (CT) (1888), president of Sewing Machine Supply Co. (MA), machine tools (MEIA)

BLENKINSOP, John, 1783–1831, English inventor; double-cylinder locomotive (1811) (also see Richard Trevithick [1771–1833] (DNB MWBD)

BLERIOT, Louis, 1872–1936, French aviator and manufacturer; monoplane, flew across English channel (1909), his plant built over 10,000 planes during WWI (HIV MMAH MWBD)

BLICKEMSDERFER, George C., 1850–1917, American inventor and manufacturer; patented innovative typewriter (1889), used cylindrical single type element, developed lightweight portable machine (approximately 10 lb.), had a scientific keyboard, a Featherweight Blich that weighed 5 lb., a Blick electric machine (1902), during WWI built military equipment (I&T NYTObit)

BLICKENSDERFER, Herman, 1897–1965, American civil engineer; in WWI with US Army (1917–1919), faculty at Valparaiso Univ. (1927–1940), mine superintendent in Peru (1940–1942), various faculty appointments (1942–1964) at South Dakota State Univ., Univ. of Texas, Univ. of Idaho, Bradley Univ., worked for US Conservation Service, airport construction and highway department (NC53)

BLOCK. Edward, 1902–1961, American chemical engineer and inventor; engineer with Superior Chemical Co. (1923ff), which became part of American Cyanide & Chemical Corp. (1930), with father and brothers formed Blockson Chemical Co. (1926), which became a part of Mathieson Chemical Co. (1955), serving in several executive positions (until 1961), held several patents which dealt with manufacture of chemicals (NC51 WWWIA)

BLOCK, Louis, 1850–1926, German American engineer; to US (ante 1882), steam heating, refrigeration machinery, separating tank for refrigeration machines (1888), brine tank for ice manufacture (1894), condenser (1918), De La Verne Co. (1882ff), serving from draftsman to president, at least 8 patents, consulting engineer (MEIA WWWIA)

BLODGETT, Katherine Burr, 1898–1979, American chemist and inventor; researcher at General Electric Co. (1918), electrical and electronic studies relating to filaments and their interaction with gases, monomolecular layer of oil for lubrication (1934) known as Langmuir-Blodgett firm that has many applications, non-reflecting glass (1938), cloud physics (1944), a major invention was invisible glass (1958) (AIE ANB MWBD NWITPS NWS WWWIA WWWIS)

BLONDEL, (Nicolas) François, 1618–1686, French military engineer and architect; fortifications, restoration of bridges and arches, port in Paris (1672), made maps (Grenada), in charge of public works in Paris (1673), lecturer at public meetings and colleges (DOSB MWBD PS WWWIS)

BLOOD, Aretas, 1816–1889, American engineer and manufacturer; locomotive builder, Amoskeag steam fire engine, organized Vulcan Works (NH) (1853) then incorporated Manchester Locomotive Works (NH) (1854) (MEIA NC7)

BOARTS, Robert Marsh, 1904–1960, American chemical engineer; worked for Detroit Water Board (1927–1929), with Comstock & Westcott, Inc. (1929–1931), faculty at Univ. of Tennessee (1934–1960), while on leave served as head of TVA Chemical Engineering Lab. (NC48 PS WWWIA)

BOBO, Melvin, 1924–1993, American mechanical engineer; conception, design, development, refinement of aircraft gas turbine engines, General Electric Co. (1949–1991), becoming chief engineer (1985), over 35 patents, NAE (1991) (MT7)

BODE, Hendrik Wade, 1905–1982, American applied mathematician and systems engineer; circuit theory, electrical network, design of electric filter, weapons systems, fire control devices (WWII), systems engineer at Bell Labs. (40 yr.), science and technology, NAS (1957), NAE (1964) (MT3 WWIE WWWIA WWWIS)

BODMER, Johann Georg, 1786–1864, Swiss mechanic and inventor; firearms (1803), interchangeable parts, textile manufacturing machinery (1824), traveling grate stoker called Bodmer stoker (EE MWBD WWWIS)

BOEING, William Edward, 1881–1956, American industrialist and manufacturer; airplane designer and manufacturer (Boeing Co.), clipper flying boat (1938), US bombers used in WWII B-17 (1935ff), B-29 (1942), first commercial jet airliner Boeing 707 (1954), built several large commercial airplanes in a 7X7 series (MWBD NC36 WPost WWWIA)

BOELTER, Llewellyn Michael Kraus, 1898–1966, American mechanical and electrical engineer; cooling of automotive engines, heat and mass transfer (including atmospheric), research led to Dittus-Boelter equation (1930) (Frederick W. Dittus [1897–1987]), air pollution, faculty at Univ. of California-Berkeley (1919–1944), dean of engineering at Univ. of California-Los Angeles (1944–1965) (AE47 HOHT L&M WWIE WWWIA)

BOGARDUS, James, 1800–1874, American inventor; watchmaker, eight-day chronometer (1828), dry gas meter (c. 1833), engraving machines, deep-sea sounding equipment, cast iron buildings, construction equipment, prolific number and variety of patents, cast iron load bearing column buildings (1848) (AIE ANB BDOACE DAB GI MEIA MWBD NC8 WWWIA WWWIS)

BOGART, John, 1836–1920, American engineer; hydroelectric development, power generation and transmission, chief engineer, caissons for dam construction, railroad and canal, bridges, tunnels, waterways, irrigation, public park planning (DAB WWWIA)

BOGUE, Virgil Gay, 1846–1916, American civil engineer; railroad engineer, building railroads over Andes in Peru, railroad across Cascade Mountain at Stampede Pass in state of Washington, consulting engineer for worldwide projects (BDOACE DAB NC13 WWWIA)

BOHR, Niels Henrik David, 1885–1962, Danish physicist; faculty at Univ. of Copenhagen (1916ff) and head of Inst. of Theoretical Physics (1920–1962), theory of atomic structures, spectroscopy, principle of complementarity (1927), in US during WWII, active in control of atomic and nuclear energy in the world, Nobel Prize (1922), FRS, fNAS (1925) (BDOS(Wi) BEST LDOS MWBD SAI-MP TBDOS(Po) TGS WWWIA WWWIS)

BOIS-REYMOND, D. H.—see DU BOIS-REYMOND

BOLLER, Alfred Pancoast, 1840–1912, American civil engineer; bridge and aqueduct construction for railways, foundations, bridge design and construction, structural engineer, consulting engineer (BDOACE DAB NC9 WWWIA)

BOLLMAN, Wendell, 1814–1884, American civil engineer; railroad survey for Baltimore & Ohio Railroad, railroad bridges, all-metal Bollman Truss of cast iron and malleable iron (1850, 1852), supervised construction of nearly 200 bridges, several bridges in South America and Mexico (BDOACE FEL NC11)

BOLTZMANN, Ludwig Eduard, 1844–1906, Austrian physicist; kinetic theory of gases, Boltzmann distribution law (1868), derived Maxwell-Boltzmann law of equipartition of energy (1871), Stefan-Boltzmann law (1884) (Josef Stefan [1835–1893]) of blackbody radiation showing that the quantity of radiation increased as the fourth power of the absolute temperature, Boltzmann constant or number, his equation relating entropy and disorder is engraved on the headstone of his grave, fNAS (1904) (BEST EOS&T GLOSB L&M ME126 MWBD WWWIS)

BONARD, Louis, 1809–1871, French American inventor; circular loom for weaving hats, brick-making machines, machine for casting iron, concerned with cruelty to animals (DAB WWWIA)

BOND, George Meade, 1852–1935, American mechanical engineer and inventor; at Harvard College Observatory, with William A. Rogers [1832–1898] developed Rogers-Bond Universal comparator (1880), standards and gauge department for Pratt and Whitney (ante 1902), weights and measures, lecturer (MEIA WWWIA)

BONWILL, William Gibson Arlington, 1833–1899, American inventor and dentist; invented diamond drill (1873, 1875), electromagnet mallet, contour filling operation, device for root canal, improved amalgam fillings, Bonwill Surgical Engine, inventions outside of dentistry included grain reaper, kerosene lamps, shoe fasteners, ideas for aerial trolley car (ANB DAB NC5 WWWIA WWWIS)

BONZANO, Adolphus, 1830–1913, German American mechanical engineer and inventor; to US (1850), with Detroit Bridge and Iron Works (MI), partner and chief engineer of Clark, Reeves & Co. and successors (1888ff), bridge construction, viaducts, invented Bonzano rail joint, turntable (1870), consulting engineer worldwide (BDOACE DAB MEIA NC16)

BOOLE, George, 1815–1865, English mathematician; linear transformation (1841), Queen's College in Cork (Ireland) (1849–1865), mathematical logic, operational calculus, laws of thought (1854), symbolic logic, method referred to as Boolean algebra (BDOS(Wi) BEST DNB DOSB MWBD RHW WWWIS)

BOOTH, Hubert Cecil, 1871–1955, English mechanical engineer; invented vacuum cleaner (1901), small electric model sweeper (1905), design of marine engines, design of bridges (DNB EAI NYPL 1000Y PA)

BORDEN, Gail, 1801–1874, American inventor; surveying making topographical map of Texas (1838), meat biscuits (1851), making condensed milk (1853, 1856) under vacuum, vacuum pan, concentrated juices (AIE ANB BEST DAB MWBD NC7 WWWIS)

BORDEN, Simeon, 1798–1856, American civil and mechanical engineer; metal and woodworking crafts, surveys for maps, surveyor and engineer for towns, railroads, computation and methods for running curves for railroads (DAB NC24 WWWIA WWWIS)

BORDEN, Thomas James, 1832–1902, American mechanical engineer; with iron works, textile mills, print works, developed advanced processes for printing fabrics, perforated pipe fire control systems that led to automatic sprinkler systems (MEIA NC29)

BORN, Ignaz von, 1742–1791, Austrian mineralogist and metallurgist; extraction of gold and silver from ores, amalgamation (MWBD WWWIS)

BORRIES, Bodo von, 1905–1956, German electrical engineer; electron microscopy, metallurgist, electron microscope especially applies to thin metal surfaces, electron optics (DOSB TFE&I)

BOSCH, Karl, 1874–1940, German chemist, chemical engineer and inventor; industrial synthesis of ammonia, led to production of fertilizers and explosives known as Haber-Bosch process, use of catalysts, helped form and became head of I. G. Farben (1925), Nobel Prize (1931) (BDOS EAI LOTOS WWWIS)

BOSCH, Robert August, 1861–1942, German engineer and industrialist; founded Robert Bosch Co. (1886), manufacturer of magnetos, igniters, lubrication devices, spark plugs, fuel injection pumps (MWBD WWWIS)

BOSE, Jagadis Chandra, 1858–1937, Indian physicist and inventor; faculty at Presidency College Calcutta (1885–1915), Bose Research Inst. founder and president (1917–1937), properties of short radio waves, radio detector, invented a form of wireless coherer, related biological responses in plants and animals to mechanical stimulus (crescograph), knighted (1917), FRS (1920) (MWBD RSObit RTFE&I TTOS WWWIS)

BOSS, William A., 1869–1945, American agricultural engineer; leader in development of profession, mechanization of seeding, mowing, threshing and germination, faculty at Univ. of Minnesota (1895–1913), formed Specialty Manufacturing Co. (1913ff), returned to Univ. of Minnesota as chair of department (1919–1938), held 17 patents (AE46 WWIE WWWIA)

BOUCICAUT, Aristide, 1810–1877, French inventor and merchandiser; invented department store known as Bon Marché selling dry goods in Paris (1838), business became the pattern for development of many department stores in western countries (EB 1000Y)

BOULTON, Matthew, 1728–1809, English manufacturer and engineer; ironmaker, developed process for making steel inlay, development and manufacture of steam engines (partner of James Watt [1736–1819]) (1775), steam press coinage machines (1786), contributed to industrial revolution (DNB GI LOTE MMAH MSAE MWBD PS TBDOS(E&I) TSOE WWWIS)

BOUDY, Ray H., 1903–1992, American chemical engineer; chemical process development, invented Dowtherm heat transfer fluid, Dow Chemical Co. (42 yr.) (1926–1968), including director of research and patents (1952) and vice president (1953), consultant (1968ff), NAE (1967) (MT7)

BOURDON, Eugène, 1808–1884, French inventor and industrialist; machine shop (1835), instrument maker, manufacture of steam engines, pressure gauge, Bourdon tube pressure gauge (1849) (CEOS&T EE FNIE MWBD TBDOS(E&I) WWWIS)

BOURGEOUYS, Marin Le, fl. early 17th century, French inventor; invented first flintlock musket (1615) (EB)

BOURN, Daniel, mid-18th century, English inventor; developed carding machines for textiles (1748) with Lewis Paul [??–1759] (BDOTHOT EB MMAH TCBE)

BOUSCAREN, Louis Frederic Gustave, 1840–1904, American civil engineer born in Guadeloupe; surveys, railroad construction in midwest and south, designed longest truss span in world at that time (1876), consultant (BDOACE WWWIA)

BOUSSINESQ, Joseph Valentin, 1842–1929, French physician, civil engineer and inventor; faculty at Gap College (France) (1866–1872), faculty of science at Lille (1872), then Academy of Science in Paris as dean, hydrodynamics, hydrostatics, mechanics, electron theory (DOTHOS WWWIS)

BOWDITCH, Nathaniel, 1773–1838, American astronomer and inventor; method of calculating longitudes, use of sextant, surveying the sea, celestial mechanics (ASAI MWBD NC6 WWWIS)

BOWEN, Edward George, 1911–1991, Australian physicist and engineer; worked on radar with British Air Ministry (ante 1944), Australian Commonwealth Scientific Research (CSIRO) (1903ff) for 25 yr. was director of radar physics, magnetron, airborne, navigational devices, radiotelescopes, experimental cloud physicist, fNAE (1977) (Bridge MT8 WWWIS)

BOWMAN, Robert A., 1907–1977, American mechanical engineer; research engineer at Westinghouse Electric Corp. (1925–1951), Bechtel Corp. (1951ff), from chief mechanical engineer to senior vice president (1970) to consultant (1973), design of heat transfer equipment, atomic power, thermal power, management, consulting engineer, NAE (1970) (MT1 WWIE WWWIA)

BOWSER, Edward Albert, 1837–1910, Canadian American civil engineer; hydromechanics, mechanics, surveying, faculty at Brooklyn Polytechnical Inst., Rutgers Univ. (1868–1904), US Coast & Geodesic Survey (1875–1895), author of several books (BDOAS NC19 WWWIA)

BOYD, Henry, 1802–1886, Black American inventor; cabinet maker, carpenter, developed new wooden bed frame called Boyd Bed at his factory (until 1863) (AAI)

BOYD, James, 1904–1987, Australian American mechanical engineer; field engineer, geology, materials, exploration, surface mining and reclamation, minerals and energy, faculty, including dean at Colorado School of Mines (1929–1947), including US Army Materiel Command (1941–1944), US Bureau of Mines (1947–1951), exploration for and vice president of Kennecott Copper Corp. (1957–1980), president to chairman of Copper Range (1960–1971), president of Materials Consultants (1973–1977), materials policy, NAE (1967) (MT5 NCH,M WWIE WWWIA WWWIS)

BOYDEN, Seth, 1788–1870, American inventor and manufacturer; improved father's machine for splitting of sheepskin and leather (1813), patent leather (1822), malleable cast iron (1826), sheet iron, cut-off governor for steam engine, numerous inventions in several subjects (DAB EB MEIA MWBD NC11 PS WWWIA WWWIS)

BOYDEN, Uriah Atherton, 1804–1879, American engineer and inventor; son of Seth Boyden [1788–1870], railroad construction, hydraulic development, design of efficient turbines—Fourneyron (1844), Boyden turbine water wheel, invented hook gauge, consulting engineer (DAB MEIA MWBD NC11 PS WWWIA WWWIS)

BOYER, Francis H., 1845–1909, American civil and mechanical engineer; millwright, architect, steamboat transportation (1863), refrigeration business, builder of refrigeration machinery, brewery refrigeration system (1884), master mechanic (1890) (MEIA)

BOYER, Marion Willard, 1901–1982, American chemical engineer; petroleum production specialist, Standard Oil of Louisiana (1927ff) for 23 yr., executive, becoming vice president of Esso Standard Oil Co. (predecessor of Exxon), general manager of Atomic Energy Commission (1950–1953) (CB NYTObit)

BOYER, Raymond Foster, 1910–1993, American physicist and engineer; high polymers, industrial development of plastics technology, radiation chemistry of polymers, light and heat stabilizers, at Dow Chemical Co. (1935ff), from research director to director (1948–1952) and plastics research (1952ff), NAE (1978) (AMOS MT7 WWWIA WWWIS)

BOYKIN, Otis, 1920–1982, Black American inventor and electrical engineer; industry research and development, precision wire round resistor known as Boykin Resistor (1961), electrical capacitor of high capacity in a small volume, high dielectric breakdown (1965), electric resistance element (1966), air filter, invented electrical devices used in guided missiles and computer heart stimulator (pace maker) (AAI BCST BDOBA BIS PS)

BOYLE, Robert, 1627–1691, British (Irish) natural philosopher; crystals, heat and thermodynamics, pneumatics, respiration, vacuum, developed air pump (1657), Boyle law (1660), barometer, hydrometer to determine density of fluids (1665), Hypatia [c. 370–417] and Francesco Folli [1625–1685] also credited with development of hydrometer, combustion, invention of match (1680), considered father of modern chemistry, FRS (1691) (BEST ChemHerit DNB DOTHOS FNIE L&M MWBD NYPL SAI TBDOS TGS WWWIS)

BOYS, Sir Charles Vernon, 1855–1944, English physicist and inventor; mining and metallurgy, quartz fibers, radio micrometer (1888), calorimeter (1905), high-speed camera (DNB PS TBDOS MWBD WWWIS)

BRADLEY, Humphrey, c. 1584–1625, Dutch drainage engineer; master of dikes (1599), drained areas in England (Fens) and France, canal projects (BDOS(Wi) PS)

BRAGG, William Henry, 1862–1942, English physicist and inventor; with son William Lawrence Bragg [1890–1971] invented x-ray spectrometer, x-ray diffraction (crystal structure), faculty at Adelaide Univ. (1885–1908), Leeds Univ. (1909–1915), Univ. of London (1915–1923), Royal Institution (1923ff), Nobel

Prize (1915), fNAS (1939) (DNB IAI MWBD TBDOS TGS WWWIS)

BRAILLE, Louis, 1809–1852, French teacher and inventor; teacher (1826), Braille raised point printing (1827) for assisting blind in reading (GI MWBD PS SAI WA WWWIS)

BRAITHWAITE, John, 1797–1870, English engineer; first practical steam engine (1829), editor of magazines dealing with steam and railroads and agriculture (DNB MWBD WWWIS)

BRAMAH (originally BRAMMER), Joseph, 1748–1814, English mechanical engineer and inventor; water closet flush toilet (1778), precision safety lock (1784), which resisted picking for 67 yr. (Englishmen Charles Chubb [1779–1845] and his brother Jeremiah Chubb [1790–1847] made improvements on locks), invented hydraulic press (1795), precision machine tools (with Henry Maudslay [1771–1831]), proposed moving ships by propeller (1785), over 18 patents (AHOT BDOTHOT DNB EAI GI GEAPIT MMAH MWBD NYPL 1000Y PS SAI TGE:B WWWIS)

BRAMANTE, Donato, 1444–1514, Italian architect and builder; designer and builder of St. Peter's Basilica, painter of murals (1000Y)

BRANCA, Giovanni, 1571–1640, Italian engineer and inventor; device in which steam diverted to vanes on a wheel—impulse steam turbine (1629) used to drive stamping mill (BDOS EE MMAH TTOS TTOT WBE)

BRANDENBERGER (BRANDENBURGER), Jacques Edwin, d. 1954, age 81, Swiss inventor; invented flexible cellophane (1908) from cellulose acetate, designed machine for continuous production of cellophane (1911), manufactured by La Cellophane Paris (1913), rights sold to DuPont, which developed moisture-proof version (1924), invented a type of artificial silk called Celta (ASAI EB NYPL NYTObit PA S&TF TBOF TTOS WBE)

BRANDT, Alfred, 1846–1899, German engineer; hydraulic drill for tunnel construction, devised double galley system for tunnel construction (MWBD)

BRANLY, Edouard Eugène Désiré, 1844–1940, French physicist and inventor; medical professor of electrotherapy (1897–1916), electricity, electrostatics, inventor of coherer (1890), established principle of wireless telegraphy (1899) later developed by Guglielmo Marconi [1874–1937] (CDOS(Mi) TBDOS WWWIS)

BRANNON, Hezzie Raymond, Jr., 1926–1994, American physicist and inventor; electronics, radiocarbon dating, spectral nuclear logging, borehole gravity meter, deepwater drilling platforms, served in WWII (1943–1946), senior research geophysicist at Humble Oil & Refining Co. (1952ff), NAE (1980) (AMOS MT8 WWWIA)

BRARD, Robert E. M., 1907–1977, French military engineer; hydrodynamics, construction and repair of ships, propeller design, probability theory, random theory function, became vice admiral in French navy, director of Paris Ship Model Research Lab. (1941–1969), director of French Maritime School (1958–1962), faculty of Univ. of Nantes (1969ff), fNAE (1976) (MT1)

BRATTAIN, Walter H., 1902–1987. NAS (1959) (BM63). See John Bardeen

BRAUN, Karl Ferdinand, 1850–1918, German physicist and inventor; crystal rectifier (1874), invented cathode ray oscilloscope (1897), research director, Nobel Prize (1909) (MMAH MWBD NC41 NYPL WWWIS)

BRAUN, Wernher Magnus von, 1912–1977, German American engineer; aerodynamicist, at Peenemunde (Germany) test facility (1932–1945), rockets, V-2 ballistic missile used in WWII, to US (1945), space exploration, Saturn series of rockets, Apollo mission, director of Marshall Space Center (1960–1970), research administrator with NASA (1960–1972) (AMOS ANB BDOTHOT BEST CDOS HIW IAI ITCOF IWAP MT1 ITCOF IWAP MT1 MWBD MWGD RHW SAI TBDOS(PO) TGS WWWIA WWWIS)

BRAYTON, George B., 1830–1892, British American mechanical engineer; steam engine, internal combustion engine, gas turbine, compressed air to start engine (1872–1873), Brayton cycle engine, engine used by George Selden [1846–1922] to power vehicle (1876) regulating supply of oil to vapor engines (1883) (ASAI E-64 EE ME125 MEIA WA WWWIS)

BREAKWELL, John Valentine, 1917–1991, Swiss American astronautics and mathematician; to US (1941), development of astrodynamics, flight trajectory optimization, faculty of Tufts Univ., North American Aviation Corp. (CA) (1949–1959), Lockheed Missiles & Space Co. (1959–1964), professor at Stanford Univ. (1964ff), editor of professional journals, NAE (1981) (AMOS Bridge MT7)

BREARLEY, Harry, 1871–1948, English metallurgist and inventor; worked on improving hardening of armor-piercing shells, various laboratories (1901–1907), Thomas Firth and Sons (1904) and John Brown Ltd. (1907ff), inventor and commercial developer of stainless steel (1913ff), alloy steels for Enfield rifle, manager of Brown Bayley steel works (1914–1925) (BDOTHOT DNB DNB(MP) EB MMAH NYPL TGE:B WA)

BRECKENRIDGE, Lester Paige, 1858–1940, American mechanical engineer; water power, machine shop, fuel tests, recording devices, smoke prevention equipment, educator, consultant (MEIA WWWIA WWWIS)

BREER, Carl, 1884–1970, American mechanical and automotive engineer; with several truck and automobile companies, streamlining cars (1927ff), consulting engineer with Fred Zeder and Owen Skelton, then to

Maxwell Motors Corp., engineering consultant (NYTObit WWWIA)

BRÉGUET, Abraham-Louis, 1747–1823, French mechanician; watchmaker, astronomical and marine clocks of great accuracy, Breguet spring for watches (EE MWBD WWWIS)

BRÉGUET, Antoine, 1851–1882, French electrical engineer; son of Louis-François-Clément Bréguet [1804–1883], electrical recording anemometer, electrical instruments (MWBD WWWIS)

BRÉGUET, Louis-Charles, 1880–1955, French manufacturer; his first airplane (1909), hydroplane (1912), gyroplane (1917), founded commercial airline that became Air France (MWBD)

BRÉGUET, Louis-François-Clément, 1804–1883, French physicist and watchmaker; grandson of Abraham-Louis Bréguet [1747–1823], electric thermometer (1840), induction coil (1842), speed of light apparatus (1843), electric clocks for transmitting time though a distance (1856) (MWBD WWWIS)

BREWER, William Henry, 1828–1910, American agriculturalist and innovator; geological survey, research administration, manufacture of glucose sugar from starch (1882), sorghum sugar industry, public health, suspension of clay in river water and river deposits, NAS (1880) (BDOAS BM12 DAB NC13 WWWIA WWWIS)

BREWSTER, Sir David, 1781–1868, Scottish physicist and inventor; optics, polarized light, invented kaleidoscope (1816), three-dimensional photographs (1849) following work of Alexander Becker, improved on stereoscope, advocated refractive light system for lighthouses, college principal, fNAS (1864) (DNB MWBD TWABOI WA WWWIS)

BREYER, Frank Gottlob, 1886–1966, American chemical engineer; New Jersey Zinc Co. (1910–1927), where he was chief of research (1917–1927), formed organization of Singmaster & Breyer (1927–1963), which entered field of design engineering (1942ff), including manufacturing plants for fertilizer, uranium, magnesium and cortisone, over 50 patents (NC53 PS WWWIA)

BRIDGMAN, Percy Williams, 1882–1961, American physicist; high-pressure reactions up to 20,000 atmospheres, design of equipment for high-pressure reactions, seals for equipment, worked with people at General Electric Co. to produce synthetic diamonds (1955), faculty at Harvard Univ. (1908–1954), NAS (1918), fRS (1949), Nobel Prize (1946) (BDOS(Po) BEST BM41 LDOS TBDOS(Wi) MWBD NC48 TGS WWWIA WWWIS)

BRIGGS, Henry, 1561–1630, English mathematician and inventor; disseminated John Napier's [1550–1617] invention of logarithm, modified Napier's logarithm into common (Briggsian) or decimal logarithm (1617) (BDOS DNB MWBD TBDOS WWWIS)

BRIGGS, Robert, 1822–1882, American civil and mechanical engineer; steam heating, constructed tube works, rolling mill, aqueduct, rotary fans, foundry, standard tapered pipe threads named after Briggs, consulting engineer (EE MEIA NC9 WWWIA)

BRIGHT, Sir Charles Tilston, 1832–1888, English electrical engineer; telegraph, transatlantic cable, standardization of electrical units (1861) (DNB DOSB MWBD WWWIS)

BRIGHT, Fred E., 1856–1925, American mechanical engineer and inventor; linotype (Typograph), patented inventions relating to ball bearings, manufacturer of bearings, co-founder of Cleveland Cap Screw Co. (OH) (1901), later Electric Welding Products Co., helped found Hess-Bright Manufacturing Co. (PA) (1904) for manufacturing ball bearings, becoming treasure and vice president (MEIA NYTObit)

BRIMACOMBE, James Keith, 1943–1997, Canadian metallurgical engineer; extractive metallurgy, process metallurgist, continuous casting of steel, faculty at Univ. of British Columbia (1970ff), where he established Center for Metallurgical Process Engineering, was on leave as president and chief executive officer of the Canadian Foundation for Innovation, several patents, worked with several industry-academic-development groups, fNAE (1997) (MT10 WWIE WWWIA)

BRINCKERHOFF, Charles M., 1901–1987, American mining engineer; block-caving mining, mineral industry, in western US and Chile with Anaconda Co. (1964–1971), including chairman and CEO (1969–1971), mining consultant (1972ff), NAE (1976) (MT4 WWIE WWWIA)

BRINDLEY, James, 1716–1772, English civil engineer; canal design, construction of several canals, used tunnels and aqueducts extensively, developed construction of canal banks by puddling clay (DNB EAI EB LOTE MWBD PS SFLOE)

BRINELL, Johann August, 1849–1925, Swedish engineer and metallurgist; hardness of metals and alloys, Brinell hardness scale (1900), nondestructive testing, engineer for the Lesjofers ironworks (1875–1882), chief engineer of Faversta ironworks (1882–1903), research at Jarnkontoret works (1903–1914), other hardness tests named after S. P. Rockwell and Frederich Mohs [1773–1839] and Vickers Co. (CEIS&T EB EE MWBD TBDOS(Po) WWWIS)

BRISTOL, William Henry, 1859–1930, American mechanical engineer and inventor; taught at Stevens Inst. of Technology (1884–1899), with brother Franklin Bristol organized Bristol Co. (1889), steel fasteners for joining leather belting (1889), recording devices for pressure of air, gas, steam (1888–1904), Bristolphone (1915), about 25 patents (DAB MEIA NC26 WWWIA WWWIS)

BROADHEAD, Garland Carr, 1827–1912, American geologist and engineer; surveyor, geologist, reconnaissance for geological features and natural resources, particularly in the Midwest (DAB NC13 WWWIA WWWIS)

BROCOT, Achille, 1817–1878, French inventor; pendulum spring system (EE)

BRO(A)HEAD, Charles C., 1772–??, American civil engineer; surveyor, charged with surveying eastern end of Erie Canal (1816), considered one of first American engineers (ASAI BDOACE)

BROGLIE, Louis Victor Pierre Raymond de, 1892–1987, French physicist; brother of Maurice de Broglie [1875–1960], showed the wavelike properties of particles (1923, 1927), work led to field of wave mechanics, Nobel Prize (1929) (GI MWBD RHW SAI WWWIS)

BROGLIE, Maurice Duc de, 1875–1960, French physicist and inventor; brother of Louis Victor Pierre Raymond de Broglie [1892–1987], photoelectric effect, established private laboratory, studied X-ray diffraction, established absorption spectra (ANB DOTHOS MWBD MWBD RHW SAI WWWIS)

BROMBERG, Robert, 1921–1999, American mechanical engineer; national defense and space systems, faculty at Univ. of California-Berkeley (1943–1936), Univ. of California-Los Angeles (UCLA) (1946–1954), Ramo Wooldridge Corp. (later TRW) (1954ff), becoming vice president (1964), energy conversion, engineering mechanics, materials, thermal engineering, instrumentation, NAE (1969) (Bridge MT9 WWIE WWWIA)

BROOKE, John Mercer, 1826–1906, American naval officer; deep-sea sounding apparatus, surveying expeditions, developed naval armor for Confederate states (BDOAS DAB NC22 WWWIS WWWIA)

BROOKS, Bryon Alden, 1845–1911, American teacher and inventor; mechanics, improvements to mill industry, typewriter improvements (acquired by Remington) (1878), over 30 patents (DAB MEIA NC3)

BROOKS, Frederick A., 1895–1967, American agricultural engineer; airplane designer for Curtiss Aeroplane and Motor Corp. (1917–1918), sales engineer at Dunlop Tire and Rubber Co. (1920–1921) and with several automotive companies (1923–1931), research engineer at Univ. of California-Davis (1931–1967) in zone climatology, forest fire research, atmospheric radiation, heat, moisture and momentum flow in the atmosphere (AE47 WWWIA WWWIS)

BROOKS, Thomas Benton, 1836–1900, American geologist and mining engineer; topographic work, surveys, New Jersey Geological Survey, designed dial compass, iron mine development (DAB NC45 WWWIA)

BROUGHTON, Donald Beddoes, 1917–1984, English American chemical engineer; US citizen (1946), petroleum refining, petrochemicals, extraction and adsorptive separation, drying, faculty at Massachusetts Inst. of Technology (1942–1949), research engineer with Universal Oil Products Co. (1949ff), 50 patents, industrial consultant, NAE (1976) (AMOS MT3 WWIE WWWIA)

BROWN, Alexander Ephrain, 1852–1911, American civil engineer, inventor and manufacturer; US Geological Survey, chief engineer of Massillon Iron Bridge Co. (OH), invented hoisting and conveying equipment for iron and coal (1879), processes for annealing malleable casting, organized Brown Hoisting Machinery Co., becoming president (1910), over 100 patents (ANB DAB MEIA NC33 WWWIA WWWIS)

BROWN, Alexander Timothy, 1854–1929, American mechanical engineer; manufacturer of farm machines, motor-driven agricultural machinery, firearms manufacture, patents pertaining to guns, typewriter (1888), numerous patents in different fields, served as president at various times of Syracuse Aluminum & Bronze Co. (NY), Globe Malleable Iron Works (NY), Pneumelectric Co. (NY) and Journal Printing and Publishing Co. (NY) (MEIA NC35)

BROWN, Charles Sumner, 1860–1926, American mechanical engineer and inventor; bolt cutter, operated gold mine, invented steam trap sold by Johns-Mansville, steam engineering, faculty at Rose Polytechnic Inst. and at Vanderbilt Univ. (MEIA NC44 WWWIA)

BROWN, Fayette, 1823–1910, American banker, inventor and manufacturer; manager of Jackson Iron Co., manufacturer of iron and steels, hoisting equipment for blast furnaces (1884), blast furnace design (1885) (DAB NC16 WWWIA)

BROWN, George Edward, Jr., 1920–1999, American applied physicist and politician; in WWII as infantry officer, California Assembly (1959–1962), US Congress (1962–1970) (1972–1999), supported congressional action of funding for science, engineering and technology and work of academia with industry, strong supporter of Office of Technology Assessment (OTA) for Congress (1972–1995) (WWWIA)

BROWN, George Granger, 1896–1957, American chemical engineer; unit operations, fractional distillation, thermodynamics, petroleum, combustion, atomic energy, during WWI in chemical warfare unit, with Aluminum Company of America (Alcoa) (1917–1918), production manager of Union Special Machine Co. (1919–1920), faculty at Univ. of Michigan (1920–1957), becoming department chairman (1942–1951) and dean of engineering (1951ff), consulting (ACCE NC43 PS WWWIA)

BROWN, George H., 1908–1987, American electrical engineer; leader in development of color television, original work on antenna theory and design, various executive positions, including vice president of Radio Corp. of America (1933–1972), 80 US patents, NAE (1965) (MT4 PS WWWIA)

BROWN, Gordon S., 1907–1996, Australian American electrical engineer; cosmic ray recording meter, servomechanisms, fire control, faculty and administration at Massachusetts Inst. of Technology (1938ff) where he founded MIT Servomechanisms Lab. (1940), became dean of engineering (1959) and Institute professor (1973), NAE (1965) (AMOS MT10 WWIE)

BROWN, James Salisbury, 1802–1879, American mechanical engineer, inventor and manufacturer; pattern making, machinist (1819), cotton machinery manufacturing, cutter for making bevel gears (1830), specialized drilling machines (1838), improvements to Blanchard lathe (1842) (Thomas Blanchard [1788–1864]), spinning mule (1876) (DAB MEIA NC11 WWWIA WWWIS)

BROWN, Joseph Rogers, 1810–1876, American inventor and manufacturer; machinist and machine tools, tower clocks (1829), watches, clocks, vernier caliper (1851), gear cutters (1855), lathes, small Hooke universal grinding machines (1877), J. R. Brown and Sharpe Co. (1853ff), numerous patents (ASAI DAB MEIA NC10 WWWIA WWWIS)

BROWN, Robert H., 1919–1994, American agricultural engineer; served in WWII with US Navy signal corps, electronics, rural electrification, solid state controls, instrumentation systems, physical properties of agricultural products, faculty and administration at Univ. of Georgia (1950ff), becoming chairman and director (Res1 WWIE)

BROWN, Sylvanus, 1747–1824, American inventor and metallurgist; construction and erection of machinery for grist mills and sawmills, machine shop and equipment for cotton manufacturing (DAB IIA NC24 WWWIA WWWIS)

BROWN, Theo, d. 1971, age 92, American agricultural engineer; agricultural mechanization, safety of farm implements, Richardson Manufacturing Co. (1901ff), design engineer then director of product development (ante 1954) of Deere & Co., 158 patents (AE52)

BROWN, Walter J., 1900–1972, English American engineer and inventor; supervisor of research and broadcasting at Vickers (1923–1929), director of engineering, electric devices, piezoelectric devises for Brush Development Co., over 150 patents in US and England (NC57 PS)

BROWN, William Henry, 1836–1910, American civil engineer; chief engineer for Pennsylvania Railroad (1881ff), stations, tunnels, stone bridges, shops, yards, piers, docks, terminals (DAB NC44 WWWIA)

BROWNE, Ralph Cowan, 1880–1960, American engineer and inventor; electric detection mines during WWI,

roentgenologist, Brown [*sic*] portable X-ray apparatus, airlift mine pumps, high-resistance transmitter, president of Brown [*sic*] Apparatus Co., machines (NC46 PS WWWIA WWWIS)

BROWNING, John Moses, 1855–1926, American inventor; breech loading rifle (1879), Colt machine gun (1890), Browning pistol (1911), Browning Automatic Rifle (1917), 128 patents on firearms and machines (CEOS&T DAB GI MEIA MPSI MWBD NC20 PS SAE-MP TGS WA WWWIA WWWIS)

BRUCE, George, 1781–1866, Scottish American engineer; type foundry (1816), typecasting machine (1814) with nephew David Bruce, Jr. [1802–1892] (MEIA NC11 WWWIA)

BRUEVICH, Nikolai Grigorevich, 1896–??, Soviet mechanical engineer; precision machines, planning and manufacturing mechanical devices and instruments, computing machines, taught aviation and military uses (1929ff), machine institute (1951), lieutenant general in engineers corps, resolving problems in aviation during WWII (SMOS WWWIS)

BRUGNATELLI, Luigo Gasparo, 1761–1818, Italian metallurgist and inventor; mineralogist and chemist, invented electroplating (1805) (MWBD WA)

BRUHN, Hjalmar Diehl, 1907–1997, American agricultural engineer; machinery and equipment for production of protein from plant juice, alfalfa cubes, aquatic vegetation harvesting, mechanical dewatering of alfalfa, conditioning and drying crops, tree planting, faculty at Univ. of Wisconsin (1933–1978) and as department chairman (1962–1966), consulting firm of H. D. Bruhn and associates (1979–1997) (AE59 WWIE WWWIA)

BRUK, Isaac Semenovich, 1902–??, Soviet electrical engineer; chief engineer then director of institute related to electrical machines and systems (1935ff), stability of power systems, mathematical machines (1936–1938) credited with first USSR machines for integrating ordinary differential equations, high-speed electrical computers (1948ff) (SMOS WWWIS)

BRUMER, Milton, 1902–1999, American civil engineer; engineer with the Port Authority of New York (1925–1938), chief engineer with Pennsylvania Turnpike Commission (1938–1940), with Ammann & Whitney as chief engineer, partner, then president (1940ff), worked with numerous turnpike projects, expressways, bridges, NAE (1969) (WWWIA)

BRUNEL, Sir Marc Isambard, 1769–1849, French British inventor and civil engineer; ship and dockyard equipment, tunneling in strata beneath water, first to use compressed air to sink caissons, sawmills, knitting equipment, tunneling shield (1818), Thames Tunnel (1825ff), first ocean-going steamship (1848), worked in US (BDOACE BDOS BDOTHOT EAI MWBD PS TGS TSOE WWWIS)

BRUNEL, Isambard Kingdom, 1806–1859, British civil engineer and inventor; son of Marc Isambard Brunel [1769–1849], suspension bridges, railroads, chief engineer of Great Western Railroad (1833ff), marine structures, tunnels, docks, steamship design, compressed air caisson (1852) (BDOS(Wi) BDOTHOT CEOS&T EOWB FEL GI HIW MWBD PS SAI TBDOS TGE:B TGS WWWIS)

BRUNELLESCHI, Filippo (aka Fille di Ser Brunellesce) (aka Brunesch), 1377–1446, Florentine architect, civil engineer and builder; goldsmith, engineer of the Grand Duke Cosimo I, formulated principles of linear proportion, established perspective (1435), design and construction of domes, cathedrals, churches, considered greatest architect and engineer of his time, built masonry domes of Cathedral of Florence (Santa Maria del Fiore) (1420–1434), design of Pitti Palace (CBD EB EOWB E-64 MWBD 1000Y PS TAE TSOE WWWIS)

BRUNTON, David William, 1849–1927, Canadian American mining and mechanical engineer and inventor; treatment of complex ores, tunnel construction, mining pump (1882), design of special machinery to handle and process ores, special ore sampler (1884), Brunton pocket transit (1904), mine manager of Silver Peak (NV) and Aspen Mines, with Anaconda Mines (1887–1901), consulting engineer (ANB DAB MEIA NCB WWWIA)

BRUNTON, William, 1777–1851, British civil engineer and inventor; revolving grate mechanical stoker, with industry shops of Boulton and Watt (1796–1818), had own shop (1816–1825), consulting engineering (1825–1835) (DNB EE WWWIS)

BRUSH, Charles Francis, 1849–1929, American mechanical engineer and inventor; gas street lighting, electric lighting, electric arc lighting, storage battery, established relationship of several chemical and physical phenomena, co-founder with Karl von Linde [1842–1934] of Linde Air Products Co. (1905), numerous patents, Brush Foundation (1928) (AIE ANB BDOAS DOSB MEIA MWBD NC21 PS WWWIA WWWIS)

BRUSH, George Jarvis, 1831–1912, American mineralogist; ability to identify essential features for recognizing mineral species, editor of technical articles, NAS (1868) (BDOS BM17 NC28 WWWIA WWWIS)

BRYANT, Gridley, 1790–1868, American civil engineer and inventor; contractor, invented portable derricks, railroad development for quarries (1826) (BDOACE DAB NC11 WWWIA)

BRYTON, G. B., 1830–1873, English inventor; Bryton thermodynamic cycle applicable to rotating turbine engine (EE)

BUCHANAN, Joseph, 1785–1829, American inventor; improvement of mills, glass of different colors, producing color symphonies with light, music of light (DAB WWWIA)

BUCHSBAUM, Solomon Jan, 1929–1993, Polish American physicist and electrical engineer; solid state physics, gaseous plasmas, research and applications, to US (1952), US citizen (1963), at AT&T Bell Labs, where he became director of electronics research lab (1958–1968), M.C. vice president of research for Sandia Labs (1968–1971), returned to Bell AT&T Labs (1971 ff) as executive director of research, 8 patents, NAE (1973), NAS (1974) (BM69 MT7 WWWIA WWWIS)

BUCK, Leffert Lefferts, 1837–1909, American civil engineer; bridge design and construction for railroads in US and foreign, viaducts (BDOACE DAB NC11 WWWIA)

BUCKHOUT, Isaac Craig, 1830–1874, American civil engineer; surveying, bridge, aqueduct, waterworks, underground railroad (NYC), executive of New York and Harlem Railroad, consultant (BDOACE DAB WWWIA)

BUCKINGHAM, Edgar, 1867–1940, American physicist and inventor; thermodynamics, dimensional analysis, dimensionless groups, scaling, Buckingham pi-theorem (1914), based on ideas presented by Englishman Thomas Simpson [1710–1761] (EB CEOS&T L&M NC29 WWWIA WWWIS)

BUCKLAND, Cyrus, 1799–1891, American inventor; cotton manufacturing plant, government armory, devised machines to manufacture locks and chambers for guns (DAB NC11 WWWIA WWWIS)

BUCKLEY, Oliver Ellsworth, 1887–1959, American physicist, engineer and inventor; research with Western Electric Co. (1914), thermionic vacuum tubes, submarine detection equipment in WWI, mercury-vapor diffusion pump, ionization manometer, developed transatlantic cable for high-speed transmission (installed in 1925), assistant director of research (1927) to director of research (1933), executive vice president (1936) to president and CEO (1940–1950) of Bell Labs., development of radar in WWII, 43 patents, NAS (1937) (ANB BM37 DAB NC53 WWWIA WWWIS)

BUCKLEY, Pade Scott, 1918–1995, American chemical engineer; development and design at Monsanto (1941–1949), at DuPont (1949ff) in several research and development activities, including chemical process control technology, statistical quality control, instrumentation, NAE (1981) (AMOS MT8 WWIE WWWIA)

BUDDING, Edwin Beard, c. 1796–1846, English engineer and inventor; invented lawn mower (1830), partnership with John Ferrabee (1832) and negotiated with manufacturer Ransomes (1832) to produce machine screw-shifting spanner, cylinders for carding machine (BDOTHOT EAI GI WA)

BUDIANSKY, Bernard, 1925–1999, American civil engineer; structural mechanics, materials behavior, aeronautical structures, shells, buckling phenomena, geophysical problems, NACA (1944–1955), Harvard Univ.

(1955ff), NAE (1976), NAS (1973) (AMOS Bridge WWIE WWWIA)

BUECHE, Arthur M., 1920–1981, American chemist and materials engineer; physical chemist, synthetic rubber, polymer and surface studies, with General Electric Co. (1950ff), including corporate vice president of research and development (1965), NAS (1971), NAE (1974) (BM56 MT2 WWIE WWWIA)

BUELL, Abel, 1742–1822, American metalworker and inventor; silversmith, engraver, first American type font (1769), map of US based on Paris peace treaty (1784) (MWBD WWWIA WWWIS)

BUICK, David Dunbar, 1854–1929, Scottish American automobile manufacturer; to US (1856), worked with plumbing supplies, engines, automobiles, Buick Manufacturing Co. (1903–1906), first Buick automobile manufactured (1903), which later became a part of General Motors Corp. under William C. Durant [1861–1947] (1908) from a holding company formed by him (AIE ANB MWBD NC34)

BULLOCK, William A., 1813–1867, American mechanical engineer, inventor and manufacturer; machine shop, hay and cotton presses (1843), developed artificial legs (1843), Bullock Press (1863–1865), invented grain drill (1850), seed planter (1854), lath-cutting machine, printing machinery (1863), web printing press (1865) (DAB MEIA NC9 WA WWWIA WWWIS)

BUNSEN, Robert Wilhelm Eberhard von, 1811–1899, German chemist and inventor; carbon electrode battery (1842), use of spectroscope for analysis, Bunsen cell (1841) to produce bright light, grease spot photometer to measure brightness (1844), Bunsen burner (1855), considered pioneer of chemical spectroscopy (1859ff) with Gustav Robert Kirchhoff [1824–1887], filter pump (1868), ice calorimeter (1870), professor at Heidelberg Univ. (1852–1889), fNAS (1864) (BDOTHOT CDOS GI MWBD RHW TBDOS WA WWWIS)

BURBANK, Luther, 1849–1926, American horticulturist and plant breeder; created more than 800 new varieties of plants by using native and foreign strains, much of work based on publications of Charles R. Darwin [1809–1882] (MWBD NC33 NYT-A 1000Y WWWIA WWWIS)

BURDEN, Henry, 1791–1871, Scottish American engineer and inventor; to US (1819), ironmaster, made agricultural implements for Townsend & Corning (NY), invented improved plow and cultivator (1820), flax and hemp machine (1822), machine for making wrought iron spikes (1825) horseshoe making machine (1835), hook-headed spikes for railroad (1840), steam navigation, Burden water wheel, established H. Burden & Sons Co. (1848) (ANB BDOACE DAB MEIA NC2 WWWIA WWWIS)

BURDEN, James Abercrombie, 1833–1906, American mechanical engineer, inventor and manufacturer; son of Henry Burden [1791–1871], was foreman and succeeded father as president of Burden Iron Works, invented devices for iron manufacture, manufacturer of horseshoes, over 20 patents (MEIA NC1 WWWIA)

BURDIN, Claude, 1790–1873, French engineer; Burdin turbine with horizontal impulse wheel with curved blades (1822), introduced word "turbine" to literature (1824) (EE WBE WWWIS)

BORGDORF, Theodore F., 1854–1905, American mechanical engineer; steam engineer, served in US Navy, inspector for navy, Univ. of Tennessee (4 yr.), fleet engineer for Caribbean Squadron (1903), commander for inspector of engineering material (1904) (MEIA)

BURGESS, Charles Frederick, 1873–1945, Canadian American chemical and electrical engineer and inventor; electrochemistry, faculty at Univ. of Wisconsin (1905–1913), manufacturer of dry cells, founded Burgess Battery Co. (1910), founded Burgess Lab. (1915), electrolytic means of purifying iron, Burgess Zeolite Co., consultant (1906ff) (ACCE NC34,C PS WWWIA WWWIS)

BURGESS, Hugh, c. 1825–1892, English American inventor and manufacturer; soda process to make paper from wood pulp (1851–1854), to US (1854), with Morris L. Keen [1820–1883] formed American Wood Paper Co. (1863) (MWBD)

BURKS, G. Edwin, 1901–1994, American automotive engineer; designer of earth-moving machinery and diesel engines, with Western Telephone Co (CA) (1920–1923), Schmeizer Co. (1923–1927), Caterpillar Co. (1927–1967), becoming vice president of research and design, NAE (1978) (MT9)

BURLS, G. A., 1866–1939, British engineer; Burls horsepower formula for internal combustion engines (1910–1911) (EE)

BURR, Theodore, 1771–1822, American civil engineer and inventor; patented arch-and-truss bridge design (1817), built at least 45 bridges of 60–367 ft. span (as of 1818), patented a water wheel, consulting engineer (BDOACE FEL PS)

BURR, William Hubert, 1851–1934, American civil engineer; built wrought iron bridges (1872ff), water works, faculty at Rensselaer Polytechnic Inst. (1876–1884), Columbia Univ. (1893–1916), design and superintendent of construction of many US bridges, docks, water systems and for several other countries, consulting engineer (NYC) (ANB BDOACE NC4 WWWIA WWWIS)

BURRELL, George Arthur, 1882–1957, American chemical engineer; chemist with US Geological Survey (1904–1908), in charge of fuel testing then in charge of research on gas, natural gas and petroleum for the US Bureau of Mines (1908–1916), during WWI worked on noxious gases while commissioned as a colonel in US Army, professional consulting engineer, worked on helium and separation of helium from natural gas, president of Burrell-Mase Engineering Co. and founded Burrell Technical Supply Co. (1923ff) (NC46 WWWIA WWWIS)

BURRISS, Stanley William, 1910–1979, American engineer and executive; US Navy in WWII (1943–1946), Los Alamos Scientific Lab. (1946–1952), Secretary of Navy (1953), Lockheed Aircraft Corp. (1956ff), becoming president (1969), NAE (1968) (MT2 WWIE WWWIA)

BURROUGHS, William Seward, 1855–1898, American inventor and mechanical engineer; mechanical adding machine (1884), calculating machine (1885), formed American Arithmometer Co., later named Burroughs Adding Machine Corp. (1905), recording adding machine (1892), wood-working machinery (AIE ANB MEIA MWBD NC27 PS WA WWWIA WWWIS)

BURROWES, Edward Thomas, 1852–1918, Canadian American manufacturer and inventor; manufacturer of screen, window screen (1878), spring-activated shades, automobile accessories, folding tables, over 40 patents (DAB WWWIA)

BURSON, William Worth, 1832–1913, American inventor, engineer and manufacturer; farm machinery, self-rake reaper (1858), twine binder for grain (1860, 1864), wire binder (1861), automatic knitting machine (1860), formed Burson Manufacturing Co. (1892) to make knitting machinery, more than 50 patents (DAB MEIA)

BURT, John, 1814–1886, American inventor and industrialist; son of William Austin Burt [1792–1858], land surveyor, built saw and grist mills, lumbering, canal superintendent, manufacture of pig and wrought iron (1869) (BDOACE DAB MEIA WWWIA WWWIS)

BURT, William Austin, 1792–1858, American surveyor, civil engineer and inventor; typographer (1829), solar compass (1836), sextant (1856), leader in Sault Ste. Marie Canal development (ANB BDOACE DAB MEIA MWBD NC18 PS WWWIS)

BURT, William Irving, 1893–1965, American chemical engineer; chemical manufacturing manager with B. F. Goodrich Co., becoming vice president then president of Goodrich-Gulf (1952), synthetic rubber industry, recognized for converting laboratory or pilot plant results to full-scale production (ACCE NYT-A)

BURTON, William Meriam, 1865–1954, American chemist and inventor; thermal cracking of petroleum to produce gasoline (1912), Standard Oil Co. (1890ff), becoming president (1918–1927) (MWBD NC41,C NYT-A WA WWWIA)

BUSCH, Paul L., 1937–1999, American civil and mechanical engineer; drinking water, toxic wastes, wastewater, radioactive wastes, consulting engineer, president and CEO of Malcolm Pirnie, Inc., NAE (1996) (WWIE)

BUSEMANN, Adolf, 1901–1986, German American mechanical engineer; to US (1947), aerodynamics, proposed swept wings for supersonic flight (1935), gas dynamics, rocketry, professor at Georgia Augusta Univ. (Göttingen) (1925–1931), Max Planck Inst. (1925), Technical Univ. Dresden (1931–1935), Braunschweig Lab. (1936–1946), NASA Langley (1947–1964), faculty at Univ. of Colorado (1964–1969), NAE (1970) (ANB DAB MT3 WWIE)

BUSH, Vannevar, 1890–1974, American electrical engineer and inventor; faculty at Tufts Univ. (1916–1919), magnetic device for detecting submarines, differential analyzer, analog computer (1928), faculty at Massachusetts Inst. of Technology (1919–1938), becoming vice president and dean of engineering (1932–1938), president of Carnegie Inst. (1939–1955), director of US Office of Research and Development (1941–1946), wrote reports leading to Manhattan Project and establishment of National Science Foundation, received 49 patents, NAS (1934) (ANB BEST BM50 CDOSB HIV IAI TBDOS MWBD NCF PS WWWIA WWWIS)

BUSHNELL, David, 1742–1824, American inventor; one-man submarine called Turtle (1776), man-propelled submarine, torpedo (1777), considered by some as the father of the submarine (ANB DAB IIAS MEIA 1000Y PS MWBD NC9 NYPL TP WA WWWIA WWWIS)

BUSIGNIES, Henri Gaston, 1905–1981, French American electrical engineer and inventor; radio navigation, antennae, direction-finding antennae, communications, moving target indicator radar, LMT (Paris) (1928–1935), International Telephone & Telegraph Lab. (ante 1975), of which he became vice president then president (1956), over 140 patents, NAE (1966) (AMOS MT2 PS WWIE WWWIA WWWIS)

BUTTERICK, Ebenezer, 1826–1903, and his wife Ellen, American inventors; standard tissue paper patterns for clothes (1859), Butterick Patterns (DAB GI MWBD NC13 WWWIA WWWIS)

BUYS BALLOT, Christoph Hendrik Diederik, 1817–1890, Dutch physicist and meteorologist; Buys Ballot law (1857) for representing wind direction and barometric pressure, professor and director of Royal Dutch Meteorological Inst. (1854ff) (DOTHOT L&M MWBD WWWIS)

BY, John, 1779–1836, English engineer; Royal Engineers (1799ff), canal design and construction in Canada (1826–1832), soil settlement (DNB MWBD)

BYRGIUS, Justus (aka Joost Bürgi), 1552–1632, Swiss mathematician and inventor; watch-maker, invented celestial globes, instruments for observation, compiled logarithms (1603–1611) but didn't publish, Scott John L. Napier [1550–1617] working independently was the first to publish logarithms (1614) (LDOS MWBD TTOS WWWIS)

BYRON, (Augusta) Ada (aka Countess of Lovelace)—see LOVELACE, Lady Ada

C

CADWELL, William, 1834–1902, American mechanical engineer; master mechanic and superintendent of National Company Cotton Mills (MA) (1866), Nashua Manufacturing Co. (NH) (1891), director of Pennichuck Water Co. (MEIA NC31)

CAESAR, Gaius Julius, c. 102–44 BC, Roman general and dictator; bridging the Rhine River, road building, military and naval operations (MWBD TEE)

CAGNIARD, de la Tour, Charles, 1777–1859, French engineer and physicist; Cagniardelle pump (1812), forced draft blower, invented siren using sound vibrations (1819) (EB EE MWBD WWWIS)

CAHILL, Thaddeus, 1867–1934, American inventor; electric typewriter (1901), electronic music production (MWBD NYPL WWWIA)

CAIRNS, Robert William, 1909–1985, American industrial chemist; with Hercules, Inc. (DE) (1934–1971) advancing from chemist to vice president of research, US Dept. of Commerce, executive director of American Chemical Society (1972ff), NAE (1969) (MT3 WWIA WWIE)

CAJORI, Florian, 1859–1930, Swiss American mathematician and engineer; to US (1875), Colorado College (1898–1918), including dean of engineering (1903–1918), Univ. of California (1918ff), physics, mathematics, history, engineering (MWBD NC27 WWWIA WWWIS)

CALDER, Alexander, 1898–1976, American sculptor; first educated as an engineer, son of Alexander Calder [1846–1923], wire sculptures, animated wire figures (1927–1930), large-scale sculptures, stabiles, mobiles, wind operated mobiles (MWBD NC61 1000Y WWWIA)

CALDWELL, Joseph M., 1911–1980, American engineer; US Army in WWII (1943–1946), civilian with Army Corps of Engineers (1933–1973) except during WWII, flood control, shore protection, navigation, hydraulics, waterways, consulting engineer, NAE (1973) (MT2 WWIE)

CALLENDAR, Hugh Longbourne, 1863–1930, British physicist and engineer; behavior of steam, platinum resistance thermometry (1886), reliable steam tables (1902), research on flow of steam through nozzles, calorimetry, faculty at McGill Univ. (1893–1898), Univ. College London (1898–1902), Imperial College of Science London (1902–1930) (CEOS&T DNB EE FNIE MWBD PS TBDOS(Po) WWWIS)

CALLOW, John Michael, 1867–1940, English American mining engineer, inventor and metallurgist; to US (1890), private practice (1901–1906), Callow flotation cell, Callow traveling belt screen for separation, invented Callow settling tank, originator of pneumatic flotation, president and general manager of General Engineering Co. (NY) (1906ff), 18 patents (CB CEOS&T WWWIA WWWIS)

CALVIN, Melvin, 1911–1991, American biochemist; first degree in engineering, plant physiology, photosynthesis, chemical biodynamics, organic and physical chemistry, faculty at Univ. of California (1937ff), government radiation laboratory, government representative on peaceful uses of atomic energy, NAS (1954), Nobel Prize (1961) (BEST BM75 NCI WWWIA WWWIS)

CAMP, Hiram, 1811–1893, American engineer; metal works, clock maker, mass production of clocks and time pieces (MEIA NC8 WWWIA)

CAMPBELL, Allan, 1825–1894, American civil engineer; railroad chief engineer, survey and construction of railroads in Chile, administration of railroads and coal company, harbor defenses of New York City, public official (DAB NC9 WWWIA)

CAMPBELL, Andrew, 1821–1890, American inventor, engineer and manufacturer; carriage builder, devised special vice for holding brushes for manufacturing (1842), printing press paper-feed mechanisms and automatic features, machines for making match and pill boxes (1850) numerous patents, over 50 patents on printing press alone (DAB MEIA NC9 WWWIA)

CAMPBELL, Andrew Chambre, 1856–1926, American mechanical engineer and inventor; son of Andrew Campbell [1821–1890], designed automatic machine for making paper flour sacks, casting of curved metal plates for rotary press (1877), gripper mechanism for printing press (1882), button-hole making machine (1884), wire-drawing machine (1892), numerous patents related to sewing, printing, hammerlock cotter pin (1912), consulting engineer (MEIA NC25 WWWIS)

CAMPBELL, Donald L., 1904–1984, American inventor; fluid analytic cracking, applied to more efficiently transform higher boiling point oils into lighter, usable products (NYT-A WPI)

CAMPBELL, George Ashley, 1870–1954, American engineer, mathematician and inventor; telephone transmission with AT&T and Bell Labs., distribution of electricity in telephone networks, repeater circuits (1912), electric wave ladder filter (1917), at least 12 patents (NC45 PS WWWIA WWWIS)

CAMPBELL, Henry, 1824–1891, American inventor; invented American type locomotive having two pairs of driving wheels and a rotatable truck (ASAI DAB)

CAMRAS, Marvin, 1916–1995, American electrical engineer and inventor; development of magnetic recording and its application, data storage, tape and wire recorders (1930s), recording heads, stereo sound, video recorders, Illinois Inst. of Technology research (1940ff), referred to as the father of modern recording, over 500 patents for devices in electrical communications, NAE (1976) (FAL HOCS&T MT8 NYT-A NYTObit TEOUT WOI WWIE WWWIA)

CANAGA, Alfred Bruce, 1850–1906, American naval engineer; chief engineer in US Navy (1895), designed machinery for several battleships, Boston Naval Yard (DAB NC22)

CANJAR, Lawrence Nicholas, 1923–1972, American chemical engineer; US Army Corps of Engineers in WWII, faculty at Carnegie Inst. of Technology (1947–1965), dean of engineering at Univ. of Detroit (1965–1972), innovations in teaching, development of core curriculum, design courses, specialty in hydraulics (NC57 PS WWWIA)

CANNING, Sir Samuel, 1823–1908, English engineer; chief engineer for laying cables across Atlantic Ocean (1865) (1869), laid cables from England to Malta and Egypt (DNB MWBD)

CAPLAN, John D., 1926–1998, American civil and mechanical engineer; identification, measurement and control of motor vehicle pollutants, at General Motors Research Lab. (1949–1987), becoming technical director (1967), NAE (1973) (MT10 WWIE WWWIA)

CAPSTAFF, John George, 1879–1960, English American inventor; invented several modifications of photographic processes, research laboratory at Eastman Kodak (1913–1954), worked on Kodochrome color process (1915), then Kodacolor process (1928), film for motion picture movie cameras, issued over 100 patents (NC49 PS WWWIS)

CARDAN, Jerome (aka Girolamo Cardano), 1501–1576, Italian mathematician and inventor; Cardan or universal joint (1545), in England this invention attributed to Robert Hooke [1635–1703] known as Hooke coupling, Cardano formula, Cardano shaft, Cardano suspension, probability (DOSB EB EE GLOSB MWBD TWABOI WWWIS)

CARLL, John Franklin, 1828–1904, American civil engineer and geologist; newspaper business publishing *Daily Eagle,* surveying, development of oil fields in Pennsylvania, invented static pressure sand pump, invented adjustable sleeve for piston rods, reported on future of oil fields of Pennsylvania (DAB NC12 PS WWWIA)

CARLSON, Chester Floyd, 1906–1968, American inventor; with Bell Telephone Co., P. R. Mallory Co., electrostatic xerography copying process (1938ff) (pat. 1940ff), licensed to Battelle Memorial Inst. (1944), first built by Haloid Co., which became Xerox Co. (1947), patented many improvements (AIE ANB CDOS IAI MWBD NYT-A NYPL 1000Y PA WA)

CARLSON, Roy W., 1900–1990, American civil engineer; construction materials, pressures, stresses, deflections, concrete technology, first with California Edison Co. (1925–1927), test engineer and director of research for Hoover Dam (1931–1934), faculty at Massachusetts Inst. of Technology (1934–1943), during WWII with Los Alamos Lab. and Radiation Lab. at Univ. of California-Berkeley, consulting engineer worldwide (1936ff), development and manufacturer of concrete-testing equipment, 5 patents, NAE (1974) (AMOS MT5 WWIE WWWIA)

CARNEGIE, Andrew, 1835–1919, Scottish American industrialist; cotton textile factory, telegraph company, railroad, iron and steel business (1865), Carnegie Steel Co. (1899), founded several companies, especially US Steel (1901), financier, several philanthropic organizations, distributed wealth for public programs devoted to education, public libraries, etc. (AIE DAB IIA MWBD NC9 1000Y WWWIA)

CARNOT, Lazare-Nicolas-Marguerite, 1753–1823, French soldier and engineer; in French army (1773ff), heat and thermodynamics, politician (DOTHOS MWBD WWWIS)

CARNOT, Sadi (Nicolas-Léonard-Sadi), 1796–1832, French physicist and engineer; son of Lazare-N.-M. Carnot [1753–1823], served in French engineer corps (1814–1828), energy, temperature, thermodynamics, heat, Carnot cycle (1824), steam engine efficiency, army engineer on fortifications, industrial development (DOTHOS FNIE IAI L&M MWBD SAI-MP TDOS TGS WWWIS)

CAROTHERS, Wallace Hume, 1896–1937, American chemist and inventor; faculty at Univ. of Illinois and Harvard Univ. (1921–1928), DuPont Co. (1928ff), invented neoprene (1934), nylon—the first synthetic fiber (1934), nylon stockings (1939), first successful synthetic rubber, NAS (1936) (AEOTHOT AIE ASAI BEST BM20 GSA HOCS&T IAI IIA I&T15 LOTCS MEI MWBD NC38 NYPL NYT-A PA WA WWWIA WWWIS)

CARPENTER, Franklin R., 1848–1910, American mining engineer; surveyor, public school teacher, semipyritic smelting, electrostatic concentration of ores, copper and nickel ore separation (DAB NC16 WWWIA)

CARPENTER, Rolla Clinton, 1852–1919, American civil and mechanical engineer; consulting engineer for cement companies, power stations, fire systems, heating and lighting, pumping stations, educator, invented furnace for boilers, calorimeter, friction testing machines, governors, writer (MEIA NC4 WWWIA)

CARRÉ, Ferdinand, 1824–1900, French inventor; introduced absorption refrigeration machine (1859–1862) using ammonia as a refrigerant and water as an absorbent, developed compression machine for refrigeration, Raoul-Pierre Pictet [1846–1929] developed compression refrigeration system (1874) using sulfur dioxide as refrigerant (S&TF WOI WWWIS)

CARRIER, Willis Haviland, 1876–1950, American engineer and inventor; temperature and humidity control systems (1902ff) using ammonia, carbon dioxide and later freon as a refrigerant, invented air conditioning (1902), standards (1911), first worked with Buffalo Forge Co., established and was president then chairman of Carrier Engineering Corp. (1915–1948), over 80 patents (AIB ANB EB GI I&T15 MWBD NCE NYPL PA PS WA WWWIA)

CARROLL, Thomas, 1888–1968, Australian agricultural engineer; helped develop the modern self-propelled combine harvester, Buckeye Harvester in Melbourne (1904–1909), with J, J. Mitchell and Co. (1909–1911), then Moore & Tudor Co. and subsequently Massey-Harris, Ltd., designed self-propelled combine (1936), hay baler, plows, planters and cultivators (AE49)

CARTER, Deane G., 1894–1980, American agricultural engineer; salesman with James Manufacturing Co., on faculty at Iowa State Univ., North Carolina State Univ., Univ. of Arkansas as department head (1922–1941), faculty at Univ. of Illinois (1941–1948), known for his work in design of farm structures and houses (AMOS AE61)

CARTWRIGHT, Edmund, 1743–1823, English clergyman and inventor; weaving machinery, power loom (1785), power weaving and mill (1787), wool-combing machine (1789), patented alcohol engine (1792) (DNB E-64 MMAH MWBD NYPL S&TF TBDOS TGE:B WA WWWIS)

CARTY, John Joseph, 1861–1932, American electrical engineer; switchboard, telephone systems, common battery for telephones, metallic or two-wire telephone transmission, chief engineer and executive at AT&T Co. (1909), signal corps in WWI, retiring as brigadier general, 24 patents, NAS (1917) (ANB BM18 MWBD NC23 PS WWWIA WWWIS)

CARVER, George Washington, 1864–1943, Black American agricultural scientist and inventor; research on use of agricultural crops such as for insulation, paving blocks, synthetic marble, made 300 different products from peanuts, made about 118 products from sweet potatoes and soybeans, with Tuskegee Inst. (1896ff), becoming director of agricultural research, over 100 US and foreign patents (ACCE AIE BDOBA BEST BI BPOSAI LOTOS NCF NYT-A SAI-MP S:TLAW TFM TGS WBEOS WWWIA WWWIS)

CASAGRANDE, Arthur, 1902–1981, Austrian American civil engineer; to US (1926), hydraulics, public roads, airfields, foundations and dams, triaxial tests, liquefaction, Panama Canal, faculty at Massachusetts Inst. of Technology (1926–1932), Harvard Univ. (1932–1974), consulting engineer (1974ff), NAE (1966) (AMOS MT2 WWIE WWWIA)

CASAGRANDE, Leo, 1903–1990, Austrian American civil engineer; faculty at Univ. of Berlin (1933) and at Technical Univ. Braunschweig (1940–1946), to England at Building Research Center (1946–1950), to US (1950), Harvard Univ. (1950–1972), foundations, design, soil mechanics, research and application in electro-osmotic soil stabilization, with technical organizations in Austria, Germany and England (ante 1950), engineering consultant with brother Arthur Casagrande [1902–1981] and son Richard Casagrande formed Casagrande Consultants (1950ff), NAE (1974) (AMOS MT5 WWIE WWWIA)

CASAMAJOR, Paul, 1831–1887, Cuban American chemical engineer; to US (1845), founded and supervised The Enterprise Mining and Boring Co. (PA) (1864–1866), American Sugar Co. (NYC) (1867), developed procedures for filtration of turbid sugar liquors (ACCE)

CASE, Jerome Increase, 1818–1891, American manufacturer; threshing business using horse treadmill, practical operating combined thresher and separator (1844), founded J. I. Case Co. (1863), later incorporated as J. I. Case Threshing Machine Co. (1880), established J. I. Case Plow Works, some attribute first stationary steam tractor to J. I. Case, a few mobile tractors followed with the first being the French Albaret locotractor (1856) (AIE DAB NC18 WOI WWWIA)

CASELLI, Abbe Giovanni, 1815–1891, Italian physicist and inventor; first person to reproduce fixed images transmitted over wire (1862), in effect television by wire or fax, credit for invention of the first fax device granted (1843) to Scot Alexander Bain [1818–1903] (GI MMAH MWBD TWABOI)

CASEY, Thomas Lincoln, 1831–1896, American military engineer; faculty at USMA at West Point, construction of forts and harbors on Delaware River and bay, constructed wagon road in northwest, fortification along Maine coast, several defense and public projects including the Washington Monument, Library of Congress, and statues as a builder, chief of engineers as brigadier general (1888), NAS (1890) (ANB BM4 NC4 PS WWWIA)

CASS, George Washington, 1810–1888, American civil engineer; topographical engineer with US Army, railroad construction, first cast iron tunnels and bridges in US (1837), transportation on rivers and railroads, was president and director of Ohio & Pennsylvania Railroad and its successors (1856–1881) (ANB DAB WWWIA)

CASSATT, Alexander Johnston, 1839–1906, American civil engineer; engineer and superintendent of Pennsylvania Railroad, traffic planning and control, improved efficiency of railroad operating systems, railroad rebate system, tunnel and terminal in NYC, president of Pennsylvania Railroad system (1861–1882) (1899–1906) (DAB NC13 WWWIA)

CASSEGRAIN, (Guillaume) N., c. 1650–1700, French physician and inventor; reflecting telescope named after him (1672), megaphone (MWBD TBDOS(Po) TTOS)

CASTIGLIANO, Carlo Alberto, 1847–1884, Italian structural engineer; Castigliano theorem (CEOS&T FNIE PS)

CATHCART, William Ledyard, 1855–1926, American mechanical engineer; engineer corps of US Navy, marine engineer, served with navy during Spanish-American War, faculty at Webb Inst. (1897) (1918) and Columbia Univ. (1899), consulting engineer (MEIA WWWIA)

CAUCHY, Augustin-Louis, 1789–1857, French mathematician, physician and inventor; professor at College of France, École Polytechnique (1816–1830) (1838–1848) and Sorbonne (1848ff), elasticity, light, hydrodynamics, hydrostatics, mechanics, refractive index (BDOS(Wi) DOI&D DOTHOS MWBD TBDOS(Po) TSOE WWWIS)

CAUX or CAUIS or CAUIX, Solomon de, 1576–1626, French engineer; landscaping, ornamental fountains, anticipated steam engine, solar-powered motor (1615) (MWBD NYPL WWWIS)

CAVE, Henry, 1874–1963, English American engineer and inventor; worked in companies making stationary steam engines, lace making machinery, engine testing with Daimler Motor Car Co., designer and inspector of motors (engines), free-lance writer who worked for several automotive companies, developed gas welding torch, consulting engineer with Fuller Brush Co., over 100 patents (NC55 PS)

CAVENDISH, Henry, 1731–1810, English physicist and chemist; developed idea of electric potential, did work in heat, meteorology, metallurgy, discovered hydrogen (1766), Cavendish experiment (1798) to determine gravitational constant, Cavendish Physical Lab. at Cambridge Univ. named after him (1871) (BEST DNB MWBD NYPL SAI TGS WWWIS)

CAXTON, William, 1422–1491, British inventor and publisher; returned to England (1496) after work in Germany and Belgium, established first press in England, printed over 100 publications (GI WWWIS)

CAYLEY, Sir George, 1773–1857, English scientist, aeronautical engineer and inventor; pioneer in aerial navigation, initiated aerodynamics, founded Polytechnic Inst. (1835), taught engineering at Royal Polytechnic Inst. (1839), originally called Regent Street Polytechnic, which he founded, formulated principles of heavier-than-air flight (1808ff), made first man-piloted glider (1853), caterpillar tractors, railway devices (BDOS(Wi) CDOS DNB EE GI ITCOF MMAH MWBD NYPL TBDOS TGE:B WA WWWIS)

CEARD, Nicolas, 1747–1821, French engineer; chief engineer for building road over Alps (TSOE)

CELSIUS, Anders, 1701–1774, Swedish astronomer and inventor; Celsius temperature scale (1742), previously called centigrade, numerous astronomical observations, particularly aurora borealis, study of falling level of Baltic Sea, professor at Uppsala Univ. (1730–1744) (BEST EB FNIE 1000Y RHW TGS WBEOS WWWIS)

CESSART, Louis-Alexandre de, 1719–1806, French engineer; military service, engineer in district of Tours, harbor work at Cherbourg, built harbor at Le Havre (TSOE)

CHADWICK, Wallace Lacy, 1897–1996, American mechanical and civil engineer; hydroelectric and thermal power, heavy engineering construction, vice president of Southern California Edison Co. (1922–1962) except for time as engineer with Metropolitan Water District of Southern California (1931–1937), consulting engineer for 30 yr., NAE (1980) (MT10 NCM PS WWIE WWWIA WWWIS)

CHAFFEE, Roger B., 1935–1967, American aeronautical engineer and astronaut; US Navy (1957) advanced to lieutenant commander, safety officer, photographic squadron, fighter pilot, combat pilot and test pilot with NASA Manned Space Flight, died in training mission (1967) along with American mechanical engineer and astronaut Virgil Ivan Grissom [1926–1967] with American engineer and astronaut Edward Higgins White II [1930–1967] (Bent CB MWBD NYT-A)

CHALMERS, Arthur Douglas, 1870–1961, Canadian American chemical engineer; with DuPont Co. (1897–1944) successively involved in management of a gun powder plant (1897), dynamite plant (1906), manufacture of dyes (1915), products for agriculture and textile business, Chambers Works at DuPont named after him (ACCE)

CHALMERS, William James, 1852–1938, American manufacturer; began business with father in Chalmers & Co. (1872), became president of company (1891), built heavy equipment, established company in England, had mining operations in South Africa, in US merged several companies making heavy machinery and farm equipment, forming Allis-Chalmers Co. (1901), retired (1905) (DAB WWWIA)

CHAMBERS, Carl Covalt, 1907–1987, American electrical engineer; electricity and magnetism, radio, inventions, faculty and administration at Univ. of Pennsylvania (1933–1975), NAE (1970) (AMOS MT5 WWIE)

CHAMPION, Albert, 1878–1927, French American aviator and inventor; bicycle and motorcycle racer, automobile racer, magnetos, sparking plugs, manufacturer of ignition and electrical devices for automobiles, founded Champion Ignition Co., which became a part of General Motors Corp. (1910), where he remained as general manager, also made spark plugs used in aircraft, made other automobile accessories (ANB)

CHANNING, William Francis, 1820–1901, American inventor; geological survey, use of electricity to improve fire alarms (1851–1859), portable electromagnetic telegraph (1877) (DAB NC23 WWWIA)

CHANUTE, Octave, 1832–1910, French American civil engineer; to US (1838), aviation pioneer, glider flight (along with Otto Lilienthal [1848–1896] and Gustav Lilienthal [1832–1910]), credited with technique of landing planes with tail low, Chanute biplane, railroad bridge across the Missouri River, aerial navigation (ANB ASAI BDOACE DAB ITCOF MWBD NC10 WWWIA WWWIA-S&T)

CHAPIN, Roy Dikeman, 1880–1936, American industrialist; automobile manufacturing (1901ff), organized Hudson Motor Co., becoming president and chairman of the board (1910–1936), manufactured Hudson and Essex automobiles, (1919ff), secretary of commerce (1932–1933) (CDOAB MWBD NCD WWWIA)

CHAPMAN, Dean Roden, 1922–1995, American aeronautical engineer; supersonic aerodynamics, recovery of vehicles from space, with NACA (which became NASA) (1944ff), including director of astronautics (1973), NAE (1975) (MT8 WWIE WWWIA)

CHAPLYGIN, Sergei Alekseevich, 1869–1942, Russian physicist and engineer; aerodynamics, fluid mechanics, fluid dynamics (CEOS&T)

CHAPPE, Claude, 1763–1805, French engineer; with brother Ignace-Urbain-Jean Chappe [1760–1829] built widely used telegraph systems, telegraph signals, semaphore (visual telegraph) (1790), signal towers between Paris and Lille (1794) (DOSB GI MWBD NYPL PS)

CHARDONNET, Louis-Marie-Hilaire Bernigaud, 1839–1924, French industrial chemist and inventor; rayon (nitrocellulose) (1878), perfected (1883), and manufactured (1889), which became a substitute for silk, first artificial fiber in common use (ASAI MWBD NYPL TBDOS WA WBEOS WWWIS)

CHARLES, Jacques-Alexandre-César, 1746–1823, French physicist and applied scientist; teacher and popularizer of science, popularized Benjamin Franklin [1706–1790], theory of electricity and lightning, constructed and used first hydrogen air balloons (1783), showed that different gases expanded the same amount with a given rise in temperature (1784) (BEST FIOTEC FNIE L&M MWBD NYPL WWWIS)

CHASE, Leon Wilson, 1878–1951, American agricultural engineer; employed by Fairbanks, Morse and Co., International Harvester Co., helped develop agricultural engineering at Univ. of Nebraska (1905–1921), crusaded efforts to establish Nebraska Tractor Test law and facility, first test in (1920) was of a Waterloo Boy tractor (1916), Claude Shedd [1884–1966] was chief engineer of operation, major in ordnance in WWI, formed and headed Chase Plow Co. (AE32,61,63)

CHASE, William Livingston, 1855–1898, American mechanical engineer; mechanics, designed and built wood working machinery, water wheels, pioneer inventor of method and apparatus for forming articles from wood pulp (1881), decimal classification for record keeping (MEIA)

CHASSEPOT, Antoine-Alphonse, 1833–1905, French inventor; invented Chassepot musket for French army (1866–1874) (MWBD)

CHELOMEI, Vladimir Nikolaevich, 1914–??, Soviet mechanical engineer; mechanics, dynamics of machinery, pneumatic and hydraulic servomechanisms, worked with several research organizations, taught at several institutes, professor at Moscow Technical College (1952ff) (SMOS)

CHENEA, Paul F., Sr., 1918–1996, American mechanical and electrical engineer; automotive, catalysts and electronic control for engine exhaust emissions and use of computerized methods, WWII service (1941–1946), faculty at Univ. of Michigan (1952–1961), at Purdue Univ. (1961–1967) as professor, department head and vice president, then research and vice president of General Motors Corp. (GMC) (1967ff), NAE (1969) (MT9 WWIE)

CHEOPS—see KHUFU-ONEKH

CHESBROUGH, Ellis Sylvester, 1813–1886, American civil engineer; railroad survey, several railroads in northeast and midwest, aqueducts and reservoirs, sewerage systems, consulting engineer (BDOACE NC9)

CHESTERS, John Hugh, d. 1994, age 83, Australian ceramic engineer; ceramic engineering, iron and steel making, fNAE (1977) (Bridge)

CHEVROLET, Louis Joseph, 1879–1941, Swiss American automobile designer and manufacturer; to US (1900), racer of automobiles, with William C. Durant [1861–1947] founded Chevrolet Motor Co. (1911), company became a part of General Motors Corp. (GMC) under William C. Durant (1915) (MWBD NC53 PS)

CHÉZY, Antoine de, 1718–1798, French engineer and mathematician; fluid stream flow, Chézy equation or formula (c. 1770), built bridge across Marne River (1758–1764), completed Pont de la Concorde, bridge designed by Hupeau [??–1763], school administrator at École des Ponts et Chaussées (CFR L&M MWBD TSOE WWWIS)

CHIAO Wei-Yo (QIAO Weiyo), fl. 10th century, Chinese engineer; developed canals and locks in China (983) (BDOTHOT GI)

CHILDE, John, 1802–1858, American civil engineer; survey, location and construction of railroad lines, harbor construction, got land donated by Congress to assist railroad development (BDOACE DAB NC16 WWWIA)

CHILTON, Thomas Hamilton, 1899–1972, American chemical engineer; research chemist, fluid flow, distillation, gas absorption, industry experiment station at E. I. DuPont (1925–1959), faculty at Univ. of California (1959ff), NAE (1966) (AMOS MT1 NCL PS WWWIA WWWIS)

CHIN, Gilbert Yukyu, 1934–1991, Chinese American metallurgical engineer; to US (1950), crystal plasticity, new magnetic alloys, industry research (1962ff) and research director (1972ff) and director of Materials Physics Research Lab. (1987ff) at AT&T Labs., NAE (1982) (MT6 WWIE WWWIA)

CHINAKAL, Nikolai Andreevich, 1888–??, Soviet mining engineer; at Dorbas mines (1912ff), improved working conditions of miners, developed plans for restoring mines, use of mechanization for handling of coal deposits, headed mechanization section (1923–1928), directed research and teaching institute in West Siberia (SMOS WWWIS)

CHIPPENDALE, Thomas, 1718–1779, English inventor; invented and developed furniture style for middle-class market, simplified styles of the times, published book and catalogue on design of furniture that made him famous (1754), succeeded by his son Thomas Chippendale [c. 1749–1822] (DNB MWBD 1000Y)

CHISHOLM, William, Sr., 1825–1908, Scottish American inventor and manufacturer; to US (1852), contractor and builder, rolling mills, manufactured horseshoes, spikes, screws and bolts (1860), methods and machinery for manufacturing spades, scoops and shovels (1879), steam engines for pumping and hoisting (1882) (MEIA NC18 WWWIA WWWIS)

CHITTENDEN, Hiram Martin, 1858–1917, American military engineer; engineering officer of US Army, road construction in Yellowstone Park, flood control, chief engineer of Corps of Engineers, reservoirs, docking and terminal facilities (Seattle) (ANB DAB NC17 WWWIA)

CHIZHIKOV, David Mikhailovich, 1896–??, Soviet metallurgist and engineer; non-ferrous metallurgy, non-chlorine method of processing tin ores, chief engineer of zinc works (1928, 1930), professor of Moscow Inst. of Non-Ferrous Metals and Gold (1933–1941), Academy of Science Inst. of Metallurgy (1939ff) (SMOS WWWIS)

CHOW, Ven Te, 1919–1981, Chinese American civil engineer; hydrosystems, hydraulic engineering, structures, water resources, academic work in China, to US (1947), at Pennsylvania State Univ. and Univ. of Illinois with several foreign assignments, NAE (1973) (AMOS MT2 WWIE WWWIA WWWIS)

CHRISTIE, James, 1840–1911, Canadian American materials engineer; to US (1856), railroad construction, machinist, construction of bridges, efficiency of labor and machinery in iron works (MEIA WWWIA)

CHRYSLER, Walter Percy, 1875–1940, American automobile manufacturer; designed and manufactured Plymouth, Dodge and Chrysler cars and Dodge trucks, previously with Buick Motor Co. (1916–1919), Willys-Overland Co. (1921), with Maxwell Motor Co. (1924), which introduced the Chrysler automobile (1924) and became Chrysler Corp., which acquired Dodge Brothers Co. (1928) and expanded the line of cars, he served as president and chairman (ante 1940) (MWBD NCD WBE WWWIA)

CHUBB, Lewis Warrington, 1882–1952, American electrical engineer and inventor; transformers (1906ff), electrochemical and magnetic research (1910), welding, radio engineering (1920), research administration at Westinghouse Co. and Radio Corporation of America, received over 200 patents (NC39 NT-CS PS WWWIA WWWIS)

CHUKHANOV, Zinovii Fedorovich, 1912–??, Soviet thermal engineer; theory of burning and vaporization of solid fuels, utilization of fuels, heat exchange and diffusion, All Union Power Engineering Inst. (1931–1934), State Inst. of Nitrogen (1932–1937), Power Engineering Inst. of USSR (1938ff) (PS SMOS)

CHURCH, George Earl, 1835–1910, American civil engineer; explorer, topographical mapping and railroad work in Massachusetts, Iowa and Argentina (ante 1857), in Civil War promoted to colonel, chief engineer, work in Mexico, Bolivia and Ecuador, writer (DAB NC13)

CHURCH, Irving Porter, 1851–1931, American civil engineer; applied mechanics and hydraulics (1876), water flow in open channels, pneumatics, faculty at Cornell Univ. (1876–1916) (MEIA NC22 WWWIA)

CHURCH, John Adams, 1843–1917, American mining engineer; mining observer with US Geological Survey (USGS) and wrote significant reports (1869–1878), faculty in mining and metallurgy at Ohio State Univ. (1878–1881), superintendent of mining company in Arizona (1880ff), operated silver mines in China (4 yr.), consulting engineer (1890ff) (ANB WWWIA)

CHURCH, William, c. 1778–1863, American inventor; invented typesetting machine, hand justification (1823) (MWBD)

CIERVA, Juan de la, 1896–1936, Spanish aeronautical engineer; autogyro-rotating wing aircraft (1923ff), which set the stage for the helicopter, he died when piloting one of his craft, first practical helicopter developed by German Heinrich Karl Foche [1890–1979] (1936), followed by Igor Ivan Sikorsky [1889–1972] (1939) (GI MWBD PopMech177 TBDOS WA)

CISLER, Walker Lee, 1897–1994, American mechanical engineer; utility power systems, atomic energy, peaceful uses of nuclear energy, Public Service Electric & Gas Co. of Newark (NJ) (1922–1941), during WWII with US War Production Board (1941–1943), from chief engineer to chairman of the board (1964) at Detroit Edison Co. (1945–1975), NAE (1964) (MT8 NCJ WWIE WWWIA)

CIST, Jacob, 1782–1825, American inventor; postal service, anthracite coal pioneer, mineral resources (geology), artist paint mixing mill (1803), printer's ink from anthracite coal (1808), stove to burn anthracite coal (1814) (ANB DAB WWWIA WWWIS)

CITROËN, André-Gustave, 1878–1935, French automobile manufacturer; munitions manufacturer (WWI), after war mass produced automobiles, company went bankrupt (1934) (MWBD WWWIS)

CLAPEYRON, Benoit Paul Émile, 1799–1864, French engineer; developed Carnot concept of function of temperature, Clapeyron law (1834), heat and thermodynamics (CEOS&T DOTHOS FIOTEC FNIE L&M PS WWWIA)

CLARK, Alvan, 1804–1897, American astronomer and inventor; patented firearms (1840), started firm Alvan Clark & Sons (1844), telescope lenses and parts, invented double eyepiece, with sons built Lick Observatory (CA) (1886) (ASAI BDOAS MEIA MWBD WWWIA WWWIS)

CLARK, Alvan Graham, 1832–1897, American inventor and lens maker; son of Alvan Clark [1804–1897], worked with his father's firm (1852ff), astronomy, casting mirrors, discovered 16 double stars, constructed 40 in. lens for Yerkes Observatory (WI), then largest in world (MEIA NC5 WWWIA WWWIS)

CLARK, Frederic Horace, c. 1860–1917, American pianist and inventor; designed and patented in Germany the Harmonie-Piano based on his earlier work on physiology and piano techniques (1913) (ANB)

CLARK, Josiah Latimer, 1822–1898, English engineer and technologist; early work as civil engineer, with telegraph company (1850), submarine telegraphy, speed of electric current pulse (1863), zinc-mercury standard cell (DNB DOSB PS WWWIS)

CLARK, Patrick, 1818–1887, Irish American inventor and civil engineer; to US (1827), lamp improvements (1834ff), dampers (1854) and water level indicators for boilers (1854), boiler safety device (1854), fan blowers (1866, 1868), dynamometer (1885) (MEIA)

CLARK, Thomas Edward, 1869–1962, Canadian American electrical engineer and inventor; equipment department of Detroit Power & Light (1889–1892), then with General Electric Co. experimental department as engineer, formed own business called Electric Service & Appliance Co. (1898–1903), electric automobiles, installation and manufacture of wireless telegraph, formed Clark Electric Engineering Co. (1903–1907) then formed Clark Wireless Telegraph & Telephone Co. (1907–1950) renamed Tesla Co., train control devices, signaling, safety, held over 70 patents (NC53 PS)

CLARK, Walter Leighton, 1859–1935, American mechanical engineer; machinist, tanning plant, tool works, rifle manufacturing, ship building (MEIA)

CLARKE, Alfred, 1848–1922, English American mechanical engineer and inventor; to US (1876), chief engineer of Bradley Fertilizer Co. (MA), superintendent of Kitson Machine Co. (MA) (1877–1885), general manager of Prospect Machine and Engine Co. (OH) (1885ff), cotton machinery, cotton opener (1881, 1882), several patents (MEIA NC20 WWWIA)

CLARKE, Edith, 1883–1959, American woman electrical engineer and inventor; supervised AT&T Transmission and Protecting Energy Department (1915–1918), turbine department at General Electric Co. (1919–1922), Central Station of Engineering at General Electric Co. (1922–1955), Clarke calculator (1921), invented voltage regulator (1925, 1927) for long-distance power lines (ANB AWIT WIS(2) WWWIA)

CLARKE, Frank Wigglesworth, 1847–1931, American chemist and inventor; at Univ. of Cincinnati (1874–1883), geologist with US Geological Survey (1883–1924), features of earth's crust, methods of evaluating, NAS (1909) (BM15 MWBD NC26 WWWIA WWWIS)

CLARKSON, Ralph Preston, 1886–1964, American mechanical engineer and inventor; machine design, US patent examiner, author of book "Practical Talks on Farm Engineering and Machines" (1915) and many other publications, hydraulic current meter, numerous patents dealing with railway track construction, air conditioning, consulting engineer (1917ff) (NC53 NUC PS WWWIA)

CLAUDE, Georges, 1870–1960, French physicist, chemist and inventor; safe transportation of acetylene (1897), produced liquid air (1902), neon light (1910), synthesis of ammonia (1917) (MMAH NYPL PA WA WWWIS)

CLAUDIUS—see APPIUS CLAUDIUS

CLAUSIUS, Rudolf Julius Emanuel, 1822–1888, German physicist and mathematician; produced second law of thermodynamics (1850), coined word "entropy," kinetic theory of gases, electrolysis, magnetism, statistical mechanics, professor at Zürich (1855–1867), Würzburg (1867–1869), and Berlin (1869ff) Univ., fNAS (1883) (BEST DOT DOTHOS FNIE L&M MWBD WWWIS)

CLAY, Lucius Du Bignon, 1898–1978, American engineer and army officer; US Army engineer, airport construction (1940–1942), wartime military procurement (1942–1945) as a general, military governor of Germany and commander of European military forces (1945–1949), federal interstate highway programs, chairman of the board of Continental Can Co. (1950–1962), headed investment firm of Lehman Brothers (1963–1973) (ANB DAB MWBD NCG,61 WWWIA)

CLEARY, Edward John, 1906–1984, American civil engineer; construction of tunnels and power plants, water sanitation, environmental health engineering, public works administration, editorial staff of *Engineering News-Record* (1935–1949), Ohio River Sanitation Commission (1949–1971), after retirement (1971) part-time at Univ. of Cincinnati, consultant, NAE (1967) (MT3 WWIE WWWIA)

CLEMENS, Ernest Victor, 1855–1893, American civil and mechanical engineer; machinist, draftsman, foundry, invented, designed and erected mining machinery, sugar plants, superintendent of National Machinery Co. (OH), established Clemens Foundry & Machine Co. (CT), superintendent of De Vergne Refrigeration Machine Co. (NY) (1888) (MEIA)

CLEMSON, Thomas Green, 1807–1888, American mining engineer; mining and mineral processing, assaying and gold mining, government service, interested in education and diffusion of knowledge, consulting engineer, estate given to establish Clemson College (now Univ.) (SC) (1889) (BDOAS DAB MWBD NC12,13 WWWIA WWWIS)

CLEON, c. 250 BC, royal engineer of Faiyum (ancient Egypt); commissioner of public works, transport of building materials, construction and repair of buildings, work followed by his assistant Theodorus (TSOE)

CLERK, Sir Dugald, 1854–1932, Scottish engineer; built successful two-stroke cycle gasoline engine (1881) called Clerk engine, author (BDOTHOT DNB EE MWBD WWWIS)

CLEVELAND, Frank Allen, 1923–1983, American mechanical and electrical engineer; turbine engines for aircraft, afterburners for turbojets, aerodynamics, aircraft design, after WWII with Lewis Research Center

(NACA) (2 yr.), then to Lockheed Co. (until 1983), becoming corporate vice president, NAE (1980) (MT3)

CLEWELL, Dayton H., 1912–1992, American physicist and petroleum engineer; new technologies for petroleum operations, exploration and production, industry executive, energy conservation, environmental improvement, seismic exploration with Mobil Oil Corp. (1938–1977), NAE (1976) (MT7 WWIE)

CLUETT, Sanford Lockwood, 1874–1968, American engineer; invented Sanforizing for mechanically preshrinking fabrics (MWBD WWWIA)

CLYDE, Arthur W., 1891–??, American agricultural engineer; serviceman for International Harvester Co. (1915–1918), engineer with Portland Cement Co. (1918–1920), faculty at Pennsylvania State Univ. (1920–1956), soil resistance encountered by tillage tools, design and operation of tractors, drying forage with natural and heated air, consulting engineer (1948ff) (AMOS)

CLYDE, George Dewey, 1898–1972, American agricultural and civil engineer; irrigation engineer, hydrology, stream flow, snow surveys, faculty at Utah State Univ. (1923–1940), including dean of engineering (1935–1945), chief of irrigation research for Soil Conservation Service (1945–1953), director of Utah Water & Power Board (1953–1957), governor of Utah (1957–1965), consulting with Richard J. Woodward [b. 1907] and Guy O. Woodward [b. 1915] (CDOS NCI WWIE WWWIA)

CLYMER, George E., 1754–1834, American inventor; designed unique plow for local conditions in Buck County (PA), pump to clear coffer dams (1801), improved printing press known as Columbian press (1817), which was quite successful in the European market (BDOTHOT DAB MEIA NC8 WWWIA WWWIS)

COATES, George Henry, 1849–1921, American inventor and manufacturer; machinist, machine design, mechanical drawing, fire kindler (1874), several hand tools, adjustable hair clipper (1876), flexible shaft (1892), with Ethan Allen Co. (1867–1875), went into business developing hair clippers and machine tools (1875ff) (DAB)

COBLE, Robert L., 1928–1992, American ceramist; materials sintering, ceramic processing, composites, industrial laboratory with General Electric Co. (1955–1960), faculty at Massachusetts Inst. of Technology (1960ff), NAE (1978) (MT7 WWIE WWWIA)

COBLENTZ, William Weber, 1873–1962, American physicist and inventor; infrared spectrometer, ice formation on plants, bioluminescence, phototherapy, ultraviolet climatology, with National Bureau of Standards (1905–1945), where he headed the radiometry section, NAS (1930) (AMOS BEST BM39 NC52,E WWWIA WWWIS)

COCHRAN, Josephine Garis, 1842–1913, American woman inventor; first commercially available dishwasher (1896, 1888, 1894) known as Garis Cochran dishwashing machine absorbed by Hobart (1926), later known as Kitchen Aid, which became a part of Whirlpool Corp. (AWIT BDOTHOT)

COCHRAINE, Edward Lull, 1892–1959, American naval officer and engineer; service and command functions with US Navy, became vice admiral, Bureau of Construction and Repair (1924–1929), submarine and general design, managed design branch, professor of naval construction and head of naval architecture and engineering at Massachusetts Inst. of Technology (1947–1950), NAS (1945) (NC46,G WWIE(I) WWWIA)

COCKCROFT, John Douglas, 1897–1967, British physicist and electrical engineer; with Irish physicist Ernest Thomas Sinton Walton [1903–1945] developed the particle accelerator (1932), radar, atomic energy, professor at Cambridge Univ. (1939–1946), director of Atomic Energy Research Establishment (UK) (1946–1958), founding master of Churchill College at Cambridge Univ. (1959), president of Manhattan College of Science and Technology (1961–1967), directed establishment of first nuclear power plant in Canada (during WWII), Nobel Prize (1951) (BEST CEOS&T DNB DOTHOS IWAP MWBD NYPL PA TBA TBDOS(Po) TGE:B TP WWWIS)

COCKERELL, Christopher Sydney, 1910–1999, British engineer and inventor; invented successful hovercraft (1955), built by Saunder-Roe Co. (1959), aircraft navigation, radar, aircraft communications, wireless telegraph, at Marconi Wireless Co. (1935–1950), worked for British government, over 36 patents on telegraph (DNB GI IAI ITTC NYPL PS TBDOS TGE:E WA)

COCKERILL, William, 1759–1832, English inventor and manufacturer; wool-carding, wool-spinning machines, significant design to streamline production, worked in Russia, Sweden and Belgium (DNB TBDOS MWBD WWWIS)

COEHOORN, Menno van Baron, 1641–1704, Dutch soldier and military engineer; invented small bronze mortar (used in 1674), artillery, fortification for level terrain (MWBD)

COFFIN, Howard Earle, 1873–1937, American engineer; automobile designer, with several automobile companies, chief engineer, founded National Air Transport (1925), the forerunner of United Air Lines Co., of which he was president (1925–1928), consulting engineer (MWBD NC30 WWWIA)

COGSWELL, William Browne, 1834–1921, American mining engineer; repair shops for navy during Civil War, introduced Solvay process into US for lead mines, developed industry to produce soda ash and soda (NY) (DAB NC13 WWWIA)

COHEN, Jerome B., 1932–1999, American materials scientist and engineer; X-ray diffraction of materials, residual stress, atomic arrangements in alloys, ceramics, phase transitions, fatigue of metals, catalysis, faculty and college administration at Northwestern Univ. (1959ff), NAE (1993) (Bridge MT9 WWIE WWWIA WWWIS)

COHEN, Mendes, 1831–1915, American civil and railroad engineer; builder of locomotives with Ross Winans [1796–1872] (1847ff), converting locomotives from wood to coal fuel, executive of railroad and navigation companies (BDOACE DAB NC13 WWWIA)

COHN, Nathan, 1907–1989, American electrical engineer and inventor; control and performance of interconnected electric power systems, Leeds & Northrop Co. (1927–1975), becoming vice president of research and development (10 yr.), vice president (1972) and corporate director (1975), 15 patents, consulting engineer as Cohn Associates (1967ff), NAE (1969) (MT4 WWIE WWWIA)

COLAHAN, Charles, 1836–1923, American mechanical engineer; wholesale importer, freight transportation, advisor to Cyrus Hall McCormick [1809–1884] in building grain harvesters and binders, machine for baling cotton (1867), mowing machine (1873), harvester rake (1883) (MEIA)

COLBURN, Allan Philip, 1904–1955, American chemical engineer; Colburn equation (1933), heat transfer, Colburn number, Colburn analogy, fluid flow, chemical processing, research at DuPont (1929–1938), professor and administrator, including acting president, at Univ. of Delaware (1938–1955) (L&M WWWIA)

COLBURN, Irving Wighman, 1861–1917, American inventor and manufacturer; leader in development of electric equipment, installation of telephones, developed continuous production of flat sheet glass, which became Libby Owens Sheet Glass Co. (DAB MWBD NC24)

COLBURN, Zerah, 1832–1870, American civil engineer; draftsman, locomotive works, editor and founder of several technical magazines in US and England (BDOACE MEIA NC7 WWWIA)

COLDING, Ludwig August, 1815–1888, Danish engineer and inventor; roads and buildings inspector (1845), technical and administrative posts in Copenhagen (1837–1865), water and sanitation systems, forces and energy concepts, professor at Polytechnic Inst. Copenhagen (1865ff), developed ideas on mechanical equivalent of heat at the same time as Julius R. von Mayer [1814–1878] (1842), James P. Joule [1818–1889] and Herman von Helmholtz [1821–1884] (DOSB MWBD PS WWWIS)

COLE, Edward Nicholas, 1909–1977, American automotive engineer; automotive industry, light tanks and vehicles, chief design engineer, chief engineer at General Motors Corp. (1940), chief engineer at Chevrolet (1952) to vice president of GMC (1956) to president and CEO of GMC (1957–1974), chairman and CEO of Checker Motor Corp. and chairman of International Husky, Inc. (1974–1977), 18 patents, NAE (1970) (MT1 WWWIA)

COLE, Francis J., 1856–1923, English American mechanical engineer and manufacturer; to US as a youth, machinist, draftsman, with numerous railroad shops and locomotive works, finally with American Locomotive Co. standard locomotive designs, worked on boiler ratios, leader in adopting superheaters, consulting engineer (MEIA)

COLE, Julian D., 1925–1999, American mechanical engineer, mathematician and aeronautics; application of mathematics to fluid mechanics, perturbation methods, shock-free transonic airfoils, aerospace, faculty at California Inst. of Technology (1950–1967), with service at Rensselaer Polytechnic Inst. and Boeing Co., Univ. of California Los Angeles (1969ff), consultant, NAE (1976), NAS (1976) (AMOS MT10 WWIE WWWIA)

COLE, J. Wendell, 1842–1921, American mechanical engineer; developed steamship machinery, chief draftsman, steam and boiler equipment, tool and drill grinders, with Fuel Saving Furnace Co., Detroit Emery Wheel Co. (1871) and William Sellers & Co. (1885ff) (MEIA)

COLLES, Christopher, 1739–1816, Irish American engineer and inventor; to US (1770), pneumatics, inland navigation canals, water supply for New York City based on reservoirs, served in artillery during Revolutionary War (1775–1777), canals and road building and maps, invented devices such as mileometer and hydrometer (BDOACE DAB MEIA NC9 WWWIA)

COLLEY, Russell, d. 1996, age 97, American mechanical engineer and inventor; employee with B. F. Goodrich Co. (1928ff) for 34 yr., developed space suits, various airplane devices, invented airplane deicer (1932), fame as an artist, 65 patents (FAL NYTObit)

COLLINS, Arthur A., 1909–1987, American electronics engineer and inventor; invented numerous electronic devices, communications, computation, control, automatic tuned multi-channel radio, founder of Collins Radio Co. (IA) (1931), becoming president and chairman (1933–1971), consultant, NAE (1969) (MT5 WWIE)

COLLINS, Edgar Vermont, 1887–1958, American agricultural engineer and inventor; at Kansas State Univ. (1915–1918), at Iowa State Univ. (1918ff), developed and improved horse-pulling dynamometer, tandem tractor, contributed to tractor testing, developed ridge farming practices, high clearance sprayer, 33 patents (AE39 T7-AP)

COLLINS, Samuel Cornette, 1898–1984, American mechanical engineer; Collins helium liquefier, faculty at Massachusetts Inst. of Technology (1945–1964), with Arthur D. Little, Inc. (1965–1967), with Naval Research Lab. (NRL) (1964–1971) as cryogenics engineer, NAS (1969) (CEOS&T MHSE PS WWIE WWWIA WWWIS)

COLLINS, W. Leighton, 1906–1999, American civil engineer; properties of materials, theoretical and applied mechanics, Holabird & Root Architects (1928–1929), faculty at Univ. of Illinois (1929–1965), executive secretary of ASEE (1955–1969), consulting engineer (WWIE)

COLPITTS, Edwin Henry, 1872–1949, Canadian American engineer; communications, research and development on telephone lines, increased distance for transmission, American Bell Telephone Co. (1899–1907), research engineer and laboratory head of Western Electric Co. (1907ff), Colpitts oscillator (1918), military communications in Europe in WWI, head of research then vice president of Bell Telephone Labs. (1933–1937), submarine detection (1940–1943), 24 patents (ANB NCE WWWIA)

COLT, Samuel, 1814–1862, American inventor and manufacturer; multishot firearm of the revolving barrel type revolver (1835ff), hand weapons, submarine battery, underwater mine explosions (1842–1845), built large armory (AIE ANB DAB GI IIA MEIA MMAH MWBD NC6 NYPL PS TBDOS WA WWWIA WWWIS)

COLWELL, Augustus, 1842–1917, American mechanical engineer; manufacture of sugar manufacturing equipment and for other products using evaporation, diffusion equipment for handling sugar cane and sorghum liquors, use of water tube boilers using bagasse as fuel (1884), vacuum pan (1885) (MEIA)

COMPTON, Arthur Holly, 1892–1962, American physicist and inventor; engineer at Westinghouse Light Co. (1917–1919), faculty at Washington Univ. (1920–1923), then chancellor (1945–1953), faculty at Univ. of Chicago (1923–1945), Compton effect (1923), with George Inman instigated the development of fluorescent lighting by General Electric Co. (1935ff) and Westinghouse Co., fluorescent light patented by Peter Hewitt [1861–1921] (1901), design of nuclear reactor (1942–1945), NAS (1927), Nobel Prize (1927) (BEST BM38 EB GSA MWBD NCG NYPL TBOF TGS WOI WWWIA WWWIS)

COMPTON, Karl Taylor, 1887–1954, American physicist and inventor; brother of Arthur H. Compton [1892–1962], served on faculties of Wooster College (OH), Reed College (OR) and Princeton Univ. (1915–1930), president of Massachusetts Inst. of Technology (1930–1948), structure of crystals, photoelectric effect, ionization, National Defense Research Committee (1940ff), involved in development of electronics that contributed to development of radar, rockets, missiles, proximity fuse, and the atomic bomb, NAS (1924) (ANB BM61 EB MWBD NCC,42 VNSE WWWIA WWWIS)

COMSTOCK, Cyrus Ballou, 1831–1910, American military engineer; engineer in Civil War, fortifications, advanced to brigadier general, wrote "Primary Triangulation of the U.S. Lake Survey" (1882) for the Corps of Engineers, NAS (1884) (BM7 NC22 PS WWWIA WWWIS)

COMSTOCK, Theodore Bryant, 1849–1915, American mining engineer and geologist; with Morgan expedition to Brazil, taught school, state geologist, faculty at Cornell Univ. (1875–1879), Univ. of Illinois (1885–1889), Univ. of Arizona (1891–1895), including president (1893–1895), consulting engineer (BDOAS NC13 WWWIA)

CONANT, Hezekiah, 1827–1902, American mechanical engineer, inventor and manufacturer; machinist, lasting pinchers for bookmakers (1852), gas check for breech loading rifle (1856), thread dressing and sewing (1859, 1860), development became part of J & P Coats Co., built several spinning mills, several patents related to thread machines, gravity escapement for clocks (1894) (DAB MEIA NC42 WWWIA WWWIS)

CONE, Hutchinson Ingham, 1871–1941, American naval officer; US Navy (1894–1922), retiring as rear admiral and engineer-in-chief, served in engineering positions on 8 ships, commanded torpedo boat (1903), served in industry and government following naval service including with the Daniel Guggenheim Fund (ante 1928), US Shipping Board (6 yr.), Moore-McCormick Line (1935ff) (ANB NC35 WWWIA)

CONE, Russell Glenn, 1896–1961, American civil engineer; bridge construction across Delaware River as assistant engineer (1923–1925), resident engineer (1925–1927), later the Ambassador Bridge (MI) (1927–1930), general manager of four-lane toll bridge over the Delaware River (1930–1933), Golden Gate Bridge (CA) (1933–1937), and various transportation, bridge and government projects (ANB NC51 WWWIA)

CONEY, Jabez, 1804–1872, American millwright and engineer; builder, machinist, machinery for US marine boats (DAB NC23 WWWIA)

CONGREVE, Sir William, 1772–1826, English rocket engineer; improved military rocket (1805), Congreve rocket, Royal artillery, founded two rocket companies (1809ff), rockets used in war against French, officer in the government of England (DNB DOSB E–64 MMAH WWWIS)

CONRAD, Frank, 1874–1941, American electrical engineer and inventor; radio experiments, radio and radio broadcasting (1919), commercial radio broadcasts (KDKA) (1920), arc lamps, electric meters, rheostats at Westinghouse Electric Co., complete automobile electrical system, including voltage regulator (MWBD NC35 TFE&I WWWIA WWWIS)

CONRAD, Pete (Charles), 1930–1999, American aeronautical engineer and astronaut; third man to walk on the moon, flew three other missions, commander of *Apollo* 12, US Navy (1953–1963), astronaut at NASA Spacecraft (1964–1974), vice president of American TV & Communications Co. (1974ff) (NYT-A WA WWWIA)

CONRADER, Rudolph, 1858–1929, American mechanical engineer and inventor; brass finishing, reamer grinding machine (1893), machine for grinding spherical surfaces (1894), steam engine governors for Erie Co. (1920–1925), head of R. Conrader Co. (1925ff), over 200 patents (MEIA)

CONSTANT, Hayne, 1904–1968, British mechanical and aeronautical engineer; vibration of crankshafts, axial flow compressors and gas turbine engine drive propellers for aircraft (1937), which became the prototype of modern jet engines and augmented work of Frank Whittle [1907–1996], 10 patents (DNB PS RS19 WWWIS)

CONVERSE, Edmund Cogswell, 1849–1921, American inventor and industrialist; lock-joint for connecting tubing (1882), involved in many manufacturing companies (DAB WWWIA)

CONWAY, Gerald R., 1915–1999, American civil engineer; Washington National Cathedral (DC) construction, field engineer, president of consulting firm dealing with construction (WPostObit(Nov))

COOK, George Hammell, 1818–1889, American geologist and engineer; glass making technology, faculty at Rutgers College (NJ) (1853–1857), New Jersey state geologist (1864–1889), New Jersey State Agricultural Experiment Station (1880–1890), director of New Jersey weather service, NAS (1887) (BDOAS BM4 DAB NC6 WWWIA)

COOK, James (aka Captain Cook), 1728–1779, English hydrographer and sea captain; surveyed St. Lawrence Channel (1759), astronomy, cartography, charted coast lines, circumnavigated the globe (MWBD WWWIS)

COOK, Neville George Wood, 1938–1998, South African American mining engineer; rock bursts, rock mechanics, design of deep mines, underground nuclear waste depositions, at Australian National Univ. at Cranberry (1962), director of Mining Research Lab. (South Africa), then administrator of Chamber of Mines (1974), faculty at Univ. of California (1903ff) for 22 yr., NAE (1988) (Bridge MT9 WWWIA)

COOKE, Morris Llewellyn, 1872–1960, American mechanical engineer; scientific management as a follower of Frederick W. Taylor [1856–1915]. engineering consultant (1905–1911), director of public works in Philadelphia (1911–1915), headed survey that emphasized support for rural electrification (1923–1925), first director of Rural Electrification Administration (1935–1937), headed several federal government programs (ANB WWWIA)

COOKE, Sir William Fothergill, 1806–1879, English electrical engineer; electric telegraph using single-needle device (1845), electric telegraph development with Charles Wheatstone [1802–1875] (1837) (DNB GI MMAH MWBD PS TGE WWWIS)

COOLEY, Lyman Edgar, 1850–1917, American civil engineer; hydraulic engineer, Mississippi and Missouri River improvements, waterways, canals, sanitary canals, writer and consultant (DAB NC9 WWWIA)

COOLEY, Mortimer Elwyn, 1855–1944, American mechanical and civil engineer; steam engineer with US Navy, steam and iron shipbuilding, faculty at Univ. of Michigan (1885–1895), appraiser of railroads, power plants, rolling stock, member of several state and federal committees and commissions (ANB MEIA NC51 WWWIA)

COOLIDGE, William David, 1873–1975, American physical chemist and inventor; ductile tungsten (1908), tungsten filament (1912), X-ray tube (1916) known as Coolidge tube, with General Electric Co. (1905–1944), including director of research (1932) and vice president (1940), NAS (1925) (AIE ANB BM53 MWBD NCG NYT-A WA WWWIA WWWIS)

COON, John Sayler, 1854–1938, American mechanical engineer; mining equipment, pumping equipment, with various mining companies (ante 1880), faculty and administration at Georgia Inst. of Technology (1889ff) for 37 yr., consultant (MEIA WWWIA)

COOPER, Alfred R., Jr., 1924–1996, American ceramic engineer; structure of glass, continuous melting of glass, field-assisted ion exchange, US Navy in WWII (1943–1946), with RCA working on cathode ray products (1948–1952), laboratory manager at Emhart Corp. (1952–1956), Massachusetts Inst. of Technology (1960–1965), faculty at Case Inst. (OH), which became Case Western Inst. of Technology (1965–1992), where he founded the ceramics program (1965ff), participated in numerous international academic programs, NAE (1996) (Bridge MT10)

COOPER, Franklin Seaney, 1908–1999, American engineering physicist; speech instrumentation, human communication, aids for handicapped, General Electric Research Lab. (1936–1939) in high voltage engineering, with Caryl Parker Haskins [1908–1958] founded Haskins Labs. (1939) and served as research director and president, during WWII headed liaison Office of Scientific Research & Development, launched research program on speech at Haskins Labs., NAE (1976) (AMOS Bridge CB WWIE)

COOPER, Hugh Lincoln, 1865–1937, American civil engineer; bridge construction as apprentice to super-

intendent (1885–1889), chief engineer for two bridges over the Mississippi River, with Chicago Bridge and Iron Works (IL) (1889–1891), with several firms designed and built hydroelectric plants worldwide, consulting, NAS (1938) (ANB BM18 NCC,33 WWWIA)

COOPER, John Haldeman, 1828–1897, American mechanical engineer; machine-making, worked on mining, agricultural and general machinery, expert in cooling of condensing water in power plants, consulting engineer (1891) (MEIA)

COOPER, Peter, 1791–1883, American engineer, inventor and manufacturer; machines for making carriage wheel hubs (1841), cloth-shearing machines (1815), glue factory, manufacturer of isinglass, built first US steam locomotive called Tom Thumb, numerous metal and wire factories, Cooper Union, an engineering school in New York City named after him, numerous patents (AIE ANB ASAI DAB MEIA MWBD NC9 PS WA WWWIA)

COOPER, Theodore, 1839–1919, American civil engineer; bridge builder, developed wheel-load analysis for railroad bridge analysis, consulting engineer (BDOACE DAB NC19 WWWIA)

COPELAND, Charles W., 1815–1895, American naval engineer; designed machinery for several government-funded ships (*Fulton, Mississippi, Missouri, Susquehanna, Saranac, Michigan*), marine ships, low-pressure steam engine (1841), consulting engineer (DAB MEIA WWWIA)

COPELAND, Norman Arland, 1915–1984, American chemical engineer; development engineer, design engineer, heat transfer beds, synthetic rubber, nylon, research engineer and chief engineer at DuPont (1937ff), NAE (1977) (AMOS MT3 WWIA)

COPERNICUS, Nicolaus (aka Mikolaj Kopernik), 1473–1543, Polish astronomer; studied orbits of sun, moon, and planets (1497ff), revolutionized astronomy by developing heliocentric cosmology system in which earth rotates daily and with other planets revolve around the sun (publ. 1543), which became known as the Copernican theory, replacing the Ptolemaic system after Clausius [c 90–170]. (BEST CDOS EB MWBD SAI-MP TBDOS(Po) WWWIS)

CORBUSIER, Le—see Le Corbusier

CORCORAN, William Harrison, 1920–1982, American chemical engineer; biomedical engineering, kinetics, processing of pharmaceuticals, transport processes, faculty and administration at California Inst. of Technology (1922ff), consulting, NAE (1980) (MT2 WWWIA)

CORE, Jesse F., 1913–1993, American mining engineer; efficient coal mine layouts and operations, improvement of health and safety in coal mines, executive with US Steel (1951ff), becoming vice president (1958), NAE (1981) (MT8 WWIE)

CORIOLIS, Gaspard-Gustave de, 1792–1843, French mathematician and engineer; mechanics, Coriolis theorem (1831), motion on a rotating surface (1835), described Coriolis forces (1835), Coriolis law (1844), defined kinetic energy, professor at École Polytechnique (1816–1838), École des Ponts et Chaussées (1832–1843) (CEOS&T DOT EE FNIE L&M MWBD TBDOS(Po) WWWIS)

CORLISS, George Henry, 1817–1888, American inventor and manufacturer; steam power, valves for steam engines, Corliss steam engine (1846ff), Corliss valve (1849ff), numerous improvements on steam engine, machine for sewing leather (1842), gear cutting machine (1856), with William A. Harris [1835–1879] formed Harris-Corliss Co. (1864) (ANB ASAI DAB EE IIA LIME MEIA MWBD NC10 PS TBDOS WWWIA)

CORNELIUS, Yngve R., 1905–1966, Swedish American metallurgical engineer and inventor; developed process for electric melting of glass, electric heating furnace design, domestic warm-air furnaces, bottle glass factories (1938), electric sodium silicate plant, consulting engineer (NC53 PS)

CORNELL, Ezra, 1807–1874, American mechanician and construction engineer; telegraph pioneer, used broken bottle tops as insulators for telegraph wires, developed telegraph lines in midwest (1856), founder of Cornell Univ. (1865) with Andrew D. White [1832–1918] (AIE ASAI IIA MEIA MWBD NC4 PS WWWIA WWWIS)

CORNU, Paul, 1881–1944, French engineer and inventor; cycle builder, built first helicopter to carry a person in flight (1907) for 20 sec. at a height of 6 ft., with numerous milestones to follow to reach commercial development and use, with a fully controllable helicopter by German and Scottish builders (1936, 1938) (MWBD TBOF)

CORRSIN, Stanley, 1920–1986, American mechanical engineer; aerodynamics, advanced aircraft, fluid dynamics, turbulence, faculty at California Inst. of Technology (1940–1947), faculty at Johns Hopkins Univ., including department head (1947ff), NAE (1980) (MT3 WWWIA)

CORRY, Andrew F., 1922–1994, American electrical engineer; electric transmission, underground electric power transmission, with US Army Signal Corps in WWII (1942–1946), electric utility industry with Boston Edison Co. (1947ff), nuclear energy research, consultant, NAE (1978) (DNB MT8 WWIE WWWIA)

CORT, Henry, 1741?–1800, British iron master and inventor; established forge factory, large-scale conversion of cast iron into wrought iron, groove rollers to replace hammering (1783), perfected puddling in manufacture of malleable iron (1784), used reverberating furnace, use of coke instead of more expensive charcoal (EOWB MMAH PS TBDOS)

CORTHELL, Elmer Lawrence, 1840–1916, American civil engineer; served in Civil War, design and construction work for railroads, chief engineer of levees along railroads, leader in international engineering education, consulting engineer in US and several South American countries (BDOACE DAB NC9 WWWIA)

COSTA, Joseph Louis, 1897–1969, English American scientist and chemical engineer; synthetic fiber development and production, filament rayon, production of cellulose material, which replaced natural casings in the meat industry, in WWI served with French legion, in WWII with US Development and Research Council, technical director of Woonsocket Rayon Co. (RI) (1927–1943), president and general manager of Transparent Package Co. (IL) (1943–1946), vice president of what became Phitec, Inc. (1946–1967), several patents (NC55 WWWIA)

COSTANO, Miguel, fl. 1769–1811, Spanish military engineer; cosmographer to US, astronomical observations, drainage of Mexico City, defending coast of US (DAB)

COTTON, Sir Arthur Thomas, 1803–1899, English engineer; irrigation in India, dams, hydraulics, founded school of hydraulic engineering in India (DNB MWBD)

COTTON, William, 1786–1866, British inventor and businessman; invented knitting machine used particularly for hosiery (1864), worked for Huddart & Co. (London) in India, wrote book on inventions of Joseph Huddart [1741–1816] primarily related to sailing, navigation and trading, contributed to schools and housing (DNB NUC TBDOS)

COTTRELL, Calvert Byron, 1821–1893, American mechanical engineer, inventor and manufacturer; machinist, improvement of machine tools and machinery, Cottrell air spring, improvement of printing press, rotary color printed press, established Cottrell & Babcock Co. (1855), over 100 patents on printing plus others (DAB MEIA NC3 WWWIA WWWIS)

COTTRELL, Frederick Gardner, 1877–1948, American chemist, metallurgist and inventor; electrostatic separator (e/s) of particles from gas named after him, formed Research Corporation (1912), development of industrial helium, petroleum dehydration, nitrogen fixation, NAS (1939) (ACCE AIE ANB BM27 NC38 NYT-A TTOS WWWIA WWWIS)

COULOMB, Charles-Augustin de, 1736–1806, French military engineer and inventor; structural mechanics, physics, chemistry, friction, electricity, Coulomb law of friction (1781), Coulomb law of electric attraction and repulsion (1785), coulomb unit of measure named after him, torsion balance (1777), developed method for calculating pressure of soil against retaining wall, magnetic needle (BEST DOT DOTHOS FIOTEC FNIE L&M MWBD NYPL TSOE WBEOS WWWIS)

COULTER, Wallace Henry, 1913–1998, American instrumentation industrialist; flow cytometry, identified blood by their signatures, method and equipment for counting and sizing blood cells and microscopic materials suspended in a fluid called Coulter Counter (1953) with younger brother Joseph Coulter, an electrical engineer, began in basement laboratory (1948), established Coulter Electronics, Inc. (1958), 74 patents, company acquired by Beckman Instruments, Inc. (1997), NAE (1990) (Bridge MT9 NYTObit WWWIA)

COUSTEAU, Jacques Yves, 1910–1997, French marine explorer and inventor; interest in diving (1936ff), gunner with French navy, with Emile Gagnan developed air lung called aqualung (1942), founded French navy underseas research group (1945), numerous sea explorations, worked with several universities in US and Europe, fNAS (1968) (BDOTHOT GI IAI NYPL SAI-MP S:TLAW SAI TGS TOIAD WWWIA WWWIS)

COWEN, Joshua Lionel (originally Joshua Cohen), 1877–1965, American inventor; batteries, fuses, model railroad trains, formed Lionel Corp. (1900) to manufacture model trains (ANB MWBD WWWIA)

COWLES, Alfred Hutchinson, 1858–1929, American metallurgist and inventor; inventions with brother Eugene Cowles [1855–1892], with Electric Smelting and Aluminum Co. (1885), pioneer in use of electric furnace, first electric smelting plant (1886), produced carborundum, graphite, phosphorus and calcium carbide, companies named after him (ACCE MWBD NC22 WWWIS)

COX, Jacob Dolson, 1852–1930, American mechanical engineer and manufacturer; marine engines, manufacture of twist drills, designed and built machinery for grinding, planing and drilling, invented grip socket, formed Cox and Prentiss Co. (1880) (MEIA WWWIA)

COX, Lemuel, 1736–1806, American bridge builder and civil engineer; wheelwright (1765), mechanic, bridge builder in Massachusetts, Maine and Ireland, inventor of first machine to cut card wire (1770) (BDOACE DAB NC24 WWWIA)

COXE, Eckley Brinton, 1839–1895, American mining engineer and manufacturer; development of coal land in US after study in Europe, dealt with issues relating to consolidation and management of holdings, leadership in preventing fires in mines, labor use, waste prevention, developed long steel tapes to measure distances in place of chains for surveying, Coxe micrometer, over 100 patents on mechanical stoking (c. 1890) (DAB MEIA NC11 WWWIA WWWIS)

CRAIG, David, 1905–1964, American chemist and inventor; research chemist, synthesis and testing antioxidants, heat stabilizer, accelerators, vulcanizing agents, manufacture of tires, synthetic rubber at B. F. Goodrich Co., over 26 patents (NC52 PS)

CRAIN, Cullen M., 1920–1998, American electrical engineer; airborne radar development, invented airborne microwave refractometer, faculty at Univ. of Texas (1943–1944) (1946–1957), Rand Corp. (1957–1988), NAE (1980) (Bridge WWIE WWWIA)

CRAMPTON, Thomas Russell, 1816–1888, English engineer; locomotion, railway construction, waterworks construction, Crampton locomotive (1843), laid telegraph cable across English Channel (1851) (DNB MWBD WWWIS)

CRANE, Charles Richard, 1858–1939, American industrialist and diplomat; plumbing supplies by Crane Co., of which he was president (1912–1914), involved in numerous political activities (MWBD NC30 WWWIA)

CRAPONNE, Adam de, 1526–1575, French hydraulic engineer; drainage of marshes of Frejus, Roman port on Mediterranean, defense works of Nice, first canal outside of Italy, called de Craponne (EB TSOE)

CRASSUS—see APPIUS Claudius

CRATES of CHALGIS, fl. 300 BC, Greek miner and tunnel engineer; made an ambitious attempt to control level of Lake Copais (Greece) (c. 325 BC), used shaft sinking and tunnel in works (EB TSOE)

CRAVEN, Alfred Wingate, 1810–1879, American civil engineer; surveys and construction of railroads, wharves, docks and basins on railroads along the Delaware River, water works, subway, consulting engineer (BDOACE NC9 WWWIA)

CRAVEN, John Joseph, 1822–1893, American inventor and physician; telegraph construction, first to use newly invented pole climbers, insulated cables for underwater installations, medical service in Civil War (DAB NC23 WWWIA)

CRAY, Seymour R., 1925–1996, American computer engineer; noted for development of computers, vice president of Control Data Corp. (CDC), which was founded by William C. Norris [b. 1911] (1957), developed supercomputer named Cray computer, used large number of integrated circuits for supercomputers, established Cray Research, Inc. (1972) (FAL IAI NT-CS NYT-A NYTObit 1000Y WWWIA)

CREDE, Charles Edwin, 1913–1964, American mechanical engineer; mechanics and vibration, shock, failure, railway equipment and rolling stock, new plant design, heat flow in refrigerated cars, shock resistant equipment for US Navy\, with USN Bureau of Ships during WWII (1942–1945), with Barry Controls (1945–1958), faculty at California Inst. of Technology (1958–1964), consulting (NC52 PS)

CREGIER, De Witt Clinton, 1829–1897, American civil engineer; ironworks, construction of engines for steamships, public works, Chicago city engineer (1880), pumping machinery for water works, manager of railroad (1887), consulting engineer (1891ff) (MEIA NC19)

CRELLE, August Leopold, 1780–1885, German engineer; road building (1816–1836), publisher (1822), founder of *Crelle's Journal* (1826) (MWBD PS WWWIS)

CRISTOFORI, Bartolomeo di Francesco, 1655–1731, Italian inventor; harpsichord instrument maker, credited for developing the hammer action for the piano (1709), invented grand piano (1709) (GI MWBD 1000Y WA)

CROCCO, Gaetano A., 1877–1968, Italian electrotechnical engineer; aerospace engineering, airship stability, design of rigid and semi-rigid airships, flight mechanics and airframe structures, founder and director of Italian Research Inst. for Aeronautics (ITCOF)

CROCCO, Luigi, 1909–1986, Italian mechanical engineer; rocketry, theoretical aerodynamics, Crocco energy integral (1931), Crocco point of gas dynamics (1937), Crocco transformation in boundary layer (1939), professor at Univ. of Rome, Princeton Univ. (1952) and École Centrale des Arts et Manufactures (1970–1977), consultant, NAE (1979) (MT3)

CROCKER, Francis Bacon, 1861–1921, American electrical engineer and inventor; with Charles E. Crocker [1822–1888] formed company specializing in patents (1883), with Charles G. Curtis [1860–1953] formed C. and C. Electric Motor Co. (1886), with Schuyler S. Wheeler [1860–1923] formed Crocker-Wheeler Co. and served as director for the remainder of his life, designed standard electric motor in use (1886), electric lighting, arc furnace, worked toward establishing electric standards, National Electric Code (NEC), with Peter Cooper Hewitt [1861–1921] experimented with helicopter (1917) and hydroplane (DAB MWBD NC12 WWWIA)

CROFTS, Freeman Wills, 1879–1957, Irish civil engineer and writer; made reputation writing detective novels (DNB MWBD)

CROMPTON, George, 1829–1886, English American engineer and inventor; to US (1839), son of William Crompton [1806–1891], took over father's loom factory, increased capacity of looms, mills, gun machinery for private and government arsenals (1861), a founder of Hartford Steam Boiler Inspection & Insurance Co., formed Crompton Carpet Co., over 200 patents (DAB MEIA MWBD NC21 WWWIA)

CROMPTON, Rookes Evelyn Bell, 1845–1940, British engineer; pioneer of the dynamo and electric lighting, military tank, electrical and mechanical standards, steam engine-driven road locomotive (DNB PS TBDOS WWWIS)

CROMPTON, Samuel, 1753–1827, English inventor; invented spinning mule (1779), by improved design could spin finer threads and a greater variety of yarn (1799), improved design based on work of Richard Arkwright [1732–1792] and James Hargreaves [1732–1792] (DNB MMAH MWBD WA TBDOS WA WBEOS WWWIS)

CROMPTON, William, 1806–1891, British American inventor and manufacturer; to US (1836), textiles, power weaving, followed by son George Crompton [1829–1886], mill machinery, loom design and manufacture (1837) (1840), made fancy woolens (1840), spinning yarns (DAB GI MEIA MWBD NC10 WWWIA)

CROOKES, William, 1832–1919, British physicist, chemist and inventor; invented cathode ray tube (1878), radiometer, discovered thallium, Crookes radiometer, Crookes tube (vacuum tube through which electricity was passed to form an ionized gas—a forerunner of cathode ray tube and television tube), printing of textiles, manufacture of sugar from sugar beets, use of manufactured fertilizer (ASAI BEST DNB EB IAI MWBD NYPL TBDOS WA WWWIS)

CROS, Charles, 1842–1888, French inventor and poet; theory of three-color photography (1869), phonograph (1877), speculated on communications with other planets (MWBD WWWIS)

CROSBY, William Otis, 1850–1925, American geologist and consulting engineer; gold mine operations, western mines, teacher, syphons, aqueducts, water supply and storage dams, construction of high dams in western US and Spain (BDOAS DAB NC22 WWWIA WWWIS)

CROSLEY, Powel, 1886–1961, American manufacturer; radio receiver, radio-frequency amplification, manufacture of household appliances, organized Crosley Radio Corp. (1921), built small Crosley automobile (MWBD WWWIA)

CROSS, Charles Frederick, 1855–1935, English chemist and inventor; with partners Edward J. Bevan [1856–1921] and Clayton Beadle invented viscose rayon process (1892), establishing artificial silk industry, pioneered development of cellulose (1908) (BDOS(Wi) BDOTHOT DNB MWBD TTOS WA WBEOS WWW WWWIS)

CROSS, Charles Robert, 1848–1921, American physicist and electrical engineer; faculty at Thayer College (1870–1876), Massachusetts Inst. of Technology (1880–1917), including director of Rogers Lab. (1885–1917), taught first course of electrical engineering in US (1882) (BDOAS NC39 WWWIA)

CROSS, Hardy, 1885–1959, American civil engineer; moment distribution for structural analysis (1930) and structural design and analysis, faculty at Brown Univ. (1911–1918), Univ. of Illinois (1921–1937), Yale Univ. (1937–1951) (MWBD WWWIA WWWIS)

CROSSFIELD, Alfred Scott, 1921–1996, American aeronautical engineer; pilot for test flight X-15 rocket plane, design and test pilot (1950s) for North American, Apollo and Saturn missions, research and development for Eastern Airlines (1967) (CB69 PS)

CROSTHWAIT, David Nelson, Jr., 1892–1976, Black American mechanical engineer; lifelong employment with Dunham-Bush Co. (IA) (1913ff), primarily in heating, ventilating and air conditioning, moved through the various positions, becoming director of research (1925) and technical director (1930–1969), notable achievements in reducing noise of steam and non-condensibles in heating systems, received 34 US and 80 foreign patents (AAISMAI ANB PS)

CROWE, Francis Trenholm, 1882–1946, Canadian American civil engineer; worked in construction of concrete water containment structures, irrigation pumping stations (1908), superintendent of construction of Jackson Lake dam (WY) (1910), then Arrowrock dam (ID) (1911), worked with numerous construction companies and projects, primarily in western US, including Hoover Dam (1931) and Lake Meade back of the dam (ANB NCD,34 WWWIA)

CROWELL, Luther Childs, 1840–1903, American inventor and mechanical engineer; manufacture of square-bottom paper bags (1873), printing press improvements, obtained over 280 US patents for printing machinery (DAB MEIA NC13 WWWIA WWWIS)

CROZET, Claude (Claudius), 1790–1864, French American military engineer; artillery, instructor at USMA (1816–1823), design and construction of railroad bridges, state engineer (VA) (1823–1832) (1839–1845), supporter of education, first president of Virginia Military Inst. (1839–1845), first railroad over the Blue Ridge range of mountains in southeastern U.S. (1849–1858) (BDOACE DAB EB MWGD NC18 WWWIA)

CRUQUIS, Nicholas, 1678–1754, Dutch drainage engineer; mapped Holland's canals, first used contour lines in mapping (TSOE)

CTESIBIUS of Alexandria, 2nd century BC (between 100–50 BC), Greek physicist, engineer and inventor; hydraulic organ, air gun, improved artillery, water clock (clepsydra), mechanics, force pump, first mechanical wind organ, founded engineering at university at Alexandria (BEST EB GI HIW MMAH MWBD PS TSOE WWWIS)

CUBITT, Sir William, 1785–1861, English civil engineer and inventor; invented self-regulating windmill sails that automatically adjusted to wind (1807), chief engineer of South-Eastern railway (England) (1812–1821), then partner (1821–1826), invented a treadmill (1818), canals, drainage structures, sluices, docks, ports, railway instrumentation (DNB EB WWWIS)

CUGNOT, Nicolas Joseph, 1725–1804, French soldier and engineer; invented three-wheeled artillery tractor (1770), earliest steam carriage (1769), believed to be earliest predecessor of automobile, invented rifle, steam-propelled tractor for hauling artillery (1779ff) (ASAI FIOTEC GI MMAH MWBD NYPL PS TBDOS WA WBE WWWIA)

CUI, Cesar Antonovich, 1835–1918, Russian military engineer and composer; fortifications, piano works, wrote songs, composed opera (MWBD)

CULLUM, A. Earl, Jr., 1909–1985, American electrical engineer; radio, television, microwave communications, broadcasting, sophisticated weaponry, formed own firm, NAE (1970) (MT3)

CULMANN, Karl, 1821–1881, German engineer; graphical statics, strength of materials, use of diagrams, Culmann diagram, professor at Zurich (1855–1881) (MWBD WWWIS)

CUMMINS, Clessie Lyle, Jr., 1888–1968, American inventor; with American Motors Co. (1904–1908), organizer and officer of Cummins Machine Co. (1912–1918), Cummins Engine Co. (OH) (1918–1954), invented first US automotive diesel-powered tractor (1930) at the same time several were produced in western Europe, invented Jake brake (1931) (LIME NYTObit TBOF WWWIA)

CURIE, Jacques (Paul-Jacques), 1855–1941, French physicist and inventor; with Pierre Curie [1859–1906] and Paul Langevin [1872–1946] developed phenomenon of piezoelectricity (1880), credited with French contribution to efforts of US and British to develop a working sonar system (GSA IAI WWWIS)

CURIE, Marie Skłodowska, 1867–1934, Polish French physicist and inventor; wife of Pierre Curie [1859–1906], together they developed extraction of radium from pitchblende ore, unit of curie named after both of them, shared Nobel Prize with Pierre Curie (1903), received second Nobel Prize for discovery of poloniom and radium and isolation of pure radium (1911) (BEST GSA IAI MWBD NWITS NWS S:TLAW TGS WIS(1) WWWIS)

CURIE, Pierre, 1859–1906, French physicist and inventor; extraction of radium from pitchblende ore, unit of curie named after both people, observed with brother Jacques Curie [1855–1941] and Paul Langevin [1872–1946] the phenomenon of piezoelectricity (1880), shared Nobel Prize (1903) (BEST GSA IAI LOTCS MWBD SAI-MP S:TLAW TGS WWWIS)

CURRY, Norval H., 1914–1995, American agricultural engineer; design of structures, design of agricultural experiment stations, facilities for crops and animals, irrigation systems, US and foreign operations, formed and operated company for consulting engineering (1959–1977), NAE (1988) (Res2 WWIE WWWIA)

CURTIS, Charles Gordon, 1860–1953, American mechanical and marine engineer; electric railway motors (1889), Curtis steam turbines (1896), first US gas turbine

(1899), steam propulsion mechanism (1904) (EE FNIE NC42 WWWIA WWWIS)

CURTIS, Francis Joseph, 1894–1960, American chemical engineer and chemist; research on explosives, inventing TNT (trinitrotoluene) at Merrimac Chemical Co. (1915–1920), production at plant (1925), then joined sales department (1926–1929) at which time the company was taken over by Monsanto Chemical Co. (1929), for which he later became director of development (1935), after which he moved to management and executive positions (ACCE EOS&T WWWIA)

CURTIS, Harry Alfred, 1884–1963, American chemical engineer; faculty at Univ. of Colorado (1908–1917), at Northwestern Univ., Clinchfield Carbocoal Corp., established department of chemical engineering at Yale Univ. (1923–1930), director of research for Vacuum Oil Co., chief engineer with Tennessee Valley Authority (5 yr.), dean of engineering at Univ. of Mississippi (1938–1949), returned to TVA working on fuels, coal tar, chemical manufacturing, fertilizer manufacturing (NC50 WWWIA)

CURTIS, Samuel Ryan, 1805–1866, American engineer, lawyer and soldier; improvements in Muskingham River, chief engineer for river improvements, city engineer, Mississippi River dikes (ANB DAB NC4 WWWIA)

CURTISS, Glenn Hammond, 1878–1930, American inventor and aviator; motorcycle manufacturer (1902), speed records (1905, 1907), aircraft, speed boats, motors for dirigibles (1907ff), seaplane (hydroplane) (1911), Curtiss pusher (1912), Curtiss Jenny (1918), Navy Curtiss flying boat (NC4) (1919), Curtiss R3C-2 (1925), won several flight records (AG AIE ANB ASAI MWBD NC22 NYPL WA WWWIA WWWIS)

CUTLER, James Gould, 1848–1927, American architect and inventor; architectural firm, office buildings and factories, patented mail chutes (1883), organized and was executive of factory (1908ff) (DAB WWWIA)

CUTTING, James Ambrose, 1814–1867, American inventor; manufacturer of bee hives, photography (1854), improvements in photolithographic process (1858) (DAB NC13 WWWIA WWWIS)

CUVIER, Georges Léopold, 1769–1832, French naturalist and inventor; at the Jardin des Plantes Museum of Natural History (Paris), invented classification system for annuals (plants), lecturer at École Centrale de Panthéon (1796), professor at College of France (1799ff), several government appointments, chancellor at Univ. of Paris (1819), president of the Committee of the Interior (1819) (BEST MWBD SAI WWWIS)

D

DAFT, Leo, 1843–1922, English American electrical engineer and inventor; power generating and transmission equipment, built early electric traction equipment called *The Ampere,* installed electric short line known as Daft System (ANB DAB EB MWBD NC18)

DAGGETT, Ellsworth, 1845–1923, American mining engineer and metallurgist; hydrometallurgist, US Geological Survey of 40th parallel, manager of smelting plant (1872), Russell system of irrigation, consultant (DAB WWWIA)

DAGUERRE, Louis-Jacques-Mande, 1789–1851, French artist and inventor; photographic device called Daguerre (1835), visual images considered a great achievement at time, with Joseph N. Niepce [1765–1833] invented photographic process named after him—Daguerreotype (1839), first consumer camera (1839) (ASAI BEST CEOS&T EOWB GEAPIT GI IIA MWBD NYPL PopSci PS SAI TGS WA)

DAHLBERG (DAHLBERGH), Eric, Count, 1625–1703, Swedish military engineer; director general of fortifications (1680ff), field marshall (1693), defense director, gained title of "Vauban of Sweden" (MWBD)

DAHLGREN, John Adolphus Bernard, 1809–1870, American naval officer, mechanical engineer and inventor; surveyor, astronomical observations, US Navy, Coast Survey, percussion lock (1835), devised Dahlgren gun (1851), served as chief of Bureau of Ordnance (1862–1863) (1868–1869), muzzle-loading gun, rifled gun, rear admiral in Union Navy (1863) (ASAI DAB MEIA MWBD)

DAILY, James Wallace, 1913–1991, American hydraulic engineer; fluid mechanics, hydraulics, cavitation, pumps and turbines, liquid-fluid flow, non-Newtonian fluid flow, hydraulics of submerged bodies, California Inst. of Technology (1936–1946), Massachusetts Inst. of Technology (1946–1955), NAE (1975) (AMOS MT6 WWWIA)

DAIMLER, Gottlieb Wilhelm, 1834–1900, German mechanical engineer, inventor and manufacturer; twin cylinder engine design, carburetor (1876), with Wilhelm Mayback [1846–1929] worked on engine compression ignition (1883), with Karl Friedrich Benz [1844–1922] developed high-speed petrol engines (1885), four-wheeled gasoline automobiles (1889), built first motorcycle (1885), founded several companies, including Daimler Motor Co. (1890), produced Mercedes automobile (1899ff) (ASAI BEST E-64 EE EIH EOWB GI IAI 1000Y MMAH MWBD NYPL PS RHW SAI S&TF TGS WA WBE)

DAKIN, Thomas Wendell, 1915–1990, American engineering consultant; electrical insulation, microwave dielectrics, electrochemistry, electric power industry, manager of Westinghouse Research Lab. (1946ff) NAE (1981) (AMOS MT5)

DALÉN, Nils Gustaf, 1869–1937, Swedish inventor; improvements in hot-air turbines, air compressors, milking machines, valve for regulating gas lights, Nobel Prize (1912) (MWBD WWWIS)

DALTON, John, 1766–1844, English chemist, physicist and meteorologist; teacher at New College of Manchester (1793–1800), president of Philosophical Society (1817–1844), meteorological studies (1787ff), built instruments for study of weather, Dalton law of partial pressures (1801), units of atomic mass once called daltons, FRS (1822) (BEST DOT MWBD SAI S:TLAW WWWIS)

DALZELL, Robert M., 1793–1873, Irish American millwright and inventor; to US (1801), water power for flour mills, construction of mills, elevator system for storing grain and meal (NY) (DAB)

DAMMOND, William Hunter, 1873–1956, Black American civil engineer; faculty of mathematics in Texas (1897), followed by joining faculty at Wilberforce Univ. (OH), railroad signaling devices (1903) and other railroad improvements (MI), bridge designer in England (1910–1916), Dammond circuit for railroad, returned to US (1916), worked for US Steel Co. (PA), then worked in Ohio, New York and Pennsylvania (UnivPitt. news release)

DAMON, Richard W., 1923–1991, American electrical engineer and physicist; microwave ferrite technology, induced conductivity in solids, electron devices, optics, electromagnetics, US Navy in WWII, General Electric Research Lab. (1951–1960), manager of microwave control devices at Microwave Assoc. (1960–1962), department director of technology, becoming corporate director of technology at Sperry Corp. (1962–1988), held several patents, NAE (1989) (AMOS Bridge MT5 WWIE WWWIA)

DANCEL, Christian, 1847–1898, German American inventor and mechanical engineer; merchant, developed machine for sewing shoes, shoe-welts, and stitching outsoles (c. 1870), machine to sew outer sole to upper shoe (1885), straight needle machines (1891), Dancel Machine Co. (c. 1895), numerous patents related to shoes and leather (DAB MEIA WWWIA WWWIS)

DANCER, John Benjamin, 1812–1887, British optician and inventor; voltaic cells, electrolysis (1830s), electrodeposition of copper, photography of microscopic objects (1840ff), microdot photography, improved Daniell cell (John F. Daniell [1790–1845]), microfilm (1852) (NYPL RHW TBDOS)

DANCKWERTS, Peter Victor, 1916–1984, British chemical engineer; liquids, continuous flow reactors, gas-liquid reactions, gas absorption, mixing, served in WWII, Atomic Energy Authority (1954–1956), Imperial College (1956–1959), Univ. of Cambridge (1959–1977), FRS, fNAE (1978) (CEOS&T DNB MT3)

DANDELIN, Germinal Pierre, 1794–1847, French Dutch mathematician and engineer; mining engineer in Liège, fortifications in Belgium, colonel in Belgian army, hyperbolic revolution, statics, algebra, probability (BDOS TBDOS PS WWWIS)

DANFORTH, Charles, 1797–1876, American inventor, mechanical engineer and manufacturer; cotton manufacturing business, improved spinning of cotton, several improvements to spinning frame known as cap spinner (1828), improved spinning of weft, manufacture of machine tools (1840) (DAB MEIA NC13 WWWIA)

DANFORTH, Richard Stevens, 1885–1962, American civil engineer and inventor; worked for industries on West Coast, developed Danforth anchor (1939), manufactured by company with his name (1939–1962), of which he was owner and president, anchor widely used in WWII for anchoring aircraft, pontoon bridges and LST boats (NC51 PS)

DANIELL, John Frederic, 1790–1845, British chemist, meteorologist and inventor; cell for producing direct current electricity called Daniell cell (1836), electrolysis, electroplating, process for sugar refining, invented a dewpoint hygrometer (1820), pyrometer (1830) for measuring high temperatures, studied behavior or earth's atmosphere, Henri-Victor Regnault [1810–1878] also developed hygrometer (1835–1838) (RHW TBDOS(Po) WWWIS)

DANIELS, Fred Harris, 1853–1913, American mechanical engineer, metallurgist and inventor; steel wire and rod products, iron works, rolling mill operations, chief engineer and executive, later became a part of American Steel and Wire, with Washburn & Moen (1876–1899), then with US Steel Corp. (1899ff), becoming president (1900), approximately 151 patents, consulting (BITAS DAB MEIA NC22 WWWIS)

DARBY, Abraham, 1678–1717, English ironmaster and inventor; founded Bristol Iron Co. (1708) to smelt iron, made castings, developed method of coking coal (1709) (BDOS BEOS EIH GEAPIT TBDOS RHW MMAH MWBD PS SAI)

DARBY, Abraham, 1750–1791, American ironmaster, engineer and inventor; constructed first iron bridge (1779) (EIH GEAPIT MMAH MWBD RHW TBOF)

DARCY, Henri-Philibert-Gaspard, 1803–1858, French hydraulic engineer; flow of water through porous media, effect of roughness on flow, water flow in pipes, water supply systems, Darcy law (1856), similar work done earlier by Frenchman Antoine de Chézy [1718–1798] with related work by John Thomas Fanning [1837–1911] and Robert Manning [1816–1897] (1888) (CFR FNIE L&M MWBD WWWIS)

DARIUS The Great, 521–486 BC, Persian builder; dug canal between the Nile to Suez, his ships sailed from the Nile through the Red Sea to Persia (EB TSOE MWBD)

DARLINGTON, Sidney, 1906–1997, American physicist and electrical engineer; electrical network theory and analysis, communications, radar, guidance systems, Darlington emitter-follower (1953), research administration at Bell Labs. (1929–1971), consultant (1971ff), 40 patents, NAS (1978), NAE (1975) (BM84 MT9 WWWIA)

DARRACQ, Pierre-Alexandre, 1855–1931, French manufacturer; automobiles, electric cars, motorcycles, racing cars, founded Gladiator Cycle Co. (1891) (MWBD)

DASSAULT, Marcel (originally Marcel Bloch), 1892–1986, French aeronautical engineer and industrialist; French Corps of Engineers in WWI, credited and produced high-performance combat aircraft, including delta wing Mirage, transport jets, director of Avions Marcel Dassault-Prequet Aviation (after WWI), member of French parliament (1951ff), fNAE (1976) (MT3 WPost1986)

DATUS, Nonius (listed by some as NONIUS), 1st century, Roman engineer; built aqueducts (TAE)

DAVENPORT, Russell Wheeler, 1849–1904, American mechanical engineer; puddling, iron and steel, chemist in steel works, introduced open hearth process, armor plate, high-grade steel for springs, axles, with Midvale Steel Works (PA) (1874–1891), supervisor at Bethlehem Iron Co. (PA) (1891ff) (MEIA NC25 WWWIA)

DAVENPORT, Thomas, 1802–1851, American inventor; direct current electric motor (1834), electric locomotive (1835), publisher (ANB BITAS DAB EIH MWBD PS WA WWWIA)

DAVEY, Henry, 1843–1928, British inventor; Davey differential valve gear for double-acting compound engine, hydraulic machinery (EE WWWIS)

DAVIDSON, Alexander, 1826–1893, American mechanical engineer and inventor; small electric cars for office and factory transport, patented paddle wheel boat (1881), keyboard for typewriter, typewriter improvements (1887) (1893), steam separator (1888), with W. D. Rutledge owned commercial school (IL) (1864), US Revenue Service (1869–1878), sold patents to Yost Writing Machine Co. (1887), typewriter (1893) (MEIA NC3 WWWIA)

DAVIDSON, Jay Brownlee, 1880–1957, American agricultural engineer; considered father of agricultural engineering in US, farm engines, tractors and machinery, faculty and administration at Iowa State Univ. (1905–1915) (1919–1956), 9 patents, on leave to China (1946–1948), consultant to US and foreign governments (AE38 NC43 NUC WWIE(I) WWWIA)

DAVIDSON, Marshall Ten Broeck, 1837–1919, American mechanical engineer; marine engine and machine shop, ocean steamers, assistant engineer US Navy (1862), chief engineer in US Army, contracting mechan-

ical engineer (1865) president of M.T. Davidson Co. (NYC), Davidson steam pump and pumping engines used by navy (MEIA)

DAVIDSON, Walter, 1876–1942, American inventor and industrialist; first president of Harley-Davidson Co. (1903), William A. Davidson [1870–1937] was works manager, Arthur Davidson [1881–1950] was secretary and sales manager, William S. Harley [1880–1943] was chief engineer and treasurer, Walter H. Davidson [1905–1992], nephew of Walter Davidson, became president of Harley-Davidson (1942) (NC47)

DAVIES, Duncan S., 1921–1987, British engineer and chemist; industry research and administrator at ICI (1945–1977), application of chemistry to industrial products and processes, design and innovation in polymer industry, manufacture of polymer reactive dyes, ceramics, chief engineer with UK Department of Trade and Industries (1977–1982), consulting, fNAE (1978) (MT5)

DAVIES, John Vipond, 1862–1939, Welsh American civil engineer; drydocks, engines, construction in coal mines, marine engineer, system for using coal dust with pitch binder to make briquettes, coal reclamation, tunnel project in NYC (1893), NY-NJ tunnel under Hudson River, Jacobs and Davies firm formed to work on tunnels, consulting engineer (ANB NC29 WWWIA)

DA VINCI—see LEONARDO

DAVIS, Arthur Powell, 1861–1933, American hydrographer and engineer; irrigation survey, reservoir sites, hydrographer (1896) for US Geological Survey on stream flows, engineer with newly established US Reclamation Service (1907), as chief engineer (1908), then director (1914), study of feasibility of Panama Canal (1909), multipurpose dams to provide electricity and irrigation in US and abroad (ANB NC24 WWWIA)

DAVIS, Edward F. C., 1847–1895, American mechanical engineer; machine works, hydraulic works, dock iron works, rolling mill, superintendent of shops for Reading Coal and Iron (PA) (1879), general manager of Richmond Locomotive and Machine Works (VA) (1890), designed compound locomotive, C. W. Hunt Co. (NYC) (1895) (MEIA NC58)

DAVIS, George Jacob, Jr., 1876–??, American surveyor and civil engineer; city engineer, faculty at Univ. of Wisconsin (1905–1912), faculty and dean of engineering at Univ. of Alabama (1912ff) (WWIE(I))

DAVIS, George Whitefield, 1839–1918, American military and civil engineer; quartermaster for US Army, assistant engineer for construction of Washington Monument (DC) (1878–1885), served in numerous military positions as instructor, on commissions and with military government, becoming major general, canal construction company executive in Nicaragua, governor of Canal Zone (DAB NC36 WWWIA)

DAVIS, Harmer E., 1905–1998, American civil engineer; concrete foundations, bridges, pavements, highway and traffic engineering, faculty at Univ. of California (1930–1959), serving as department chairman (1954–1959), NAE (1969) (WWIE)

DAVIS, Phineas, 1800–1835, American inventor and manufacturer; iron foundry, machine shop for manufacture of steam engines, steam engine York, boat builder, manager of machine shops of Baltimore & Ohio Railroad (MD) (1831ff) (DAB MEIA WWWIA)

DAVIS, Sir Robert Henry, 1870–1965, English inventor; invention of breathing, diving and escape apparatus, first patented oxygen-breathing equipment for mine rescues, oxygen-breathing equipment and heated apparel for high-altitude flying (BDOTHOT GDIMS PS)

DAVIS, Stephen Smith, 1910–1977, Black American mechanical engineer and inventor; at Howard Univ. (1938ff), HVAC specialty, during WWII (1943–1945) with National Bureau of Standards working on guided missiles, National Ordnance Lab. (1953–1963), consulted on wind tunnel construction while at Howard Univ., head of department (1962) and dean of engineering (1964–1970), continued teaching and consulting engineering (AAISMAI WWWIA)

DAVIS, William Augustine, 1809–1875, American inventor and postal authority; postmaster (MO), devised system of sorting mail, originated railway postal service (1862) (DAB MWBD NC16 WWWIA)

DAVISON, Gregory Caldwell, 1871–1935, American naval officer and inventor; communications, weapons, ordnance for navy, commander of torpedo boats, with Guglielmo Marconi [1874–1937] developed ship-to-shore communication transmission, designed and constructed the Davy Jones gunsight, officer with Electric Boat Co., developed recoilless aircraft gun (ANB NC26)

DAVY, Sir Humphry, 1778–1829, British chemist and inventor; electrochemistry, electrolysis, physiological effects of gases, arc lamp (1808), Davy miners safety lamp (1816), work led to electric welding, FRS (1801) (ASAI EE EIH EOWB GEAPIT GI MMAH MWBD NYPL SAI SAI-MP S:TLAW TBDOS TEIH TGE:B WA WWWIS)

DEAN, Ernest Woodward, 1888–1959, American chemical engineer; instructor at Hobart College (NY) (1908–1911), petroleum research at US Bureau of Mines (1914–1922), inspector then director of laboratory for Standard Oil Co. (1922–1952), retired (1953), helped set quality standards related to cracking petroleum fractions, introduced Dean-Stark method for removing water from petroleum, during WWI worked with US Shipping Board, helped develop viscosity index and kinematic viscosimeter (NC50 PS)

DEAN, Francis Winthrop, 1852–1940, American mechanical engineer; stationary and locomotive steam power, internally fired large-capacity boilers, high-pressure steam engines, vertical engine, Dean uniflow locomotive (1934), with Charles T. Main [1856–1943] (1893–1907), consulting engineer (MEIA WWWIA)

DEANE, Charles P., 1845–c. 1902, American mechanical engineer and inventor; formed Deane Steam Pump Co. (MA) (1867), direct acting engine (1870), steam engine valve (1871), reciprocating engine (1872), steam pump and pumping engine (1876), steam condenser (1882) (MEIA)

DE BRAHM, William Gerard (aka John Gerar William), 1717–c. 1799, American military engineer; surveyor of south district of colonies (1764), fortification of Charleston (1755), hydrography, defense of Savannah (1757), mapped Florida and Georgia (1765ff) (ANB BDOACE BDOAS BITAS DAB PS)

DEBUS, Kurt Heinrich, 1908–1983, German American electrical and mechanical engineer; surge voltages, development of V-2 rockets and rocketry in WWII, to US (1945), space vehicle launch operations, administration of development of space systems, missiles aircraft for US Army then NASA (ante 1975), NAE (1975) (MT2 PS WWWIA)

DECAUVILLE, Paul, 1846–1922, French industrialist and inventor; manufacture of equipment for narrow gauge railroad called Decauville (MWBD)

DECROW, David Augustus, 1859–1923, American mechanical engineer; mechanical draftsman for Holly Manufacturing Co. (NY), built large water works pumping engines, design engineer and chief engineer of commercial companies, designed unaflow pumping engine, manager of waterworks (1916) for Worthington Pump and Machinery Corp., which succeeded International Steam Pump Co. and other companies (MEIA)

DE DION, Albert, 1856–1946, French inventor; invented high-speed light oil engine named after him (EE WWWIS)

DEERE, John, 1804–1886, American blacksmith and industrialist; blacksmith, steel plows that would scour (1833) (1838), plows with wrought iron landside, steel share, and steel or iron plow with wooden moldboard, farm machinery, cultivators, formed Deere & Co. (IA) (1868) manufacturing company (AIE ASAI DAB GI IIA MEIA MWBD NC20 NYT-A PS WWWIA)

DEERING, William, 1826–1913, American industrialist; manufactured farm harvesting equipment, twine baler (1879ff), formed William Deering & Co. (1883), later merged with International Harvester Co. (IHC) (1902) (MWBD NC11 WWWIA)

DE FOREST, Erastus Lyman, 1834–1888, American mathematician and innovator; gold mines, contributed to mathematics of probability and errors, linear compounding (BDOAS WWWIA)

DE FOREST, Lee, 1873–1961, American scientist and inventor; served in Spanish American War, formed De Forest Wireless Telephone Co. (1902), wireless telegraphy, rectifier (1904), radio tube triode (1906), radio amplifier (1906), radio telephony (1909), feedback circuit (1912), telephone amplifier (1912), radio tube oscillator (1915), loudspeaker (1916), sound pictures, grid in electric tube, sold radio tube patents to American Telephone and Telegraph Co., over 300 radio patents (AIE ANB ASAI BEST CEOS&T EIH GI HOCS&T IAI IIA TBDOS TFE&I MMAH MWBD NC17 NC58 PA PS S:TLAW WA WWWIA)

DEGENKOLB, Henry John, 1913–1989, American structural engineer; earthquake engineering, seismic codes, seismic safety, chairman of board of H. J. Degenkolb consulting engineering firm, NAE (1977) (MT4 WWIE WWWIA)

DE HAVILAND, Sir Geoffrey, 1882–1965, British aeronautical engineer; pioneered in use of pusher propellers and rotary engines, designed all-wood plane, commercial airplane manufacture, founded De Haviland Aircraft Co. (1920), developed jet engines (DNB MWBD TBDOS)

DE HONECOURT, Wilars, fl. c. 1230, French craftsman; surveyor, crossbow, how to cut stone arches (TSOE)

DE LA CIERVA, Juan, 1896–1936, Spanish aeronautical engineer; autogyro (1920), exhibited and tested (1928) (MWBD WA)

DE LACY, Walter Washington, 1819–1892, American soldier and engineer; map of Bitterroot Mountain, railroad construction engineer, surveys, gold mine operation (DAB NC3 WWWIA)

DELAFIELD, Richard, 1798–1873, American military engineer; developed ordnance stores, defenses of New York harbor, fortification of Hampton Roads (VA) and several eastern defenses, advanced to rank of major general (DAB NC11 WWWIA)

DELAMATER, Cornelius Henry, 1821–1889, American mechanical engineer and manufacturer; iron boats, boilers and engines for sidewheel steamers, steam fire engines, ships for Civil War, company became known at Delamater Iron Works, propellers, air compressors, built submarine (1881) for John P. Holland [1840–1914] (DAB MEIA MWBD WWWIA)

DELANY, Patrick Bernard, 1845–1924, Irish American electrical engineer and inventor; telephone operator (1863–1876), invented multiplex telegraph, telegraph tape perforator, submarine detection devices (MWBD NC13 WWWIA)

DE LA RUE, Warren, 1815–1889, English astronomer and inventor; envelope-making machine (1851), devised method for daily photographing of the sun (1858),

studied electric discharge through gas (1868), invented silver chloride electric cell, constructed reflecting telescope (DOSB MWBD TBDOS)

DELAUER, Richard David, 1918–1990, American aeronautics; space, defense, electronic systems, experimental engineer, missile, aerodynamics, intercontinental ballistic missiles, unmanned space exploration, nuclear propulsion, TRW (1958ff), becoming executive vice president (1968) and chairman (1973–1974), government service and administration, NAE (1969) (IWAP MT5 WWIE WWWIA)

DE LAVAL, Carl Gustav Patrik, 1845–1912, Swedish engineer; with Klosters-Bruck Steel Worksteel (1872ff), centrifugal cream separator (1878), high-speed turbine (1882), founded De Laval Manufacturing Co., developed divergent steam nozzle, gears for converting high-speed steam turbines to slower speed shafts, vacuum milking machines (1913) (BDOTHOT E-64 EE EIH FNIE MWBD PS TBDOS WA WWWIS)

DE LA VERGNE, John Chester, 1840–1896, American mechanical engineer and inventor; oil seal vertical ammonia compressor, joint and flange connection for ammonia piping and fittings, direct piping for cooling agent, started De La Vergne Refrigerating Machine Co. (1880) for manufacturing refrigerating machinery and ice-making plants (MEIA NC2)

DE LESSEPS, Ferdinand-Marie, Vicomte de, 1805–1894, French engineering, construction and diplomat; diplomatic service (1825–1837), Suez Canal (1832), construction of canals (1859–1869), formed companies for construction of Panama Canal (1881–1888) (FEL MWBD PS)

DEL MAR, Alexander, 1836–1926, American mining engineer and economist; mining, wrote history of precious metals, director of Bureau of Statistics (1866–1869), engineer for US Monetary Commission, author (DAB NC4 WWWIA)

DEMING, William Edwards, 1900–1993, American statistician and engineering consultant; quality control, sampling, physical materials, depreciation methods, management consultant for engineering and industrial organizations, NAE (1983) (AMOS MT8 WWWIA)

DEMOCRITUS (DEMOCRITOS), c. 460–c. 370 BC, Greek natural philosopher; postulated atomic theory of matter, devoted considerable attention to structure of human body, provided a collection of ethical precepts, credited with 72 written works, known as laughing philosopher. Modern atomic theory is credited to John Dalton [1766–1844] (c. 1800) (EB DOSB MWBD OED RHW WWWIS)

DEMOUSTIER, Pierre-Antoine, 1735–1796, French civil engineer; hydraulics, bridge design and construction (TSOE)

DEMPSTER, Arthur Jeffrey, 1886–1950, Canadian American physicist and inventor; first mass spectrometer (1918), faculty at Univ. of Chicago (1919–1950), NAS (1937) (BM27 MWBD NC38 WA WWWIA)

DEN HARTOG, Jacob Pieter, 1901–1989, Dutch American engineering mechanics; at Westinghouse (1920s), US citizen (1929), at Harvard Univ. (1932–1942), with US Navy Reserve in WWII, at Massachusetts Inst. of Technology (1945ff), mechanics, elasticity, dynamics, mechanical vibrations, NAS (1953), NAE (1975) (AMOS BM67 Bridge MT5 WWWIA)

DENNER, Johann Christoph, 1655–1707, German instrument maker; manufacturer of woodwind musical instruments, including recorders, flutes, oboe, invented Klarinette, which was forerunner of the clarinette (c. 1700) (GI MWBD)

DENNISON, Aaron Lufkin, 1812–1895, American mechanical engineer; cobbler, watchmaker, first firm to manufacture watches with interchangeable parts, after being a part of several companies became American Waltham Co. (1859) variously in Switzerland, England and US (DAB MEIA WWWIA)

DENT, Frederick J., 1905–1973, English engineer; gas production research (1929–1941), joint assistant director of Gas Research Board (1941–1952), Gas Council Midlands Research Station (1952–1967) (RS20 WWW)

DEPREZ, Marcel, 1843–1918, French engineer and pioneer electrician; long-distance transmission of electric power, worked on motors and friction, innovator of steam engine valves and indicators, ballistics, railroad engineering, industrial use of direct current electricity (1881–1886), alternating current transmission of electricity with transformers (1881) (BDOS CDOSB MWBD PS WWWIS)

DERBY, George Horatio, 1823–1861, American soldier, humorist and engineer; topographical engineer, exploration and mapping, construction of harbors, lighthouses in Florida, military roads, mapped mining areas in California, published humorous writing under pseudonyms of Squibob and John Phoenix (ANB DAB NC5,11 WWWIA)

DERBYSHIRE, William H., 1859–1907, American mechanical engineer and inventor; development and management of machine tool works, smith shop and boiler works, president of Chambersburg Engineering Co. (1897), quick-acting hydraulic riveting machine (1900, 1901), hydraulic machine and valve (1904) (MEIA)

DE ROCHAS—see BEAU DE ROCHAS

DEROSNE, Louis-Charles, 1780–1846, French chemist and inventor; pharmacist, production of acetone (1807) with brother Jean-François Derosne, new techniques and equipment in sugar technology, improved methods of beet sugar manufacture (1811), continuous distillation process (1817), made commercial equipment (DOSB PS WWWIS)

DERRINGER, Henry, 1786–1869, American inventor and manufacturer; gun maker, developed military rifles, pistols, short-barrel pistol (c. 1852) known as derringer (MWBD)

DESCARTES, René (aka Renatus Cartesius), 1596–1650, French mathematician and philosopher; applied mechanistic view, laid foundation for analytic geometry, developed system for plotting of equations based on Cartesian coordinates, combined geometry and algebra to solve problems, known for statement "I think, therefore I am," started use of constants (a, b, c) the letters at the beginning of the alphabet and variables (x, y, z) the letters at the end of the alphabet, introduced exponentials to indicate powers of numbers (BDOS(Wi) BEST EIH GI MWBD SAI-MP TBDOS(Po) TGE WWWIS)

DESCHAMPS, Claude, 1765–1843, French engineer; bridge building, using caissons to support bridges (TSOE)

DESCHAMPS, Georges A., 1911–1998, French American electrical engineer; served in French Army, to US (1937), taught mathematics and physics at Lycée Français de New York (10 yr.), project engineer for International Telephone & Telegraph Co. (1947–1958), faculty at Univ. of Illinois (1958–1982), microwave engineering, direction finding, radio navigation systems, formulated diagram factor, laser beam propagation, electromagnetic scattering, several patents, NAE (1978) (Bridge MT10)

DESGULIERS, John Theophilus, 1683–1744, English inventor; mechanics, optics, improvements to steam engine (MWBD)

DE SEVERSKY, Alexander Procofieff, 1894–1974, Russian American aeronautical engineer and aviator; served in WWI in Russian air force, to US (1918), developed pursuit planes, bombsight, air pollution control, founder and president of Seversky Aircraft Co. (1931–1939), which was predecessor to Republic Aviation Corp. (1939), founded several companies related to aeronautical work, including Seversky Electronation Corp. (1952–1968), consultant to education, industry and government institutions (ANB MWBD NCF WWWIA)

DESSAUER, John H., 1905–1993, German American engineer and inventor; research with Agfa Co. (1929–1935), inventions and developments in xerography, company executive positions with Xerox (1935–1973), becoming vice president, NAE (1967) (Bridge MT7 WWIE WWWIA)

DETMOLD, Christian Edward, 1810–1887, German American civil engineer; surveyor, railroad survey work, manufacture of iron, author (DAB WWWIA)

DEVEREUX, John Henry, 1832–1886, American civil engineer; railroad engineering, building railroads, railroad executive with New York Central Railroad (NYC) (DAB NC12 WWWIA)

DE VRIES, John, 1853–1932, Dutch American mechanical engineer and manufacturer; mechanical drawing, machine tool manufacturing (1872), repair work (1875), erecting and testing iron works and steam engines (1876), first adjustable cut-off for low- and intermediate-cylinders on compound and triple-expansion engines, with brother formed Novelty Iron Works (NJ) (1892), built De Vries feedwater heater and purifier, manufactured machinery for variety of applications—tin cans, textiles, artificial leather, chemicals, rubber, matches (IAI MEIA)

DEVYATKOV, Nikolai Dmitrievich, 1907–??, Soviet electrical engineer; ultra-high frequency electronics, radio engineering, construction of gaseous discharge devices to protect communication lines, devices for generating and converting frequencies of electromagnetic oscillation, administrative positions, professor at Moscow Inst. of Energy (1944ff) (PS SMOS)

DEWAR, James, 1842–1923, Scottish scientist and inventor; invented double-walled glass vessel with vacuum between (1892) named after him, thermos flask (1904), first to produce liquid nitrogen (1898), invented cordite with Frederick August Abel [1827–1902] (1889), faculty at Cambridge Univ. (1875–1923) and Royal Inst. (1887–1923), fNAS (1907) (CDOS GI MWBD PA RHS TGE:B WWWIS)

DEWITT, Simeon, 1756–1834, American surveyor and civil engineer; in Continental Army (1778–1780), geographer, surveyor general (NY), published map of New York State, surveyed canal routes, NY-PA boundary disputes, Univ. of State of New York (1798–1834), serving as chancellor (1829–1834) (BDOACE WWWIA)

DIADES, fl. 330 BC, Greek military engineer; developed siege engines, invented flying bridge (GEAPIT TAE)

DIBNER, Bern, 1897–1988, Russian American electrical engineer; design engineer, invented solderless connectors, electrical apparatus, founder (1924) and executive of Burdy Corp. (1947–1972), Burdy Library (1935), author, historian of science, served on many public service and industry boards, Dibner Library at Smithsonian Inst. (DC) (1975), 25 US patents (WWIE WWWIA)

DICKEMAN, Raymond Louis, 1922–1983, American nuclear engineer; nuclear reactor technology, production reactor, at General Electric Co. (1948–1967), including general manager of Hanford project (WA), chief executive officer of Exxon Nuclear Co. (1968–1978), duel purpose N-reactor, consultant (1978ff), NAE (1978) (MT2 WWIE WWWIA)

DICKIE, George William, 1844–1918, Scottish American engineer and shipbuilder; to US (1869), design and construction of gas plant for Pacific Gas Co., marine compound engines, battleships and cruisers for navy, superin-

tendent of Risdon Iron Works (1870), manager of Union Iron Works (CA) (1883), merchant marine ships for US government in WWI (DAB MEIA NC22 WWWIS)

DICKINSON, John, 1782–1864, English inventor; papermaker, papermaking machine (1809), continuous papermaking, a machine to paste layers of paper to make cardboard, machine to add threads to paper to change texture, identification and strength (1829) (BDOTHOT RS WA)

DICKSEE, Cedric Bernard, 1888–1981, British engineer; compression ignition engine (diesel) development, engine designer (1928), combustion chamber design (WWII) for diesel engines and jet engines, industry product development (TBDOS)

DIESEL, Rudolf, 1858–1913, German mechanical engineer (born in Paris); internal combustion engines, machine factory, invented diesel engine with high-compression ignition (1892), built first successful diesel engine (1896), founded Diesel manufacturing company in Germany (1899) (BEST CEOS&T E–64 EB EIH EOWB FNIE GI IAI MMAH MWBD NYPL PS SAI SAI-MP S:TLAW TBDOS TBOF TGS WA WWWIS)

DIETRICH, Joseph Robert, 1914–1982, American physicist and nuclear engineer; reactor physics, nuclear reactor design, development of nuclear reactors, NACA (1943–1946), Oak Ridge then to Argonne National Labs. (1947–1956), executive of General Nuclear Engineering Corp. (1956ff), NAE (1975) (AMOS MT2)

DIETZ, Daniel N., 1914–1988, Dutch American engineer; engineering physics, hydrologic survey, underground water management, reservoir engineering, with Shell Co. (1945–1973), professor at Delft Univ. of Technology (1973–1984), consulting engineer, fNAE (1988) (MT4)

DIETZ, Johann Christian, 1773–1849, German engineer; musical instruments, melodeon (1805), claviharp (1814) (MWBD)

DIKUSHIN, Vladimir Ivanovich, 1902–??, Soviet engineer; machine tool design, bases of metal-cutting machine design, assembly-line work, standard machines for rotating bodies in automatic lines (SMOS)

DILLARD, Joseph K., 1917–1988, American electrical engineer; electric power systems, underground transmission, pioneer work in applying digital computers to technical and economic analysis of power generation and transmission systems, Westinghouse Electric Co. (1950–1982), retiring as senior consultant, NAE (1975) (Bridge MT5 WWIE)

DIOPHANTUS of Alexandria, fl. 250, Greek mathematician; earliest writing on algebra, set out rules for multiplication and division, use of minus sign, use of symbols to represent unknown quantities, theory of numbers, indeterminate equations. See René Descartes (CDOS GI MWBD RHW)

DISSTON, Henry, 1819–1878, English American mechanical engineer; to US (1833), saw manufacturing business (1840), utilized steam power (1844), converted waste steel into usable ingots, manufactured high-grade crucible steel during Civil War, manufacturer of saws and other steel tools (DAB MEIA NC6 WWWIA)

DITTUS, Frederick William, 1897–1987, German American mechanical engineer; heat transfer, turbulent flow in a smooth tube, Dittus-Boelter equation (c. 1930), pipeline expansion joint, chief draftsman for Standard Oil. Co. of California (1916–1922), faculty at Univ. of California (1922–1927), research, design and chief engineer with Standard Oil of California (1927–1944), chief engineer at California Tex Oil Co. (1944ff), becoming vice president (1957) (L&M WWIE(I))

DIVINI, Eustahio, 1610–1685, Italian optics and inventor; maker of clocks and lenses (1646), innovative compound microscope (1648), long telescopes, made numerous observations of moon, Saturn and Jupiter (DOSB PS)

DIXON, Charles A., 1856–1900, American mechanical engineer and inventor; builder of steam engines, pattern maker, draftsman, syphon-pipe heater (1887), steam safety valve (1888), Dixon Steam Engine Works (with others) (1893), engineering consultant (1895ff) (MEIA)

DIXON, George Edward, 1850–1904, English American mechanical engineer and inventor; to US (1881), iron works, officer in army, heating business (1885), founded George E. Dixon and Co. (1893) with two sons, patented furnace heating devices, hot-water heating, Dixon radiator shield (1893), firebox dimensions (MEIA)

DIXON, Jeremiah, 1733–1779, English surveyor; with Charles Mason [1728–1786] determined the boundary between Maryland and Pennsylvania (1763–1768) known as the Mason-Dixon line (DNB MWBD)

DIXON, Joseph, 1799–1869, American mechanical engineer, manufacturer and inventor; casting (1850), wood-planing machine (1866), printing color-fast calico, formed Joseph Dixon Crucible Co. (1867), machine for cutting files (1820), superheated steam engine to power boat, credited with originating lithography, graphite crucible to withstand high heat (1826), making currency to reduce counterfeiting (1832), with I. S. Hill made Babbitt metal, invented by Isaac Babbitt [1799–1862] (1839) for anti-friction bearings, galvanic batteries (1866), obtained numerous patents (DAB GSA MEIA MWBD NC22 WWWIS)

DIXON, Robert Munn, 1860–1918, American mechanical engineer and inventor; engineer (1887) to president of Safety Car Heating and Lighting Co., harbor and coast lighting, received over 50 patents in heating and lighting railroad cars (MEIA WWWIA)

DOANE, Thomas, 1821–1897, American civil and mechanical engineer; surveyor for railroads in northeast US, built tunnels using machinery and explosives, chief engineer of Burlington & Missouri Railroad, pioneer in use of compressed air in US, consulting engineer (BDO-ACE DAB MEIA NC25 WWWIA)

DOANE, William Howard, 1832–1915, American mechanical engineer; manufacturer of woodworking machinery working with J. A. Fay & Co. as bookkeeper (1851), from partner to president (1861), also president of Central Safe & Deposit Co. (OH), composer of music, over 70 patents of woodworking machines (MEIA NC41 WWWIA)

DOBSON, William John Marshall, 1847–1932, English American mechanical engineer and inventor; to US (ante 1876), designed and constructed large chemical plant for McKesson & Robbins Co. (1876), chemical appliances, plans for factories, power plants, chemical works, wallpaper mills, invented auger guide, tension indicator, drying system for wallpaper and skins, consulting engineer (1886ff) (MEIA)

DOD, Daniel, 1778–1823, American inventor and engineer; watchmaking, instrument making, surveying, mathematical instruments, steam engine building, ferry service, cotton machinery company (1799), boilers, condensers for steamboats and mills (1811–1812) based on principle of James Watt [1736–1819] engine, Daniel Dod and Aaron Ogden [1756–1839] were partners (1812–1818), steamboat manufacturing (NY) (1820) (DAB GEAPIT MEIA NC24)

DODGE, Barnett Fred, 1895–1972, American chemical engineer; worked on explosives for DuPont (1917–1920), faculty at Yale Univ. (1925–1964), serving as dean of engineering (1960–1962), concerned with thermodynamics, phase equilibrium, reaction rates at high pressures, water pollution control (ACCE)

DODGE, Grenville Mellen, 1831–1916, American army officer and engineer; railroad development in US and Cuba, railroad construction in Midwest, Mississippi and Missouri, served with distinction in Civil War, worked with railroad as chief engineer (1871), president of Texas and Pacific Railway (1880) (ANB BDOACE DAB MWBD NC16 WWWIA)

DODGE, James Mapes, 1852–1915, American mechanical engineer and inventor; foreman and superintendent of erection of shipbuilding works, partner of company to manufacture mining equipment (1876), superintendent of company to manufacture flat detachable chain links, Link-Belt Machinery Co. formed (1880), of which he was president then chairman of the board (1906), president (1892) of two companies to store and handle anthracite coal, over 200 patents covering link-belts, moving stairs, toys, etc. (MEIA NC29 WWWIA)

DOLBEAR, Amos Emerson, 1837–1910, American physicist and inventor; writing telegraph (1864), magnetic telephone (1864) (1870), static telephone (1879), ammeter (1889) photography of moving waves (1893) and numerous other inventions (BDOAS NC9 WWWIA)

DOLLEZHAL, Nikolai Antonovich, 1899–??, Soviet power engineer; thermopower installations including chief engineer (1932–1942), including Kiev plant, director of Inst. of Chemical Machine Building (Moscow) (1942–1953), taught at Inst. of National Economy and Moscow Higher Technical School (1923ff), compressor for chemical industry, nuclear energy development, atomic energy power plant, USSR Academy (1953) (SMOS)

DOLLOND, John, 1706–1761, English optician and inventor; with brother Peter Dollond [1730–1820] worked on heliometer (1754), achromatic lens for eyeglasses (1758), triple achromatic lens (1765), George Dollond [1774–1852] took over family business (1819) (MWBD NYPL WA WWWIS)

DONKIN, Bryan, I, 1768–1855, English mechanical engineer and inventor; Donkin pantograph engraving machine, machine for making paper (1803), rotating printing machine, built automatic papermaking machine (1804), preservation of meat and vegetables by canning in tinned iron containers (1812) and with John Hall [fl. 1800s] formed Donkin and Hall, helped Fourdrinier Brothers develop making paper in continuous rolls, acquired rights to patent by Peter Durand [fl. early 1800s] (1810), Bryan Donkin II [1809–1893] and Bryan Donkin III [1835–1902] were also engineers who worked in paper mills and papermaking, FRS (1838) (BDOTHOT GEAPIT MWBD TDDOS TBOF)

DONOVAN, Allen Francis, 1914–1995, American aeronautical engineer; space reentry vehicles, space operations, applicability of space to military missions, supersonic control, Curtiss-Wright (NY) (1936–1955) for the laboratory that was given to Cornell Univ., missile and rocket design for Ramo-Wooldridge Corp. (1955–1960), Aerospace Corp. (CA) (1960–1978), consulting engineer, NAE (1969) (AMOS Bridge MT8 WWWIA)

DONOVAN, Marion O'Brien, 1917–1998, American woman inventor; sold patents to others to develop, invented new type of baby diaper (1947, 1949), invented forerunner of disposable diapers, developed vinyl and synthetic products, a hanger for holding up to 30 shirts, device to help floss teeth, started her own company and sold household items to Saks Fifth Ave. (1949), sold right to products to Procter & Gamble, for which design engineer Victor Mills [1897–1997] created Pampers, disposable diapers that were mass produced, 12 patents (AIE AWIT FAL GI NYTObit WI WWWIA)

DOOLITTLE, James Harold, 1896–1993, American aviator and army officer; air corps test pilot (WWI until 1930), commanded bombing mission in Asia in WWII with B-25s (1942), promoted to lieutenant general, NAS (1959) (IWAP MWBD NCG WWWIA)

DOREMUS, Robert Ogden, 1824–1906, American chemist and inventor; arc light, use of Bunsen cells, street lighting, perfected compressed granulated gunpowder, blasting, first large-scale use of chlorine gas for cholera treatment, fire protection, lecturer (ANB DAB NC28 WWWIA WWWIS)

DOREY, Stanley F., 1891–1972, British mechanical engineer; marine engineer, with British Royal Navy in WWI (1912–1919), chief engineer and surveyor for Lloyd Ltd. (1919, 1932–1956) (PS RS19 WWW)

DORNBERGER, Walter Robert, 1895–1980, German American engineer; in charge of rocket research in Germany (1932–1945), developed V-2 German rocket with Wernher von Braun [1912–1977] (1936ff), to US (1947), Bell Aircraft Corp. (1950ff) (MWBD TTOS)

DORNIER, Claudius, 1884–1969, German airplane builder; first all-metal airplane (1911), large airplanes, bombers, reconnaissance planes (MWBD)

DORODNITSYN, Anatolii Nekseevich, 1910–??, Soviet hydrodynamicist; dynamic meteorology, aerodynamics, influence of uneven land surfaces on air stream, research (1941) and professor (1947), boundary strata in compressible gas, taught and worked in institutions of higher education (1936ff) in Moscow and Leningrad, Central Aerodynamics Inst. (1941), Mathematical Inst. of USSR Academy of Sciences (1944–1955), director of Computer Sciences (1955ff), member of USSR Academy of Sciences (1953) (SMOS)

DORR, John Van Nostrand, 1872–1962, American chemical engineer and inventor; extractive metallurgy, incorporated Dorr Co. (1910), involved primarily in developing and marketing of Dorr classifier, thickener, Dorr agitator, ore dressing, Dorr Co. merged to form Dorr-Oliver, Inc. (1971) (ACCE MWBD NC48 PS WWWIA)

DORRANCE, John Thompson, 1873–1930, American businessman and inventor; chemist, food processing with Joseph Campbell Preserve Co. (NJ) (1869ff), formed by Joseph Campbell and Abram Anderson, condensed soup developed and marketed (1899), officer of company as vice president (1900) and general manager (1910), then sole owner and president (1915–1930), renamed Campbell Soup Co. (1922), company remained in private control (until 1954) (AIE ANB NYTObit WWWIA)

DOUGLAS, Clifford Hugh, 1879–1952, English engineer and economist; social economist, economic theory, reconstruction in Canada (1935) (DNB MWBD)

DOUGLAS, Donald Wills, 1892–1981, American engineer and industrialist; aircraft, civilian and military airplanes, was chief engineer for Glenn L. Martin Co. (1915–1920), established Douglas Aircraft Co. (1920), of which he was president (1928–1957) and chairman (1957–1967), which merged to form McDonnell Douglas Co. (1967), designed and built the widely used commercial plane the DC-3 (1936ff), NAE (1967) (ANB EOWB MT2 NCF MWBD WPost WWWIA)

DOUGLAS, James, 1837–1918, American metallurgist and mining engineer; copper extraction (1873), Hunt-Douglas (Charles W. Hunt [1841–1911]), ore processing, established several mines (BITAS DAB DOSB MWBD NC23 WWWIA)

DOUGLAS, Walter Spalding, 1912–1985, American civil engineer; design and construction, bridges, military-type construction, Bay Area Rapid Transit (CA) for mass transit, system-wide analysis and study of transportation systems, in WWII with US Navy construction unit (1942–1945) as lieutenant commander, chairman of the board of Parsons Brinckerhoff Quade & Douglas (1939–1977), consulting engineer, NAE (1967) (MT3 WWIE WWWIA)

DOUGLASS, David Bates, 1790–1849, American engineer; military engineering officer, survey for defenses of Long Island and Niagara, consulting engineer (DAB EIH)

DOW, Herbert Henry, 1866–1930, Canadian American chemist, inventor and manufacturer; electrolytic extraction of bromide (1889), chlorine extraction (1895), founded Dow Chemical Co. (MI) (1897), over 100 patents (ACCE BDOTHOT HOCS&T MWBD NC24 NYT-A WWWIA)

DOW, Lorenzo, 1825–1899, American inventor; ordnance, waterproof cartridge (1861), steamboat service in Colombia (1866–1870), mining in Venezuela (DAB WWWIA)

DOW, William Gould, 1895–1999, American electrical engineer; space research, radio research, vacuum tube development, engineering research, electric welding, controlled nuclear fission, atmospheric research with rockets, microwave electronics, Westinghouse Electric & Manufacturing Co. (1920–1926), faculty and department administration at Univ. of Michigan (1926ff) except during WWII, when he was at Harvard Univ. (AMOS NCG WWIE WWWIA)

DOWNING, C. Glen E., 1914–1999, Canadian agricultural engineer; research engineer at Canada Research Station in farm power and machinery (1939–1942), captain in Canadian Royal Electric Mechanical Corps (1942–1945), faculty and administration at Univ. of Guelph as head of engineering science department (1946–1963) and director of the School (1964–1967), mechanization and automation in agriculture, systems engineering and analysis (AE Res7 WWIE)

DRAKE, Edwin Laurentine, 1819–1880, American engineer; pioneer oil well driller with Rock Oil Co. based on methods proposed by George H. Bissell

[1821–1884], first got oil by drilling at Titusville (PA) (1859) along with William A. Smith, conceived use of pipe extended to bedrock to protect well hole, 4-cent US postage stamp in his honor (1959) (ASAI BDOS EIH GI 1000Y PS MWBD NYPL NC26 WWWIA)

DRAPER, Charles Stark, 1901–1987, American aeronautical engineer; instrumentation, gyroscope (1920s), guidance systems, inertial guidance for air, marine navigation and guided missiles, Draper Lab. (MA) named after him, International Draper Award of National Academy of Engineering (NAE) named in his honor, NAS (1957), NAE (1965) (ANB BM65 IWAP MT4 MWBD NYT-A WWWIA)

DRAPER, George, 1817–1887, American mechanical engineer and inventor; manufacturer of textile machinery and operator of mills, originally E. D. & G. Draper then changed to George Draper & Sons (1877) and later to Draper Co. (MA) (1897), incorporated improvements to hold cloth taut and built spinning machinery, made Northrop loom (MEIA NC12 WWWIA)

DRAPER, Ira, 1764–1848, American inventor and manufacturer; textile manufacturing machinery, mechanical improvement of threshing machine and road scraper, contributed to building of Draper Co. (DAB MEIA NC23 WWWIA)

DRAPER, William Franklin, 1842–1910, American manufacturer and inventor; textile machinery, served in Civil War, becoming brigadier general (1865), became president of Draper Co. (until 1907), invented many improvements to weaving machinery, company perfected Northrop loom (1892), held over 200 patents (ANB NC32 WWWIA)

DREBBEL (or DRUBBEL), Cornelis Jacobszoon, c. 1572–1633, Dutch engineer and inventor; designed water supply systems for towns (1597ff), moved to England (c. 1604) developed microscope, experimented with submarine (1620) along lines proposed (c. 1578) by Englishman William Bourne [??–1583], possibly developed torpedo, incubator, magic lantern, improved thermometer, machine activated by atmospheric pressure (BDOS(Wi) EB GI MWBD NYPL TSOE WWWIS)

DREW, Charles Richard, 1904–1950, Black American physician and inventor; blood plasma development and blood bank (1940), storing, shipping and protecting stored blood supplies, faculty at Howard Univ. (DC) (BDOBA MWBD WWWIA)

DREW, Richard Gurley, 1899–1980, American mechanical engineer and inventor; with 3M Co. (1921–1962), developed masking tape (1925), cellophane tape that became known as Scotch tape (1930), technical director of 3M Co. fabrication laboratory (1943–1962), consultant (1962–1967) (ANB)

DREW, Thomas Bradford, 1902–1985, American chemical engineer; alloys, flow of fluids, heat transfer, diffusion, combustion, nuclear energy, faculty at Drexel Univ., Massachusetts Inst. of Technology (1927–1934) (1965ff), DuPont (1934–1940), Columbia Univ. (1940–1965), including department head, NAE (1983) (AMOS MT3 WWWIA)

DREYSE, Johann Nikolaus von, 1787–1867, German inventor; multiple-loading guns (1827), breech-loading gun (1836), needle guns (MWBD WWWIS)

DRIFTMIER, Rudolph H., 1898–1979, American agricultural engineer; served in WWI with Army Expeditionary Force (1917–1918), faculty at Kansas State Univ. (1920–1930), faculty and administration at Univ. of Georgia (1930–1965), including department chairman (1950–1965), electricity for agriculture, university system head of plant operations, consultant (1965ff) (AE60,61 WWIE WWWIA)

DRINKER, Philip, 1894–1972, American chemical engineer, industrial hygienist and inventor; US Army in WWI (1917–1919), engineer with Buffalo Foundry and Machine Co. (1920–1921), ventilation and illumination specialty, Harvard Univ. School of Public Health (1922–1960), with his brother Cecil Kent Drinker [1887–1956], invented the iron lung (first used in 1928) (GI TIY(2) WWWIA)

DRIPPS, Isaac L., 1810–1892, Irish American engineer and inventor; locomotive construction popularly known as John Bull (1831), fitting of machinery on steamboats, partner in machine works, management of shops, superintendent of motive power, gauge of rail tracks, efficiency of locomotives on grades and curves with Pennsylvania Railroad (DAB MEIA)

DRUMMOND, Thomas, 1797–1840, British military engineer; British Royal Engineer (1815ff), ordnance, survey of Great Britain (1820), developed Drummond light, government service in Ireland (1835–1840) (GEAPIT MWBD PS WWWIS)

DRYDEN, Hugh Latimer, 1898–1965, American physicist and aeronautical engineer; aerodynamics, mechanics, sound, radar honing missile (WWII), weather and communications satellites, National Bureau of Standards (1918–1947), including director of wind tunnel (1920), director of research and director of NACA (1947–1958), NAS (1944), NAE (1964) (BM40 IWAP MT1 MWBD PS)

DUANE, James Chatham, 1824–1897, American civil and military engineer; chief engineer in army, service in Civil War, fortifications, lighthouses, river and harbor improvements (DAB NC10 WWWIA)

DUBOIS, Augustus Jay, 1849–1915, American civil and mechanical engineer; faculty at Lehigh Univ. (1875–1877), faculty at Yale Univ. (1877ff), including department head, graphical statics, author of numerous engineering references (BITAS DAB MEIA)

DU BOIS-REYMOND, Emil Heinrich, 1818–1896, German physiologist; investigation in animal electricity, physiology of muscles and nerves, electrophysiology, Univ. of Berlin (1855–1896), fNAS (1892) (BDOTHOT CDOS MWBD RHW WWWIS)

DUBRIDGE, Lee Alvin, 1901–1994, American physicist; faculty of Washington Univ. (St. Louis) (1928–1934), research in areas of biophysics, nuclear disintegration, photoelectrons, Univ. of Rochester (1934–1940), director of Radiation Lab. at MIT (1940–1945), president at California Inst. of Technology (1946–1969), advisor and consultant to many government committees, NAS (1943) (BM72 CB NCI NYTObit TNAS WWWIA)

DU BUAT, Pierre-Louis-Georges, 1734–1809, French hydraulic engineer; military engineer (1761–1791), fortifications, developed empirical formula for discharge of fluids from pipes and flow in channels (1779), author (MWBD WWWIS)

DUDDELL, William du Bois, 1872–1917, English electrical engineer; developed an oscillograph, high-frequency generator, thermometer (EB MWBD WWWIS)

DUDLEY, Charles Benjamin, 1842–1909, American chemist and inventor; increased efficiency of Pennsylvania Railroad (1875–1909), founder (1898) and president (1902–1909) of American Society for Testing Materials (BDOAS MWBD NC12 WWWIA)

DUDLEY, Dud, 1599–1684, English ironmaster; first to use coal and coke for iron smelting (1619); process was refined by Englishman Abraham Darby [1678–1717] (1709) (EIH GEAPIT GI MMAH MWBD)

DUESENBERG, Frederick Samuel, 1877–1932, German American manufacturer; engine design, racing cars, luxury engines, Mason engine (1903), Duesenberg engine (1913), chief engineer of Duesenberg Motors Co. (1917ff), luxury automobiles (MWBD NC16)

DUFFEE, Floyd W., 1893–1982, American agricultural engineer; farm machinery, farm power, performance of ensilage cutters, design and performance of small hammer type feed mills, use of electricity in agriculture, faculty at Univ. of Wisconsin (1918–1963), including department chairman (1937–1962) (AE61 WWIE WWIE(I))

DUFOUR, Guillaume-Henri (aka Wilhelm Heinrich), 1787–1875, Swiss military engineer; returned to Switzerland (1817) after serving in the Napoleonic Army (1807–1814), topographical survey of Switzerland (1842–1864), Swiss army general (1847), efforts lead to formation of Red Cross (1864), presided over international conference that framed the Geneva Convention for treatment of prisoners of war (1864) (EB MWBD PS)

DUHAMEL DU MONCEAU, Henri-Louis, 1700–1782, French naval engineer and agriculturalist; naval architect, plant and animal physiologist, experimented on strength of wood, inspector general of marine operations (EB MWBD WWWIS)

DUKLER, Abraham E., 1925–1994, American chemical engineer; fluid mechanics, heat transfer, turbulent flow transfer, gas-in-liquid flow, distillation, Rohm & Haas Co. (1945–1948), Shell Oil Co. (1950–1952), faculty and administration at Univ. of Houston (1952–1982), including department chairman (1967) and dean (1976), NAE (1977) (AMOS Bridge MT8 WWIE WWWIA)

DUKOV, Niklolai Leonidovich, 1904–??, Soviet mechanical engineer; design engineer, new designs in two different power plants in Leningrad (1931) and Chelyabinsk (1941) (SMOS)

DULONG, Pierre-Louis, 1785–1838, French chemist; with Frenchman Alexis Thérèse Petit [1791–1820] enunciated Dulong and Petit law of gram-atomic heat (1819), with Swede Jöns Jakob Berzelius [1779–1848] worked on fluid densities (1820), specific heat (1829), with François Arago [1786–1853] on the elasticity of steam at high temperatures (1830), at École Polytechnique (Paris) (1820–1838), including director (1830–1838) (BEST DOSB DOT EB EDOP MWBD RHW WWWIS)

DU MONT, Allen Balcom, 1901–1965, American engineer; manufacture of vacuum tubes, perfected first commercially successful cathode ray tubes, oscilloscope, chief engineer, Westinghouse Lamp Co. (1924–1928), DeForest Radio Co. (1928–1931), founded Du Mont Labs. (1931), television receiver (1937), broadcasting network (1946) (EB MWBD WWWIA)

DUNBAR, Robert, 1812–1890, Scottish American engineer and inventor; designed and built mechanical aspects of shipyards, rebuilt old flour mills, designed and erected grain elevators with C. W. Evans (1838–1853), designed, built and installed conveyors (DAB WWWIA)

DUNCAN, William Jolly, 1894–1960, British mechanical engineer; professor of aeronautics and fluid mechanics at Univ. of Glasgow (1950ff), service in WWI, with marine engineering firm (1919–1926), National Physical Lab. (1926–1934), with Royal Aircraft Establishment (ante 1945) in WWII, senior partner of Ross and Duncan engineers (1945ff), government assignments, FRS (1947) (PS RS7 WWW)

DUNLAP, Matthew Elbridge, 1891–1971, American architectural engineer and inventor; US Forest Products Lab. (WI) (1916–1952), where he served in department of wood preservation and timber physics, including engineering of forest products (1920–1924), moving from engineer to general engineer (1927–1952), vapor barriers, insulation, moisture movement, heat losses, advances in glue, fireproofing of wood, invented duff hygrometer (1937) for moisture measurement, high-frequency curing of wood, 5 patents (NC56 PS)

DUNLOP, John Boyd, 1840–1921, Scottish veterinarian and inventor; pneumatic tires (1888), tricycles with rubber tires, Dunlop Rubber Co. formed (1889) by William H. Du Cros [1846–1918] and John Boyd Dunlop [1840–1921], pioneered in producing rubber tires (BDOTHOT DNB EIH MMAH MWBD PS SAI TBDOS(Po) TGE:B WA WWW WWWIS)

DUNN, Gano Sillick, 1870–1953, American electrical engineer; improved efficiency of electric motors and dynamos, obtained patents on frames, automatic brakes, speed controls, voltage and current regulators, electromagnetic circuit breakers, electric hoist for elevators, presided over firm of J. G. White Co., which had major construction projects worldwide, NAS (1919) (ANB BM28 PS WWWIA)

DUPONT, Eleuthère Irénée, 1771–1834, French American industrialist; to US (1799), established manufacturing plant for gunpowder (1802), beginning E. I. DuPont de Nemours & Co. (DE), expanded to a variety of chemicals, succeeded as president by his son Alfred Victor DuPont [1798–1856] (1834–1850) (MWBD NC6 TodChem11 WWWIA)

DUPONT, Francis Irénée, 1873–1942, American chemist and inventor; helped develop smokeless power, led development of experimental station within DuPont Co. (est. 1903) for advanced research and became its director, established company with a broader range of research, including petrochemicals and synthetic rubber (1936), nitrocellulose, TNT, started brokerage firm (1931), over 100 patents (AIE ANB WWWIS)

DUPORTAIL, Louis Lebeque dePresle, Chevalier, c. 1736–1802, French military engineer; sometimes referred to as the father of the US Army Engineers, planned defenses of Valley Forge (1778) during the War for Independence under the direction of Gen. George Washington [1732–1799] (BDOACE MWBD)

DUPUY DE LOME, Stanislas-Charles-Henri-Laurent, 1816–1885, French naval engineer; first French screw-driven steamship, armored vessels, senator in French government (1877ff) (MWBD WWWIS)

DURAND, Cyrus, 1787–1868, American engraver and inventor; machines for carding and weaving hair for carpets, machines for engraving, printing presses (DAB WWWIA)

DURAND, Peter, fl. early 1800s, English merchant and inventor; use of tin-plated iron cans for food canning (1810), sold idea to Bryan Donkin [1768–1855], an English mechanical engineer and inventor, and John Hall, who opened first food factory firm of Donkin and Hall (1811) (BDOTHOT EB GI MMAH TBOF WOI)

DURAND, William Frederick, 1859–1958, American mechanical engineer; Bureau of Steam Engineering in US Navy, organized department of mechanical engineering at Michigan State Univ. (1887–1891), Cornell

Univ. (1891–1904), Stanford Univ. (1904–1924), specialist in hydraulic machinery, marine and aircraft propellers (1914), mechanics, consultant, served on federal committees, NAS (1917) (ANB BM48 DAB MWBD NCD PS WWWIA)

DURANT, William Crapo, 1861–1947, American industrialist; organized carriage company (1886), began automobile manufacture, Buick Motor Car (1905), Chevrolet (1915) with Louis Chevrolet [1879–1941], Durant Motors (1921), president of General Motors Co. (1916–1920) (AIE MWBD NC34,36 WWWIA)

DÜRER, Albrecht, 1471–1528, German artist and designer; copper and iron engraving, designed first flying war machine (1522) (HIV MWBD WWWIS)

DURFEE, William Franklin, 1833–1899, American engineer and inventor; city surveyor, manufacture of horseshoe nails, ores for manufacture of steel, works for manufacture of steel rails (1856), refining copper (DAB NC6 WWWIA)

DURFEE, Zoheth Sherman, 1831–1880, American inventor, engineer and manufacturer; blacksmith, manufacture of steel from pig iron, helped in development of Bessemer converter (c. 1850) with William Kelly [1811–1888] of Eddyville (KY), took over operation (1861), joined in forming Pneumatic Steel Corp. (1886), developed cupola for processing iron, 16 patents (DAB MWBD NC6 WWWIA)

DURHAM, Caleb Wheeler, 1848–1910, American engineer and inventor; house drainage, developed Durham system, established business to manufacture and install system (DAB NC16 WWWIA)

DURYEA, Charles Edgar, 1861–1938, American inventor and manufacturer; built gasoline engine with brother James Frank Duryea [1869–1967] (1893), believed to be first American-built car, which was called buggyaut, automotive racing, Duryea Motor Wagon Co. (1885), Duryea Power Co. (1900), limousine development, brothers separated and formed new companies (AIE ANB ASAI DAB EIH MWBD NCD PA WA WBE WWWIA)

DURYEA, James Frank, 1869–1967, American inventor; with brother Charles Edgar Duryea [1861–1938] built successful gasoline automobile (1893), involved in automobile racing, brothers dissolved business (1898), James Frank Duryea formed Steven-Duryea Motor Car Company, which built luxury cars, sold factories to Westinghouse Corp. (1918), used for military vehicles, 5 patents (AIE ANB ASAI DAB IAOSTAT MWBD PS WBE)

DUSHMAN, Saul, 1883–1954, Russian American chemist, electronics and inventor; worked on electrochemistry in Canada, physical chemistry, electron emission from hot filaments with General Electric Co., Richardson-Dushman equation (1923) (Owen W. Richardson [1879–1959]), several different vacuum tubes, vac-uum pumps, vacuum research, consultant (ACCE DOT EDOP PS WWWIA)

DUWEZ, Pol E., 1907–1984, Belgian American physicist and metallurgical engineer; materials, structures and properties of alloys, glass alloys, magnetic and superconducting alloys, research with government of Belgium, US citizen (1944), California Inst. of Technology (1940–1978) serving with the Jet Propulsion Lab. (JPL) (CA) (ante 1954), NAS (1972), NAE (1979) (MT3 WWIE WWWIA)

DVORAK, August, 1894–1975, American inventor; developed typewriter keyboard (1932) that had vowels and most commonly used constants on the same row, a competitor to the Qwerty keyboard developed by Christopher L. Sholes [1819–1890] (1867), served in WWI and WWII, elementary and high school teacher (1911–1923), faculty at Univ. of Washington (1923–1956), specialist in time and motion study, artificial limbs, visual education (WWWIA)

DYER, Joseph Cheesborough, 1780–1871, American inventor; settled in England, invented roving frame for cotton cloth manufacture (1825, 1830, 1835), steel engraving (1809), nail-making machine (1811), carding machine (1811, 1825), machine for spinning long fibers of hemp, flax, grasses (1813) (BDOTHOT)

DYMOND, John, 1836–1922, Canadian American sugar-planter and inventor; to US (ante 1880), sugar cane planter (1868ff), development and use of processing equipment for sugar and molasses, labor-saving devices to harvest sugar cane, including grinding and rolling cane, multiple-effect evaporation, vacuum boiling, president of Louisiana Sugar Planters' Assoc. (1887–1897), editor and publisher, politician (DAB MEIA NC25 WWWIA)

E

EADS, James Buchanan, 1820–1887, American civil engineer and inventor; diving bell, ironclad ships, first steel arched bridge over the Mississippi River (at St. Louis, MO) (1867–1874), jetty system for navigation channel for the Mississippi River (1879), NAS (1872) (ANB BDOACE BITAS BM3 DAB EIH EOWB FEL MWBD PS TEIH WWWIA)

EARLE, Pliny, 1762–1832, American inventor; machinist, cotton machinery manufacturer, carding machine (1791), machine to prick leather (1803) (DAB WWWIA)

EARLY, Earl Joshua, 1875–1961, American inventor and educator; founded Philadelphia Drawing Instrument Co. (1906–1916), art instructor in high school (1916–1942), taught engineering graphics at Drexel Univ. (PA) (1943ff), invented numerous devices related to drafting, display devices, storage units, Lin-o-graph, Kip-tack, umbrella, perpetual calendar (NC50 PS)

EASTER, Everette C., 1893–1979, American agricultural engineer; faculty at Auburn Univ. (6 yr.), in management positions from chief agricultural engineer to executive vice president of Alabama Power Co. (1927–1958), rural electrification and management (AE61 WWIE)

EASTMAN, George, 1854–1932, American inventor and industrialist; dry film photography (1880), with Henry Reichenbach made transparent film (1889), with William Walker devised flexible film and film holder (1882), transparent photofilm (1884), Kodak box roll film camera (1888), organized Eastman Kodak Co. (1892), introduced Brownie camera (1900), roll film, motion picture film (AIE ASAI BEST CDOAB EOWB IAI IIA ME126 MWBD NC23,26 NYPL 1000Y PA PopSci PS SAI TBDOS WA WWWIA)

EASTMAN, William Reed, 1835–1925, American engineer and clergyman; enlargement of Erie Canal (1854–1857), railroad construction in Michigan, Indiana and Mexico (ANB DAB WWWIA)

EASTWOOD, Sir Eric, 1910–1981, British electronics engineer; Royal Air Force (WWII), electric and electronic signals, telecommunications, radar development, meteorological phenomena, flight of birds (1967), with English Electric Co. (1945), with Marconi Research Lab. (1948) and then director of research in what became General Electric Co. (1962–1974), FRS (DNB TBDOS(Po))

EATON, Benjamin Harrison, 1833–1904, American irrigation engineer; pioneer in irrigation of western land (NM and CO), built reservoirs and ditches, large land holder, land development, town of Eaton (CO), governor of Colorado (1895–1897) (CDOAB DAB NC6 WWWIA)

EATON, Cyrus Stephen, 1883–1979, Canadian American industrialist; to US (1905), organized gas and electric companies, United Light and Power Co. (1923), Continental Shares (1926), formed Republic Steel Corp. (1930), involved in international activities (MWBD NCC WWWIA)

EBERHART, Howard Davis, 1906–1993, American civil engineer; gas and electric firms, structural analysis, photoelasticity, artificial limbs, organized several companies, faculty of Univ. of California (1936ff) interspersed with other assignments, consulting, NAE (1977) (AMOS MT7)

ECCLES, William Henry, 1875–1966, English physicist, engineer and inventor; with F. W. Jordan invented flip-flop or binary circuit (1918) using valves known as Eccles-Jordan circuit (1919), professor at London Technical College and University College London, design of mechanical structures, graphic design, wireless transmitters, FRS (1921), consulting engineer (1926ff) (BDOTHOT DNB EB EOS&T MMAH RS17 WWW WWWIS)

ECKART, William Roberts, 1841–1914, American civil engineer; Putman flour mills, improved water wheels, naval and marine service, propeller design, steam locomotives, consulting engineer (DAB)

ECKENER, Hugo, 1868–1954, German aeronaut; zeppelin (1908), aerial navigation (1911), with Zeppelin works (1917–1937) and officer of company (1924ff), first to fly across Atlantic (1924), polar flight (1931) and a flight that circled the earth (1929) (MWBD WWWIS)

ECKERT, John Presper, Jr., 1919–1995, American electrical engineer and inventor; worked with John William Mauchly [1907–1980] in radio, design and construction of electronic digital computing devices, faculty at Univ. of Pennsylvania (1943–1946), formed and was officer in electronic companies (1944ff), co-inventor of Electronic Numerical Integrator and Compiler (ENIAC) (1947), formed Eckert-Mauchly Corp., which was acquired by Remington Rand (1950), which became Sperry Rand Corp. (1963–1982), which merged with Burroughs Corp., forming Unisys (1986); at least 87 patents, NAE (1967) (AMOS FAL IAI MT10 NYPL NYTObit PS TBDOS)

EDDISON, William Barton, 1899–1980, American mechanical engineer and inventor; in WWI as naval officer, design of fuel oil engines, faculty at Columbia Univ. (1913–1915), design of rotary blowers and compressors, invented aspects of stokers, clock, consulting mechanical engineer (1915–1921), chief engineer of Surface Combustion Engineering Co. (1921ff), devices for gas and oil combustion, approximately 20 patents (ANB NYTObit WWWIA)

EDDY, Harrison Prescott, 1870–1937, American sanitary engineer; city sewer department at Worcester (MA) from chemist to superintendent (1892), sewage treatment plants for Louisville (KY), chemical treatment plants, sand filter, stream pollution studies, consulting engineer with Metcalf & Eddy (1907–1937), retained by at least 125 cities (ANB NC28 NYTObit WWIE WWWIA)

EDDY, Henry Turner, 1844–1921, American civil engineer and mathematician; faculty at Cornell Univ. (1889–1873), Princeton Univ. (1943–1974), faculty and administration at Univ. of Cincinnati (1874–1889), including acting president, Rose Polytechnical Inst. (1891–1912), including president (1891–1894), civil engineering astronomy, machines and mechanisms (BDOAS NC15 TEIH WWWIA)

EDGAR, Charles, 1862–1922, American inventor; lumber industry, formed company and was leader in lumbering in Michigan, Wisconsin and Minnesota, developed improvements in tools, including band saw (1894, 1895) (DAB)

EDGAR, Charles Leavitt, 1860–1932, American electrical and mechanical engineer; central power station equipment (1884), small central power stations (NYC), Edison Electric Illuminating Co. superintendent (1887) and general manager (1889), president of several power companies in New York and Boston area, district steam heating service (MEIA NC27 WWWIA)

EDGERTON, Harold Eugene, 1903–1990, American electrical engineer and inventor; flash lighting, stroboscopic phenomena, strobe photography (1931), transients of synchronous machines, inventor of high-speed photography, electron strobe for night reconnaissance photography in WWII, faculty at Massachusetts Inst. of Technology (1942ff), NAS (1964), NAE (1966) (AIE AMOS ANB MT4 NYT-A PS WWWIA)

EDGEWORTH, Richard Lovell, 1744–1817, British inventor and educator; telegraph communication, invented semaphore, velocipede, pedometer land-measuring device, carriage improvements (MWBD)

EDISON, Thomas Alva, 1847–1931, American inventor and electrical engineer; operator at Western Union (1868), mimeograph (1875), phonograph (1877), electric pen (1877), telegraphic devices, incandescent electric lamp (1897), stock ticker (1870), carbon filament for light bulb (1880), system of cable and wiring (1880, 1881), Edison screw thread for light bulb (1881), electric meter, three-wire electric power system (1882), carbon brush for electric generators (1883), electric valve (1883), Edison effect (1883), wax cylinders for recording (1888), kinescope (1889), talking motion pictures (1913), storage battery, numerous electrical devices, 1,093 US patents, NAS (1927), honored on 3-cent US postage stamp (1947) (AG AIE ANB ASAI BEST BM15 CEOS&T EE EIH EOWB FAL GI H-BW IAI IIA1000Y MEIA MMAH MWBD NC5,25 NYPL SAI S:TLAW TBDOS TEIH TGS TNAS TP PA PS WA WWWIA)

EDLUND, Milton C., 1924–1993, American physicist and nuclear engineer; nuclear reactor theory, development and design of power racers, research with Union Carbide Co. (1948–1955), research and administration with Babcock & Wilcox Co. (1955–1966), Atomic Energy Commission (1966–1970), faculty at Virginia Polytechnic Univ. (1970ff), consultant, NAE (1976) (WWIE WWWIA)

EDMINSTER, Talcott White, 1920–1980, American agricultural engineer; soil and water conservation, government agricultural research and administrator with the US Department of Agriculture (1944ff), becoming administrator (1971–1980) (AE61,62 WWIE WWWIA)

EDSON, Jarvis B., 1845–1911, American mechanical engineer and inventor; Edson time and pressure recording steam gauge, manufactured engineering instruments, worked on engines for US ships during Civil War (MEIA)

EDWARDS, Oliver, 1835–1904, American soldier and inventor; partner in Neberling, Edwards & Co. (1856–

1861), officer during Civil War, retiring as brigadier general (1861–1866), Florence Machine Co. (MA) (1867ff), becoming superintendent, several improvements to sewing machine, invented Florence oil stove, general superintendent of Gardner Machine and Gun Co. (1867ff) (DAB NC27)

EDWARDS, William, 1719–1789, Welsh bridgebuilder and inventor; built bridge over River Taff (SE Wales) (1754)—longest single span at time (DNB GEAPIT LOTE MWBD MWGD SFLOE)

EDWARDS, William, 1770–1851, American inventor; built tannery (1780), patented equipment for preparing leather (1812), hide mill for softening dry leather, improved sole leather tanning process (1812), recognized as founder of hide and leather industry in US (MEIA NC11)

EGAN, Thomas P., 1847–1922, Irish American mechanical engineer and inventor; machinist, designed woodworking machinery, opened business (1874) that later merged with J. A. Fay & Co. (1893) (MEIA NC12)

EHBETS, Carl J., 1845–1925, German American mechanical engineer and patent authority; gunsmith (1864), to US (1868), worked on Gatling gun at Colt Patent Fire Arms Co. (1833), handled patents for Samuel Colt [1814–1862] and John M. Browning [1855–1926], assembled one of most complete firearms patent libraries in existence (MEIA MWBD)

EHRENSVÄRD, Count Augustin, 1710–1772, Swedish military engineer; fortifications of Sveaborg (1749), his son was Count Carl August Ehrensvärd [1745–1800], who was an admiral in the navy (MWBD)

EHRICKE, Krafft Arnold, 1917–1984, German American aeronautical engineer and physicist; during WWII developed German rocketry (V-2) and propulsion, to US (1946), designed and developed Atlas rocket for US (1947–1952), advisor to Rockwell International Corp. (ANB MWBD PS WWWIA WWWIS)

EICKEMEYER, Rudolf, 1831–1895, Bavarian American inventor and manufacturer; to US (1850), railroad construction, small machine repair shop (1854), invented machines related to hat manufacture, hat blocking (1865) and hat stretching (1869), invented differential gear for mowing and reaping machine (1870), electric power plant, high-potential phenomena, alternating electric current machinery, business merged with General Electric Co. (1892), was first employer in US of Charles P. Steinmetz [1865–1923], granted about 150 patents (ANB DAB MEIA MWBD NC1,11 WWWIA WWWIS)

EIFFEL, Alexandre-Gustave, 1832–1923, French engineer; bridges, iron bridge (1858), viaducts (1882), Eiffel tower (1899), framework for Statue of Liberty, aerodynamics and wind tunnel (1912), meteorology; as result of work of Gustave-Auguste Ferrie [1868–1932] the Eiffel Tower was used as a radio station (1904)

(EB EIH EOWB FEL ITCOF MWBD TBDOS WA WWWIS)

EIMBECK, William, 1841–1909, German American geodetic engineer; coastal survey, geodesic survey (1869), solar eclipse study, triangulation, survey of western continental divide, survey of 39th parallel (DAB WWWIA)

EINSTEIN, Albert, 1879–1955, German American physicist and inventor; patent examiner, relativity (1905), theory of photoelectric effect (1905), theory of heat capacity (1907), Einstein shift (1911), Einstein-Stark (Johannes Stark [1874–1957]) law (1912), general theory of relativity (1915), Einstein-Szilard (Leo Szilard [1898–1964]) pump (1931), gyroscope, gravitation (1929), Nobel Prize (1921), to US (1933), Inst. of Advanced Studies at Princeton Univ. (1933–1955), US citizen (1940), fNAS (1922), NAS (1942) (BEST BM51 CDOS DOT EDOP IAI MWBD NC30,F,G NYPL RHW SAI SAI-MP TGS TP WWWIA WWWIS)

EINTHOVEN, Willem, 1860–1927, Dutch physiologist and inventor; string galvanometer (1903), which served as the basis of electrocardiograph (1903) based on work of French engineer Clément Ader [1841–1926] (1897), Nobel Prize (1924), FRS (1926) (GI IAI MWBD SAI WA WWWIS)

EISENBERG, Phillip, 1919–1984, American civil and structural engineer; hydraulics, model testing, mechanics, WWII in US Navy (1945–1946), David Taylor Model Basin (DC) (1942–1944) (1946–1953), US Office of Naval Research (1953–1959), founder and president of Hydronautics (1959ff), NAE (1974) (MT3 WWIE)

EISENBUD, Merril, 1915–1997, American electrical engineer and industrial hygienist; control of environmental hazards, radioactive pollution, Liberty Mutual Insurance Co. (1936–1947), US Atomic Energy Commission (1947–1957), faculty at New York Univ. Medical Center (1959–1985), adjunct professor at Univ. of North Carolina (1985ff), NAE (1977) (WWIE WWWIA)

ELDER, John, 1824–1889, Scottish marine engineer; shipbuilder, invented compound reciprocating steam engine (1854) (MWBD)

ELIASSEN, Rolf, 1911–1987, American environmental engineer; sanitary engineering, man and environment, methods of water and sewerage treatment, design engineer, industry waste treatment, design engineer with J. N. Chester Engineers (1935–1936), Dorr Co. (1936–1939), with faculty of Illinois Inst. of Technology (1939), New York Univ. (1940–1949), service in US Corps of Engineers in WWII, Massachusetts Inst. of Technology (1949ff), consulting engineer (1940ff), NAE (1971) (AMOS MT9 WWWIA)

ELION, Gertrude Belle, 1918–1999, American woman chemist and inventor; human physiology and nucleic acids, invented target drugs, with Burroughs Wellcome Co. (1944–1999), serving as head of experimental therapy (1967ff), National Inventors Hall of Fame (1991), Nobel Prize (1988) (AIE I&T21 LOTCS PA WWWIA)

ELKINGTON, George Richards, 1801–1865, British inventor; use of electroplating for finishing metal objects, plating of base metals with silver and gold (1840), persons in employ also issued related patents (PS TBDOS)

ELLET, Charles, Jr., 1810–1862, American civil engineer; wire suspension bridges (1842ff), built first bridge over Niagara River (1847), built longest single-span bridge in the world that at the time was over the Ohio River (1849), designed and built ram warships (1861, 1862) (ANB BDOACE BITAS DAB FEL MWBD NC4 PS TBDOS TEIH WWWIA)

ELLICOTT, Andrew, 1754–1820, American surveyor; surveyed state of Maryland (1775), extended Mason-Dixon line (1784), boundaries between states—OH-PA (1785), PA-NY (1786) (1790), DC (1791–1792), US-FL (1796–1800), GA-SC (1811), instructor at West Point (1813–1820), Ellicott plan for Washington (DC) (BDOACE BDOAS MWBD NC13 TEIH WWWIA)

ELLICOTT, Joseph, 1760–1826, American engineer; surveying, located western border of Pennsylvania, laid out the city of Buffalo (NY) (1793) that was incorporated (1810), land development, agent for Holland Co. (DAB EB MWBD NC13 WWWIA)

ELLIOTT, Ezekiel Brown, 1823–1888, American inventor; telegraphic work, statistics and electricity, several inventions, including dynamos and motors, telegraph insulator, work on actuaries for government and insurance companies (BDOAS(El) PS)

ELLIOTT, John F., 1920–1991, American metallurgical engineer and metallurgist; process and extractive metallurgy, heat corrosion, WWII with US Navy (1942–1946), US Steel and Inland Steel Co. (ante 1955), faculty of Massachusetts Inst. of Technology (1956ff), NAE (1975) (AMOS MT6 NC2 WWIE WWWIA)

ELLIOTT, Martin Anderson, 1909–1988, American mechanical engineer; engine performance, fuels, research with Consolidated Gas and Electric Co. (MD) (1934–1938), research with US Bureau of Mines (1938–1952), faculty at Johns Hopkins Univ. (1930–1934), director of Inst. of Gas Technology (1956ff), energy consultant, NAE (1976) (AMOS MT4 WWIE WWWIA)

ELLIS, Carleton, 1876–1941, American chemist and inventor; developed paint and varnish remover, formed applied research organization to discover new products and processes, hydrogenation of vegetable oils (1912), coatings, resins in paint industry, first durable lacquer for automobile paint (1925), soap, distillation and cracking of crude oil, dyes, explosives and fertilizer, obtained over 753 patents (ANB NC32 WWWIA)

ELLIS, Francis Cutler, 1890–1957, American engineer, inventor and manufacturer; first worked with Electric Bond and Share Corp., then with Consolidated Utah Power and Light Co., then Idaho Power and Light, served in WWI, founded Ellis Research Lab., where he developed and manufactured electrical devices for medical use, studied electricity in the human body, made various testing devices, best known of which was the Ellis microdynameter, received 21 US patents (NC48 PS)

ELLISON, Walter D., 1898–1970, American agricultural engineer; research engineer with US Department of Agriculture, investigations in drainage, first project supervisor of Coshocton Watershed Project (OH) (1936–1949) related to runoff, hydrology, forestry, meteorology, soil conservation and erosion control for US Navy (1949ff) (AE51)

ELLSWORTH, Lincoln, 1880–1951, American explorer and engineer; aviation, Andes mountain exploration (1924), transpolar flight (1926), transarctic submarine expedition (1931), airplane flight across Antarctica (1935), author (MWBD NC39,E PS WWWIA)

ELLWOOD, Isaac Leonard, 1833–1910, American industrialist; worked with Joseph F. Glidden [1813–1906] in manufacture and sale of barbed wire (1870s), sold interest to Washburn & Moen Manufacturing Co. (1876), formed I. L. Ellwood & Co. followed by Ellwood Manufacturing Co., then Ellwood Wire and Nail Co., both of which were sold to American Steel & Wire Co. (1898) (MWBD WWWIA)

ELMEN, Gustav Waldemar, 1876–1957, Swedish American electrical engineer; to US (1893), magnetism, developed materials such as permalloy (1916), employed by several organizations, including Western Electric (1901–1925), Bell Telephone Labs. (1925–1941), US Naval Ordnance Lab. (1941–1956) (MWBD NC43)

ELMES, Charles F., 1845–1904, American mechanical engineer; machinist in father's firm Elmes & Son, which became Charles F. Elmes Co. (1877), designed pumps, propelling machinery for fireboats (MEIA)

ELMORE, Francis Edward, 1864–1932, English inventor; with brother Alexander Stanley Elmore [1867–1944] developed flotation process of ore separation and recovery (1898) (MWBD)

ELMS, James Cornelius, 1916–1993, American physicist and engineering electronics; aeronautics, seismic studies, space, strategic defense initiative, armament control, radar, US Army Air Force (1942–1946), with several aeronautical industries on West Coast, vice president of Avco Corp. (1950–1962), National Aeronautics and Space Agency (1962), Department of Transportation Systems Center (1970–1975), consultant (1975ff), NAE (1974) (AMOS MT7 WWIE WWWIA)

ELSTER, Johann Philipp Ludwig Julius, 1854–1920, German physicist and inventor; with Hans Friedrich Geitel [1855–1923] developed first practical photoelectric cell (1895), photometer, a Tesla transformer, studied atmospheric electricity, radioactivity (MWBD NYPL WA WWWIS)

ELSTON, Charles W., 1914–1989, American mechanical engineer; energy conversion, electric power generation, design and manufacture of large steam turbine generators, technology assessment, with General Electric Co. (1937ff), becoming manager of turbine engineering (1949), NAE (1967) (MT5 WWIE)

ELYUTIN, Vyacheslav Petrovich, 1907–??, Soviet metallurgist; researcher then director of Machine Inst. of Steel (1930–1950), professor (1947), ferrous alloy production, reaction of different components (SMOS WWWIS)

EMANUELLI, Pio, 1888–1946, Italian astronomer and inventor; invented celestial maps (DOSB WA)

EMBANK, Thomas, 1792–1870, English American inventor and manufacturer; manufacture of copper, lead and tin tubing, developed improved methods for tinning lead (1823) (1832), improved safety valves (1831), commissioner of patents (1849) (ANB BITAS DAB NCI PS WWWIA WWWIS)

EMEL'YANOV, Vasilii Semenovich, 1901–??, Soviet metallurgist; developed new grades of steel, produced armor, role of gases in steel and ferroalloys of manganese steel, new electric furnaces (SMOS)

EMERSON, James Ezekiel, 1823–1900, American machinist and inventor; carpentry and construction, developed automatic machine to bore, turn and cut heads on wooden spools and bobbins used in cotton mills, established and supervised sawmills (CA) (1852ff), organized company to manufacture edge tools (NJ) (1859), manufactured circular saws, invented removable tooth saw, established Emerson, Ford & Co. (PA) (1871) (DAB MEIA NC12 WWWIA)

EMERSON, Ralph, 1831–1914, American inventor and manufacturer; hardware business, reaper business developed on patents of John H. Manny [19th century inventor], patented tongue, caster wheel (1851) and lever board and improved guards (1862) for reapers and mowers, managed several companies (DAB NC12 WWWIA)

EMERSON, Ralph Waldo, 1803–1882, American philosopher and writer; minister of Unitarian Church in Boston (1829–1832), traveled and lectured widely, preached and wrote on the transforming power of technology, realized that artifacts and machines were value-laden, in his philosophy accommodated environmentalism and materialism as represented by scientist, engineer and inventor (EB HBW MWBD)

EMERY, Albert Hamilton, 1834–1926, American civil and mechanical engineer; invented cheese press (1859), window sash fastener (1859), ordnance projectiles, fortifications (1861), cannon foundry, percussion fuses, scales, testing machines for tension and compression (1873–1878), hydraulic pressure measuring devices, established Emery Scale Co. (1882), some inventions manufactured by Yale and Towne Manufacturing Co. (CT), numerous patents (DAB MEIA NC12 WWWIA)

EMERY, Charles Edward, 1838–1898, American civil engineer; steam expansion studies, tested stationary engines, chief engineer and manager of New York Steam Co. (1879ff), central steam plant, consulting engineer and patent authority (1888ff) (BITAS DAB MEIA NC9 WWWIA)

EMMET, William LeRoy, 1859–1941, American electrical and mechanical engineer; improved apparatus used by electric street railways such as controllers, armatures, swivel trolley, insulation materials and devices, worked in various industries, electric power development, particularly alternating current systems, worked with Charles G. Curtis [1860–1953] (1869) developed first successful steam turbines to produce electricity, NAS (1921) (ANB BM22 NC30,D WWWIA)

EMMONS, Howard Wilson, 1912–1998, American mechanical engineer; aerodynamics of combustion, supersonic flight, gas dynamics, numerical solutions, fire safety research, development of turbines at Westinghouse Electric and Manufacturing Co. (1937–1939), faculty at Univ. of Pennsylvania (1939–1940), faculty at Harvard Univ. (1940–1983), some of the time with Massachusetts Inst. of Technology, consulting engineer, NAE (1966), NAS (1966) (AMOS MT10 WWIE WWWIA)

EMMONS, Samuel Franklin, 1841–1911, American mining engineer and geologist; associated with Eckley B. Coxe [1839–1895] and surveyed 40th parallel in western US, statistics of precious metals, origin of ore deposits, NAS (1892), consulting engineer (ANB BDOAS BM7 DAB NC10 TNAS WWWIA)

EMORY, William Hemsley, 1811–1887, American topographical engineer; in military service engaged in civil engineering, survey of California-Mexico border (1848–1853), compiled useful astronomical, meteorological and ethnographical data (BDOAS NC4 PS WWWIA)

ENDICOTT, Mordecal Thomas, 1844–1926, American naval engineer; railroad extension projects, consulting engineer in charge of US Navy engineering projects (1890), directed several navy projects, such as Dewey floating dry dock, rear admiral (DAB NC15 WWWIA)

ENGEL, Godfrey, 1860–1936, Prussian American mechanical engineer; erection of machinery in sugar refineries (1882), worked with several companies involving sugar processing, worked on beet sugar processing and cane sugar processing, design and construction with several companies, consulting engineer (1921ff) (MEIA)

ENGELBRECHT, Richard S., 1926–1996, American environmental engineer; sanitary science, environment, health, waste water, water quality and treatment, faculty at Massachusetts Inst. of Technology (1950–1954), faculty at Univ. of Illinois (1954ff), NAE (1976) (AMOS MT9)

ENGELHARD, Charles Philip, 1867–1950, German American chemical industry entrepreneur; to US (1891), US citizen (1906), founded American Platinum Works (1903), reorganized Baker and Co., which refined precious metals, business carried on by son Charles William Engelhard [1917–1971] as Engelhard Minerals and Chemicals Corp. (ACCE NC41,F PS)

ENGLEHARDT, Orton, 1901–1989, American rancher and inventor; invented the world-famous Rain Bird Sprinkler, which had a horizontal arm impact device (1935), developed and manufactured by Rain Bird Sprinkler Co. (1935) by Clement M. La Fetra [d. 1963] (1935ff), purchased Thermal Hydraulic Co. (1967) (AE(supp))

ENGSTROM, Elmer William, 1901–1984, American electrical engineer; high-power radio transmitter, radio tubes, television, General Electric Co. (1923–1930), executive with Radio Corp. of America (1930–1955), executive with National Broadcasting Co. (1958ff), NAE (1964) (AMOS ANB MT3 WWWIA)

ENTWISTLE, James, 1837–1910, American naval engineer; engineer on gunboats in Civil War, inspector of machinery (1877), fleet engineer to Commodore George Dewey [1837–1917] (1897), retired as rear admiral (DAB MWBD WWWIA)

EPIMACHOS, ??–305 BC, Greek engineer; developed belfry to protect attacking troops (TAE)

ÉRARD, Sébastien, 1752–1831, French inventor and manufacturer; musical instruments, mechanical harpsichord (c. 1775), French square piano, piano key mechanism (1809), double action harp (1911), followed by Pierre Érard [1796–1855] (MWBD)

ERCKER, Lazarus, c. 1530–1594, German metallurgist; analytical and metallurgical chemistry, assay master of Saxony (1555), warden of mint to Duke (1558–1566), mine master, consultant, author (GEAPIT MWBD WWWIS)

ERICSSON, John, 1803–1889, Swedish American marine engineer and inventor; in Swedish army (1816–1827), with Francis Pettit Smith [1808–1874] worked on steam engine improvement, surveyor (1817), locomotive (1829), first successful screw propellers for ships (1837), designed Monitor, torpedo launched underwater (1878), worked on numerous warships, received over 36 US patents (ANB ASAI BEST BITAS DAB EE EIH FNIE I&T MEIA MMAH MWBD NC 4,9 NYPL RHS SAI TBDOS PS WWWIA WWWIS)

ERLENMEYER, Richard August Carl Emil, 1825–1909, German chemist and inventor; began as pharmacist, professor at Polytechnikum Munich (1868–1883), invented conical flask named after him (1861), synthesized several organic compounds (MWBD WWWIS)

ERNST, Oswald Herbert, 1842–1926, German American soldier and engineer; fortifications during Civil War, developed deep sea channel to Galveston, developed plans for Panama Canal, river and harbor improvements in Baltimore (1900) and Chicago (1901), became major general (1916) (DAB NC4 WWWIA)

ERSKINE, Robert, 1735–1780, Scottish American mechanical engineer; invented centrifugal pump (1763), continuous steam pump (1766), managed several iron works, manufactured cannon shot and war supplies for colonies in Revolutionary War, geographer and surveyor general to Continental Army (1777) (MEIA WWWIA)

ESMAY, Merle L., 1920–1990, American agricultural engineer; teaching, research and extension specialist in design of farm structures for storage and animal housing, environmental control, wood trusses, in US Army in WWII, Iowa State Univ. (1947–1951), Univ. of Missouri (1951–1955), Michigan State Univ. (1955ff), international education activities (10 yr.) (AE(supp) AMOS WWIE)

ESNAULT-PELTERIE, Robert-Allen-Charles, 1881–1957, French aviator and inventor; ailerons, monoplane with engine in front (1907), invented improved fuel pump, coined term "astronautics," space exploration, rockets for studying upper atmosphere (MWBD WWWIS)

ESSELEN, Gustavus John, 1888–1952, American chemist and chemical engineer; research for General Electric Co. (1912–1914), Chemical Products Co. (1914–1917), Arthur D. Little, Inc. (1917–1921), fabricated silk from collagen, curing agents for epoxy resins, improvement of safety glass, founded G. J. Esselen Co. (1920), which became Esselen Research Corp. (1921–1949), which merged with United States Testing Co. Inc., reserve officer in US Army Chemical Warfare Service (ACCE NC42 PS WWWIA)

ESTCOURT, Vivian Fitzgeorge, 1897–1985, English American mechanical and electrical engineer; fossil and nuclear power generating, steam power, air pollution control, control and safety in industry, US citizen (1921), manager of generation for Pacific Gas & Electric Co. (1923–1964), consulting engineer for Bechtel Power Corp. (1963–1984), consultant, NAE (1981) (MT3 WWIE WWWIA)

ESTERLY, George, 1809–1893, American inventor and manufacturer; horse-pushed harvester (1844), agricultural machinery business for mowing machine, plow, hand rake reaper, sulky cultivator (1856), reaping machine (1857), established factory in Whitewater (WI) (1858), seeder (1865), grain-harvesting machine (1882),

automatic twine binder (1884) (ANB DAB MEIA WWWIA)

ESTES, Elliott M., 1916–1988, American mechanical and automotive engineer; with General Motors Corp., in engine development then assistant chief engineer of Oldsmobile division, chief engineer of Pontiac division (1956–1981), general manager of Chevrolet division (1965–1969), president of General Motors Corp. (1974ff), NAE (1976) (MT4 WWWIA)

ETCHEVERRY, Bernard Alfred, 1881–1954, American irrigation and civil engineer; faculty at Univ. of Nevada (1903ff) after 1 yr. at Univ. of California, land drainage and flood protection, author on subject (WWWIA)

ETHERINGTON, Harold, 1900–1994, English American mechanical engineer; nuclear power plants, nuclear safety, metallurgical engineering, Allis-Chalmers Manufacturing Co. (1937–1946), director of nuclear energy for power at Oak Ridge National Lab. (1946–1947) and Argonne National Lab. (1948–1950), vice president for nuclear products of ACF Industries (1954–1959), general manager of Atomic Energy Division (1959–1963), consulting engineer (1963ff), NAE (1978) (MT8 WWIE)

EUCLID, fl. 325–300 BC, Greek mathematician and philosopher; founded school in Alexandria, geometry, synthesized work of people before him, major work is *Elements* (BDOA(Wi) BEST EIH HOFL MWBD TBDOS(Po) TEIH TGS SAI-MP WWWIS)

EUDOXUS of Cnidus, c. 400–347 BC, Greek mathematician, astronomer and inventor; mechanics, proportion, meteorology, volume of cones, spherical astronomy (BDOS(Wi) EB TBDOS TAE TEE TGS WWWIS)

EULER, Leonhard, 1707–1783, Swiss mathematician and inventor; laws of motion for engineering, Euler ratio of length to diameter or width, buckling of columns, one of first crude forms of turbine (1750–1754), at St. Petersburg (Russia) (1727–1744) (1766ff), Berlin Academy of Science (1744–1766) (BDOS(Wi) BEOS DOSB EB FIOTEC FNIE GEAPIT HOFL MMAH MWBD TBDOS(Po) TGS WWWIS)

EUPALINUS of Megara, 6th century BC, Greek surveyor and engineer; tunneling through mountain starting from two headings, tunnel for water supplies on Island of Samos, said to be first civil engineer known by that title (AIn EB E-64 EIH GEAPIT PS TAE TEIH TGS TSOE WWWIS)

EUSTIS, Henry Lawrence, 1819–1885, American engineer and soldier; US Corps of Engineers (1842–1849), Boston sea wall construction, Newport engineering operations, faculty and dean at Harvard Univ. (1849–1885), Civil War service as a brigadier general of volunteers (BDOAS BITAS DAB NC22 PS WWWIA)

EVANS, Anthony Walton Whyte, 1817–1886, American civil engineer; enlargement of Erie Canal, international railroad building in Chile and Peru, agent for US companies, consulting engineer (DAB NC10 WWWIA)

EVANS, David C., 1924–1998, American electrical engineer and computer scientist; interactive computer operation, computer graphics, Bendix Corp. (1953–1962), faculty at Univ. of Utah (1965–1977), founder of Evans & Sutherland Computer Corp. (1968), computer-aided design, NAE (1978) (FAL NYTObit WWIE WWWIA)

EVANS, James Carmichael, 1900–1992, Black American engineer and inventor; aide to secretary of war in WWII, patented device to utilize exhaust gases to prevent icing on aircraft (AAI WWWIA)

EVANS, Oliver, 1755–1819, American inventor and mechanical engineer; crusade to build self-propelled steam-powered carriage (1772–1777), machine to make textile cards (1777), water power for automatic grain and flour mill (1785), boring machine, invented method of making millstones (1796), invented mechanical devices for handling grain—elevator, screw conveyors, hoppers, etc., Evans hopper bay to cool and dry flour, first steam traction (1801), first high-pressure steam engine (1802), amphibious dredge (1802), established Mars Iron Works (1807), patents for improved sawmill (1811) (AIE ANB ASAI ASMENews20 BOAC BITAS CDOAB DAB E-64 EIH GEAPIT IAI IIA MEIA MWBD NC6 NYPL 1000Y TBDOS TEIH WWWIA WWWIS)

EVE, Joseph, 1760–1835, American inventor and scientist; invented machine to separate seed from cotton (1787), manufacturer of gins (1800), cottonseed huller (1803), metallic bands and belts for power transmission (1828) (DAB GEAPIT MEIA PS WWWIA)

EVERETT, William L., 1900–1986, American electrical engineer and physicist; electronics and communication engineering, radio engineering, served in WWI and WWII, faculty and department chairman at Ohio State Univ. (1926–1949), faculty and dean of engineering at Univ. of Illinois (1949–1968), editor of engineering journals, NAE (1964) (MT4 WWIE)

EVINRUDE, Ole, 1877–1934, Norwegian American inventor; outboard marine engine (1909), with his wife Bessie Evinrude founded Evinrude Motors (WI) (1911) (MWBD PA)

EWER, Roland Gibbs, 1848–1902, American mechanical engineer; machine shop superintendent of Charles Pratt's Oil Works (NY) (1869–1884), with Henry R. Broad established Progressive Iron Works (NY) (1884), superintendent (1888) and general manager (1894) of salt manufacturing plant, which became known as Michigan Alkali Co. (MI), manager of Progressive Iron Works (1897ff) (MEIA)

EWING, Alfred (James Alfred), 1855–1935, Scottish engineer and physicist; magnetic properties of iron and steel, Ewing extensiometer to test elongation of a test piece, named phenomenon of hysteresis, professor at Univ. of Tokyo (1878–1883), at Dundee College (1883–1890), at Cambridge Univ. (1890–1903), director of naval education (1903–1916), principal and vice chancellor at Univ. of Edinburgh (1916–1929) (DNB DOT EE MWBD WWWIS)

EYTELWEIN, Johann Albert C., 1764–1848, German hydraulic engineer; surveyor (1786), civil engineer (1790), planned harbors, designed structures for water control, developed education program for engineers (BDOS GEAPIT PS WWWIS)

EYTH, Maximilian von (aka Edward Friedrich Maximilian), 1836–1906, German engineer and inventor; power machinery for plowing, agricultural machinery, earth-moving equipment, developed irrigation, as did John Fowler [1826–1864], to advance use of power equipment, internationally known Max Eyth medal for engineers awarded by Verein Deutscher Ingenieur (VDI) named after him (MWBD WWWIS)

F

FABER, Johann Lothar von, 1817–1896, German manufacturer; took over family business (1839) originally established (1761) to make pencils, put rubber tip on pencil, made writing and drawing instruments, brother John Eberhard Faber [1822–1879] represented business in US (1948) and who established an independent business in US known as Eberhard Faber Pencil Co. (NY) (1861) (MWBD)

FABRY, Charles, 1867–1945, French physicist and inventor; optics, Fabry-Perot interferometer (1896), set a series of standard wavelengths for vapors and gases, professor at Univ. of Marseilles (1894–1920), at Sorbonne (1920–1945) (MWBD TBDOS)

FAGE, Arthur, 1890–1977, British aeronautical scientist and engineer; aerodynamic research (1912ff), British National Physical Lab. (1914–1953), becoming laboratory superintendent (1946–1952), FRS (PS RS24 WWWIS)

FAHRENHEIT, Daniel Gabriel (or Gabriel Daniel), 1686–1736, Polish Dutch German physicist and inventor; instrument maker (1701ff), alcohol-filled thermometer (1709), first accurate thermometer (1714), changed from alcohol- to mercury-filled (1714), introduced Fahrenheit temperature scale of 32 to 212 deg. F. (1717), manufactured meteorological instruments, developed improved hydrometer, FRS (1724) (BDOS BEST DOSB EB FNIE FETCTW GI MWBD NYPL 1000Y TBDOS WA WWWIS)

FAIR, Gordon Maskew, 1894–1970, South African American sanitary engineer; water purification, sewage treatment, health, windmills for water supply, served in WWI with Canadian forces, faculty and dean of engineering at Harvard Univ. (1918–1945), Rocke-feller Foundation (1945–1948), consultant, NAE (1967) (AMOS MT1 WWWIA WWWIS)

FAIRBAIRN, Sir Peter, 1799–1861, Scottish engineer; with brother William Fairbairn [1789–1874] formed company to build textile machinery and machine tools (DNB MWBD)

FAIRBAIRN, Sir William, 1789–1874, Scottish civil and mechanical engineer; brother of Peter Fairbairn [1799–1861], riveting machine, introduced shipbuild-ing (1835–1849) and use of wrought iron for ships and railway bridges (1846), machinery for cotton mills, self-controlled planing machines (1862), Fairbairn coupling (half-lap shaft) (BDSE BDOS DNB EE EIH GEAPIT MWBD PS TBDOS TEIH TGE:B WWWIS)

FAIRBANKS, Frank Latta, 1888–1939, American ag-ricultural engineer; successively worked for Franklin Au-tomobile Co., Pendleton Automobile Co., Morse Chain Co., Cornell Univ. (1917ff), first in mechanical engineer-ing then in rural engineering, research on animal temper-ature control, electric brooding, poultry house lighting, tractors, rubber tires for implements (AE20)

FAIRBANKS, Henry, 1830–1918, American inventor, engineer and clergyman; with his father, Thaddeus Fair-banks [1796–1886], invented scales to weigh grain as it went into hopper (1868), paper pulp business, alter-nating electric current generator (1897), management of family scale business (DAB MEIA NC10 WWWIA)

FAIRBANKS, Thaddeus, 1796–1886, American man-ufacturer and inventor; with his brother Erastus Fair-banks [1792–1864] established E. & T. Fairbanks Co. (1824), cast iron plow (1826), compound lever platform scales (1831), formed Fairbanks Scale Co. (1834) to manufacture all-purpose platform scales, Franklin Fair-banks [1828–1895] became partner (1855) (EE DAB MWBD NC10 WWWIA)

FAIRCHILD, Sherman Mills, 1896–1971, Ameri-can engineer, inventor and manufacturer; flash camera (1916), aerial photography, high-speed camera, orga-nized Fairchild Aerial Camera Co., organized Fairchild Aviation (1924), innovative airplane, organized air trans-port companies (NC58 WWWIA)

FANNING, John Thomas, 1837–1911, American civil engineer; served in Civil War, water treatment, water supply system, city engineer (1862–1870), chief en-gineer of St. Anthony Falls Water Power Co. (1886ff), Fanning friction factor (1882), related to work of Rob-ert Manning [1816–1897] and Antoine de Chézy [1718–1798], consulting (BDOACE CEOS&T DAB FNIE L&M NC9 WWWIA WWWIA WWWIS)

FARADAY, Michael, 1791–1867, British physicist and inventor; high-steel alloys (1818), electric motor (1821), at Royal Inst. (1825ff), dynamo (1831), invented trans-former, magneto hydrodynamic electric generation

(1840ff), RS (1824), fNAS (1864) (BDOS EIH EOWB FNIE GEAPIT GI MMAH MWBD NYPL 1000Y SAI-MP S:TLAW TBDOS TGE:B TGS WWWIS)

FARMER, Moses Gerrish, 1820–1893, American electrical inventor; with William F. Channing [1820–1901] invented the electric fire alarm system (1851), du-plex and quadraplex telegraphy (1855), wet cell battery (1859), dynamo (1866) platinum incandescent light, at US Torpedo Station (RI) (1872–1881) (ANB ASAI DAB MWBD NC7 WWWIA WWWIS)

FARNSWORTH, Philo Taylor, 1906–1971, American engineer and inventor; electronics, television (1928) us-ing image dissector camera, nuclear fission, established laboratory (CA) (1926–1931), moved to Philadelphia, where he worked for several companies, including Ra-dio Corp. of America and International Telephone and Telegraph Co., established several companies using his name, consulting engineer, over 300 US and foreign pat-ents (AIE ANB MWBD NYT-A PS WA WWWIA)

FARNUM, Henry, 1803–1883, American civil engi-neer; surveyor for Erie Canal (1821–1824), chief engi-neer and superintendent (1827) of canal system, railroad layout and construction, chief engineer of various rail-roads, railroad expansion to west (BDOACE WWWIA)

FARRALL, Arthur W., 1800–1986, American agricul-tural engineer; director of spray drying laboratory for Douthitt Co. (1929–1932), research engineer for food processing equipment with Creamery Package Co. (IL) (1932–1945), developed continuous flow ice cream and food processing equipment, faculty at Univ. of Califor-nia (1922–1929), faculty and administration at Michi-gan State Univ. (1945–1968), including department chairman (1945–1964), 7 patents, author, consultant (AE(supp) NCM WWIE WWWIA)

FARRARE, Jean de and GOZZO, Jacques de, 14th cen-tury, Italian Renaissance builders (before Italy became a nation); probably built Ponte Castelvecchio bridge, in-cluding an arch of 160 ft. span, at Verona (1354–1356), Della Navi bridge (c. 1374) (TSOE)

FARREN, Sir William Scott, 1892–1970, English aero-nautical engineer; Royal Aircraft Factory and its succes-sors (1916ff), head of aeronautics department, designed combat aircraft, in charge of production of flying boat (1917), with Armstrong, Whitworth Aircraft (1918–1937), lecturer and researcher (1922–1931), wind tunnel instrumentation, Air Ministry (1937–1941), returned to Royal Aircraft Establishment (1941ff) and other compa-nies in England (DNB PS RS17)

FEDDERSEN, Berend Wilhelm, 1832–1918, Ger-man physicist and inventor; electric spark discharges, electric waves, relationship of discharge and fre-quency, author of biographical literature (1898) (TFE&I WWWIS)

FEDER, Gottfried, 1883–1941, German civil engineer and economist; civil engineer by profession, Ministry of Economics (1933), state housing director (1934), politician, German government official (MWBD)

FEELY, Frank J., Jr., 1918–1995, American engineer; known for strengthening national engineering standards, developing and using petroleum technology, understanding brittle fracture of steel, Standard Oil of New Jersey (later named Exxon) (1940ff), became vice president of engineering (1961), vice president of Exxon Research and Engineering (1974ff), NAE (1979) (Bridge MT8 WWIE)

FELKER, Jean Howard, 1919–1994, American electrical engineer and inventor; designed first transistorized digital computer, digital systems, served in signal corps in WWII (1942–1945), research administrator at AT&T Bell (1945ff), becoming assistant chief engineer for transmission engineering, NAE (1974) (AMOS Bridge MT8 WWIE)

FELT, Dorr Eugene, 1862–1930, American inventor; desk-type mechanical adding machine called Comptometer (1886), formed Felt & Tarrant Co. (1889), of which he was president (1889–1930) (MWBD NC40 NYPL WWWIA WWWIS)

FELTON, Samuel Morse, 1809–1889, American civil engineer; constructed railroads for specialized uses, transportation of troops during Civil War, management of railroad systems (DAB NCB WWWIA)

FENSKE, Merrell Robert, 1904–1971, American chemical engineer; petroleum processing, catalytic reactions, distillation, extraction and separation, preparation of hydrocarbons, faculty at Pennsylvania State Univ. (1929ff), including director of Petroleum Research Lab., NAE (1967) (AMOS MT1 WWIE WWWIA WWWIS)

FENTON, Frederick Charles, 1891–1981, American agricultural engineer; economics of using farm machinery, design of rural structures, grain storage, extension work at Iowa State Univ. (1914–1916) and Univ. of Missouri (1916–1917), WWI in France as officer, faculty at Iowa State Univ. (1919–1928), faculty and head of department at Kansas State Univ. (1928–1956), consultant in India, retired (1961) (AE61 AMOS NUC WWIE(I) WWWIA)

FERENCE, Michael, Jr., 1911–1996, American physicist and automotive engineer; physics and electronics for industry and government, experimental hydrodynamics, microwave properties, radar, faculty at Univ. of Chicago (1937–1946), meteorologist with Evans Signal Lab. (1946–1953), research manager for Ford Motor Co. Scientific Labs. (1953–1962), including vice president for research (1962ff), NAE (1971) (AMOS Bridge MT9 WWIE)

FERGUSON, Harry George, 1884–1960, Irish industrialist and manufacturer; airplane design, farm machinery design and manufacture, tractor design and manufacture with automatic draft control known as the Ferguson system (1935), merged with Massey-Harris to form Massey-Ferguson Co. (1953), then manufactured in US by the Ford Motor Co. and called Ford Ferguson system (MWBD NC37 TBDOS TBOF WOI WWWIA WWWIS)

FERGUSON, Thomas Barker, 1841–1922, American soldier and inventor; Confederate soldier entering service as engineer, organized and member of Maryland Fish Commission (1870ff), invented improvements to the apparatus for incubating fish eggs, patented improved coffee pot (DAB PS)

FERGUSON, Phil Moss, 1899–1986, American civil engineer; frame analysis, reinforced concrete, testing, faculty at Univ. of Texas (1928–1957), including department chairman, American Concrete Inst. (1957–1959), consulting (1959ff), NAE (1973) (AMOS MT3 WWIE WWWIA)

FERMI, Enrico, 1901–1954, Italian American physicist and inventor; to US (1938), US citizen (1944), professor at Columbia Univ. (1939–1942), with others developed first controlled nuclear chain reaction (1942–1945), associated with Univ. of Chicago (1946–1954), worked on beta decay of radioactive materials, development of atomic bomb, professor at Inst. of Nuclear Studies (1945–1954), Nobel Prize (1938), NAS (1945), Enrico Fermi Award established by US Department of Energy (1956) in his honor (ASAI BEST BM30 CB GI IAI MMAH NC40 NYT-A NYPL 1000Y SAI S:TLAW TBDOS(Po) TBE TNAS TP WWWIA WWWIS)

FERRANTI, Basil Reginald Vincent de, 1930–1988, British mechanical engineer and scientist; son of Sir Vincent Ziani de Ferranti [1864–1930], worked in Glasgow engineering firm, served in political office, officer in International Computer and Tabulators (1964ff), Ferranti Ltd. (1984), then was president of Ferranti International (1988) (DNB)

FERRANTI, Sebastian Ziani de, 1864–1930, British electrical engineer and inventor; centralized power station, developed Ferranti alternator, pioneered high-voltage alternating current electric power generation and transmission, high-tension lighting cables, circuit breakers, turbines, obtained 176 patents (BDOS DNB MMAH PS TBDOS TGE:B)

FERRANTI, Sir Vincent de, 1893–1980, British electrical engineer; served in WWI in British Royal Navy (1914–1919), in WWII (1939–1944), electrical equipment manufacturing, active in professional societies (WWW)

FERRARIS, Galileo, 1847–1897, Italian physicist and electrical engineer; rotary magnetic field (1885), work led to polyphase electric motors, hydroelectric industry in Italy, transformer for alternating current electricity, established first electrical engineering school in Italy (1886) (BDOS MWBD PS WWWIS)

FERRI, Antonio, 1912–1975, Italian American aeronautical and electrical engineer; racer for seaplanes, wind tunnel, aeronautical research in Italy (1935–1943), NACA (US) (1944–1951), US Gasdynamics Lab. at Langley (VA), aerodynamics of high-speed flow, astronautics, mechanics, faculty at Polytechnic Inst. of Brooklyn (1951ff), including department head (1957), NAE (1967) (AMOS MT1 WWWIA)

FERRIS, George Washington Gale, 1859–1896, American engineer and inventor; railroad and bridge firm (1881ff), built Ferris wheel (1893) (BDOACE DAB MWBD NC13 WWWIA)

FERRIS, Robert E., 1907–1970, American agricultural engineer; most of career with Starline Co., which was formed by his grandfather (1929), serving in progression of offices, becoming chairman of the board (1958), during WWII with Office of Scientific Research and Development, 80 patents related to agricultural engineering, 9 patents on missiles and projectiles (AE51)

FESSENDEN, Reginald Aubrey, 1866–1932, Canadian American radio engineer and physicist; radio compass, electrolytic detector (1900), amplitude modulation, early microphone, transmission of voice by radio (1900), loop antenna radio compass, high-frequency alternator (1902), radio telephony (1906), invented FM radio, sonic depth finder, submarine signalling devices, heterodyne (1913), radio-television system, faculty at Univ. of Pittsburgh (1893–1900), head of National Signalling Co. (1902–1910), over 500 patents (ANB ASAI CDOS EIH GI IAI MMAH MWBD NC15 NYPL PA PS RHW TBDOS TFE&I WA WWWIA WWWIA-S&T WWWIS)

FESSENDEN, Thomas Green, 1771–1837, American inventor and journalist; development of hydraulic devices, new type of grain mill, patents on heating devices, established and edited magazine *New England Farmer* (1822–1837) (ANB DAB MWBD NC7 WWWIA)

FETTERS, Karl Leroy, 1909–1990, American metallurgical engineer; steel-making authority, slag metal reactions, ingot structure, electrolytic tin plate, corrosion, manager of research at National Tube Co. (1933–1936), with Youngstown Sheet & Tube Co. (OH) (1936ff), becoming vice president (1943), faculty at Carnegie Inst. of Technology (1941–1943) (AMOS Bridge MT6)

FEYNMAN, Richard Phillips, 1918–1988, American physicist; during WWII worked on nuclear energy as related to atomic bomb project at Princeton Univ. and Los Alamos Lab. (1942–1945), Cornell Univ. (1945–1950), California Inst. of Technology (1951ff), NAS (rejected), Nobel Prize (1965), explained behavior of electrons in high-energy collisions, developed theory of liquid helium, quantum mechanics, Feynman diagrams, honored on US 37-cent postage stamp (2005) (ANB CB MWBD SN168 WWWIA)

FIEDLER, Albert G., 1897–??, American civil engineer; water resources division of US Geological Survey (1918–1961), ground water studies, well-drilling, inspection of gauging stations, studies became basis of codes for ground water use (HOWRD)

FIELD, Charles William, 1828–1892, American soldier and civil engineer; served in US Army (1849–1861), joined Confederate Army (1961), superintendent of Hot Springs Indian Reservation (1885–1889), advanced to major general (DAB WWWIA)

FIELD, Lester Marshall, 1918–1997, American electrical engineer; microwave tube theory and design, electron optics, electromagnetic theory, plasma physics, faculty at Stanford Univ. (1939–1944) (1946–1952), Bell Telephone Labs. (1944–1946), faculty at California Inst. of Technology (1952–1955), Hughes Research Labs. (1966–1977) to vice president and chief scientist, NAE (1967) (AMAWOS Bridge)

FIELD, Stephen Dudley, 1846–1912, American electrical engineer and inventor; telegraphy, electrical improvements for telegraph and fire alarms, developed dynamo and system sold to Western Union Telegraph Co., electric locomotion and railway, over 100 patents covering many fields of electrical engineering (AI DAB WWWIA WWWIS)

FINK, Albert, 1827–1897, German American engineer; to US (1849), designed Fink truss (1952), truss for railroad bridges, railroad economics, Louisville & Nashville Railroad (1857–1875), including vice president (1869) (ANB BDOACE FEL MWBD NC9)

FINK, Donald G., 1911–1996, American electrical engineer; television development, stereo broadcasting, standardization activities, research director, general manager and vice president of Philco Corp. (1952–1962), director of IEEE (1963–1974), NAE (1969) (Bridge WWIE MT9 WWWIA)

FINLEY, James, 1762–1828, American engineer and inventor; bridge designer, cast iron eye-bar chain suspension bridge built (1801) and patented (1908), 40 bridges built in US (EIH GEAPIT WWWIA)

FINSTERWALDER, Ulrich, 1897–1988, German structural engineer; construction, pioneer work in reinforced concrete, leader in prestressed concrete, highways, bridges, technical manager with Dyckerhoff & Widmann, Inc. (1923ff), consulting engineer, fNAE (1976) (MT4 WWIE)

FIORAVANTE, Neri de, fl. 14th century, Florentine mason and builder; credited with designing the Ponte Vecchio at Florence (1341–1345) (TSOE)

FIRESTONE, Harvey Samuel, 1868–1938, American industrialist; tire innovations, nonskid tires, balloon tires, rubber plantation (1924), organized (1900) and president (1903–1932) of Firestone Tire and Rubber Co., son Harvey Samuel Firestone, Jr. [1898–1973] followed in the business (AIE MWBD NC1,5,C,J WWWIA)

FISHER, Alva J., ??–1947, American inventor; invented electric washing machine (1906) (some references have 1901), called The Thor manufactured by Hurley Machine Corp., Chicago (1907); American James T. King invented the washing machine using a manually operated cylinder (1851). See Vincent H. Bendix. (GI NYTObit SATF TBOF TTOT TWABOI WOI)

FISHER, Chester G., 1891–1965, American chemist, inventor and manufacturer; took over Pittsburgh Testing Lab. and formed Scientific Materials Co. (1902), which he renamed Fisher Scientific Co. (1925), marketed Fisher burner designed by brother Edwin Fisher, issued landmark Fisher catalog (1936), became chairman of company (1949), company purchased by Allied Corp. (1981) and returned to name of Fisher Scientific Co. (ACCE LOTCS ChemHerit20)

FISHER, Clark, 1837–1903, American naval engineer and inventor; served in Civil War, research for US Navy to establish oil as a fuel, was chief engineer for the US Navy (1871), developed numerous patents for Eagle Works, such as railroad spikes (1874), cast iron anvil, rail joints, hydropneumatic engine, approximately 20 patents (DAB MEIA)

FISHER, David A., Jr., 19th century, Black American inventor; invented carpenter's joiners clamp (1875), furniture caster (1876) (AAI)

FISHER, Ronald Aylmer, 1890–1962, British statistician; Rothampstead Experiment Station (1919), geneticist Univ. College London (1933–1943), Cambridge Univ. (1943–1957), developed techniques for design of experiments, originally applied mainly to design of experiments in agriculture, now applied to a variety of experiments, contributed to statistics in analysis of variance, consultant (CDOS DNB ME126 MWBD RHW WWWIA WWWIS)

FISK, James Brown, 1910–1981, American physicist and electrical engineer; communications research and development, radio, faculty at Massachusetts Inst. of Technology (1936–1938), Univ. of North Carolina (1938–1939), research and administration at Bell Labs. (1939ff), becoming president (1959), NAS (1954), NAE (1966) (AMOS BM56 CB MT4 WWWIA)

FISKE, Bradley Allen, 1854–1942, American naval officer, engineer and inventor; boat detaching apparatus (1878), mechanism for electric turning of turrets (1884), communication in ships, electric range finder (1889), navy operations, rear admiral in US Navy (1911), over 60 patents (ANB MEIA MWBD NC31,B WWWIA WWWIS)

FITCH, John, 1743–1798, American mechanical engineer and inventor; in charge of gun factory during War of Independence, surveyed Ohio River territory (1780), application of steam to transportation, invented steamboat with rotating side paddles (1785), navigation pioneer (1785), steamboat building and operating (1787ff), suggested jet propulsion of boat (1790) (AIE ANB BEST DAB EIH EOWB GEAPIT GI IIA MEIA MMAH MWBD NC6 NYPL PS WA WWWIA WWWIS)

FITZ, Henry, 1808–1863, American optician and inventor; printer, locksmith (1827), reflecting telescope (1835, 1838), photography (1840), made lenses and telescopes (BDOAS DAB MEIA NC4 PS WWWIA)

FITZGERALD, Desmond, 1846–1926, British American civil and hydraulic engineer; various railroad positions, sanitary protection of water supply, reservoirs, study of organisms in drinking water, worked to eliminate pollution of water supplies, consulting engineer (BDOACE DAB NC39 WWWIA)

FLAD, Henry, 1824–1898, Bavarian American civil engineer and inventor; construction of railroads in New York, East and Midwest, improvement of railroad communications, assistant to James B. Eads [1820–1887] on Eads Bridge, which was the first bridge across the Mississippi River (1867–1874), water supply for St. Louis, numerous patents covering filters, gauges, hydraulic elevator, pile driver, consulting engineer (BDOACE DAB MEIA MWBD NC12 WWWIA WWWIS)

FLATHER, John Joseph, 1862–1926, American mechanical engineer; machine shop practice in US and Europe, thermodynamics, worked for several companies, power plant development, tall chimney design, designer and foreman for two companies, faculty at Lehigh Univ. (1888–1891), Purdue Univ. (1891–1898), faculty and department head at Univ. of Minnesota (1898ff), consulting engineer (DAB NC22 WWWIA)

FLEISCHMANN, Charles Louis, 1834–1897, Hungarian American manufacturer and inventor; developed yeast product (1866), distilling business (1866), baker's yeast, compressed yeast based on patent of brother Henry Fleischmann, inventions include plow, improved cotton gins, cottonseed oil extraction, distillation equipment, sewing machine improvements, 13 patents between 1866 and 1888 (AIE ANB DAB NC22 WWWIA WWWIS)

FLEMING, Alexander, 1881–1955, Scottish bacteriologist; discovered lysozyme (1921), along with others discovered penicillin (1928), Nobel Prize shared (1945) (CDOS GI MWBD 1000Y PA RHW)

FLEMING, Sir Arthur Percy Morris, 1881–1960, English electrical engineer; development of radio, radio vacuum tube diode (1905), development of radar, director of research at Metropolitan-Vickers Co. (1931–1954) (MWBD NYPL SAI-MP TGS WA)

FLEMING, Sir John Ambrose, 1849–1945, English electrical engineer and physicist; application of electricity to telephony, radio rectifier (1904), light bulb and lighting, telegraphy, radio tube diode (1905), magnetic field, right hand rule (generators) and left hand rule (motors), thermionic valve, professor at Univ. College London (1885–1926), NAS (1938) (BEST BM39 CEOS&T DNB DOT GI IIA MMAH MWBD PS TBDOS TFE&I WA WWWIS)

FLEMING, Sir Sandford, 1827–1915, Scottish Canadian engineer; to Canada (1845), chief engineer of Inter-Colonial Railroad (1867–1876) and Canadian Pacific Railway (1872–1880), developed scheme for standard time (1884), telegraph communication system (DNB MWBD)

FLETCHER, Andrew, 1829–c. 1905, Scottish American mechanical engineer; machinist, built steamboats, erected and operated sugar plant in Havana, built and repaired steamboats, formed commercial firms of Fletcher, Harrison & Co. and North River Iron Works (1853), firms combined and reorganized as W. & A. Fletcher Co., of which he became president (MEIA WWWIA)

FLETCHER, James Chipman, 1918–1991, American physicist and electronics engineer; space and missile systems, communications, underwater acoustics, guidance, control, systems laboratory for guided missiles at Hughes Aircraft Co. (1948–1958), chairman of the board and president of Space Electronics Corp. (1958ff), consulting, NAE (1970) (AMOS Bridge MT6 WWWIA)

FLETCHER, Leonard J., 1891–1983, American agricultural engineer; power and machinery for agricultural and industrial uses, faculty and administration at Washington State Univ. (1915–1916) and Univ. of California (1916–1926), Caterpillar Co. from sales engineer to vice president (1927–1964) (AE64 WWIE(I))

FLETTNER, Anton, 1885–1961, German American engineer and inventor; to US (after WWII) established Berlin Aircraft Co. (1926), rotor aircraft (1926), aircraft control, marine rudder, helicopter, established Flettner Aircraft Co. (NYC) after WWII (MWBD)

FLIPPER, Henry Ossian, 1856–1940, Black American soldier and engineer; first Black American to graduate from USMA (1877), with US military involved in drainage of swamps, building wagon roads, installing telegraph lines, as a civilian (1882ff) involved in mining, engineering, surveying, cartography, resident engineer for a number of mining companies, including Alaska railroads and a Venezuelan oil company (ANB CBB)

FLOSDORF, Earl W., 1904–1958, American bacteriologist and inventor; with Stuart Mudd [1893–1975] developed method of dehydrating blood plasma and food and with George S. Sperti [1900–1991] further developed freeze-drying or lyphilization of juices, coffee and thinly sliced food (ANB GSA WOI WWWIA WWWIS)

FLÜGGE-LOTZ, Irmgard, 1903–1974, German American woman aeronautical engineer; developed Lotz model for calculating the lift of airplane wings (1931), automatic flight control, head of theoretical aerodynamics at Göttingen Univ. (1938–1948), to US (1948), faculty at Stanford Univ. (1948ff) (AMOS MADOI WWWIA)

FOCKE, Heinrich Karl Johann, 1890–1979, German aircraft designer; aircraft designer and builder (1908ff), monoplane (1912ff), founded Focke-Wulf Co. [Georg Wulf] (1924), developed helicopter with counter-rotating rotors (FW 61) (1936) (MWBD NYPL)

FOGARTY, Charles Franklin, 1917–1981, American mining engineer; petroleum geology and sulfur mining, exploration, served in WWII (1942–1946), geologist with Socony-Vacuum Oil Co. (1946–1950), research and executive with Texas Gulf Sulfur Co. (1952ff), NAE (1976) (AMOS MT2 WWWIA)

FOKKER, Anthony Herman Gerard, 1890–1939, German Dutch American aircraft designer and inventor; designed pursuit planes (WWI), founded manufacturing plants in Germany and Netherlands, developed Fokker D-7 (1917), invented apparatus to shoot through propeller, to US (1922), founder and president of Fokker Aircraft Corp. of America (1922) (ASAI MWBD TTOS WWWIA WWWIS)

FOLGER, Walter, 1765–1849, American inventor; expert watch and clock maker, Folger astronomic clock (1890), built scientific apparatuses, land surveyor, taught navigation, developed process for annealing wire (1812), telescope (1821), thermometer (1823), served in government (BDOAS(El) BITAS DAB PS WWWIA)

FOLLI, Francesco, 1623–1685, Italian physician and inventor; invented hygrometer (1664), early proponent of blood transfusions (NYPL TOIAD WWWIS)

FOLMER, William Frederic, 1861–1936, American inventor; gas burners, photographic equipment, formed Folmer & Schwing Co. (1887), which manufactured Speed Graphic camera, firm acquired by Eastman Kodak Co. (1905), aerial photography, over 300 patents (MWBD NC33)

FOLSOM, Richard Gilman, 1907–1996, American mechanical engineer; pumps, fluid metering, fluid flow and heat transfer, fluid mechanics, faculty and administration at Univ. of California (1933–1953), faculty and director of engineering research at Univ. of Michigan (1953–1958), president of Rensselaer Polytechnic Inst. (1958–1971), consulting engineer (1971ff), NAE (1965) (AMOS WWIE)

FONTAINE, Hippolyte, 1833–1917, French engineer; designer and builder, becoming head of Cail & Co. (1857), dynamo, electric motor, electric energy transmission (1873), founder and editor of *Revue Industrielle* (1870ff) (MWBD WWWIS)

FONTANA, Domenico, 1543–1607, Italian architect and engineer; chief architect of Vatican (1585–1592), directed moving of Egyptian obelisk from Heliopolis (originally built c. 40 AD) to Rome, domes in Vatican, aqueducts, fountains, Vatican library (1587–1590), built drainage canal in Naples, planned port improvements, built coast road (EIH GEAPIT MWBD TEIH TSOE WWWIS)

FONTANA, Mars Guy, 1910–1988, American metallurgical engineer; corrosion, oxygen and nitrogen in iron and steel, vacuum fusion, equilibrium of iron-oxygen-hydrogen, research at E. I. DuPont (1934–1945), faculty and department chairman at Ohio State Univ. (1945–1975), consulting, NAE (1967) (AMOS MT5 WWWIA WWWIS)

FÖPPL, August Otto, 1854–1924, German physicist and engineer; railroad engineering, bridge engineer, statics, strength of materials, electromagnetism, agricultural machinery and forestry (Leipzig), engineering mechanics, taught at trade school (1877–1892), Univ. of Leipzig (1892–1894), Univ. of Munich (1894–1922) (DOSB PS WWWIS)

FORBES, Robert Bennet, 1804–1889, American marine engineer and inventor; importer and exporter primarily with China, constructed many trading vessels, early use of metal hulls, screw propeller, maritime safety (EOWB WWWIA)

FORCHHEIMER, Philipp, 1852–1933, Austrian hydraulic engineer; consulted and worked for several companies, Technical Univ. of Aachen, Technical Univ. Graz, ground water flow, use of LaPlace equations for studies (MWBD WWWIS)

FORD, Hannibal Choate, 1877–1955, American mechanical and electrical engineer and inventor; design of subway systems in New York City and Philippines, invented safety devices for control of trains (1906), typewriter and office machine design, assisted Elmer A. Sperry [1860–1930] in development of gyrocompass (1909ff), formed Ford Instrument Co. to manufacture his ideas such as systems and analog computers to control guns aimed at airplanes flying over ship, company became a part of Sperry Corp. (ANB NCF)

FORD, Henry, 1863–1947, American automobile manufacturer; machinist, chief engineer of Edison Illuminating Co. (ante 1899), designed first Ford automobile (1896), racing automobiles, transmission mechanism, organizer and president of Ford Motor Co. (1903–1919) (1943–1945), standardized parts, moving assembly line production (1908), lightweight engine, model-T automobile (1910–1927), tractor manufacture (1917ff), trimotor all-metal airplane (1925ff), model-A automobile (1927–1933), V-8 engine (1932), 12-cent US postage stamp issued (1968) in his honor (AIE ANB BEST EIH H-BW HIW GI IAI IIA MMAH ME125 MWBD NC38,E NYPL 1000Y SAI SS1660 TEIH TIA WWWIA WWWIS)

FORD, Jon Baptiste, 1811–1903, American inventor and manufacturer; foundry and rolling mill, iron products for railroad, steamboat building, plate glass, inverted glass tube, founded company that became Pittsburgh Plate Glass Co. (DAB)

FOREST, Fernand, 1851–1914, French inventor; introduced internal combustion engine cooled by air, first 4-cycle 4-cylinder engine (1891), invented magneto ignition (1897), engine used in motorcycles and some airplanes and automobiles (TTOT WWWIS)

FORNEY, Matthias Nace, 1835–1908, American mechanical engineer and inventor; draftsman for Baltimore & Ohio Railroad (1856ff), designed and patented Forney engine (1866), superintendent of locomotive building (1865), author, editor of journals relating to railroads, argued against narrow gauge railroad, obtained 33 patents related to railroads (DAB MEIA NC22)

FORREST, John S., 1907–1992, British electrical engineer; research, electrical power and transmission systems, insulators on power lines, lightning protection, environmental effects, with Central Electricity Board of London (1931–1973), member of Royal Society, fNAE (1979) (Bridge MT7)

FORSYTHE, Alexander John, 1769–1843, Scottish minister and inventor; minister (1790ff), invented percussion lock for firearms (MWBD PS WA)

FORTEN, James, 1766–1842, Black American inventor; invented several sail-handling devices, particularly for sailing on rough seas, owner of sail-making business (1798), taken prisoner by British in Revolutionary War (AAI AIE BI)

FORTIN, Jean-Nicolas, 1750–1831, French inventor; scientific instruments, devised precision balance (1778), developed cup barometer (c. 1800) known as the Fortin barometer, clocks (BDOS(Wi) DOSB MWBD PS TBDOS(Po) WWWIS)

FOSTER, Charles F., 1852–1910, American mechanical engineer; rodman, leveler and transitman, water works (1872), superintendent of St. Louis Cotton Factory (MO) (1876), helped set up world's fair in Chicago (1893), chief operating engineer of Universal Exposition, St. Louis (1904) (MEIA NYT-A)

FOSTER, John Gray, 1823–1874, American soldier and engineer; commissioned in Corps of Engineers (1846–1867), served in war with Mexico, assistant engineer of Engineering Bureau in Washington (DC), coastal surveys and harbor defenses, Civil War fortifications, superintendent of engineers for river and harbor improvements, submarine operations, taught at USMA (ANB NC10 WWWIA)

FOTTINGER, Hermann, 1877–1945, German inventor and engineer; engine development, automatic transmission using hydrokinetic drive called Fottinger coupling (1905), at Technical Univ. Danzig (1909), Univ. of Berlin (1924ff) (EE WA WWWIS)

FOUCAULT, Jean-Bernard-Léon, 1819–1868, French inventor and physicist; action of pendulum (1851), gyroscope (1852), velocity of light, Foucault prism (1857), testing surfaces of telescope mirrors (1859) (GI IAI MWBD NYPL 1000Y SAI TBDOS WA WWWIS)

FOURDRINIER, Henry, 1766–1854, English papermaker and inventor; with his brother Sealy Fourdrinier [d. 1847] invented making paper using wood pulp, with the aid of Bryan Donkin [1768–1855] invented and developed a continuous papermaking process (1802, 1807) (MMAH MWBD)

FOURIER, Jean-Baptiste-Joseph, 1768–1830, French physicist and mathematician; equations developed that are widely used in heat transfer and other numerical applications called Fourier equations, law of thermal conduction (1822), at École Polytechnique (1794–1798) and other institutions (FNIE L&M MWBD WWWIS)

FOURNEYRON, Benoît, 1802–1867, French hydraulic engineer; with Franche-Compte Iron Works, first practical radial outward flow water turbine (1827), improved water wheel turbine (1855), sold throughout the world, called Fourneyron wheels (EE EIH FNIE MWBD NYPL PS TBDOS WBE WWWIS)

FOWKE, Sir George Henry, 1864–1936, British engineer; military service, becoming lieutenant general, known for tunneling under German lines in WWI (E-64 WWW)

FOWLER, George L., 1855–1926, American mechanical engineer; axle company, building excavating and traveling cranes (1879–1881), worked for A. F. Bartlett Co. (1886) building sawmill machinery and steam engines, worked for railway brake company, railway wheel company, consulting engineer (1892ff) (MEIA WWWIA)

FOWLER, Sir John, 1817–1898, English engineer; railways, railway bridges, pioneer in underground railways, engineer for London Metropolitan Railway (1853ff), consultant, advisor to Egypt, worked with Englishman Sir Benjamin Baker [1840–1907] (DNB EIH MWBD PS TGE:B WWWIS)

FOWLER, John, 1826–1864, English engineer and inventor; invented mechanical system for drainage (1847), plowed field using steam power (1854), steam-powered plow developed (1856) with Jeremiah Head, invented seed drill and reaping machine (1850–1864), founded Steam Plow Works (1862) in partnership with William Watson Hewitson (BDOTHOT DNB MMAH MWBD WWWIS)

FOWLER, Joseph William, 1894–1993, American admiral and builder; built warships, built theme parks, including Disneyland (CA) and Disney World (FL) (WA WWWIA)

FOX, Sir Charles, 1810–1874, English engineer; with his son Sir Douglas Fox [1840–1921] in similar work, construction of railways in several countries, buildings, railway switches, waterworks (MWBD TGE:B WWWIS)

FOX, Daniel Wayne, 1923–1989, American chemist and inventor; developed high-performance engineering plastics, insulation materials, pigments, in US Army Air Force in WWII (1943–1946), manager of research and product development at General Electric Co. (1953ff), NAE (1984) (AMOS Bridge MT4 WWWIA)

FOYN, Svend, 1809–1894, Norwegian inventor; invented harpoon gun for whaling (1946), began modern whaling industry (BDOS PS S&TF)

FRANCESCO DI GIORGIO MARTINI (aka Francesco Maurizio di Giorgio Martini), 1439–1502, Italian engineer and architect; furniture, military construction, fortresses, drawbridges, earthen embankments, inventor of mines of siege (1495), built churches and municipal buildings, manuscript translation and illumination (GEAPIT MWBD TAE)

FRANCIS, James Bicheno, 1815–1892, English American hydraulic engineer and inventor; draftsman (1834), manufacture of locomotives, erection of cotton mills (1837), water power facilities, chief engineer of locks and canals (1845), built Francis mixed-flow turbines (1849) (1855), flow of water through various devices, chief engineer and general manager of organization called the Proprietors of Locks and Canals, fire protection system, consulting hydraulic engineer (1855ff) (ANB BDOACE BITAS DAB EE FNIE IIA MEIA MWBD NC9TBDOS(Po) TEIH WWWIA WWWIS)

FRANCIS, Joseph, 1801–1893, American inventor and manufacturer; made unsinkable boats using cork, portable boats, made lifeboats, first hydraulic press to make ships, cargo boats, use of corrugated metal for boats (1850–1855), floating docks, surf boats, decorated for his work in US and abroad (ASAI DAB NC10 WWWIA WWWIS)

FRANK, Charles Frederick, 1911–1998, British engineer; behavior dislocation of solids, fNAE (1980) (Bridge WWW)

FRANK, Richard S., 1914–1986, American mechanical engineer; advanced product technology, lubrication equipment, heavy equipment design and manufacture, reliability of equipment, company executive at Caterpillar Co. (1936–1976), NAE (1980) (MT4)

FRANKLIN, Benjamin, 1706–1790, American inventor and physical scientist; writer, printer and publisher, conservation of electric charge, lightning and electricity (1750), kite experiment (1752), lightning rod (1752), electrical phenomena, public official, plenipotentiary to France (1778), bifocal lens for eyeglasses (1780), wrote *Poor Richard's Almanac* (1782ff), Franklin stove, street lighting (AIE ASAI BEST DAB EIH EOS&T GEAPIT GI IIA MMAH MWBD NC1 NYPL 1000Y SAI SAI-MP TBDOS TBOF TEE TGS WWWIA WWWIS)

FRANKLIN, William Buel, 1823–1903, American soldier and civil engineer; Corps of Topographical Engineers, Great Lakes survey, Mexican War (1846–1847), served in Washington (DC), where he supervised construction of iron dome over rotunda of Capitol and on other federal projects, in Civil War commanded large units, general manager of gun manufacturing company (1866ff) (ANB NC4 WWWIA)

FRASCH, Herman, 1851–1914, German American chemical engineer and inventor; petroleum industry, opened chemical laboratory (1877), developed process for refining petroleum, founded Empire Oil Co. in Canada, which later became part of Standard Oil Co., Frasch process for desulfurizing Canadian and US oils, sulfur mining (1891) (AIE ANB DAB NC19 WWWIA)

FRAUNHOFER, Josef von, 1787–1826, German physicist and inventor; worked for Optical Institute (1806ff), spectroscope, heliometer, micrometer, Fraunhofer lines for identification of materials and rays of sun (1814), diffraction grating (1814), optic spectrum relationships, made improvements in microscope and telescope (ASAI BDOS(Wi) GI MWBD SAI TBDOS WWWIS)

FRAZIER, John Earl, 1902–1985, American chemical and ceramic engineer; glass technology, use of electricity for glass melting, Owens-Illinois Co. (1924–1926), Simplex Engineering Co. (1926–1938), Frazier-Simplex Co. (1938ff), becoming secretary and president (1966), consulting engineer, over 50 patents in glass technology, NAE (1978) (AMOS MT3 WWIE WWWIA)

FREEMAN, Benjamin William, 1890–1963, American mechanical engineer, inventor and manufacturer; founded Benjamin W. Freeman Heel Co. (1913) and was general manager (ante 1916), served in WWI, joined his father's firm Louis G. Freeman Co. (1919ff) in manufacture of shoe-making machinery (1919–1962) first as designer then as president (1930), patented more than 60 devices (NC50 PS)

FREEMAN, John Ripley, 1855–1932, American civil and hydraulic engineer; fire inspector for insurance company, helped standardize fire protection apparatuses, hydraulics, city water supply study (NYC) (1899), consulting engineer for Panama Canal, US, and Chinese government, NAS (1918) (ANB BM12 MEIA NC36,C WWWIA WWWIS)

FREEMAN, Sir Ralph, 1880–1950, English civil engineer; steel structures, bridges in Australia, Victoria Falls, and five bridges in the Rhodesias, ship building, consulting firm of Freeman, Fox & Partners (1901ff) (DNB MWBD)

FREEMAN, Thomas, ??–1821, Irish American civil engineer and surveyor; one of surveyors of Washington (DC) (1794–1796), construction of Ft. Adams (1798–1800), boundary between US and Spanish Florida (1796–1811), exploration of newly acquired Louisiana territory, mapped Red River, several public buildings (1804ff), mapped borders between some states (ANB BDOAS BITAS DAB NC24 PS WWWIA)

FRENCH, Aaron, 1823–1902, American inventor; wagon builder (1845), railroad shop supervisor (1854), invented coil and elliptic railroad car springs, formed A. French Spring Co., which merged with Railway Steel Spring Co. (DAB MEIA MWBD NC24 WWWIA)

FRENCH, Orval C., 1908–1999, American agricultural engineer; development engineer with Black-Sivells & Bryson Co. (KS) (1930), faculty at Univ. of California-Davis (1931–1947), during WWII at Univ. of California Radiation Lab. (1942–1945), professor and head of agricultural engineering at Cornell Univ. (1947–1972), consulting engineer (AE Res6 WWIE)

FRESNEL, Augustin-Jean, 1788–1827, French physicist, engineer and inventor; civil engineer for French government, optics, wave nature of light, development of dioptric systems for lighthouses (1819), Fresnel lens (1820), development based on work of Comte Georges Louis Leclerc de Buffon [1707–1788] (1748) and Dominique-François-Jean Arago [1786–1853] (BEST GEAPIT NYPL TBDOS WWWIS)

FREUD, Benjamin Ball, 1884–1955, American chemical engineer; graduated as a chemist and became a chemical engineer, on faculty of Armour Inst. of Technology (IL), served in France in WWI as a captain with chemical warfare unit, faculty at Illinois Inst. of Technology (1927–1947) as department chairman, dean of engineering, then as professor (until 1952), worked on monomolecular films (ACCE PS WWWIA)

FREUDENTHAL, Alfred Martin, 1906–1977, Austrian American engineer; structural and consulting engineer in Prague, Warsaw and Tel Aviv (1930–1938) and with Palestine Transjordan British Forces (1938–1945), to US (1947), served on faculties at Hebrew Inst. of Technology (1938–1949), Univ. of Illinois (1947–1950), Columbia Univ. (1949–1969), George Washington Univ. (1969ff), NAE (1976) (MT1 WWIE WWWIA WWWIS)

FREUNDLICH, Herbert Max Finlay, 1880–1941, German American physical chemist and inventor; Freundlich isotherm (1909), Kaiser Wilhelm Inst. in

Berlin (1914–1933), Cambridge Univ. (1933–1938), to US (1938), Univ. of Minnesota (1938ff), viscosity, elasticity, introduced the term "thixotropy," developed non-drip paints (L&M RHW RSObit WWWIS)

FREYSSINET, Marie-Eugène-Léon, 1879–1962, French civil engineer; bridge and highway engineer, reinforced concrete bridges, developed pre-stressed concrete (1928) used worldwide (1938ff), with consulting firm (1919–1928) (EIH FEL MWBD)

FRICK, Abraham O., 1852–1934, American mechanical engineer; improved designs for Frick portable steam engines and boilers, developed steam traction engine (1876–1880s), improved balance slide valve, refrigerating and ice-making machines (1882), became president of Frick Co. (1904), consulting engineer, nine patents (MEIA NC24)

FRIEDRICH, Hans Rudolf, 1911–1958, German American rocket engineer; selenium rectifiers, with General Electric Co. in Germany, co-developer of V-2 missile, supervisor of electronic and industry related to V-2, after WWII to US (1946), US citizen (1955), worked on automatic control, flight mechanics, design of guided missiles, with US government at Ft. Bliss (TX) followed by Redstone Arsenal (AL) (1946–1951), with Convair of General Dynamics Co. (1951ff) (NC46 PS)

FRIESE-GREENE, William, 1855–1921, English photographer and inventor; motion photography (1889), with engineer Mortimer Evans invented motion picture camera (pat. 1890), photographic typesetting (1895), stereoscopic color motion picture making (DNB MWBD PS TBDOS TGE:B WWWIS)

FRIIS, Harald Trap, 1893–1976, Danish American electrical engineer; radio research engineer, first in Denmark then in US, at Western Electric Co. (1919) and Bell Telephone Labs, (1925) studies related to radio reception, antenna development, vacuum tube technology, ship-to-shore reception, radio-telephone transatlantic links, director of research in high frequencies and electronics (1952–1957), microwave radio links, consultant, held 31 patents (ANB)

FRISCH, Otto Robert, 1904–1979, Austrian British physicist and inventor; worked in England, Denmark and US, cloud chamber (1933), with Lise Meitner [1878–1968] (1939) found that neutron could be used to split uranium, Frisch-Peierls Memorandum (Rudolf Peierls [1907–1995]) (1940), industrial laboratory research, diffraction of atoms by crystal surfaces, invented an instrument to analyze photographs of particle paths, formed Laserscan Co. to produce instrument, Los Alamos Lab. (1943–1945), faculty at Cambridge Univ. (1947–1974) (DNB GLOSB IAI MMAH RS27 WWWIS)

FRITTS, Charles E., 19th century, American inventor; invented working photovoltaic cell using selenium (1883, 1886) called solar cell, which was developed by

Bell Labs. (1954); photoconductivity effect of selenium first reported in Great Britain by Willoughby Smith [1828–1891] (1893) and voltaic effects between selenium and various metals reported by Americans W. G. Adams and R. E. Day (1877), Alexander Becquerel [1820–1891] constructed a selenium photoelectric cell (1883) (DNB EB GI PA TPA)

FRITZ, John, 1822–1913, American mechanical engineer; rolling mill operation, blast furnace operation using anthracite coal instead of coke, rail mill, superintendent and chief engineer of Bethlehem Iron Co. (PA) (1860–1892), open-hearth furnaces, forging processes, manufacture of armor plate (1897), John Fritz Medal named in his honor (ANB BITAS DAB MEIA MWBD NC13 WWWIA)

FRIZELL, Joseph Palmer, 1832–1910, Canadian American hydraulic engineer; began hydraulic work under James B. Francis [1815–1892], fortifications on Gulf Coast during Civil War, consulting engineer, patented air compressor (1878), hydraulic investigations on head waters of the Mississippi River (BITAS DAB MWBD NC23 WWWIA)

FROCHT, Max Mark, 1894–??, Polish American mechanical engineer; faculty of Carnegie Inst. of Technology (1922–1946), research professor and director of stress analysis at Illinois Inst. of Technology (1946ff), photoelasticity, strength of materials, numerical methods (AMOS WWIE WWWIS)

FROEHLICH, Jack Edward, 1921–1967, American aeronautical engineer; high-pressure gas reactions, test pilot, US Marine Corps in WWII (1942–1946), with Jet Propulsion Lab. at California Inst. of Technology (1949–1959) for space missiles and satellites, electronic systems, fuels and lubricants, officer and executive at Collins Radio Co. (1959ff) (AMOS CB59 PS WWWIA)

FROELICH, John M., 19th century, American farmer and inventor; got Van Duze Gas and Gasoline Engine Co. (OH) to build what became recognized as the first practical gasoline-powered tractor later produced by Waterloo Gasoline Tractor Co. (1892); another reference states that the first gasoline-powered vehicle in America (1889) was the Burger tractor manufactured by Charter Engine Co. (IL), industrial-scale production began with Hart-Parr Co. (IA) (1901), with the company incorporated (1903); the first mobile tractor was developed in France and called the *Albaret locotractor* (1856) (GI IAOSTI MISAT TAT TBOF WOI)

FROLICH, Per Keyser, 1889–1977, Norwegian American chemical engineer; to US (1922), citizen (1929), research in applied chemistry at Standard Oil Co., polymer and copolymer research, research laboratory director (1929, 1933), issued approximately 75 patents (NC60 PS WWWIS)

FRONTINUS, Sextus Julius, 35–c. 103 AD, Roman engineer; military tactics, hydraulics, director of water supplies (97 AD), aqueducts, measurement of flow of water, water commissioner of Rome (EB EIH EITAW E-64 GEAPIT PS TAE TEE TEIH TSOE WWWIS)

FROUDE, William, 1810–1879, English engineer and naval architect; hydrodynamicist, bilge keels, scale model experiments with ships, Froude law of skin friction (1852), dynamometer called Froude brake (1858) for marine engines, Froude number (1869) (CEOS&T EE FNIE L&M MWBD PS TBDOS WWWIS)

FRUDDEN, Conrad Erwin, 1887–1971, American agricultural engineer; helped survey site and development for Hart-Parr Co. (C. W. Hart [1872–1937] and C. H. Parr [1868–1941]), which became a part of Oliver Corp., worked for Duda Co., which became a part of Allis-Chalmers Manufacturing Co., chief engineer of the tractor division (1929), served in WWI and WWII, at least 12 patents (AE52 GTHP NC56)

FRY, Alfred Brooks, 1860–1933, American mechanical engineer; engineer for Corliss Engine Co., inspection engineer (1886), chief engineer for US Treasury, US buildings and facilities engineering operations, served in US Navy as acting chief engineer (1898), consulting engineer (MEIA WWWIA)

FRY, Joshua, c. 1700–1754, English American civil engineer and surveyor; college teaching in natural philosophy and mathematics (1731), established boundaries of Lord Fairfax grant by King George II of England (1746), surveyed inhabited part of Virginia-Carolina boundary (1749), treaty with Indians (1752) (BDOACE BDOS DAB WWWIA WWWIS)

FRYE, John Chapman, 1912–1982, American geologist and engineer; physiography, Cenozoic geology, ground water, metallurgical and petroleum engineering, US Geological Survey and Kansas State Geological Survey (ante 1954), faculty at Univ. of Kansas (1942–1954) and Univ. of Illinois (1963–1974), executive director of Geological Society of America (1974–1982), NAE (1971) (AMOS MT2 WWWIA WWWIS)

FTELEY, Alphonse, 1837–1903, French American civil engineer; water works systems, including drainage, water supply, reservoirs, dams, aqueducts, gauging stream flow, consulting engineer (BDOACE EIH NC13 WWWIA)

FU, King-Sun, 1930–1985, Chinese American electrical engineer; network theory, communication theory, pattern recognition, image processing, control system, Taiwan Power Co., Boeing Aircraft Co., faculty at Seattle Univ., and Purdue Univ. (1960–1985), NAE (1976) (AMOS MT3 WWWIA)

FUBINI, Eugene Ghiron, 1913–1997, Italian American electrical engineer and inventor; airborne instruments, radio, electronic devices, antennas, microwave devices, Columbia Broadcasting Co. (1938–1942), at Harvard Univ. (1942–1944), US Army Air Force (1944), Airborne Instruments Lab. Co., becoming vice president (1945–1961), director of research in DOD Defense Research and Engineering (1961–1965), vice president of IBM (1965–1969), several patents, consultant (1969ff), NAE (1966) (Bridge MT10 WWIE WWWIA)

FUERTES, Estevan Antonio, 1838–1903, Puerto Rican civil engineer; public works for roads, bridges, harbors, waterworks, faculty and administration at Cornell Univ., initiated engineering laboratories (BITAS BDOACE DAB NC4 WWWIA)

FULLER, Calvin Souther, 1902–1994, American chemist and inventor; chemist at AT&T Lab. (37 yr.), development of synthetic rubber, drying oils, co-inventor of solar cell with Gerald L. Pearson and Daryl M. Chapin, development of semiconductor devices, 33 patents (AMOS WWWIS)

FULLER, George A., 20th century, American builder and inventor; built first Quonset building (1941) (pat. 1946), named after Quonset Point Naval Air Station (RI), principal manufacturer was Great Lakes Steel Corp., widely used during and after WWII for storage, services, businesses and housing (LATI(1991) OED RG)

FULLER, George Warren, 1868–1934, American sanitary engineer; chemist, studied public health and water pollution in Europe, Massachusetts State Board of Health (1890–1895), known for his work in water purification and sewage treatment, rapid-sand filter or mechanical filtration ahead of treatment, advised many cities throughout US, public works projects, an award named after him (ANB EIH NC13 WWWIA)

FULLER, Levi Knight, 1841–1896, American inventor and manufacturer; machine shop, woodworking machinery, sewing machines, Estey Organ Co. engineer and executive, governor of Vermont, about 50 patents (DAB NC8 WWWIA)

FULLER, Richard Buckminster, 1895–1983, American inventor, engineer, architect, idea person; Dymaxion house (1927), housing design, Dymaxion car (1932), world map (1943), geodesic dome (1947), world game (1969), buckyball or buckminsterfullerene (a 60-atom carbon molecule) named after him because of its geodesic dome structure, over 2,000 patents worldwide (AIE CB76 IG MWBD 1000Y PS SAI-MP S:TLAW TGS WWWIA WWWIS)

FULLER, Robert Mason, 1845–1919, American inventor, pharmacist and physician; toxology, forensic medicine, microphotography, use of photography to study dermatology, healing of wounds (DAB NC12)

FULTON, Robert, 1765–1815, American artist, engineer and inventor; power shovel (1795), canals (1796), saw for cutting marble, machine for twisting hemp into rope, navigation, submarine (1804), designed cast iron bridges, iron aqueduct, double-incline plane for raising and lowering canal boats, excavating machines, marine torpedo (1804), pioneer in building the first economically successful steamboat for transportation called the Clermont (1807) preceded by experimental steamboat (1803), first steam-powered warship (1814) (ANB ASAI BDOACE BEST BITAS DAB EIH GEAPIT GI IAI IIA LAI LOTE 1000Y MEIA MMAH MWBD NC3 NYPL PS SAI SFLOE TBDOS(Po) TEIH TGS WA WWWIA WWWIS)

FURER, Julius Augustus, 1880–1963, American naval officer and inventor; US Navy Construction Corps after graduation from USNA, assistant naval contractor for Brooklyn Navy Yard (1905), installed ship machinery and repaired ships at Charleston (1907), became industrial manager (1909), subsequent service at Philadelphia Navy Yard (1910), Honolulu (1913), submarine recovery, head of supply division of Navy Bureau (1915), designed WWI submarine chasers, coordinator of research and development for US Navy (1941) as rear admiral (ANB WWWIA)

FURLOW, Floyd Charles, 1877–1923, American mechanical engineer and inventor; consulting engineer, especially in construction, chief engineer to president of Otis Elevator Co., engineering research on electric and hydraulic elevators, steam-power engineering, machine design, 25 patents on elevators between 1906 and 1922 (DAB NC20 WWWIA)

FURNAS, Clifford Cook, 1900–1969, American chemical and metallurgical engineer; chemical and physical reactions, metallurgical processes, heat transfer, carbon combustion, reaction kinetics, worked on metallurgical processes for US Bureau of Mines (1926–1931), faculty at Yale Univ. (1931–1942), research for Curtiss-Wright Corp. (NY) (1942–1946), faculty and vice president at Cornell Univ. and director of Cornell Aeronautical Lab. (1943–1954), chancellor at Univ. of Buffalo (1954–1966) except for two years to serve as assistant secretary of defense, impact of science on society, consulting (1966ff), NAE (1967) (AMOS E-64 MT1 NC56 WWWIA WWWIS)

G

GABOR, Dennis, 1900–1979, Hungarian British electrical engineer, physicist and inventor; invented holography (1947), stereoscopic cinema (based on Soviet engineer and inventor Ivanov) (1939), color television tube of greatly reduced depth (1958), high-speed oscillograph, communications, laser beam, over 100 inventions, Nobel Prize (1971), Imperial College of Science and Technology in London (1949–1961), CBS Labs. (CT) (1967ff), fNAS (1973) (BEOS CB72 CDOS EAI GI GLOSB MMAH MWBD TBDOS(Po) PS RHW WBE WWWIS)

GABRIEL, Jacques II, 1667–1742, French engineer; first engineer in engineer corps of French Army, built several public buildings, inspector general of public construction, major bridge of his works—the Bridge of Bois—was destroyed (GEAPIT TSOE)

GABRIELLI, Giuseppe, 1903–1987, Italian aeronautical and mechanical engineer; faculty at Polytechnic Inst. of Turin (1927–1973), while also being chief designer at Piaggio Co. in Genoa, involved with aircraft design, metal airplane structures, flying boat made of aluminum alloy, fNAE (1983) (MT4)

GAEDE, Wolfgang, 1878–1945, German inventor; Gaede pump, mercury vapor diffusion pump (1913) (EE)

GAGARIN, Yuri Alekseyevich, 1934–1968, Soviet engineer and astronaut; pilot, first human to orbit the earth (1961) on Soviet spacecraft *Vostok I*, elected to Supreme Soviet (1962), died in plane crash while training for *Soyuz* 3 (1968) (CB62 EOWB GI MMAH MWBD PS RHW TBOF)

GAGGE, Adolf Pharo, 1908–1993, American physicist and aeronautics; basic principles of air conditioning, bioastronautics, bioengineering of heat transfer in humans, environmental physiology, Pierce Lab. associated with Yale Univ. (1933–1941), with USAAF Aero Medical Lab. (OH) (1941–1950) as colonel, Yale School of Medicine (1963–1976) and Pierce Lab. (1963–1968), consultant, NAE (1979) (AMOS Bridge MT7)

GAGNEBIN, Albert P., 1909–1999, American mechanical engineer; co-invented Ductil iron, invention of process for making ductile iron, development of spheroidal cast iron, nickel, research and executive of International Nickel Co. (1930–1974), serving in several offices and becoming president (1967) and chairman (1972), NAE (1974) (Bridge WWIE)

GAILLARD, David Du Boise, 1859–1913, American military officer; dredging and excavation of Panama Canal (1907), excavation at Culebra cut through the Panama continental divide (later named after Gaillard) of the Panama Canal (1908–1913) (AmSci81 ANB DAB MWBD WBE)

GAILLARD, Peter, 19th century, American inventor; first US patent on a mowing machine (1812) (S&TF)

GAIUS, Plinius Secundus (aka Pliny the Elder), 23–79 AD, Roman military officer; served as military commander in Africa and Germany, procurator in Spain (c. 72), known for works in history, writing, natural science and military tactics, died as result of eruption of Mt. Vesuvius (EITAW MWBD)

GALILEO (usual name), often referenced as Galilei, 1564–1642, Italian mathematician and inventor; swinging pendulum principle (1581), hydrostatic balance (c. 1586), irrigation machine (1594), thermometer called thermoscope (1607), pendulum clock (1641), hydrostatic balance, suction pump theories used by Evangelista Torricelli [1608–1647], proportional compass (1606), constructed astronomical telescope (1608), an improvement over others, studied stresses on beams, earth-sun relationship, sunspots, professor at Pisa (1589–1591) and Padua (1592–1610), under house arrest (1631ff) (EIH GEAPIT HIW HOFL 1000Y MMAH MWBD RHW SAI SAI-MP TAE TEIH TGS WA WBE WP WWWIS)

GALIN, Ledv Aleksandrovich, 1912–??, Soviet mechanical engineer; mechanics, theory of elasticity, elastic-plastic problems in unsettled filtration of liquids, first with USSR Academy Inst. of Mechanics and then faculty at Moscow Univ. (1965ff) (PS SMOS)

GALLAGHER, Richard H., 1927–1997, American civil engineer; design of aircraft and aerospace structures, finite element theory and use, research, development and design methods at Bell Aerospace Corp. (1955–1967), faculty and administration at Cornell Univ. (1967–1978), faculty and dean of engineering at Univ. of Arizona (1978–1984), faculty at Worcester Polytechnic Inst., president of Clarkson Univ. (1988ff), NAE (1983) (Bridge WWIE)

GALLOWAY, Donald F., 1913–1996, English consulting engineer; theory and practice of ship berthing, mooring and maritime structural design, Dunlop Rubber Co. (1937–1939), director of research for Inst. of Production Engineers (1930–1946), organized and founded Production Engineering Research Associates (1946–1978), consulting engineer, fNAE (1984) (Bridge MT10)

GALLY, Merritt, 1838–1916, American clergyman and inventor; engraver, printer, patented platen job-printing press (1860) sold under name of Universal, composing machine for linotype (1872), firearms manufacturer for Colt Firearms Co. (ante 1886) (Samuel Colt [1814–1862]), over 50 patents (DAB MEIA MWBD NC4 PS)

GALTON, Francis, 1822–1911, English scientist and inventor; invented silent dog whistle, weather map, teletype printer, fingerprinter, designed several instruments to plot meteorological data (MWBD NYPL RHW WWWIS)

GALVANI, Luigi, 1737–1798, Italian (Bologna) physiologist and inventor; galvanic electricity in frog leg (1780ff), work led to Alessandre Volta's [1745–1804] invention of electric cell, name associated with galvanizing and galvanometer, faculty at Univ. of Bologna (1768–1798) and Inst. of Science—Bologna (1782–1798) (E-64 EIH FNIE RHW 1000Y WBE WWWIS)

GAMBEY, Henri Prudence, 1787–1847, French instrument maker; at École des Arts et Métiers, established own shop in Paris for manufacturing precision instruments for physicists and government, built first cathetometer for Pierre-L. Dulong [1785–1838] and Alexis Petit [1791–1820], a heliostat for Augustin Fres-

nel [1788–1827], and an improved compass for Charles A. Coulomb [1736–1806] (CDOSB DOSB MWBD PS RHW WWWIS)

GAMBLE, James Norris, 1836–1932, American chemist and inventor; son of William Gamble, who with Alexander Procter [1817–1874] established Procter and Gamble Co. (P&G) (OH), invented Ivory soap (1878), was vice president of P&G Co. (1890ff), built business on use of animal remains of slaughtering industry (ANB NC20,25 NYTObit TIA)

GANGUILLET, Émile Oscar, 1818–1894, Swiss engineer; worked with German engineer Wilhelm Kutter [1818–1888] in developing their formula for open channel and stream flow (CFR DOSB NUC)

GANTT, Henry Laurence, 1861–1919, American mechanical engineer; scientific methods of industrial management, periodically worked with Frederick W. Taylor [1865–1915], established office as consulting engineer (1902), also known for worker management and leadership development (DAB NUC TEIH WWWIS)

GARAND, John Cantius, 1888–1974, Canadian American engineer and inventor; toolmaker, light machine gun, US Ordnance (1919–1953), US citizen (1920), developed semi-automatic rifle (M-1) 0.30 cal. (1929) adopted by US Army (1936) to replace the Springfield rifle (adopted 1903) and the American Enfield rifle, used in WWI (ANB EB MWBD WBE WWWIS)

GARDINER, James Terry, 1842–1912, American topographical engineer and inventor; civil engineering works, Brooklyn Water Works (NYC), inspector with US Ordnance Corps (1861–1862), earthworks around San Francisco harbor, survey of California, US Geological Survey of 40th parallel (1867–1873), survey of territories (1873–1875), director of New York State Survey (1876–1886), consulting engineer, director and president of street railroad, coal companies and related enterprises (ANB BDOAS(El) BITAS PS)

GARDINER, Lion, 1599–1663, British military engineer; as a colonist in North America built fortifications, defenses for settlements, fought native Indians (CT, MA), later lived peacefully with Indians on Long Island (DAB)

GARDINER, William Cecil, 1904–??, Canadian American chemical engineer; US citizen (1939), Mathieson Chemical Corp. then Olin Mathieson Corp. (1929ff), electrochemical engineer, several patents dealing with chlorine metallurgy, mercury cells, brine treatment, metal brine (NCJ)

GARNERIN, André-Jacques, 1769–1823, French aeronaut and physicist; worked with his brother Jean-Baptiste Olivier Garnerin [1766–1849] in development of parachute, invented and personally demonstrated parachute (1797), made jump of 8,000 ft. (1800) (EB GI MWBD)

GARRETT, Joseph, 1857–1910, English American mechanical engineer; to US (1879), foreman and pattern maker for O. A. Pray & Co., manufacturers of flour and sawmill machinery (1879ff), later chief engineer and partner of Twin City Iron Works (MN), company absorbed by Minneapolis Steel Machinery Co. (1902), of which he later became director (MEIA)

GARRETT, William, 1843–1903, Welsh American engineer and inventor; worked in and then supervised rolling mill in Scotland (1854–1868), to US (1868), where he was foreman at Cleveland Rolling Mill Co., then superintendent of rod mill at American Wire Co. (OH), head of Garrett-Cromwell Engineering Co., several patents on iron and steel products (MEIA)

GARRY, Frederick W., 1921–1993, American mechanical engineer; design and development of gas turbine engines, aircraft, corporate management of engineering and manufacturing at General Electric Co. (1951–1974) (1981), president of Rohr Industries (1974–1980), transportation, NAE (1982) (Bridge MT7 WWIE)

GASCHE, Fred, 1903–1966, German American mechanical engineer and inventor; engineer with several companies, manager and chief engineer with company making naval auxiliary equipment, founded Autoclave Engineers, Inc. (1945), where he was chairman of the board and president, company devoted to design and manufacturing high-pressure equipment and systems, held several patents (NC53 PS)

GASCOIGNE, William, 1612–1644, English optician and inventor; invented micrometer (1638), introduced cross-hairs to telescope, invented instruments to measure small angles (CDOSB DOSB WA WWWIS)

GASKILL, Harvey Freeman, 1845–1889, American engineer and inventor; invented revolving hay rake (1858), manufactured a patent clock at Penfield, Martin & Gaskill Co. (1866), developed planing mill and sash and blind factory, invented horse-drawn hay rake, developed pumping machinery for waterworks (1882), steam pumps, developed Gaskill engine for Holly Manufacturing Co. (NY) (1873ff), official of Holly Co., including vice president (1885ff) (DAB MEIA NC23)

GATLING, Richard Jordan, 1818–1903, American inventor; screw propeller for ships (1839), agricultural implements for sowing cotton seeds (1844), machine for thinning cotton, hemp-breaking machine (1850), steam plow (1857), hand-cranked rapid-fire machine gun named Gatling (1862), adopted by US Army, marine steam ram (1862), motor-driven plow (1900) (ANB ASAI BEST DAB EAI MEIA MMAH MWBD NYPL NYTObit PS RHW WA WBE WWWIS)

GAUDI, Antonio (aka Antonio Gaudi y Cornet), 1852–1926, Spanish architect and builder; skilled use of masonry, thin-shell construction for houses, parks, churches (MWBD WBE)

GAUDIN, Antoine Marc, 1900–1974, Turkish American metallurgist and mineral engineer; rock and ore crushing, mineral dressing, adsorption, flotation, leaching of ores, radioactive tracers, faculty at Columbia Univ., Univ. of Utah, Colorado School of Mines (1929–1939), Massachusetts Inst. of Technology (1939ff), consulting metallurgist, NAE (1964) (AMOS MHSE MT1 NC58 WWWIA)

GAUGER, Alfred William, 1892–1963, American chemical engineer and inventor; chemist with Great Western Sugar Co. (1914–1915), Bureau of Mines (PA) (1915–1917), captain in chemical warfare during WWI, research with several chemical companies, director of division of mines and mining experiments at Univ. of North Dakota (1926–1931), patented several processes for recovery of saline from brines, faculty at Pennsylvania State Univ. (1931–1952) working on fuel technology and mineral industries, consultant (1952ff) including universities in South America (NC50)

GAUSS, Johann Carl Friedrich, 1777–1855, German mathematician and inventor; work led to many engineering uses of mathematics, theory of numbers, normal distribution curve, optics, errors, crystallography, mechanics, capillarity, measured magnetic phenomenon (1832), magnetic unit gauss named after him, with Wilhelm Edward Weber [1804–1891] invented electromagnetic telegraph (1833), director of Göttingen Observatory (1807ff) (BEST FNIE GEAPIT GI MWBD NYPL 1000Y SAT-MP RHW TGS WBE WWWIS)

GAUTHEY, Emiland-Marie, 1732–1806, French civil engineer; design and building canals, particularly for connecting Loire and Saône Rivers (France) (1783–1789), built churches (1773–1791), streamlined piers, bridges characterized by design decorations (BDOS(Wi) GEAPIT LDS MWBD MWGD TSOE)

GAUTIER, Henri, 1660–1737, French doctor and engineer; Languedoc provincial engineer (1688–1716), design, building and inspection of bridges and highways (1716ff), worked with brother Hubert Gautier, author of books describing foundations, scaffolding, arches, machines and floods, books led to structural mechanics as a specialty (BDOS(Wi) GEAPIT LDS MWBD TSOE)

GAUTIER, Paul Ferdinand, 1842–1909, French instrument maker; astronomical instruments, including reflecting telescopes, equatorial telescopes, transits (1876ff), astrographs, reflectors (1887), equipment used in leading observatories in Europe, Algeria, Argentina and Brazil (DOSB PS)

GAUTREAUX, Marcelian F., Jr., 1930–1994, American chemical engineer; development of chemical processes for production of synthetic linear alcohols, olefins and others, industry research and development and president of Ethyl Corp. (1951–1955) (1958ff), faculty at Louisiana State Univ. (1955–1958), NAE (1977) (Bridge MT8 WWIE)

GAUVIN, William Henry, 1913–1994, Canadian engineer and inventor; new technologies, specialist in spray drying and plasma processing, heat and mass transfer, fluid mechanics, invented atomized suspension technique (AST) to pulverize waste from pulp and paper, president of technology company named after him, advanced from manager to director of Quebec Research Inst., faculty at McGill Univ. (1947–1952), founded and was director of research and development of Noranda Research Centre (1961–1983), fNAE(1987) (Bridge MT8 WWIE)

GAYLEY, James, 1855–1920, American metallurgist, engineer and inventor; cooling of blast furnaces, dry-air blast (1894), management of operation of blast furnaces, bronze cooling-plate for furnace walls (1891), casting apparatus (1896), vice president of US Steel (1901–1909), which absorbed Carnegie Steel Co. (1897), numerous patents (AIE ANB BITAS DAB MWBD)

GAY-LUSSAC, Joseph-Louis, 1778–1850, French chemist and physicist; studied upper atmosphere (1804), composition of water (1805), law of combining volumes (1809), improved commercial processes for making sulfuric acid, fermentation (1811–1815), professor at Sorbonne (1808–1832), Jardin des Plantes (1832ff), followed by Henri-Victor Regnault [1810–1878] at École Polytechnique (ANB BEST DOSB MWBD RHW TGE WBE WWWIS)

GED, William, 1690–1749, Scottish goldsmith; invented stereotyping (1725) molding a whole page of type to make a printing plate, process developed and patented by Scott Alexander Tilloch [1759–1825] and his partner printer Andrew Foulis (DOSB MWBD WWWIS)

GEDDES, James, 1763–1838, American civil engineer and lawyer; surveyor, one of principal engineers for Erie Canal and other canals, ran survey from Great Lakes to Hudson River (1808) along with Charles Bro(a)dhead [1772–??] and Benjamin Wright [1770–1842], called by some the first American engineer (ANB ASAI BDOACE DAB MWBD TEIH)

GEIGER, Johannes Hans Wilhelm, 1882–1945, German physicist and inventor; device for measuring radiation (1908), alpha particles, artificial radiation, cosmic rays, Geiger counter for alpha particles (1913), a more sensitive radiation Geiger counter for alpha particles developed with Walther Müller (1928), Geiger-Nuttall rule (John M. Nuttall [1890–1958]) (CEOS&T EDOP HIW MWBD PA RHW TTOS WA WWWIS)

GEISSLER, Johann Heinrich Wilhelm, 1815–1879, German inventor; glass blower, manufacture and sale of scientific instruments in his shop (c. 1852ff), air pump without moving parts (1855), Geissler vacuum tubes (BEST DOSB MWBD PS WWWIS)

GEIST, Jacob M., 1921–1991, American chemical engineer; cryogenics, industrial gases, distillation, heat transfer membranes, safety, US Army in WWII, with Air Products & Chemicals (1955–1982), had own company, Geist Tech (1982ff), for consulting engineering, NAE (1980) (MT6 WWIE)

GÉRARD, Pierre Simon, 1765–1836, French hydraulic engineer; strength of wood as structural material (1790), studied Nile River surface elevation and bed characteristics, director of Paris water supply (ante 1831), ship canal to serve Paris (1813–1820), recovered and initiated use of studies by Antoine de Chézy [1718–1798] (DOSB PS WWWIS)

GERBER, Heinz Joseph, 1924–1996, Austrian American aeronautical engineer and inventor; computer automation company, president and chairman of Gerber Scientific and subsidiaries (1948–1996), imaginative inventor, laboratory instruments and supplies, NAE (1982), holder of 677 US and foreign patents (Bridge MT10 WWWIA)

GERHARDT, Karl Jakob Christian Adolf, 1833–1902, Alsatian physician, chemist and inventor; synthesized acetylsalicylic acid (1853), named "aspirin," which was produced by Felix Hoffmann [1868–1946] of Bayer AG in Germany (1897) and commercially sold in powder form (1899) and marketed extensively by Bayer (1905) (EB TBOF WWWIS)

GERMER, Edmund, 1901–1987, German American inventor; fluorescent light, high-pressure mercury vapor, safety of lighting (NYT-A)

GERNSBACK, Hugo, 1884–1967, American inventor and publisher; improved dry battery, radio and electronic devices, over 80 patents (MWBD)

GERSTNER, Franz Anton, Ritter von, 1796–1840, Czech civil engineer; developed railroads for internal improvement of Czechoslovakia, mechanics publication (1829), project for railroad in Russia from St. Petersburg to Moscow, which was later built by American engineers Whistler and McNeill (1834), studied construction, management and fiscal state of US railroads and canals (BDOACE)

GEYER, John Charles, 1906–1995, American civil engineer; sanitary engineering, water filtration, nuclear wastes, ground water, runoff, cooling water, US Navy in WWII (1943–1946), faculty at Univ. of North Carolina (1934–1937), faculty and administration at Johns Hopkins Univ. (1937–1943) (1946–1976), World Health Organization (WHO) (1954–1955), NAE (1970) (Bridge MT9 WWIA WWIE)

GHERARDI, Bancroft, 1873–1941, American mechanical engineer and inventor; telephone engineer with metropolitan system in Manhattan, developed system of communication wires to permit underground cables, developed system of handling toll cable and designed expandable switching equipment, engineer and assistant chief engineer of telephone company, invented interoffice lamp-indicator system for multiplex telephone circuits (1915–1918), signaling systems, grounding systems, transcontinental telephone system, NAS (1933) (ANB BM30 PS)

GHIBERTI, Lorenzo, c. 1381–1455, Italian goldsmith, architect and builder; known for a variety of works of sculpture, goldsmith (1409), painter (1423), stonemason (1427), commissioned to redo the doors of castle of Carlo Malatesta (1403, 1407), east doors of Baptistry of Florence (1425), many bronze works (EOBD)

GIAMBATTISTA—see PORTA

GIBB, Sir Claude Dixon, 1898–1959, English Australian engineer and inventor; firm of C. A. Parsons (1924ff), chief engineer (1929), general manager (1937), joint managing director (1943), Ministry of Supply (1940–1945), formed company and collaborated in first designs for gas-cooled nuclear power plants (DNB PS RS5)

GIBBON, John Heysham, 1903–1973, American surgeon and inventor; invented heart-lung machine (1953), NAS (1972) (BM53 NYPL RHW)

GIBBS, George, 1861–1940, American mechanical engineer and inventor; superintendent of meters for electric generation station, chemist, materials testing laboratory for railroad company (1885) as well as in charge of machinery and rolling stock, steam heating to eliminate coal stoves, fire extinguisher for railroad cars (1888), shields and reflectors for incandescent lamps (1889, 1891), manufacture of industrial electric motors with brother Lucius Gibbs, consultant (ANB NCE,12 PS WWWIA WWWIAH-S&T)

GIBBS, James Ethan Allen, 1829–1902, American inventor; designed and built economically unsuccessful wool carding machine, shuttle sewing machine (1856), lock stitch sewing machine (1857), twisted loop rotary hook machine (1857), formed Willcox & Gibbs (James Willcox) sewing machine company (1859), served in Confederate Army as lieutenant in ordnance, sewing machine was manufactured by Brown and Sharpe Manufacturing Co. (RI), clutch-driven bicycle, 25 patents for improvement of sewing machines (DAB MEIA NC19 WWWIAH-S&T WWWIS)

GIBBS, Josiah Willard, 1839–1903, American physicist and engineer; first doctor of philosophy degree in mechanical engineering in US (1863) with dissertation covering teeth on spur gears at Yale Univ. (CT), thermodynamics, Gibbs adsorption, phase rule (1875), adsorption theorem (1878), energy equilibrium relationships, faculty at Yale Univ. (1871–1903), NAS (1879) (ASAI BEST BITAS BM6 CEOS&T DOT NC4 1000Y TBDOS RHW TGS WBE WBEOS WWWIS)

GIBBS, William Francis, 1886–1967, American naval architect and marine engineer; ship designs and ship building, Gibbs Brothers, Inc. (1922), then Gibbs & Cox

(1929), served in government during WWII, designed Liberty ship (WWII), NAS (1949), NAE (1965) (BM42 MHSE MT1 MWBD NC53 PS WWWIA WWWIS)

GIESE, Henry, 1890–1981, American agricultural engineer; background in architectural engineering, taught in public school (1912–1916), served in WWI, Iowa engineering and extension service (1916–1923), faculty in farm structures and mechanics at Iowa State Univ. (1923–1966), originated Midwest Plan Service (1929), design of animal structures, economics of farm structures (AE63 WWIE WWWIA)

GIFFARD, Henri, 1825–1882, French aeronautical engineer and inventor; steerable balloon flight, first practical steam engine-powered airship (1852), water injector for steam boiler (1858) (BDOS EAI EE GI MMAH MWBD TBDOS PS WBE WWWIS)

GIFFORD, Richard P., 1922–1976, American electrical engineer; communications, including mobile, microwave and carrier systems, radar, US Navy in WWII (1943–1946), manager of several laboratories dealing with communication systems at General Electric Co. (1946ff), NAE (1973) (MT1 WWIA)

GILBERT, Rufus Henry, 1832–1885, American physician and inventor; skilled mechanic, medical service in Civil War, elevated railway system (NYC and Chicago) called el (1870), Gilbert Railway Co. (1872) (DAB MWBD WBE)

GILBRETH, Frank Bunker, 1868–1924, American engineer; efficiency studies and work expert, time and motion study, therblig for work units (reverse spelling of his name), contractor (1895), consulting firm (1911), well known for book *Applied Motion Study* (1917), worked with his wife, Lillian E. (Moller) Gilbreth [1878–1972], who continued the work (HIW MWBD NC26 1000Y WBE WWWIS)

GILBRETH, Lillian Evelyn (or Moller), 1878–1972, American engineer; work efficiency expert, time and motion study, fatigue study, took over business on work efficiency when husband Frank B. Gilbreth [1868–1924] died, NAE (1965) (MADOI MT1 MWBD NWS TEIH WBE WCBE WWWIS)

GILCHRIST, Percy Carlyle, 1851–1935, British metallurgist and inventor; with Sidney G. Thomas [1850–1885] developed use of phosphorus-bearing ores in steel making (1875ff), developed steel-making process and Thomas steel, removal of phosphorus to make steel, work led to basic Bessemer process, FRS (1891) (BDOS MMAH PS RHW RSObit TBDOS WIS(2) WWWIS)

GILKESON, Robert Fairbairn, 1917–1993, American electrical engineer; development and commercial application of advanced steam cycles, nuclear power generation, research and development manager, chief executive officer and chairman of Philadelphia Electric

Power Co., except for service in WWII in US Army, NAE (1978) (Bridge MT7 WWIA)

GILLESPIE, William Mitchell, 1816–1868, American civil engineer; land surveying, road building, educator, proponent of liberal education for engineers (DAB NC23)

GILLETTE, King Camp, 1855–1932, American inventor and businessman; invented safety razor (1895, pat. 1901) and disposable blades (1903), organized and was president of Gillette Safety Razor Co. (1901–1932) (AIE ANB GI MWBD NC10 1000Y PA RHW WA WWWIS)

GILLILAND, Edwin R., 1909–1973, American chemical engineer; fractional distillation, research and development of rubber (NDRC), faculty and department head at Massachusetts Inst. of Technology (1934ff), government advisor, NAS (1948), NAE (1965) (BM49 MT1 WWWIA)

GILLMORE, Quincy Adams, 1825–1888, American military engineer; considerable service as officer of northern troops in Civil War, artillery use, testing of wrought iron cannon, chair of Mississippi River Commission (1879), retired as major general (1865) (ANB BDOACE DAB)

GINSBURG, Charles P., 1920–1992, American electrical engineer and inventor; industry research and development, engineer for radio stations (1942–1952), research and advanced development, becoming corporate officer with Ampex Corp. (1952ff), recording and transmitting radio signals, video magnetic recording, wideband data recording, NAE (1973) (Bridge MT7 WWIE)

GINZTON, Edward Leonard, 1915–1998, Soviet (Ukrainian) American engineer, physicist and inventor; to US (1929), microwave tube development, linear electron accelerator, circuit development, radar, research engineer with Sperry Gyroscope Co. (1940–1946), faculty at Stanford Univ. (1946–1968), including director of the microwave laboratory (1949–1959), Varian Assoc. (1959–1972), including CEO, president and chairman of the executive committee, NAE (1965), NAS (1966) (AMOS MHSE MT10 PS WWWIA)

GIORDANA, Friar, 13th century, Pisan inventor; eyeglasses (c. 1286), indexed under Friar by some authors (TAE)

GIORGI, Giovanni, 1871–1950, Italian electrical and civil engineer and physicist; steam-generated electrical traction, innovations in urban trolley systems, hydroelectric installations, electric power network distribution, director of Technology Office for Rome (1906–1923), professor and administrator at Univ. of Rome (1913–1939), instituted systemization of electrical units (1903), Giorgi International System led to international system of units based on meter-kilogram-second-ampere units (CEOS&T DOSB)

GIRARD, Philippe Henri de, 1775–1845, French mechanical engineer and inventor; improved method for spinning flax (1810), supervised mechanized spinning factory in Warsaw, desk lamp, achromatic lens using liquid, equipment for extracting and evaporating beet juice, machine for manufacture of rifle stocks (GEAPIT PS WWWIS)

GIRAUD, André Y., 1925–1997, French engineer; energy technology, development of commercial nuclear breeder reactors, administrator general of French Atomic Energy Commission (1970–1978), French minister of defense (1986–1988), consultant, fNAE (1977) (Bridge MT10)

GLASER, Edward L., 1929–1990, American computer engineer; time-shared computer systems, software and hardware, computer-aided design, with several companies, including International Business Machine (1951–1955), Burroughs Corp. (1955–1963), faculty at Massachusetts Inst. of Technology (1963–1967), faculty and administration at Case Western Reserve Univ. (1967–1975), Systems Development Corp. (1975–1982), with Ray Sanders founder and served as chief technical officer of Nucleus International Corp. (1982ff), NAE (1977) (Bridge MT7)

GLENNAN, Thomas Keith, 1905–1995, American electrical engineer; talking motion pictures, operations and studio manager for Paramount (5 yr.) and Samuel Goldwyn Studios (2 yr.), Atomic Energy Commission (AEC), director of US Navy underwater sound laboratory (1942–1945), president of Case Inst. of Technology (OH) (1947–1966), first administrator of NASA (previously NACA) (1958–1961), NAE (1967) (IWAP MT8 WWWIA)

GLIDDEN, Carlos, 1834–1877, American inventor; shared in inventing typewriter (1868) with co-inventors Christopher L. Sholes [1819–1890] and Samuel W. Soule [19th century] with rights sold to Remington Arms Co. (1873), which became the basis of the Remington typewriter, developed spading machine for agricultural use (MWBD WBE)

GLIDDEN, Joseph Farwell, 1813–1906, American farmer, rancher and inventor; first commercially successful barbed wire for fencing, process of attaching steel points to wire (1873, 1874), with Isaac L. Ellwood [1833–1910] formed Barb Fence Co. (IL) (1874) then sold interests (1876) (AIE ANB ASAI DAB FFF GI MWBD NC23 1000Y WA WBE WWWIA WWWIS)

GLUSHKO, Valentin Petrovich, 1908–??, Soviet power engineer; numerous aspects of power engineering, USSR Academy of Sciences (1953), becoming academician (1958) (SMOS)

GODDARD, Calvin Luther, 1822–1895, American inventor; taught school in NYC (1845), method of cleansing wool, designed and patented burring picker, several patents on process, organized company to manufacture equipment (NYC) (1862) (DAB MEIA NC41)

GODDARD, Robert Hutchings, 1882–1945, American physicist and inventor; taught at Princeton Univ. (1911) then Clark Univ. (1914ff), developed theory of rocket action, rocketry with solid fuel (1914), power rockets used in space exploration (1919), liquid-fueled rockets (1926), reaction motor, rocket motors for jet-assisted takeoff of airplanes, instruments and camera carried by rockets (1929), 214 patents, 8-cent US postage stamp (1964) (ANB DAB E-64 EAI EOWB GI HOCS&T I&T16 ITCOF MMAH NC35 NYT-A 1000Y RHW SAI S:TLAW TGS WA WBE WWWIS)

GODDU, Louis, 1837–1919, Canadian American inventor; to US (1858), developed shoe-making and sewing machinery, McKay Metallic Association (MA) (1856), nailing shoe soles by machine, use of tacks, nails and wire with first machine (1876), oil burner for power plants, obtained about 300 patents, of which 137 were for shoe production (DAB MEIA NC13)

GODEFROY, Maximilian, fl. 1806–1824, French American military engineer; building design and construction, devised fortifications for Baltimore, built battle monument (DAB WWWIA WWWIS)

GODWIN-AUSTEN, Henry Haversham, 1834–1923, English geologist and surveyor; topographical drawing, trigonometrical survey in India, mapped mountains, including portions of Himalayan Mountains, his name given to mountain in Kashmir (K2), retired (1877) after service in India (DNB WBE WWWIS)

GOETHALS, George Washington, 1858–1928, American civil engineer; army officer, superintendent of Panama Canal construction and locks design and construction (1907ff), dams at Muscle Shoals (TN), governor of Panama Canal Zone (1914–1916), consulting engineer, NAS (1914) (AmSci81 ANB BDOACE EIH FEL MWBD NC24 TGS WWWIS)

GOLAND, Martin, 1919–1997, American mechanical engineer; thermodynamics, aerodynamics, applied mechanics, faculty at Cornell Univ. and with Curtiss Wright Corp. (1942–1946), Midwest Research Inst. (MO) (1946–1955), Southwest Research Institute (TX) (1955–1997), becoming president (1959), NAE (1967) (Bridge MT9 WWIA WWIE)

GOLAY, Marcel J. E., 1902–1989, Swiss American electrical engineer and physicist; Bell Labs. (US) (1924), US Army Signal Corps Lab. (25 yr.), Perkin-Elmer Corp. (1955–1961), Technical Univ. of Eindhaven (1961–1963), code systems for computers, computerization of analytical instruments, infrared detection, over 58 patents (LOTCS)

GOLDBERG, Rube (aka Reuben Lucius), 1883–1970, American engineer and cartoonist; city engineer for San Francisco (CA), newspaper work, cartoonist who de-picted inventions to achieve simple tasks using complex mechanical devices (1921–1964), Pulitzer Prize (1948) (ET22 WBE MWBD SciAm278)

GOLDMARK, Henry, 1857–1941, American civil engineer; construction of railroad bridges, metallurgy, advised on use of metals to strengthen bridges for heavier loads, design engineer for building, steel dam, concrete arch over Rock Creek River (DC) (1897), with US Board of Engineers for deep waterways, locks and canals, Panama Canal, consulting engineer (1914–1928) (ANB DAB NC38)

GOLDMARK, Peter Carl, 1906–1977, Hungarian American electrical engineer; color television (1940), miniature television, 33 1/3 rpm phonograph record (long playing) (1948) (RCA), engineer (CBS), recording broadcasting and reception, NAE (1967), NAS (1972) (AIE AMOS ANB BM55 MT1 MWBD NC60 NYPL 1000Y PS WA WBE WWIA WWWIS)

GOLDTHWAIT, Abel G., 1837–1907, American mechanical engineer and inventor; mowing machine development (1861), pattern maker, draftsman and designer, advancing to chief engineer in shops of William H. Tolhurst (1863ff), assisted in design of Hartshorn window shade roller, designed paper-box machines, Magee hot-air furnace, laundry machine, including Tolhurst hydro-extractor, milling apparatus, first successful design of universal railroad car coupler (MEIA)

GOLUBTSOV, Vyacheslav Alekseevich, 1894–??, Soviet power engineer; built and operated a series of power stations, chief engineer of power plants (1925–1934), professor (1944ff), USSR Academy of Sciences (1953), emphasis on water preparation for boilers and water softening, air preheating, deaeration, dust control, utilization of fuels in power stations (1955ff) (SMOS)

GOOCH, Sir Daniel, 1st baronet, 1816–1889, English engineer and inventor; locomotive construction, stationary link motion (1843), superintendent of Great Western Railway (1837–1864), recognized as one of greatest locomotive designers, laying telegraph cables, first transatlantic cable (1866) (EAI EE MWBD WWWIS)

GOOD, John, 1841–1908, Irish American inventor; ropemaking, machinist, machinery to replace hand-making of rope, patented breaker, nipper, spreaders, regulators for rope-making, John Good Cordage & Machine Co. (1893–1898), organized John Good & Jennings Patent Machine Cordage Co. (1898), over 100 patents (DAB MEIA)

GOODING, Robert Carpenter, 1918–1999, American naval officer and engineer; maintenance and logistics for navy, technical director of shipbuilding and repair, US Navy, advancing to vice admiral (1941–1972) at various command posts, Naval Sea Systems Command (1974), Polaris/Poseidon weapon systems, NAE (1976) (Bridge WWIA)

GOODRICH, Benjamin Franklin, 1841–1888, American industrialist; rubber goods manufacturer, formed B. F. Goodrich Co. (OH) (1870) (1880), company produced rubber tires for tractors (1931), educated as a physician (ANB CDOAB DAB MWBD NC28 TOIAD WBE WWWIA)

GOODRICH, Simon, 1773–1847, British engineer; designed arrangement for emptying naval locks (EE)

GOODWIN, Hannibal Williston, 1822–1900, American clergyman and inventor; flexible photographic film (1887), communications, motion pictures (DAB MWBD WBE)

GOODYEAR, Charles, 1800–1880, American inventor; successful hardware firm (1821–1836), developed rubber coating (1837), use of sulfur and heat to form vulcanized rubber (1839) (1844), welt shoe sewing machine, belting, hoses, company made automotive tires named after him, which led to numerous rubber products, had about 60 patents (AIE ANB ASAI BEST BITAS CDOAB DAB EAI GI I&T15 MMAH NC3 NYT-A 1000Y PS RHW SAI-MP TBDOS TGS WA WWWIA WWWIS)

GOOGIN, John Milton, 1922–1994, American chemist, technologist and inventor; separation of isotopes, uranium isotopic enrichment, lithium processing, purification of metals by solvent extraction and ion exchange, metallurgy, ceramics, Tennessee Eastman Corp. (1944–1947), Union Carbide Corp. (1947ff) advancing to executive positions, NAE (1988) (AMS Bridge MT8 WWIA)

GORBACHEV, Timofei Fedorovich, 1900–??, Soviet mining engineer; worked in coal industry, chief engineer for group of coal enterprises, USSR Academy of Sciences (1958) (SMOS)

GORDON, George Phineas, 1810–1878*, American printer and inventor; improved presses for card printing called Yankee job press (ante 1851), followed by invention and manufacture (RI, NJ) of the Turnover press, the Firefly press (1854), and the Franklin press (1858), subsequently known as the Gordon press, over 50 patents for invention of presses (*DAB MEIA WWWIS)

GORGAS, William Crawford, 1854–1920, American army officer and physician; US Army (1880–1898) medical corps, sanitary office in Cuba (1898–1902), Panama Canal Commission (1904–1913), became surgeon general with the rank of major general (1915), NAS (1914) (EIH MWBD NC32 WWWIA)

GORINOV, Aleksandr Vasil'evich, 1902–??, Soviet railway engineer; surveying new railroads, chief construction engineer, professor at several locations, scientific research organizations, USSR Academy of Sciences (1939), work regarding complex design of railroads, inertia calculations, use of train kinetic energy, classification of railroads (SMOS)

GORNOWSKI, Edward John, 1918–1983, American chemical engineer; catalytic cracking, reforming, coal gasification, fluid mechanics, hydroformer, faculty at Villanova Univ. and Univ. of Pennsylvania (1939–1942), Standard Oil Co. (1942–1945), research and administration at Esso Research & Development Co. (1945ff), NAE (1971) (AMOS MT2)

GORRIE, John, 1803–1855, American physician and inventor; expansion system refrigeration cycle (installed for space cooling in 1944), patented 1851, used for ice making, credited with first US patent on mechanical refrigeration, invented artificial eye (AIE ANB DAB MWBD NC15 1000Y WBE)

GORYACHKIN, Vasilii Prokhorovich, 1868–??, Russian agricultural engineer; founder of agricultural mechanics, construction of agricultural machinery, first major publication on moldboard plow (1898), followed by books on threshing, grading, dryers, and harvesters, teacher at the Agricultural Inst. of Moscow (1894ff) machine testing station, instrumental in founding All-Union Inst. devoted to mechanization and electrification of agriculture (from book edited by N. D. Luchinskii (1968))

GOSS, Floyd L., 1907–1980, American electrical engineer; electric power transmission line vibration, regional and national power transmission and power supply planning, reliability, design and operation, engineer and manager for city of Los Angeles (CA) (1933–1972), NAE (1979) (MT2)

GOSS, William Freeman Myrick, 1859–1928, American mechanical engineer; taught practical mechanics and experimental engineering at Purdue Univ. (1879–1907), including dean of engineering (1900–1907), then dean of engineering at Univ. of Illinois (1907–1913) (1915–1917), president of Railroad Car Manufacturer's Assoc. (1917–1925) (MEIA WWWIA)

GOSSET, William Sealey, 1876–1937, British industrial scientist; statistics, analysis of normal distribution, errors, sampling, analysis of variance, with Guiness firm (1899–1937) manufacturing of ale first in Dublin then in London (RHW TBDOS(Po))

GOTAAS, Harold Benedict, 1906–1977, American civil and sanitary engineer; bridge designer, chief draftsman, sanitary engineer in several positions, US Army in WWII (1942–1946), faculty at Univ. of North Carolina (1937–1942), Univ. of California (1946–1957), faculty and dean of institute at Northwestern Univ. (1957ff), numerous activities in South America with World Health Organization, consulting engineer, NAE (1967) (AMOS MT1)

GOUGH, Herbert John, 1890–1965, English engineer and inventor; metal fatigue, National Physical Lab. (England) (1914–1938), serving as superintendent (1930ff), fatigue and plastic deformation, design of metal structures to avoid fatigue, chief engineer of Unilever, Ltd. (1945–1955) (DNB PS RS13)

GOULD, James P., 1923–1998, English American consulting engineer; application of geotechnical theory to design and construction of underground works and foundations, worked on embankment dams with US Bureau of Reclamation (1950–1953), partner in Muesser Rutledge Consulting Engineers (1953–1994), consulting engineer (1994ff), NAE (1988) (Bridge MT9 WWIA)

GOURDINE, Meredith Charles, 1929–1998, Black American physicist, engineer and inventor; research in electrogas dynamics, acoustic imaging, electrostatic precipitation, development of devices, aerospace industries with Jet Propulsion Lab., Plasmadyne Corp., chief scientist of Curtiss Wright aerodynamics division, founder and president of Gourdine Systems, Inc. (1964–1973), president of Energy Innovations (1974ff), 70 patents, NAE (1991) (AAFIS&T AAI Bridge DAAS MT9 S:TLAW WWWIA)

GOZZO, Jacques de—see FARRARE

GRAFF, Frederic, 1774–1847, American civil engineer; chief engineer, city water department, city reservoir for Philadelphia, established park on Schuylkill River, designer and manufacturer of water works equipment for Worthington Co., consulting engineer (BDO-ACE DAB)

GRAHAM, Charles Kinnaird, 1824–1889, American civil engineer; surveyor, construction engineer for Brooklyn navy yard, Union officer as brigadier general (1862) in Civil War, engineer on several projects in New York City (DAB)

GRAHAM, George, 1673–1751, British mechanician and instrumentation; worked with English clockmaker Thomas Tompion [1643–1713], invented mercurial pendulum, built first orrery (mechanical representation of solar system), produced astronomical instruments, improved clock mechanism (1715) (DOSB MMAH MWBD PS WWWIS)

GRAHAM, Jackson, 1915–1985, American civil engineer and military officer; served with corps of engineers combat units in WWII in ETO, commanded two engineer aviation groups in Korean conflict, with Army Corps of Engineers (31 yr.), district engineer in Portland (OR), in charge of all civil and military construction for 14 states as brigadier general, retired as major general (1967), first general manager of the Metro Transit Authority of Washington (DC) (1967–1976) (Oregon State Univ. press release W Post (1975) W Post (1985) (WWIA)

GRAHAM, James Duncan, 1790–1865, American topographical engineer; served in military Corps of Topographical Engineers, US-Texas border (1839–1840), commissioner of survey for Maine border (1840–1843), head of scientific corps and principal US astronomer for demarcation of border between US and British prov-

inces, resurvey of Mason-Dixon line (1849–1850), US-Mexican border (1850–1851), Great Lakes and Atlantic harbor improvements and construction (1854–1865) (ANB BDOAS(El) BITAS DAB PS)

GRAMME, Zénobe Théophile, 1826–1901, French Belgian electrical engineer; invented direct current (DC) ring dynamo (generator) (1870), commercially successful DC electric motor (1873), with French engineer Hippolyte Fontaine [1833–1917] (1878) developed electric alternator (AIE E–64 EIH MMAH MWBD NYPL PS WA WBE WWWIS)

GRANGER, John V. N., 1918–1997, American electrical engineer; communications entrepreneur, National Defense Research Council (1944–1945), Stanford Research Inst. (1949–1956), National Science Foundation (1975–1977), US Department of State (1971–1975) (1977–1981), consulting with Granger Associates (1956–1970), NAE (1975) (Bridge MT10 WWIE)

GRANT, Eugene Lodewick, 1897–1996, American civil engineer; statistical quality control, economics of engineering depreciation, industrial engineering, water resources, US Navy in WWII, faculty at Montana State Univ. (1920–1930), Stanford Univ. (1930–1956), consulting, NAE (1987) (AMOS Bridge)

GRANT, George Barnard, 1849–1917, American mechanical engineer and inventor; developed Grant's Difference machine for calculations (1872) (1876), gear-cutting business, established gear works in Boston, Philadelphia, and Cleveland), considered one of the founders of gear-cutting industry in US (DAB MEIA)

GRASHOF, Franz, 1826–1893, German mechanical engineer; application of mathematics to engineering, Grashof number (1875), fundamentals of elasticity, served in navy, faculty of Royal Industrial Inst. (1854), director of Standards Bureau (1856ff), Technical Univ. Karlsruhe (1863–1893) (CEOS&T DOSB L&M WWWIS)

GRASSELLI, Eugène Ramiro, 1810–1882, French American chemist and chemical manufacturer; to US (1837), formed E. Grasselli & Son (1873), which became Grasselli Chemical Co. (OH) (1885), son Caesar Augustin Grasselli [1850–1927] became president, developer of instruments, manufacture of sulfuric acid, soda ash, artificial alkali, product uniformity (ACCE BITAS PS WWWIA WWWIS)

GRATIOT, Charles, 1788–1855, American civil engineer; army service in engineers, engineering duty in Missouri territory (1807–1808), construction of defenses at Charleston (1808–1818), chief engineer under Gen. William H. Harrison [1773–1841], Ft. Meig (OH) defense (1813), fortification on Delaware River and later Hampton Roads (VA) (1815), chief engineer in charge of Bureau of Engineering (DC), brevet brigadier general (1818) (BDOACE)

GRAY, Elisha, 1835–1901, American inventor; inventions in telegraphy, contested Alexander G. Bell's claims as inventor of telephone, invented several communication aids and devices such as harmonic telegraph for simultaneous transmission of messages, with Eno M. Barton [1844–1916] formed Gray and Barton Co. (1872–1880), which developed into Western Electric Co., faculty at Oberlin College (OH) (1880–1891) (AIE ANB BITAS DAB EIH IAI MWBD NC4,30 PS WW-WIS)

GRAY, J. McFarlane, 1832–1908, English mechanical engineer; steam steering gear in steamship (1867) called Gray hunting gear (EE)

GRAY, John Edmund, 1922–1997, American nuclear and chemical engineer; nuclear materials production, nuclear power programs, environmental safeguards, formulation of national energy policy, numerous positions beginning with Westinghouse Electric Corp. (1943–1946), General Electric Co., US Navy, atomic power station with Duquesne Light Co. (PA) (1954–1960), president of NUS Co. (MD) (1960–1972), Edison Electric Inst. (NYC) (1974–1977), chairman and CEO of International Energy Assoc. Ltd. (DC) (1975–1985), NAE (1992) (Bridge MT10)

GRAY, Roy Burton, 1884–1975, American agricultural engineer; farm machinery and equipment for tillage and harvesting, author of popular book *The Agricultural Tractor,* with International Harvester Co. (1910–1920), faculty and head of department at Univ. of Idaho (1921–1925), US Department of Agriculture (1925–1954) (AE35,56 NUC)

GRAY, Stephen, 1666–1736, English electrical experimenter; conduction of electricity in glass tube (1729), work resulted in identifying some materials as conductors and others as insulators, as described further by French English physicist John Théophile Desaguliers [1683–1744] (BEST EB EIH MWBD WBE WWWIS)

GREEN, George, 1793–1841, English mathematician; formulated mathematical relationships of electricity and magnetism (1828), Green formula (1828), coined term "potential," hydrodynamics, introduced Green functions, which showed how volume integrals could be reduced to surface integrals (DOSB DOT EDOP MWBD WWWIS)

GREENE, Charles Ezra, 1842–1903, American civil engineer; assistant engineer of railroad, assistant engineer on federal and harbor improvements (ME, NH), city engineer, faculty and administration at Univ. of Michigan, his son Albert Emerson Greene followed him as chairman of department at Univ. of Michigan, consulting engineer (BDOAC BITAS DAB NC26)

GREENE, Francis Vinton, 1850–1921, American civil engineer; son of George Sears Greene, Sr. [1801–1899], served in artillery and Corps of Engineers, made study of Russo-Turkish War, public works in Washington (DC), president of Baker Asphalt Paving Co. (1886), police commissioner of New York City (1903) (DAB)

GREENE, George Sears, 1801–1899, American civil engineer; instructor at USMA (1823–1827), construction of railroads in northeast US and Maryland, mining (1836–1856), aqueduct construction (NYC), water supply systems, including reservoir and piping, brevet major general in US Volunteers (1865–1866), consulting engineer, numerous projects in water supply, chief engineer of public works in Washington (DC) (1871–1872) (ANB BDOACE DAB)

GREENE, George Sears, 1837–1922, American civil engineer; son of George Sears Greene [1801–1899], assistant engineer of Croton Aqueduct, topographical surveys in New York (Westchester Co. and Long Island), improved surveying instruments, engineer-in-chief of New York City docks (1875–1897), sea wall construction, underwater use of concrete, consulting engineer (DAB NC1 WWWIA)

GREENER, William, 1806–1869, English gunmaker and inventor; self-expanding bullet for muzzle rifles (1836) (MWBD)

GREENWOOD, Henry Burgess, 1897–1961, American electrical engineer, inventor and manufacturer; with F. X. Hooper Co., later Koppers Co. (MD) (1918ff) as draftsman, chief engineer, vice president (1918–1945), manufacturer of corrugating machinery, with his brothers formed Greenwood Engineering Co. (1945ff) for manufacture and installation of oil burners in addition to manufacture of bags and printing machinery, 20 US patents (NC49 PS)

GREGG, John Robert, 1867–1948, Irish American inventor and publisher; shorthand system of writing and taking notes known as the Gregg System (1888) (AIE ANB MWBD)

GREGORY XIII, Pope (Ugo Buoncompagni), 1502–1585, Italian pope; pope of Roman Catholic Church (1572–1585), introduced Gregorian calendar (1582), instituted colleges and seminaries under Jesuits (GI MWBD)

GREGORY (GREGORIE), James, 1638–1675, Scottish mathematician and inventor; earliest form of reflecting telescope (1663), number theory, geometric figures, areas and volumes, professor at Univ. of Saint Andrews (1669–1674) and Edinburgh Univ. (1674–1675) (MWBD WWWIS)

GREIST, John Milton, 1850–1906, American manufacturer and inventor; sewing machine salesman (1864ff), organized Greist Manufacturing Co. (1870ff), sold some of patents to Singer Manufacturing Co (1883), managed attachment division of Singer, held approximately 50 patents on sewing machine attachments (DAB MEIA)

GRESLEY, Nigel, 1876–1941, British engineer and inventor; steam locomotive designer, superintendent of carriage works, invented Gresley conjugated valve gear (1922), chief engineer of railway companies (EAI)

GRIDLEY, Richard, 1711–1796, American military engineer; built breastworks in Revolutionary War and fortification for Boston harbor, first for British then for Colonial Army, chief engineer and colonel of artillery with rank of major general, engineer general (1777–1780) (ANB BDOACE DAB MWBD)

GRIFFIN, Eugene, 1855–1907, American electrical engineer and manufacturer; in military service as civil engineer (1855–1907) and instructor, electricity as a motive power for railroads (1888), assistant to engineer (DC) (1886–1888), with electric railroad industry as executive, president of Thomson-Houston Electric Co. (1889), which became consolidated with General Electric Co. (1892) (DAB NC2)

GRIFFIN, Robert Stanislaus, 1857–1933, American mechanical and naval engineer; Bureau of Steam Engineering (1885), fleet engineer of North Atlantic Fleet (1904), chief of Bureau of Engineering (1913), rear admiral (1916), service in WWI, consulting engineer (1921ff) (MEIA)

GRIFFIN, Roger Castle, 1883–1956, American chemical engineer; with Arthur D. Little, Inc. (1909–1949), where he served as analyst, laboratory director, treasurer and vice president, specialized in procedures for testing particularly for wood and wood products (ACCE AMOS PS)

GRIFFITH, Alan Arnold, 1893–1963, English aeroengineer and inventor; graduated in mechanical engineering, heat flow between metal and gases, collaborated with Geoffrey I. Taylor [1886–1975], theory of rupture, propellers and turbine blades, gas turbines as engines for aircraft, research engineer at Rolls-Royce, vertical takeoff (1957–1960), use of jet engines (DNB PS RS WWWIS)

GRIFFITHS, John Willis, 1809–1882, American naval architect; clipper ship design, with the first being the Rainbow (1845), design of steamships, bilge keel, armored gunboat called the *Pawnee* (1858), twin and triple screw propellers (ASAI MWBD WBE)

GRIGOLYUK, Eduard Ivanovich, 1923–??, Soviet mechanical engineer; construction, theory of shells, elasticity and plasticity, mechanics, USSR Academy of Sciences (1958) (SMOS)

GRINNELL, Frederick, 1836–1905, American civil and mechanical engineer and inventor; industrialist with New Jersey Locomotive Works, Burlington & Missouri Railroad (1855–1860), treasurer and superintendent of Corliss Steam Engine Works (1860), founded by George H. Corliss [1817–1888], locomotive building as superintendent and general manager with Atlantic and Great Western Railroad (1865–1869), controlling interest in fire-extinguishing apparatus built by Providence Steam and Gas Co. (RI), Grinnell automatic sprinkler (1881), invented several types of sprinklers and fire alarm systems, approximately 40 patents (DAB MWBD WWWIS)

GRINTER, Linton E., 1902–1993, American civil engineer; engineer and designer for Standard Oil Co. of Indiana (1923–1928), faculty at Texas A&M Univ. (1928–1937), at Illinois Inst. of Technology (1937–1952), including dean of graduate school, at Univ. of Florida (1952–1970), including dean of graduate school, leader in engineering education (ACE WWIA WWIE)

GRISSOM, Virgil Ivan, 1926–1967, American mechanical engineer and astronaut; made suborbital space flight (1961) with Alan B. Shepard [b. 1923], Grissom was killed in space flight training accident along with Edward H. White [1930–1967] and Roger B. Chaffee [1935–1967], served in Korean War as a combat pilot, test pilot for the Air Force (MMAH MWBD NYT-A RHW WBE WWWIA)

GROPIUS, Walter Adolph, 1883–1969, German American architect, inventor and builder; influenced by the production philosophy of Frederick W. Taylor [1856–1915] and Henry Ford [1863–1947] as applied to construction, founder and director of Bauhaus in Germany (1919–1928), to England (1928), to US (1937), he greatly influenced engineering design, was considered an innovator in design and construction of buildings, leader in custom-built houses with variety, faculty and administration at Harvard Univ. (1938–1952), consulting (1953ff) (AG MWBD NC63,N TP WWWIA)

GROSS, Eric Taras Benjamin, 1892–1988, Austrian American electrical engineer; circuit analysis, grounding electrical systems, non-conventional energy conversion, voltage protective devices, industry in Austria and England, to US (1939), faculty at Cornell Univ. (1943–1945), Illinois Inst. of Technology (1945–1962), faculty and administration at Rensselaer Polytechnic Inst. (1962–1973), NAE (1978) (AMOS MT4 WWWIA)

GROVE, Sir George, 1820–1900, English engineer and writer; designer and builder of lighthouses in Jamaica and Bermuda (MWBD)

GROVE, Sir William Robert, 1811–1896, British physicist; first useful fuel cell developed (1839), Grove gas voltaic battery (1839), invented first demonstration on dissociation of water, originated theory of material convertibility (1846), professor at London Inst. (1841–1846), judge (1871ff) (CDOB DOSB MWBD GI NYPL WWWIS)

GROVES, Leslie Richard, Jr., 1896–1970, American army officer and engineer; managed Manhattan atomic bomb project during WWII, retired as lieutenant general (1948) (ANB MWBD NC56 WBE WWWIS)

GRUBENMANN, Hans Ulrich, 1709–1783, and GR-UBENMANN, Johannes, 1707–1771, Swiss builders; brothers who were carpenters and who built three covered timbered bridges, area churches, incorporated arch in bridge design (1766) (EIH FEL MWBD WWWIS)

GRUMMAN, Leroy Randle, 1895–1982, American industrialist; built aircraft, fighter-bombers, developed wing retraction system, retractable landing gear, founded Grumman Aircraft Engineering Corp. (1929–1966), built military and civilian aircraft, trucks, trailers and boats (MWBD WBE)

GRUSON, Hermann August Jacques, 1821–1895, German industrialist; shipyard, chill-casting, railway equipment, armor plate, turrets (MWBD)

GUERICKE, Otto von, 1602–1686, German physicist, engineer and inventor; engineer in army (1631–1635), invented air pump (1650, 1654), air vacuum pump (c. 1654), machine to generate static electricity (1663), barometer to forecast weather (1660), electroluminescence (1672) (CDOS E–64 EE EIH GEAPIT GI MMAH MWBD NYPL RHW TGS WA)

GUGGENHEIM, Daniel, 1856–1930, American industrialist; son of Meyer Guggenheim [1828–1905], developed copper industry and expanded into gold and tin mines, developed rubber plantations, numerous philanthropic activities (ANB CDOAB DAB MWBD NC22,34 WWWIA)

GUGGENHEIM, Meyer, 1828–1905, Swiss American industrialist; to US (1847), copper refining, formed Philadelphia Smelting and Refining Co. (1888), then combined other operations with help of seven sons to form American Smelting and Refining Co. (1901) (ANB DAB MWBD NC12)

GUGLIELMINI, Domenico, 1655–1710, Italian mathematician and hydrologist; stream flow measurements and analysis, resistance to water flow in streams, founded Italian School of Hydraulics (MWBD WWWIS)

GUILLAUME, Charles Édouard, 1861–1938, Swiss French physicist and inventor; with International Bureau of Weights and Measures (1883–1936), becoming director (1915), known for development of standards for meter, kilogram and liter, expansion of metals, invention of nickel-steel alloys, nickel-chromium-steel alloys, Nobel Prize (1920) (BEOS BEST MWBD TGA WWWIS)

GUILLEMIN, Ernst Adolph, 1898–??, American electrical engineer; communications, electrical networks, network analysis and synthesis, faculty at Massachusetts Inst. of Technology (1922ff), including work in radiation laboratory (1940–1944) (AMOS MSAE PS WWIE WWWIS)

GUILLOTIN, Joseph-Ignace, 1738–1814, French physician and inventor; a proponent of capital punishment, inventor of modern device for beheading people (1792), device and procedure named after him called guillotine, politician (EB GI MWBD WWWIS)

GUINAND, Pierre L., c. 1744–1824, Swiss inventor; invented method of making improved optical glass, bell foundry, related techniques of stirring process for making bells to making improved glass, worked for German manufacturer making lenses (1805–1814) (BDOS(Wi) PS WWWIS)

GULDBERG, Cato Maximillian, 1836–1902, Norwegian chemist and mathematician; law of mass action (1864) in collaboration with Peter Waage [1835–1900] called Guldberg-Waage law, reported on importance of concentration in chemical reactions, formulated Guldberg law of boiling point (1890), professor at Univ. of Christiana, now Univ. of Oslo (1869ff) (BEST CDOS(Mi) DOT L&M MWBD WWWIS)

GULDNER, Hugo, 1866–1926, German inventor; built early German Diesel engine, built four-stroke engine (1903) (EE)

GUNN, James Newton, 1867–1927, American industrial engineer; Library Bureau of Boston (c. 1890), developed use of commercial card indexes, tab type and vertical file, business systemization, production engineering, organized consulting firm (1901), helped several diverse industries improve businesses and productivity, the latest being Lockwood, Greene & Co. (DAB)

GUNNISON, John Williams, 1812–1853, American military engineer; first was artillery then ordnance officer (1838–1839), topographical engineering in army, survey in Georgia and northeast US (1840–1849), surveyed western route across country (ANB DAB)

GUNTER, Edmund, 1581–1626, English mathematician and inventor; devised Gunter chain (c. 1620), with a chain as 1/10 furlong (for linear measurements in which 10 sq. chains equal 1 acre and 80 chains equal 1 mile), quadrant (1606), invented surveying instruments, introduced terms of "cosine" and "cotangent," first successful analog device (1620), graphical logarithm scale, which was forerunner of slide rule (DOSB EIH GI HIW LDS MWBD RHW TSOE WWWIS)

GURNERY, Sir Goldsworthy, 1793–1875, British inventor; invented oxyhydrogen blowpipe, high-pressure steam jet, Drummond light, six-wheeled steam-powered carriage (1830s), boiler with wrought iron pipes bent in shape of hairpin (1826) (EE EIH GEAPIT MWBD WBE)

GUTENBERG, Beno, 1889–1960, German American seismologist and engineer; professor at Frankfurt (1926–1930), to US (1930), citizen (1936), California Inst. of Technology (1930–1957), earthquake waves, earthquake energies, postulated core in center of earth called Gutenberg continuity (1913), NAS (1945) (BEST BM76 IAI MWBD WWWIS)

GUTENBERG, Johannes Gensfleisch (aka Johannes Gens-fleisch zur Laden), c. 1398–1468, German printer and inventor; with business partner Johann Fust [c. 1400–1460] developed movable metal type printing (1436), reused type, known for printing 300 copies of the Gutenberg Bible (1456ff**), of which 47 known copies survive, man of the millennium* (BDOS(Wi) BDOTHOT BEST CEOS&T EAI EB GEAPIT GI MMAH MWBD NYPL *1000Y PS **RHW SAI-MP TBDOS(Po) TGS WBE)

GUTHRIE, Alfred, 1805–1882, American civil engineer; hydraulic works for Illinois and Michigan Canal (1848), studied explosions, did study and got legislation passed to reduce accidents (1852), consulting engineer (DAB)

GUY, Sir Henry Lewis, 1887–1956, British mechanical engineer; engineer for centrifugal pumps and turbocompressors at Westinghouse Co. (1915), chief engineer of mechanical department at Metropolitan-Vickers Ltd. (1918–1941), research and invention directed to improvement of steam power, research and development of armaments, aircraft, gun design, ordnance, chemical warfare, numerous government and professional activities, helped define mechanical engineering research (DSIR), RS (1936) (DNB PS RS WWW)

GZOWSKI, Casimer Stanislaus, 1813–1898, Russian American civil engineer; considerable practice of engineering in Canada, commissioned in engineering branch of Imperial Guard (1830), assistant engineer on Pennsylvania and Erie Canal, construction of New York and Erie Railroad from Montreal to Portland (ME), consulting engineer, established Toronto Rolling Mills to manufacture railroad equipment (1857), harbor improvements for Montreal (BDOACE)

H

HAASEN, Peter, 1927–1993, German materials science engineer; metal physics, physical metallurgist, plastic deformation of solids, fiber reinforcement, dislocation of effects of high pressure, fracturing, ductile brittle transition fatigue, crystallization, Haasen-Kelly effect, at Georg August Univ. of Göttingen (1959–1992), fNAE (1981) (Bridge MT7 WWWIS)

HABER, Fritz, 1868–1934, Polish German chemist, chemical engineer and inventor; industrial process for producing ammonia named Haber process (1909), with Carl Bosch [1874–1940] developed Haber-Bosch process, commercial cracking of oil, technical gas reactions, electrochemistry, demonstrated that oxidation and reduction occur at electrodes, explosives from nitric aid, gas masks, at Karlsruhe Technische Hochschule (1896–1911), director of Kaiser Wilhelm Inst. for physical chemistry in Berlin, Nobel Prize (1918), fNAS (1932),

left Germany (1933) (EOWB LOTCS MWBD RHW TBDOS(Po) TGS WBE WWWIS)

HABER, Fritz, ??–1988, German American scientist and inventor; to US after WWII as a part of operation paperclip, worked on aero-medical problems, first at Randolph Air Force Base (TX), left US government service (1954), to Avco Lycoming Co., where he helped develop gas turbine engine (I&T15)

HACKMUTH, Karl Henry, 1904–1966, American chemical engineer and inventor; chemist and assistant engineer with Grand Rapids Gas Light Co. (MI) (1920–1925), research, development, manufacture, and management with Phillips Petroleum Co. (OK), primarily involved with hydrocarbons, held more than 20 patents (NC23)

HADLEY, John, 1682–1744, English mathematician and inventor; optical instrumentation, first reflecting telescope (1719–1721) that had been introduced by Isaac Newton [1642–1727] (1668) then further developed by Hadley (1719), improved reflecting quadrant called Hadley quadrant (1730) (BEOS DOSB MWBD PS WBE WWWIS)

HADRIAN (aka Publius Aelius Hadrianus), 76–138, Roman emperor and builder (called mighty builder); reigned 117–138, builder of large structures, architectural monuments, rebuilt Pantheon, built villas, built Hadrian's Wall in Great Britain (c. 125), one of his engineer architects was Decrianus (AIn EB MWBD TAE TEIH TSOE)

HAFSTAD, Lawrence Randolph, 1904–1993, American physicist and electrical engineer; radio wave propagation, high voltage, geophysics, ordnance, nuclear reactor development, participant in several government activities, developed million-volt electron tube (with Merle A. Tuve [1901–1982] and Ole J. Dahl [b. 1933]), military ordnance, radio proximity fuse in WWII, industry research and administration as vice president of General Motors Corp. (15 yr.), NAE (1968) (AMOS Bridge MT7 WWIE)

HAGEN, Gotthilf Heinrich Ludwig, 1797–1884, German hydraulic engineer; hydraulics, laminar and turbulent flow, Hagen-Poiseuille law (1839, 1840) of flow in circular pipe, which was independently developed by Jean-Léonard-Marie Poiseuille [1799–1869], dikes, harbors, fortifications (CEOS&T HOHT L&M MWBD WWWIS)

HAGGERTY, Patrick Eugene, 1914–1980, American electrical engineer; electronics, astronautics, defense issues, industrial research administration and executive at Texas Instrument Co. (1945ff), becoming president (1958) then CEO (1967ff), NAE (1965) (MT2 WWIA)

HAGUE, James Duncan, 1836–1908, American geologist and mining engineer; geologist for US Geological Survey of fortieth parallel with Clarence Rivers King

[1842–1901], mining, consulting engineer (BDOAS(El) BITAS DAB)

HAHN, Otto, 1878–1968, German chemist and inventor; Kaiser Wilhelm Inst. (1912–1944), becoming director (1928) then with its successor, the Max Planck Inst. in Berlin (1946–1960), with Fritz Strassman [1902–1980] developed use of uranium atom for release of nuclear energy (1938), Nobel Prize (1944) (BEST IAI LOTCS MMAH RHW SAI TGS WWWIS)

HAIDER, Michael Lawrence, 1904–1986, American chemical and petroleum engineer; petroleum production at Carter Oil Co., chief engineer, engineering research and manager for Standard Oil Co. and affiliates (1929ff), president of International Petroleum Co. Ltd. (1954–1964), CEO of Exxon (1965ff), NAE (1964) (MT5 WWIA WWIE)

HAISH, Jacob, 1826–1926, German American contractor, inventor and manufacturer; to US (1836), farmer, building contractor, lumber business, invented S-barbed wire (1875), built barbed wire manufacturing machines, worked with Isaac Leonard Ellwood [1833–1910], sold interests to Washborn & Moen Manufacturing Co. (MA) (1876), manufactured barbed wire, plain wire, nails, staples, woven wire fence and agricultural implements, legal battle over patent rights (1876–1892) with Joseph F. Glidden [1813–1906] and Isaac L. Ellwood [1833–1910], over 20 patents (ANB DAB WA WWWIA WWWIA-S&T)

HALDANE, John Scott, 1860–1936, Scottish engineer; work related to medicine, mechanics of respiration, caisson disease, mine ventilation, industrial diseases, explosions, toxicity of carbon dioxide, director of Mining Research Lab. (1913–1928), affiliated with Birmingham Univ. (1921ff), fNAS (1935) (MWBD RHW WWWIS)

HALDANE, Thomas Graeme Nelson, 20th century, British engineer and inventor; invented first heat pump (1927) based on concepts credited to William Thomson (aka Lord Kelvin [1824–1907]), demonstrated practical use of heat pump for heating and cooling using vapor compression cycle (GSA IAI S&TF WOI)

HALE, George Ellery, 1868–1938, American astronomer and inventor; invented spectroheliograph (1899), established largest telescope at that time at Mt. Wilson (CA) (1917), NAS (1902) (ASAI BM21 RHW)

HALE, William Jay, 1876–1955, American chemist and inventor; patented processes for manufacture of acetic acid, phenol, and aniline and their derivatives, known as the father of chemurgy (MWBD WBE)

HALL, Albert Carruthers, 1914–1992, American electrical engineer; servomechanisms, spacecraft, boosters, national security, research administrator, faculty at Massachusetts Inst. of Technology (1939–1950), Bendix Aviation Corp. (1950–1957), Martin Co. (1958–1963), Martin-Marietta Corp. (1965–1971), research and development at Department of Defense (1971ff), NAE (1970) (Bridge MT7 WWIA)

HALL, Albert F., 1845–1907, American mechanical engineer and inventor; with George F. Blake Manufacturing Co. (1868ff), patented several devices related to steam engines, designed large pumps, 10 patents (MEIA)

HALL, Charles Martin, 1863–1914, American chemist and inventor; electrolytic process of making aluminum (1886) (also independently by French metallurgist Paul L.-T.-Héroult [1863–1914]), by adding cryolite to aluminum found that the cost of processing was greatly decreased (1914), electric arc furnace, work led to formation of Pittsburgh Reduction Co., which became Aluminum Company of America (Alcoa), of which he was vice president (1890ff) (AIE ANB BEST BITAS EAI MMAH MWBD NC13 RHW WWWIA WWWIS)

HALL, Chester Moor, 1703–1771, English scientist and inventor; invented achromatic lens (1729), achromatic telescope (1733) (MWBD WWWIS)

HALL, Edwin Herbert, 1855–1938, physicist and inventor; voltage versus magnetic field around a conductor (1879) known as Hall effect, work led to additional findings by others such as quantum Hall effect, thermoelectric phenomena in metals, faculty at Harvard Univ. (1888–1921), NAS (1911) (BM21 DOT EDOP L&M MWBD NC39)

HALL, John H., 1788–1841, American inventor; gunmaker, made flintlock pistols for US government at Harpers Ferry armory (WV), built machine tools to provide interchangeable parts (IIA PS WOI)

HALL, Lloyd August, 1894–1971, Black American chemist and inventor; during WWI worked on explosives, developed improved methods of preserving meat and food produce, antioxidants for food industry (1961), Chicago Department of Health Labs. (1916–1919), John Morrell & Co. (IA) (1919–1921), president and director of Chemical Products Corp. (IL) (1922), chief chemist and director of research at Boyer Griffin Labs. (1922–1959), 105 US and foreign patents (AAI BI BPOSAI CBB WWWIA)

HALL, Samuel, 1781–1863, English engineer and inventor; hot gassing of lace fibers (1817, 1823), surface condenser for steam marine engine boilers called Hall condenser (1834, 1838), which led to tubular condensers for cooling purposes, 20 patents (DNB EE EIH GI MWBD)

HALL, Thomas, 1834–1911, American inventor and patent attorney; mechanics, writing machine (1867), a pioneer typewriting invention, one-keyed typing machine (1880) called Hall typewriter (1881), invented improvements to sewing machines (DAB MEIA)

HALL, Thomas Seavey, 1827–1880, American manufacturer and inventor; textile industry (ante 1866), woolen manufacturing, invented system of signals to prevent railroad accidents (1867) also applied to drawbridges (1871) and highway crossings (1879), numerous patents (ANB DAB WA)

HALL, Wilfred McGregor, 1894–1986, American civil engineer; consulting work on civil engineering projects, hydroelectric projects, utility components, member of consulting engineering firms, manager and officer of Charles T. Main Co. (1916–1917) (1920–1922) (1941ff), becoming president and CEO (1953–1972), NAE (1983) (MT3 WWIA WWIE WWWIA)

HALL, Wilfred Newman, 1908–??, Canadian chemical engineer; chemical process industries management, Canadian Industries, Ltd. (1929–1945), Standard Chemical Co. (1945–1950), Dominion Tar & Chemical Co. (1950ff) in management and executive positions, becoming president (1957) (AMOS)

HALLIDIE, Andrew Smith (originally Andrew Smith), 1836–1900, English American engineer and inventor; to US (1953), wire suspension bridges, rigid suspension bridges (1867), invented cable street railway used in San Francisco particularly for pulling loads uphill (1869), father was Andrew Smith, who invented wire rope in England (1835) (BDOACE CDOAB DAB LIME MWBD NC7)

HALSEY, Frederick Arthur, 1856–1935, American mechanical engineer; machinist, draftsman, engineer for Rand Drill Co. (1880–1890), where he designed slugger rock drill, general manager of Canadian Rand Drill Co. (1890–1894), where he introduced incentive plan for workers, editor and publisher of magazines (DAB MEIA)

HALSEY, James Taggart, 1854–1915, American mechanical engineer and inventor; invented new signal devices for railroad, with Talbot Works (VA) (7 yr.), opened shop in Philadelphia making portable machine tools, invented Halsey motor truck for trolleys (MEIA)

HAMILTON, Schuyler, 1822–1903, American soldier and engineer; infantry soldier, was wounded, aide-de-camp (1847), administration quicksilver mines (1854–1858), major general of volunteers in Civil War, engineer for Department of Docks (NYC) (DAB)

HAMLIN, Emmons, 1821–1885, American inventor and manufacturer; maker of organs and pianos, improved reeds in melodeons, founded Mason & Hamlin Co. (1854), violin maker (DAB)

HAMMER, William Joseph, 1858–1934, American electrical engineer; established first central incandescent electric lighting system, chief engineer of Edison companies in US, England and Germany, consulting engineer (1890ff) (ANB MWBD)

HAMMERSTEIN, Oscar, 1846–1919, German American inventor and composer; to US (c. 1863), designed and patented machine for spreading and shaping tobacco leaf, instituted *US Tobacco Journal*, wrote operetta works and music, built theaters and opera houses in New York City, Philadelphia and London, his grandson Oscar Hammerstein II [1895–1960] became one of greatest contributors to American music (ANB DAB MWBD NC17,45)

HAMMING, Richard W., 1915–1998, American mathematician and engineer; research at Los Alamos (1945–1946), Bell Labs. (1946–1976), faculty at Naval Postgraduate School (CA) (1976ff), digital computer development, error detecting and correcting codes, automatic coding systems, NAE (1980) (AMOS Bridge FAL MT10 NYTObit WWIE)

HAMMOND, George Henry, 1838–1886, American meat packer and inventor; credited with foreseeing possibilities of refrigerated railroad cars built by William Davis (1867), George K. Wood (1868), and Joel Tiffany (1877) (ASAI WWWIA)

HAMMOND, James Bartlett, 1839–1913, American inventor and manufacturer; wrote literature and religious articles, designed typewriter (1880), Hammond typewriter (1884), formed a manufacturing company of which he was president (NYC) (25 yr.) (ANB)

HAMMOND, John Hays, 1855–1936, American mining engineer; inspected gold fields for US Geological Survey (1880), reopened North Star Mine (CA) (1885), took over Bunker Hill Mine (ID) (1889), resource development in Africa, consultant to Cecil J. Rhodes [1853–1902] (1893ff), Guggenheim Exploration Co. (1903–1907), authority on gold mining, introduced railroads to mining industry, irrigation, coal industry survey, consultant (ANB BDOACE MWBD WWWIA)

HAMMOND, John Hays, Jr., 1888–1965, American electrical engineer and inventor; son of John Hays Hammond [1855–1936], radio engineer, devices for radio remote control, used in numerous applications from radio-controlled torpedo for coastal defense to home use to missile control, for telegraphy, variable pitch ship propeller, president of Radio Engineering Co. of New York, established Hammond Radio Research Corp. (MA) (1911), developed organs, consultant (ANB DAB MWBD 1000Y PS WWWIA)

HAMMOND, Laurens, 1895–1973, American inventor; developed Hammond electronic organ (1933), developed musical instruments such as Hammond Novachord (MWBD WWWIA)

HAMMURABI, c. 2000 BC, Mesopotamia; director of engineering, developed astrolabe (angle measuring instrument for surveying), building codes named after him (TSOE)

HAMPSON, William C., 1854–1926, English chemical engineer and inventor; worked on inert gases, which led to discovery of neon, invented device for producing liquid air (1895) using cascade effect, invention used by British Oxygen Co. (DOSB MWBD WWW WWWIS)

HANCOCK, Thomas, 1786–1865, English inventor and manufacturer; shredded rubber scraps and compressed into blocks with a machine called a masticator, waterproofed fabrics, adopted heat process of vulcanization based on work of Charles Goodyear [1800–1860], made improvements to rubber manufacture, at least 17 patents (EAI PS RHW MWBD)

HANCOCK, W., 1799–1852, English inventor; developed Hancock cellular boiler for steam carriage (1827) (EE)

HANDLEY PAGE, Frederick, Sr., 1885–1962, British aeronautical engineer; established private company (1908), first large bomber (1915), designed the Halifax bomber used in WWII, built Hercules—a 40-seat plane with four engines (1930), after WWII built 4-engine jet plane named Victor (1952), in civil aviation his planes led to what became a part of British Airways Ltd. (EAI DNB RHW)

HANEY, Paul D., 1911–1990, American civil engineer; research and design of water supply systems, wastewater treatment, water quality, water pollution, hydrology, faculty at Univ. of Kansas and engineer for Kansas State Board of Health (1937–1947), commission in US Public Health Service (1948–1954), partner in consulting engineers-architects firm of Black & Vetch, Inc. (1954–1978), NAE (1974) (Bridge MT5 RHW TBDOS(Po) WWIE)

HANFORD, William Edward "Butch," 1908–1996, American chemist and inventor; synthetic organic chemistry, high-pressure reactions, developed polyurethanes (1941), worked for several chemical companies, including research chemist with DuPont (1935–1942), vice president of Olin Mathieson Corp. (NY) (1957–1973), important inventions for manufacturing, over 120 patents, consultant (1974ff) (AMOS FAL NYT-A NYTObit TTOS WWIA)

HANNAY, Norman Bruce, 1921–1996, American chemist, physicist and engineer; chemical and physical aspects of electronic materials, materials science, mass spectrometry, dipole moments, thermionic emission, semiconductors, with Bell Labs. (1944ff), first in research then as vice president (1972–1982), NAS (1977), NAE (1974) (AMOS Bridge MT10 WWIA WWIE)

HANS, Edmund (Ewald), 1886–1959, American inventor and manufacturer; automobile business (1908–1910), experimental work on farm tractors, oil and water gauges for gasoline-powered vehicles, established National Gage Co. (1914), manufactured an agricultural tractor (1917) and numerous devices for automobiles such as heaters, dashboard gauges, cooling systems, contracts with government during WWII, several patents (NC49 PS)

HANSEN, William Webster, 1909–1949, American physicist, electrical engineer and inventor; invented klystron (1937), studied microwaves, electromagnetic radiation, quantum mechanics, nuclear induction, at Stanford Univ. (1934–1941), with Sperry Gyroscope Co. (1941–1945), developed electron accelerator, director of microwave laboratory at Stanford Univ. (1945ff), co-founder of Varian Assoc. (1948), NAS (1949) (ANB BM27 DOSB TFE&I WWWIS)

HARDING, Frank Welland, 1855–1938, American engineer; worked in approximately a dozen shops (1876–1883), master mechanic at Cleveland Rubber Works (OH) (1891), New York Belting & Packing Co. (1892) (NJ) designing fabric cloth, rubber and hose machines, Quaker City Rubber Co. (PA) (1903), Russell Engine Co. (OH) (1905), Canadian Consolidated Rubber (1907), consulting engineer, retired (1922) (MEIA)

HARGRAVE, Lawrence, 1850–1915, English Australian aeronautical pioneer; to Australia (1866), models of monoplane, model of rotary airplane engine (1889), box kite (1893) (MWBD)

HARGREAVES, James, 1720–1778, English inventor; textile equipment, co-invented wool carding machine (1760), spinning jenny (1764ff), which allowed several spindles to be controlled by one person (EAI GEAPIT GI MMAH MWBD NYPL RHW SAI TEIH TGS WA WBE WWWIS)

HARLEY, William S. 1880–1943, American engineer and manufacturer; draftsman, studied engineering, designed V-twin engine, worked at Barth Manufacturing Co. (WI), built loop motorcycle frame with pattern maker Arthur Davidson [1881–1950], with brothers Walter Davidson [1876–1942], who was the first president, and William A. Davidson [1870–1937], who was works manager, formed Harley-Davidson Motor Co. (1903) and began with a single cylinder 2 hp. engine for motorcycle, William S. Harley was followed as chief engineer by his son William A. Harley (1943), Walter H. Davidson [1905–1992] became the second president (1942ff). Harley-Davidson Co. was owned by American Machine and Foundry (1969–1981) (NC47 S&TF)

HARPER, John Dickson, 1910–1985, American electrical engineer; design engineer, power manager, satellite communications, industry executive at Alcoa, Inc. (1933–1975), becoming president (1963) and CEO (1970), consultant, NAE (1971) (MT3 WWIA WWIE)

HARPER, John Lyell, 1873–1924, American mechanical and electrical engineer; electrician, operating and construction engineer (1898) with Twin City Rapid Transit Co., in charge of construction of hydroelectric plants, chief engineer and vice president of electric power companies in Niagara (NY) area, vice president of Harper-Taylor Co. consulting engineers (DAB)

HARRINGTON, John Lyle, 1868–1942, American civil engineer; consulting engineer on bridges, 74 bridges built under his direction, entered into partnership with John Alexander Low Waddell [1854–1938], specialized in movable bridges (1907–1914), established firms (1914, 1928) (ANB)

HARRIS, Daniel Lester, 1818–1879, American civil engineer; worked with several railroad companies, contractor, member of a contracting company for road building, formed firm of Roody, Stone & Harris (1845), railroad and bridge construction, part owner of Howe Truss patent, management and executive of railroad companies (DAB)

HARRIS, Milton, 1906–1991, American chemist and inventor; formed Harris Research Labs., Inc., research on textiles, cellulose and polymers, work resulted in cold permanent wave, shrinkproof wool, razor blades with polymer-coated edges, research and development officer for Gillette Co. (1957ff), NAE (1976) (AMOS Bridge MT6)

HARRIS, William Andrew, 1835–1879, American manufacturer; steam engine builder, with Corliss Steam Engine Co. (1856–1864), designed improvements on engine, founded William A. Harris Co. (1864), manufactured Harris-Corliss (George H. Corliss [1817–1888]) engines (1864–1879), used worldwide (LIME MEIA SciAm)

HARRISON, John, 1693–1776, English instrument maker and inventor; compensated clock pendulum (1726), clock development (1728ff), gridiron pendulum, marine chronometer (1728), inspired establishment of Greenwich Observatory (BEST GEAPIT MMAH MWBD PS RHW WA WWWIS)

HARRISON, Joseph LeRoy, 1810–1874, American mechanical engineer and inventor; apprenticed to build steam engines, manufactured small lathes and presses, designed locomotives, partner in what became Eastwick & Harrison Co. (1837ff) (Andrew M. Eastwick [1810–1879]) for building locomotives, worked on railroad development in Russia, patented sectional Harrison steam boiler (1859), safety of steam boilers (BITAS DAB MEIA WWWIA WWWIA-S&T)

HARROD, Benjamin Morgan, 1837–1912, American civil engineer; construction of lighthouses and forts along Gulf of Mexico in office of US Engineers, officer in Confederate Army, reconstruction after Civil War, hydraulic engineer, levees to control flood of Mississippi River, city engineer for New Orleans (1888) (DAB)

HARROLD, Lloyd Larren, 1907–1981, American civil and agricultural engineer; hydraulic engineer with US Geological Survey (1930–1935), hydrology and hydraulics, manager of research for US Department of Ag-

riculture (1930–1974), retired as director of North Appalachian Experimental Watershed (OH) (1974), water resources, soil conservation, professor at Ohio State Univ. (1962ff), consulting engineer (AE55,62 AMOS WWIE)

HART, Charles W., 1872–1937, American inventor and manufacturer; with Charles H. Parr [1868–1941] credited with first industrial-scale mobile gasoline tractor (1902) in US, formed Hart-Parr Tractor Co. (IA), devoted exclusively to manufacturing tractors (1903), credited with use of "tractor" as a replacement for "gasoline traction engine," references vary in crediting John M. Froehlich as responsible for the first mobile gasoline tractor in US (1889) and with the first gasoline-powered vehicle in US as being the Burger tractor (1889) manufactured by the Charter Engine Co. (IL) and a production model marketed as the Sterling tractor (1893), Frenchman Alabaret is credited with the first locotractor (1856), the Hart-Parr Co. was consolidated with the Oliver Farm Equipment Co. and other companies (1929) and became White Co. (1974) (FFF GTHP MISAT S&TF TAT WOI)

HARTLEY, Fred Lloyd, 1917–1990, Canadian American chemical engineer; instrumentation, process design, engineering economics, technology sales, oil shale, geothermal energy, industry manager and executive of Union Oil Co. (1939ff), becoming vice president (1960), NAE (1980) (AMOS Bridge MT6 EEIA WWIE)

HARTLEY, Ralph Vinton Lyon, 1888–1970, American electrical and mechanical engineer and inventor; radio communications, invented Hartley oscillator (1914) with Edwin H. Colpitts [1872–1949], carrier telephony at Western Electric Co., followed by research at Bell Labs. (1913–1929), followed by consulting with Bell Labs. (1939–1950), Hartley principle, information theory, servomechanisms (BDOTHOT CEOS&T DAB EDOP WWWIS)

HARTMANN, Carl Friedrich Alexander, 1796–1863, German mining and metallurgical engineer and mineralogist; commissioner of mines in Brunswick (1829–1841), recognized for contributions to technical publications on mining (DOSB)

HARTMANN, Georg, 1489–1564, German instrument maker; sundials, caliper gauge for artillery (1540), discovered magnetic dip (1544), timepieces, astrolabes, globes, quadrants, star altimeter (DOSB PS WWWIS)

HARTNESS, James, 1861–1934, American engineer and inventor; invented and developed turret lathe (1889, 1893), automatic die, Hartness chuck, optical screw thread comparator, governor of Vermont (1921–1923) (ANB EE MWBD)

HARTNETT, Laurence John, 1898–1986, English Australian engineer and industrialist; designed and built the Holden automobile, the first mass-produced automobile in Australia (RHW)

HARVEY, Hayward Augustus, 1824–1893, American steel manufacturer and inventor; with New York Screw Co., organized by his father, William Harvey (1840), hay cutter, grip bolt (1874), carbonizing process for steel surface used for armor plate, corrugated staple, wire mill, peripheral grip bolt (1874), founded Harvey Steel Co. (1886), developed machine for rolling thread on stock (1880), steel alloy (1891), over 125 patents (DAB MWBD NC13 WA WWWIA)

HASELWANDER, Friedrich August, 1859–1932, German inventor; invented three-phase dynamo (1887), compressionless diesel engine (1897), oil engine with special fuel injection (1899) (EE MWBD)

HASKINS, Caryl Davis, 1867–1911, American electrical engineer and inventor; with Ferranti (1888), General Electric Co. (1889ff), specializing in metering of electricity, head of metering department, invention of meters, steel refining, transformers, switches, circuit breakers, time relays, and auto dirigible torpedo (1906), his son was Caryl Parker Haskins [1908–2001], who was cofounder and director of Haskins Lab., Inc. (1935ff) (BITAS NC28,E WWWIA)

HASKINS, John Ferguson, 1833–1893, American mechanical engineer, inventor and manufacturer; manufactured Ericsson (Swedish engineer John Ericsson [1803–1889]) caloric engine (1858), Shanghai Navigation Co. (China) (1863ff), Burleigh Rock Drill Co. (MA), manufactured Haskins steam engine, Stowe Flexible Shaft Co., Globe Horse Nail Co. (1878), later with Stearns Manufacturing Co., senior partner of John F. Haskins Co. (IL) (MEIA MWBD)

HASLAM, Robert Thomas, 1888–1961, American chemical engineer and inventor; research chemist to production engineer to assistant superintendent for National Carbon Co. (OH) (1912–1920), faculty at Massachusetts Inst. of Technology (1920–1927), vice president and director of Esso Research and Engineering Co. (NJ) (1927–1942), developed a hydrogenation process for lubricating oil, held 16 patents for Standard Oil Co. (1926–1939), vice president and director of Standard Oil Co. of New Jersey (1935–1950), consultant (1950ff) (NC55 WWWIA)

HASSELBLAD, Victor, 1906–1978, Swedish inventor; developed camera for Swedish Air Force (1941), sold as Hasselblad, single lens reflex camera, interchangeable lens (1948) (MWBD)

HASSLER, Ferdinand Rudolf, 1770–1843, Swiss American civil engineer; geodetic field work in Switzerland, improvement of surveying instruments, superintendent of Coast Survey (1816) (1830–1832), superintendent of weights and measures (1830–1832), author (BDOACE DAB)

HASSLER, Francis Jefferson, 1921–1996, American agricultural engineer; US Army in WWII, faculty and department administration at North Carolina State Univ. (1950ff), frost control with infrared, fundamentals of tobacco curing, bulk curing of tobacco, water resources (AMOS Res2 WWIE)

HASTINGS, Warren W., 1911–1988, American chemist and agriculturalist; first worked with agriculture department in Arizona and Maryland (1935–1937), US Geological Survey (1937–1973), laboratory studies on water quality with responsibilities covering western states, to Washington (DC) headquarters (1948), became heavily involved in administration and was area hydrologist, continued work after retirement (until 1982) (HOWRD)

HASWELL, Charles Haynes, 1809–1907, American engineer; with Allaire Works (NYC) (1828ff), helped introduce steam power to navy, designed and supervised building of machinery for naval vessels, served as chief engineer of US Navy, consulting engineer (BITAS DAB)

HATHAWAY, Gail A., 1895–1979, American hydraulic engineer; construction engineer, field engineer, chief engineer, streamflow, hydrology and hydraulics, reservoir, US Corps of Engineers (1928–1956), several assignments in foreign countries, models, maximum flood flows, airport drainage, river sedimentation, large dam design, water resource development, consultant, NAE (1979) (AMOS MT2)

HATVANY, Jozsef, 1926–1987, Hungarian mechanical engineer; manufacturing engineering, computer and automation, sensors, computer programming, machine tool controls, Inst. of Hungarian Academy of Science (1946ff), including head of department, fNAE (1985) (MT4)

HAUKSBEE, Francis, 1688–1763, English physicist and instrument maker; opened own shop, performed experiments for Royal Society, worked on reflecting telescope with John Hadley [1682–1744] (GEAPIT DOSB MWBD PS WWWIS)

HAUPT, Herman, 1817–1905, American civil engineer and inventor; railway surveys, bridge construction, tunnel, military railroad transportation, railroad operation and management, in Civil War, development of pneumatic drill for tunneling, chief engineer for Pennsylvania Railroad (1853–1856), Hoosac railroad tunnel (MA) (1856–1862) (finished in 1875) (ANB BDOACE BITAS DAB EIH FEL MWBD MWGD TEIH)

HAUSSMANN, Baron Georges-Eugène, 1809–1891, French municipal engineer; responsible for inauguration and development of huge municipal systems in Paris, new water supply and sewage system, new streets, landscape gardening parks in France and Luxembourg (EIH MWBD)

HAVILAND, John, 1792–1852, English American civil engineer and architect; designed and constructed several buildings, architectural drawing school, prison architecture copied by many (BDOACE DAB)

HAWKINS, George Andrew, 1907–1978, American mechanical engineer; heat transfer, thermodynamics, instrumentation, corrosion of steel and alloys, interior ballistics, US Army Ordnance Experiment Station (1941–1951), faculty and administration at Purdue Univ. (1932ff), becoming dean (1953–1971) retiring (1974), NAE (1967) (AMOS MT1)

HAWKINS, Walter Lincoln, 1911–1992, American Black chemist, chemical engineer and inventor; chemistry of cellulose, alkaloids, deterioration of polymers, research and research manager at AT&T Labs. (1939–1976), co-invented antioxidant additives that made inexpensive telephone wire insulation, director of Plastics Inst. of America (1976–1982), 18 US patents and 129 foreign patents, NAE (1975) (AAISMAI AMOS Bridge MT7 WWIE)

HAWKSHAW, Sir John, 1811–1891, English civil engineer; underground railroad, tunnels, bridges, chief engineer of railway (1845–1850), consulting engineer (1850ff) (MWBD)

HAXO, François Nicolas Benoit, 1774–1838, French military engineer; served in Napoleonic armies, director of siege of Antwerp (1882), inspector-general of fortifications (MWBD WWWIS)

HAYASHI, Tsuyoshi, d. 1998, age 86, Japanese engineer; computational studies of aircraft materials and structures, development of composite aircraft, university teaching and research, fNAE (1987) (Bridge)

HAYDEN, Arthur Gunderson, 1875–1964, American civil engineer; structural details and bridge designer (1901–1904), then designer and assistant engineer for New York Department of Public Works (1904–1920), chief engineer for commissions in New York (1920–1937), private consulting (1937–1960), publications on bridges and structural engineering (NC51)

HAYDEN, George Dickerson, 1878–1962, American inventor and manufacturer; with numerous companies as draftsman, National Twist Drill & Tool Co. as a machine designer (1918–1932), Ace Drill Co. as officer (1933–1951), with his son Howard Albert Hayden organized the Hayden Twist Drill Co. (1951), at least 12 patents on metal cutting and drilling (NC51 PS)

HAYDEN, Hiram Washington, 1820–1904, American manufacturer and inventor; worked in manufacture of buttons (1838), die maker (1841), machinery for making kettles (1851), organized company to make sheet brass into finished articles (1853), several inventions (DAB)

HAYDEN, Joseph Shepard, 1802–1877, American inventor and manufacturer; father of Hiram Washington Hayden [1820–1904], maker of brass and gilt buttons,

machines for making cloth-covered buttons, machine design, constructed engine-lathe, several patents (DAB)

HAYDEN, Sophia Georgia, 1868–1953, Chilean American architect; teacher of mechanical drawing, winner of design competitions, first woman to obtain architecture degree at Massachusetts Inst. of Technology (1890) (AWIT AWOA)

HAYFIELD, Sir Robert Abbott, 1858–1940, English metallurgist and inventor; manager of steel foundry, high manganese steel (1883), silicon steel, alloy steel, invented nonrusting stainless steel containing chromium and nickel (1913), developed other ferrous alloys (BEST EAI MMAH MWBD RHW WA)

HAYFORD, John Fillmore, 1868–1925, American civil engineer; US Coast Guard (1889–1895) (1898–1909), faculty at Cornell Univ. (1895–1898), faculty and director of engineering at Northwestern Univ. (1909–1925), theory of isostasy, calculated dimensions of the earth, international spheroid of reference (1924), NAS (1911) (BM16 DAB MWBD WWWIS)

HAYNES, Elwood, 1857–1925, American inventor; tungsten chrome steel (1881), horseless carriage (1893–1894), alloys such as Cr-Ni (1897), Co-Cr (1900) and stainless steel (1911), formed company that became Haynes Stellite Works (1912–1920), which was sold to Union Carbide & Carbon Co., formed a company that became Haynes Automobile Co. (1902–1925), which made one of the first automobiles in US built by Haynes-Apperson Co. (1898) (Elmer Apperson [1861–1920]) (ANB MEIA MWBD NC24 WWWIS)

HAYWARD, Nathaniel Manley, 1808–1865, American inventor and manufacturer; development of India-rubber products, India-rubber fabrics (1834), rubber fabric (1870), developed company to make rubber shoes and sold business to Charles Goodyear Co. (DAB)

HAZELTINE, (Louis) Alan, 1886–1964, American electrical engineer; neutrodyne circuit, radio, organized Hazeltine Corp. (1924), faculty at Stevens Inst. of Technology (1907–1924) (1933–1960), in WWI was consultant to US Navy, in WWII in OSRD, consulting engineer (MWBD WWWIAH-S&T WWWIS)

HAZEN, Elizabeth Lee, 1885–1975, American bacteriologist and inventor; invented first nontoxic artifungal antibiotic, called nystatin, used to treat art objects against molds as well as other uses (AWIT MOI NYT-A MI WIS)

HAZEN, Richard, 1911–1990, American consulting engineer; water supply, hydrology, sewage and industrial waste disposal, wastewater treatment, partner and consulting engineer with Hazen & Sawyer (1951ff), worked in local communities, metropolitan areas and overseas, NAE (1974) (Bridge MT6 WWIE)

HAZEN, Richard Ray, 1925–1990, American electrical engineer; radio engineering, vacuum tube design, transistor, US Navy (1943–1946), faculty at Univ. of Dayton (1953ff), including chairman of electrical technology (1958) (AMOS)

HEALD, Henry Townley, 1904–1975, American civil engineer; structural engineer, bridge design, engineering education and administration, including dean and president (1938) at Illinois Inst. of Technology (1927–1952), chancellor of New York Univ. (1952–1956), president of Ford Foundation (1956–1966), consulting firm (1966ff), NAE (1965) (AMOS ANB MT1 WWWIA WWWIS)

HEATHCOAT, John, 1783–1861, British inventor; master mechanic (1803), lace-making machine (1807), net and ribbon-making machines, self-narrowing stocking frame, with Henry Handley developed steam plow (1832) (EAI EB PS RHW MWBD)

HEAVISIDE, Oliver, 1850–1925, English physicist and electrical engineer; electrical circuits, added feature of inductance to electric circuits, wave motion, electromagnetic theory, Kennelly-Heaviside electrically charged layer in atmosphere (1902) (Arthur E. Kennelly [1861–1939]) (BEST MWBD PS RHW TFE&I WW-WIS)

HEBERT, Louis, 1820–1901, American engineer and soldier; chief engineer of Louisiana (1855–1860), distinguished soldier in Confederate Army, retiring as a brigadier general (DAB)

HEDEFINE, Alfred, 1906–1981, American civil engineer; design engineer, project engineer, bridges and heavy structures, strands for suspension bridge cables, associate with Wadell & Hardesy (1948), partner with Parsons, Brinckerhoff, Quade & Douglas (NY) (20 yr.), including president (1965), NAE (1973) (MT2 WWIE)

HEDLEY, William, 1779–1843, English coal mine operator and inventor; built Puffing Billy (1813), use of steam locomotives in mines and to move coal to trucks on docks (GEAPIT MWBD)

HEFNER, Hubert, 1924–1975, American electrical engineer; radio engineering, traveling wave tubes, electron beams, solid state microwave amplifiers, US Army Signal Corps (1944–1946), Bell Labs. (1942–1954), faculty at Stanford Univ. (1954ff), NAE (1971) (AMOS MT1)

HEFNER-ALTENECK, Friedrich Franz von, 1845–1904, German electrical engineer; improved drum armature (1872), candle luminosity, German unit of light intensity, Hefner lamp, Hefner candle (1884), signal and telemetering devices, employed by Siemens Co. (1867–1890) (CEOS&T E-64 MWBD PS WWWIS)

HEGGEM, Charles Oliver, 1851–1939, Norwegian American mechanical engineer; to US (1875), Buckeye Engine Works (OH), from foreman to company vice president (1890s–1924), which manufactured tractors, sawmills, and threshers, patents on engine and tractor components (MEIA)

HEINEMAN, Daniel Webster, 1872–1962, American chemical engineer; returned to Germany, the place of his mother's birth, worked with firms associated with General Electric Co. (1895–1905), converted city transit systems to electricity in several European cities, joined as managing director of Belgian company of Sopina (1905–1955), acquired other firms throughout the world and returned to New York City (ANB)

HEINEMANN, Edward H., 1908–1991, American engineer; designer and developer of airplanes, project engineer, chief engineer, vice president of Douglass Aircraft Co. (1926–1960), with Guidance Technology (1960–1962), president of consulting company of Heinemann Associates (1973ff), NAE (1965) (Bridge DOSB MT6 WWIE)

HEINKEL, Ernst Heinrich, 1888–1858, German engineer, aircraft designer and industrialist; founded Heinkel-Flugzeugwere (1922ff) to design and build aircraft, a pioneer of jet aircraft (1939ff), built a civilian aircraft that became the HE 111 as a medium bomber during WWII, after the war he established a company to build bicycles, motorcycles, and small automobiles (BDOTHOT MWBD RHW)

HEISENBERG, Werner Karl, 1901–1976, German physicist; professor at Leipzig (1927–1941), director of Max Planck Inst. (first known as Kaiser Wilhelm Inst.) (1945–1976), quantum mechanics, founded uncertainty or indeterminacy principle (1927), Nobel Prize (1932), fNAS (1961) (CB L&M MWBD NYPL PA SAI-MP TGS WWWIS)

HELMHOLTZ, Hermann Ludwig Ferdinand von, 1821–1894, German physiologist, physicist and inventor; revived 3-color light theory, principle of conservation of energy (1847), speed of nerve impulse (1850), invented ophthalmoscope (1851), ophthalmometer (1855), acoustics (1862), professor at Potsdam (1843–1848), Königsberg (1849), Bonn (1855), Heidelberg (1858), Berlin (1871), fNAS (1883) (MWBD NYPL RHW SAI-MP TGS WBE WWWIS)

HENCK, John Benjamin, 1815–1903, American civil engineer; worked with consulting firm, formed consulting partnership of Whitwell & Henck (Willam S. Whitwell), waterworks, early street railways, development of Charles River (MA) basin and Back Bay (MA) (BITAS BOACE DAB WWWIA)

HENDERSON, Alexander, 1832–1901, American mechanical engineer; served as engineering officer on several navy projects (1851–1894), in Civil War was chief of blockading fleet (1861) and fleet engineer during Spanish War after retiring (MEIA)

HENLE, Robert A., 1924–1989, American electrical engineer; computer research and development with International Business Machines (1952–1987), solid state technology, advanced silicon technology, semiconductor devices for computers, computer systems, NAE (1982) (Bridge MT4)

HENNEBIQUE, François, 1842–1921, French structural engineer; made reinforced concrete, design and construction using reinforced concrete (FEL MWBD TEIH)

HENNING, Gustavus Charles, 1855–1910, American mechanical engineer; worked on construction of foundations, shops and track of New York Elevated Railroad (1876), the Brooklyn Bridge (1877–1882), then East Baltimore Machine & Boiler Works (MD) and Beaver Wire Mill (PA), designed testing machines for materials and extensiometers, Yale and Towne Manufacturing Co., consultant, several patents (MEIA)

HENNY, David Christiaan, 1860–1935, Dutch American civil engineer; hydraulic engineering, general manager and chief engineer of Excelsior Wooden Pipe Co. (1892–1902), then with Redwood Manufacturing Co. (1902–1905), US Reclamation Service (1905–1909), construction of dams in western US, consultant (1910ff) in land reclamation, power, flood control, and related subjects (ANB DAB)

HENRY, Beulah Louise, 1887–1939, American woman inventor; known as Lady Edison, over 100 inventions, of which 52 were patented, including umbrella improvements, telephone index, snap-on wigs, toys, and dolls (MOI NWS NYTObit WI)

HENRY, Joseph, 1797–1878, American physicist and inventor; magneto (1828), invented electromagnetic motor (1829), telegraphic apparatus, self-induction (1835), electromagnets, electric relay, transformer, galvanometers, faculty at Princeton Univ. (1832–1848), first director of Smithsonian Inst. (1846), initiated weather reporting system, NAS (1863) (cofounder), helped found Philosophical Society of Washington (1871), unit of electrical inductance named henry after him (1893) (BEST BITAS BM5 EIH FNIE GEAPIT GI IIA I&T15 MWBD NYPL SAI TGS WWWIS)

HENRY, William, 1729–1786, American mechanical engineer and inventor; formed partnership with Joseph Simon to manufacture firearms (1750ff), later purchased by William Henry (1760), experimented with steam power, invented screw conveyor, built steam heating system and controls (BITAS DAB MEIA NC11 NC15)

HENSCHEL, Georg Christian Carl, 1759–1835, German manufacturer; founded manufacturing firm of Henschel & Sohn (1810), became largest European locomotive manufacturer (MWBD)

HENSON, William, 1812–1888, English American engineer; recommended use of airscrew for thrust, suggested tricycle landing gear, his design of Arial (1843) with many features but never flew (ITCOF)

HERDIC, Peter, 1824–1888, American lumberman and inventor; built and operated shingle mill, steam sawmill, vehicle running gear, induced several manufacturers to locate in Williamsport (NY), formed Herdic Coach Co. (1880) (DAB)

HERING, Carl, 1860–1926, American electrical engineer and inventor; studied in Germany and worked with German company manufacturing electrical machines (1885–1886), in US studied regeneration of battery solutions and patented several improvements, electric furnaces (1900ff), founded and edited various technical journals (1892ff) (DAB)

HERING, Rudolph, 1847–1923, American civil engineer; sanitary engineer, bridges, water supply, sewage disposal, drainage system for Chicago, hydraulics, studies for numerous North American cities, translated from German to English and introduced Kutter formula (William Rudolf Kutter [1818–1888]), consulting engineer (BDOACE DAB NUC)

HERMAN, Robert, 1914–1997, American physicist and civil engineer; statistical mechanics, transportation, vibration-rotation spectra, infrared spectroscopy, solid state physics, high-energy electron scattering, faculty at City College of New York City, at Carnegie Inst., at Univ. of Maryland (1940–1956), research administration at General Motors Research Lab. (1956–1979), faculty at Univ. of Texas (1979ff) on traffic science, NAE (1978) (AMOS MT9 WWIE)

HERMANY, Charles, 1830–1908, American civil engineer; city engineer of Cleveland (1853–1857), assistant then chief engineer and superintendent of water work design and construction, including pumping stations, reservoirs, distribution, water filtration, use of coagulants to clean water in reservoirs (c. 1857), consulting engineer (BDOACE)

HERO of Alexandria (Heron in Greek), c. 20–c. 70, Greek engineer and scientist, inventor and mathematician; geometry, mechanics, machines (lever, pulley, wheel, inclined plane, screw, wedge), law of action and reaction, syphons, primitive steam turbine (aeolipile) (c. 50), gear wheels, techniques of measurements, mechanics, use of wind power, fire engines, wrote several famous books (BEST CEOS&T DOSB E-64 EIH EITAW FINE GEAPIT ME125 MMAH MWBD PS S&TF TAE TEIH TSOE)

HERODOTUS (or HERODOTOS) of Halicarnassus, D., c. 484–425 BC, Greek historian and natural philosopher; traveler and historian, measurements and dimensions of features of city of Babylon and Euphrates River, settled in Italy (DOSB MWBD TEE TEIH TGS WWWIS)

HÉROULT, Paul-Louis-Toussaint, 1863–1914, French metallurgist and inventor; produced aluminum by electrolytic method using direct current (also done by Charles Martin Hall [1863–1914]) (1886), aluminum making, electric arc Heroult furnace, electric arc making of steel, aluminum alloys (1888) (AIE BEST CEOS&T EAI MMAH MWBD RHW WWWIS)

HERR, Herbert Thacker, 1876–1933, American engineer; locomotive air-brake equipment, double-heading device, improvement in turbine and oil and gas engines, with railroad companies (ante 1908), with George Westinghouse companies (1908ff), vice president of Westinghouse Electric and Manufacturing Co. (1917–1933) (MWBD)

HERRESHOFF, James Brown, 1834–1930, American inventor; manufacturing chemist with Rumford Chemical Co. (1855–1862), manufacturer of fish oil and fertilizer, sliding seat for rowboats (1860), invented oil press (1862), racing shells, developed thread tension regulator for sewing machines (1866), gasoline motor bicycle (1870, 1872), improved marine boiler working with his brother John B. Herreshoff [1850–1932], called Herreshoff boiler (1874), developed steam engine to use superheated steam up to 800 deg. F. (1879) (ANB DAB EE MEIA MWBD NC12 WWWIA WWWIS)

HERRESHOFF, John Brown Francis, 1850–1932, American chemist, chemical engineer and inventor; worked as chemist for several companies in New York City, became plant superintendent for G. H. Nichols & Co. (1876ff), then vice president of Nichols Chemical Co. (1890) and Nichols Copper Co. (1899), consulting engineer for its successor, the General Chemical Co. (1924), worked primarily in manufacturing and extraction of ores, Herreshoff furnace (1876), held several patents (ACCE NC24)

HERRESHOFF, Nathanael Greene, 1848–1938, American inventor and manufacturer; founded Herreshoff Manufacturing Co. (1878) with brother James Brown Herreshoff [1834–1930], superintendent and president (1881–1924), steam and hydraulic equipment, fin keel (1891), seagoing torpedo boat, design and building of yachts, developed universal rule for measurement of racing yachts (1903) (MEIA NC30 MWBD WWWIA)

HERRICK, James Amory, 1850–1920, American mechanical engineer and inventor; built open hearth furnaces and rolling mills, built first mechanically operated Pernot furnaces in US (1874), engineer with H. A. Gadsden and Co. (1882ff), patented core for casting steel (1888), established own business as mechanical engineer, over 40 patents MEIA)

HERSCHEL, Clemens, 1842–1930, American hydraulic engineer; bridge engineering (continuous revolving drawbridge), venturi tube studies and use, flow of water in pipes, venturi meter (pub. 1888) based on work of Giovanni B. Venturi [1746–1822], water turbines, worked with seven water power companies, consulting engineer (BDOACE DAB MWBD)

HERSCHEL, Sir (Frederick Wilhelm) William, 1738–1822, German British astronomer and inventor; to England (1757), built reflecting telescope (c. 1773), appointed court astronomer in England (1782), discovered planets (1787ff), with sister Caroline Lucretia Herschel [1750–1848] developed superior lens grinding, son John Frederick William [1792–1871] invented equipment for astronomers (c. 1837) (BDOA(Wi) BEST IAI MWBD SAI SAI-MP S:TLAW TBDOS(Po) WWWIS)

HERSHEY, Milton Snavely, 1857–1945, American industrialist; candy and chocolate manufacturer, established company named after him (1903), built town of Hershey (PA), founded school for orphan boys (1909) (AIE ANB CDOAB DAB MWBD NC33 NYTObit)

HERSTEIN, Karl Marx, 1896–1961, American chemical engineer; chemist with army, navy, and several companies, formed partnership of Kenny-Herstein, Inc. (1935), formed own firm of Herstein Lab. (1937), research in textiles, paving materials, tobacco and colloids, wrote about wine (ACCE)

HERTZ, Heinrich Rudolf, 1857–1894, German physicist and mechanician; demonstrated existence of electromagnetic waves (1886ff), proved similarity of radio, light, and heat waves, mechanics (1890), many inventions of others based on his work—radio, wireless telegraphy, television, radar, satellite communication, microwave ovens (EIH FNIE IAI 1000Y RHW MMAH MWBD SAI SAI-MP TGS WBE WBEOS WWWIS)

HERWALD, Seymour W., 1917–1998, American electrical engineer; industry research and development, automatic control, theory and development of servomechanisms, airborne military electronic group, Westinghouse Electric Corp. (1939ff), becoming vice president in various capacities (1959–1975), consultant, NAE (1967) (Bridge WWIE)

HERZBERG, Gerhard, 1904–1999, German Canadian scientist and inventor; founding father of molecular spectroscopy, Darmstadt Inst. of Technology, to Canada (1935), Univ. of Saskatchewan, Univ. of Chicago (1945–1948), National Research Council of Canada (1948–1994), Nobel Prize (1971), fNAS (1968) (CBY NYObit WWWIA WWWIS)

HEWITT, Peter Cooper, 1861–1921, American electrical engineer and inventor; mercury vapor electric lamp (1903) (1912), rectifier, vacuum tube amplifier, wireless receiver, based on the work of others, such as Antoine-Henri Becquerel [1852–1908], developed fluorescent lamp (1901), which reached the market in the 1930s by General Electric Co. and Westinghouse Corp., with the advice of Arthur H. Compton [1892–1962] (AIE ANB DAB MWBD PS WA WOI WWWIS)

HEXAMER, Charles John, 1862–1921, American civil engineer; worked as area engineer in Philadelphia (1882–1917), specialty in fire hazards and combustion, active in promoting German-American Alliance (DANB) (1901–1917) (ANB)

HEYROVSKY, Jaroslav, 1890–1967, Czech chemist and inventor; development of polarography (1925), approach can be used for analysis of environmental trace metals and medical diagnostics, professor at Charles Univ. (1921ff), director of Polarographic Inst. of the Czechoslovak Academy of Sciences in Prague (40 yr.), Nobel Prize (1959) (LOTCS MWBD RHW WWWIS)

HEYWOOD, Levi, 1800–1882, American manufacturer and inventor; manufacturer of wooden chairs (1826), sawmill specializing in veneers, invention of machinery for making chairs (1841), including tilting chair, machines for bending wood, established American Rattan Co., manufacturer of rattan furniture (1876ff), built foundry for iron parts (DAB MEIA)

HICKENLOOPER, Andrew, 1837–1904, American soldier and engineer; city surveyor of Cincinnati (1859), in various engineering positions with Union Army, was brevetted brigadier general (1865), city engineer of Cincinnati (1871), executive of Cincinnati Light & Coke Co. (1871–1879) (DAB)

HICKS, Beatrice Alice, 1919–1979, American chemical engineer and physician; research, design, development, and manufacture in electro-mechanical area, international management, world economics and production, faculty of Newark College of Engineering (NJ) (1939–1942), Western Electric Co. (NJ) (1942–1945), Newark Controls Co. (1945–1967), including president (1955–1967), head of Rodney D. Chipp & Associates (1967ff), NAE (1978) (MT2 WWWIA)

HICKS, William Cleveland, 1829–1885, American mechanical engineer and inventor; worked in shop of Woodruff & Reach, assistant superintendent of Colt armory, city engineer of Hartford (CT), superintendent of Volcanic Firearms Co. (1855), numerous patents covering railway switches, steam engine valves, sewing machines (1861), drop press (1863), Volcanic gun later known as Winchester rifle (1864), Hicks engine (MEIA)

HIENTON, Truman E., 1898–1986, American agricultural engineer; research, extension, and teaching in universities, rural electrification (1925ff), director of federal programs at US Department of Agriculture on use of electricity in rural areas (1921–1968), served in WWI and WWII (AE49 WPost1986 WWIE)

HIGGINS, Andrew Jackson, 1886–1952, American industrialist and inventor; timber business (early 1900s), boat builder, beach landing assault boats (LCI, LCT, LCVP) in WWII known as Higgins boats (1942ff), PT boats in WWII, founder of Higgins Industries (LA) (1930), 30 patents (I&T19 VFWVetMag WPost2000 WWWIA)

HIGGINS, Milton Prince, 1842–1912, American mechanical engineer; considered pioneer in technical education as shop foreman at what became known as Worcester Polytechnic Inst. (1869–1896), with George I. Alden [1843–1926] organized Norton Emery Wheel Co. (1885), organized shop at Georgia Inst. of Technology (1889), organized Plunger Elevator Co. (1904), which was sold to Otis Elevator Co. (BITAS MEIA)

HIGONNET, René Alphonse, 1902–1983, French inventor; developed photo composing machine (1946) with Louis Marius Mayrould, film composition replacing type set from metal (NYT-A)

HILBERT, David, 1862–1943, German mathematician; one of greatest mathematicians, Hilbert's basis theorem (1888), theory of invariants, theory of number fields, foundation of geometry, Hilbert's program (1920), FRS (1928), faculty at Univ. of Göttingen (1895–1930), fNAS (1907) (BEST CBDOS DOT HOFL RSObit WWWIS)

HILDEBRANDT, Johann Lukas von, 1668–1745, Austrian architect and engineer; engineer in Austrian army (1695–1701), public works in Vienna, Hungary, Bamberg, Salzburg (MWBD)

HILGARD, Julius Erasmus, 1825–1891, Bavarian American civil engineer; surveying, US Coast Survey, became superintendent of US Coast & Geodesic Survey (1881), worked at introducing metric system, NAS (1863) (BDOACE BM3 DAB)

HILL, Clair Ashcroft, 1909–1998, American consulting engineer; water resources, wastewater treatment, photogrammetry, bridge engineer for California state highways, ordnance officer as lieutenant colonel in US Army (1941–1945), formed Clair A. Hill and Assoc. (prewar and 1945ff), which merged to form internationally known CH2M-Hill (1971), NAE (1992) (Bridge MT10 WWIE(I))

HILL, Louis Clarence, 1865–1938, American civil and electrical engineer; assistant engineer and engineer, becoming division engineer for railroads (1886–1888), faculty and administrator of hydraulic engineering and electrical engineering at Colorado School of Mines (1890–1903), US Bureau of Reclamation (1903–1914), developed policy on hydroelectric power and building of dams, private engineering practice (1914ff), main projects on flood control, water distribution, and hydropower (ANB DAB)

HINDLEY, Henry, 1701–1771, British inventor; Hindley worm gear made on his wheel-cutting machine (BDOS(Wi) EE)

HINE, Charles De Lano, 1867–1927, American industrial engineer; served in military (1898), WWI, motor transport corps, railroading, inspector of safety devices for government (1900), advocate of unit system of management, organization specialist for railroads (DAB)

HINTON, Sir Christopher, 1901–1983, British mechanical and nuclear engineer; chief engineer at Imperial Chemical Industries (UK) (1930–1940), director of ordnance and construction at Ministry of Supply (1940–1942), returned to ICI at end of war, worked in public sector on British atomic projects, harness of nuclear fission for useful purposes, first large atomic power plant to produce electricity (Calder Hall) (1954), fNAE (1976) (BEST DNB MSAE MT4 PS TEIH)

HIPPARCHUS of Nicaea, c. 190–c. 120 BC, Greek astronomer, mathematician, and inventor; invented trigonometry, calculated the lengths of the solar year and lunar month, catalogue of about 850 stars (BEOS DOSB MWBD RHW)

HIPPODAMUS of Miletos, 5th century, Greek engineer; father of city planning, known for checkerboard pattern, developed port facilities (E-64 EB EIH MWBD TSOE GEAPIT)

HIRSH, Leonard Frederick, 1901–1962, American chemical engineer; founder of Philadelphia Rust-Proof Co. (1923–1962), serving as officer, founded Chrome Gauge Corp. (PA) and Lustrik Co. (PA), serving as president, specialized in cadmium, chrome, and zinc plating (NC50)

HIRT, Louis Joseph, 1854–1933, French American mechanical engineer and inventor; to US (1880), previously worked in France and England in marine shop, where he invented twine baling machine, nail and tack machine, railroad cars, friction clutch, and gas engine among other items (MEIA)

HITCHCOCK, Lauren Blakely, 1900–1972, French American chemical engineer; faculty at Univ. of Virginia (1928–1935), Univ. of Buffalo (1963–1972), with Hooker Electrochemical Co., Quaker Oats Co., and National Dairy Products Corp., Los Angeles Air Pollution Federation (1954–1957), consulting engineering firm (NY) (1957–1963) (ACCE NYTObit)

HJORTH, Søren, 1801–1870, Danish engineer and inventor; railroad development in Denmark, built electric motor (1851), self-exciting dynamo (1855), electric machines (BDOTHOT)

HOADLEY, John Chipman, 1818–1886, American civil and mechanical engineer and manufacturer; various positions with survey crew for enlargement of Erie Canal (1836–1844), located, constructed, and installed new textile mills, with Gordon McKay [1821–1903] formed McKay & Hoadley Manufacturers (MA) (1848) as engineers of mill machinery, entered locomotive building (1852–1857), consulting engineer (1873ff) (DAB)

HOBBS, Alfred Charles, 1812–1891, American mechanical engineer, inventor, and manufacturer; glass cutter (1836), lock development and manufacture, formed company of Jones & Hobbs, locks and fireproofing safes, later Hobbs, Ashley & Co. (London), recognized as leading lock expert, recognized by civil engineers, patented process in carriage-making machinery and machine tools, superintendent and mechanical engineer for Union Metallic Cartridge Co. (1866–1890), numerous patents (DAB)

HOBGOOD, Price, 1911–1999, American agricultural engineer; public high school teacher and superintendent (7 yr.), faculty and administration at Texas A&M Univ. (1942, 1946ff), USAAF in WWII (1942–1946), chairman of Hakco, Inc., structures for farms and agricultural enterprises, engineering consultant (Res6 WWIE)

HODGE, Raymond J., 1922–1990, American civil engineer; naval civil engineering corps, transportation systems, airport development and construction, economic development, senior partner of industry consulting business of Tippetts-Abbett-McCarthy-Stratton, NAE (1983) (Bridge MT5 WWIE)

HODGES, Harry Foote, 1860–1929, American military engineer; assigned to Corps of Engineers, instructor at West Point (1888–1892), supervised engineering projects in Ohio, Missouri, and upper Mississippi Rivers, chief engineer of Department of Cuba, assistant chief engineer for Panama Canal (1902), served in WWI, retired as major general (1921) (DAB WWWIA)

HODGKIN, Dorothy Mary Crowfoot, 1910–1994, British woman chemist and computer specialist; structure of biological substances, X-ray crystallography (1932ff), use of computing machines, structure of vitamin B_{12} (1956), Oxford Univ. (1936–1977), chancellor at Univ. of Bristol (1970ff), Nobel Prize (1964), fNAS (1971) (HOCS&T LOTCS PharmCent TCBE WWW)

HODGKINS, George Sherwood, 1859–1919, Canadian mechanical engineer; locomotive designer for Canadian Locomotive & Engine Co. (1882ff), locomotive inspector for Canadian Pacific Railroad, chief engineer for Canadian Locomotive Works (1899), inspector for Pressed Steel Car Co. (PA), Richmond Locomotive Works (VA), edited several railroad magazines (1915ff) (MEIA)

HODGKINSON, Eaton, 1789–1861, English civil engineer and inventor; devised experiments to measure the strength of materials, neutral line in section of fracture, traverse strain, and strength of materials (1822), bridge construction, Hodgkinson beam (DOSB EAI GEAPIT RHW PS WWWIS)

HODGKINSON, Francis, 1867–1949, English American mechanical engineer and inventor; engineer for Clayton & Shuttleworth Ltd. (1882–1885), an agricultural engineering firm in England making farm implements and steam engines, to US for design and construction of Parsons turbines for Westinghouse (1896–1936), where he was in charge of numerous turbine and related projects, issued 101 patents (ANB WWWIA)

HOE, Richard March, 1812–1886, American engineer, inventor, and industrialist; son and successor (1830) of Robert Hoe [1784–1833], high-speed rotary press (1846) printing on a continuous roll (1865), with Stephen D. Tucker [1818–1902] developed web press improvements (1871), which superseded rotary press, and Hoe Perfecting Press, patented several improvements in printing equipment (ANB CBD DAB EAI EOWB IIA MMAH MWBD NC7 NYPL PS RHW TBDOS (Po) WA WWWIA)

HOE, Robert, 1839–1909, American industrialist; succeeded Richard Hoe [1812–1886] and entered family firm of R. Hoe & Co. (1886), developed type revolving presses, rotary art press (1890), color presses (MEIA MWBD NC7 PS WWWIA)

HOFF, Nicholas John, 1906–1997, Hungarian American aeronautical engineer; applied mechanics, stability of thin-walled structures, airplane stress analysis, shell theory, high-temperature effects on structures, design in airplane industry in Hungary (1929–1938), faculties of Brooklyn Polytechnic Inst. (1940–1957), department head at Stanford Univ. (1957–1971), visiting professor at Rensselaer Polytechnic Inst. (1971–1979), NAE (1965) (AMOS Bridge MT10 WWIE)

HOFFMANN, Felix, 1868–1946, German chemist, pharmacist and inventor; worked for Bayer AG, produced aspirin (acetylated salicylic acid) (1897) in quantities based on work of Alsatian chemist Karl Gerhardt (1853) and procedures developed by Adolph Wilhelm Hermann Kolbe [1818–1884], first introduced as a medicine by Hermann Dreser (1893), first sold as a powder (1899) and then as a tablet (1900) (C&EN75 EB EO&T GI TBOF WWWIS)

HOFMANN, August Wilhelm von, 1818–1892, German chemist and inventor; exploited uses of coal tar for dyes, Hofmann reaction, medicines, worked for William Henry Perking [1838–1907], first director of Royal College of Chemistry London (1845–1865), professor in Berlin (1865–1892) (IAI MWBD WWWIS)

HOLBROOK, George Edward, 1909–1987, American chemical engineer; detergents, textile, leather, paper, petroleum, tetraethyl lead, combustion, research and company officer with DuPont (1933–1976), NAE (1964) (AMOS Bridge MT5 WWIE)

HOLCOMB, Amasa, 1787–1873, American scientist and inventor; taught courses in civil engineering, surveying, astronomy, manufacture of reflecting telescopes (20 yr.), civil engineering instruments, water and its use for power, dam construction (BITAS NC3 PS WWWIA WWWIS)

HOLDEN, Sir Isaac, 1807–1897, British inventor; devices for carding, combing, preparing yarns, new

methods developed with Samuel Cunliffe Lisper [1815–1906] (1847), established factory in France (1848) then moved to England (1864) (DNB MWBD)

HOLIDAY, William Marion, 1901–1958, American automotive engineer and inventor; production and utilization of gasoline, developed refining operations, defense research and development, Standard Oil Co. (1927–1937), Socony-Vacuum Oil Co. (1947–1956), with NASA and DOD (AMOS CB58 PS)

HOLLAND, Clifford Millrun, 1883–1924, American civil engineer; built four double subway tunnels under East River (NYC) (1914–1919), Holland vehicle tunnel under Hudson River (NYC) (completed 1927), tunnel engineer for Public Service Commission of New York City (1914) (EB DAB FEL MWBD)

HOLLAND, John Philip, 1840–1914, Irish American inventor; to US (1872), designed and launched first submarine in Hudson River (1891) using electric motor for submerged submarine, formed J. P. Holland Torpedo Boat Co., which became Electric Boat Co., US purchased submarine (1900) (ANB BITAS DAB EOWB MEIA MWBD NC15 PS WA WWWIS)

HOLLERITH, Herman, 1860–1929, American inventor and consulting engineer; developed punched cards (1899) for storing and reading information (used for US Census of 1890), reading punched cards, used for tabulating and calculating machines, established firm called The Tabulating Machine Co. (1911), which became a part of International Business Machines (IBM) (1924) (AIE E-64 EAI IAI IIA MMAH MWBD NYPL NYT-A RHW WA)

HOLLEY, Alexander Layman, 1832–1883, American mechanical engineer, metallurgist, and inventor; invented steam engine cut-off (1852), experimental locomotive equipped with Corliss valve gearing, redesign of locomotive (1861), published *Holley's Railroad Advocate*, development of Bessemer process in US (1865), technical writer for *New York Times* (1858–1875), obtained 15 patents (ANB ASAI BDOACE BITAS DAB EIH MEIA NC11 NC13 MWBD WWWIA)

HOLLEY, George Malvin, 1878–1963, American manufacturer and inventor; bicycle builder, motorcycle builder, automobile accessories, aviation, formed Holley Carburetor Co. (1918ff), becoming president (NC53 WWWIA)

HOLLIS, Ira Nelson, 1856–1930, American naval engineer; assistant engineer on ship (1880), marine engineer, became assistant chief of Bureau of Steam Engineering in US Navy, faculty at Harvard Univ. (1892–1913), president of Worcester Polytechnic Inst. (1913ff) (ANB DAB MEIA NC22 WWWIA)

HOLLIS, Mark D., 1908–1998, American civil engineer; sanitary engineering, environmental engineering, US Public Health Service (1931–1961), retiring as major general (1961), air-water-land pollution, World Health Organization (1961–1972), NAE (1967) (Bridge WWIA WWIE)

HOLLISTER, Solomon Cady, 1891–1982, American civil engineer; structural engineer, materials, mechanics, reinforced concrete, structures, masonry, consulting engineer (1920–1930), faculty and administrator at Purdue Univ. (1930–1959), including dean of engineering (1937–1959), NAE (1973) (AMOS MT2 WWIA)

HOLLOMAN, John Herbert, 1919–1985, American physicist and engineer; metallurgist, nucleation kinetics, ferrous metallurgy, corrosion, manager of research at General Electric Co. (1946–1960), assistant secretary of US Department of Commerce (1968–1970), president at Univ. of Oklahoma (1970–1972), faculty and administration at Massachusetts Inst. of Technology (1972ff), public policy, safety, consumer affairs, NAE (1964) (AMOS Bridge MT5 WWIA WWIE WWWIA WWWIS)

HOLLOWAY, Frederic Ancrum Lord, 1914–1990, American chemical engineer; absorption tower performance, research and management at Esso Standard Oil Co. (1937–1960), general superintendent and manager of Humble Oil & Refinery Co. (1961–1964), president of Exxon Corp. (1964ff), previously called Standard Oil of New Jersey, NAE (1965) (AMOS MT6 WWIA WWIE)

HOLLOWAY, Josephus Flavius, 1825–1896, American mechanical engineer; mechanics, repair of watches and clocks, machinist for Cuyahoga Steam Furnace Co., worked with E. H. Reese in designing screw propeller boat (1848), worked with several companies in steamboat design and manufacturing, was vice president and treasurer of H. R. Worthington Co. (NYC) (1887ff), consulting (DAB MEIA NC12 WWWIA)

HOLLOWAY, Marshall Gleckler, 1912–1991, American nuclear physicist and engineer; charged particle nuclear reactions, management of atomic bomb development, testing of nuclear weapons, with Manhattan Project and Los Alamos Lab. (1943–1955), Lincoln Lab at MIT (1955–1957), president of ACF Industries (1957–1959), vice president of Budd Co. (1959ff), NAE (1967) (AMOS Bridge MT6 WWIA)

HOLLY, Birdsill, 1822–1894, American hydraulic engineer and inventor; organized firm of Silsby, Race & Holly (NY) for manufacture of hydraulic machinery, formed Holly Manufacturing Co. (1859) primarily to build water works pumps and equipment, then formed Holly Steam Combination Co. (NY) (1877), which led to central heating and became American District Steam Co., about 150 patents, consulting engineer (1888ff) (MEIA NC26)

HOLM, LeRoy Wallace, 1923–1989, American chemical engineer; petroleum and gas production, US Navy

in WWII (1943–1946), research engineer with Pure Oil Co. (1952ff), NAE (1986) (AMOS MT4)

HOLMAN, Minard Lafevre, 1852–1925, American mechanical and civil engineer; water works engineer, assistant engineer for US Treasury, architect for Washington (DC) (1874–1876), worked with various agencies, water commission of St. Louis (1877–1887), private practice as Holman & Laird (MO) (1904–1920) (DAB MEIA NC14 WWWIA)

HOLME, Thomas, 1624–1695, English Colonial surveyor; surveyor-general of Province by William Penn (1682ff), laid out part of city of Philadelphia, surveyed southeastern section of three counties of Pennsylvania and issued map, served on government committees dealing with Indian affairs and boundary disputes with Maryland (BDOACE BITAS DAB)

HOLMES, Isaac V., 1835–1906, American mechanical engineer; Novelty Iron Works (NYC) (1849–1969) with interruption of service to design and erect machinery for iron mines near Lake Champlain (NY), superintendent and engineer for John Casper Co. (OH) (1869ff), worked on power plants and factories (MEIA)

HOLMES, Joseph Austin, 1859–1915, American geologist and mining engineer; test laboratories, accident prevention and safety, faculty at Univ. of North Carolina (1882–1892), US Geological Survey (1904–1910), director of US Bureau of Mines (1910–1915) (BITAS DAB MWBD)

HOLMQUIST, Nils T., 1918–1969, Swedish agricultural engineer; Swedish Inst. for Farm Buildings (Lund) (1940ff), for which he was director (1964–1969), except for government service on building research and with the Agricultural School at Alnarp, becoming professor of the college that absorbed the institute (AE50)

HOLT, Benjamin, 1849–1920, American inventor and manufacturer; formed Stockton Wheel Co. (CA) (1884), developed combine harvester (1886), built steam tractors (1890s), formed Holt Manufacturing Co. (CA) (1904–1908), crawler tractor (1904), crawler or track tractors known as Caterpillars (1904–1908), exporter of heavy machinery, used in WWI for tanks and artillery, joined with Daniel Best [1838–1923], who built steam tractors (1914), formed Caterpillar Tractor Co. (1925), Alvin O. Lombard [1856–1937] of Maine manufactured an early steam power track tractor primarily used for logging, David Roberts (UK) built a crawler tractor with diesel engines (1904), tracks used for development of tanks in WWI (BDOTHOT GI I&T93 LIME NC NYPL Res8 S&TF TBOF TOIAD WA WWWIA)

HOLZMAN, Albert George, 1921–1985, American industrial engineer; operations research, statistics, mathematical programming, game and queuing theory, systems engineering, mine appliances, USAAF in WWII (1942–1945), Bethlehem Steel Corp. (1949–1951), faculty at Univ. of Pittsburgh (1951ff), consultant, NAE (1984) (AMOS MT3 WWIE)

HONDA, Kotaro, 1870–1954, Japanese metallurgist and inventor; addition of cobalt to tungsten steel to produce a more powerful magnet (1916), magnetic alloys, work led to production of alnico, faculty at Tohoku Univ. (1911ff), director of Materials Research Inst. (1922ff), president of Tokyo Univ. of Science (1948ff), research in iron and steel led to development of steel then known as the strongest (BEST GLOSB TTOS WWWIS)

HONDA, Soichiro, 1906–1991, Japanese engineer; established Honda Motor Co. (1947), first built small engines fitted to bicycles, motorcycles, entered automobile market (1961), produced Honda Civic (1972), was first Japanese car to meet US pollution standards, retired (1973) (GLOSB)

HONISS, William Henry, 1858–1940, American mechanical engineer and inventor; machinery for manufacturing paper bags, several patents on paper bag making (1885ff), patents for voting machines, streetcar fare registers, sealer for glass jars, type justifying devices, automatic glass feeder (MEIA)

HOOD, Washington, 1808–1840, American topographical engineer; surveyed and made maps for US government, with Robert E. Lee [1807–1870] (1835), worked on location of boundary between Ohio and Michigan, made map of passes through Rocky Mountains to the West Coast (DAB)

HOOKE, Robert, 1635–1703, English scientist and inventor; outstanding experimenter, originated Hooke law (1676), elasticity, compound microscope, spirit level, marine barometer, part of universal joint called Hooke coupling (known in Europe as Cardan joint after Italian Girolamo Cardano [1501–1576]), balance wheel and hair spring clock (1658), sounding device, optics (BEST DOT EDOP EE EIH EOP EOWB FNIE GEAPIT GI L&M MMAH RHW WWWIS)

HOOKER, Elon Huntington, 1869–1938, American chemical engineer, manufacturer, and inventor; civil and hydraulic engineering, founded Hooker Electrochemical Co. (1909), and later as Hooker Chemical Co. and other names, manufacture of bleaching powder, caustic soda, hydrochloric acid, and numerous other commercial products, consulting engineer (ACCE AIE NC28 PS WWWIA)

HOOKER, Stanley George, 1907–1984, English engineer; anti-aircraft rocket development, aircraft engine with turboprop (1957), aircraft engine with turbojet (1958), improvements to Frank Whittle [1907–1996] turbojet engine, vectored thrust engine for Harrier VTOL military aircraft, fNAE (1981) (DNB EAI MT3 RHW)

HOOVEN, Frederick Johnson, 1905–1985, American engineer and inventor; first radio compass (1936), bombing intervalameter (1944), self-employed inventor (1927–1957), faculty at Dartmouth Thayer School of Engineering (1967ff), consultant (1967ff), 38 US patents, NAE (1979) (MT3 WWWIA)

HOOVER, Herbert Clark, 1874–1964, American mining engineer and politician; worldwide mining activities, world relief, politician, secretary of US Department of Commerce (1921–1929), president of US (1929–1933), headed several government study commissions, Boulder Dam (726 ft. high) (1935) later named Hoover Dam (1947) (AZ, NV) with man-made Lake Meade behind the dam, NAS (1922), Hoover Medal est. (1929) (ACE ANB BM39 CB DAB EB MWBD NC56,C OED TEIH WWWIA)

HOOVER, William Henry, 1849–1932, American industrialist; tannery and harness, cleaning and sweeper equipment initially for domestic use, formed and was president of Electric Suction Sweeper Co. (1908), changed to Hoover Suction Sweeper Co. (1910), of which he was president (1908–1922), chairman of successor, Hoover Co. (1922–1932) (MWBD NC27 RHW WWWIA)

HOPE-JONES, Robert, 1859–1914, British American engineer and inventor; to US (1903), organ builder, electric organ development for Hope-Jones Organ Co. (1907–1910), which became Wurlitzer Co. (1910ff) (MWBD)

HOPKINS, Samuel, 1765–1840, American inventor; issued first US patent (1790) for improvement of potash manufacture, was basis of making US the leading producer and exporter of potash (I&T6)

HOPKINSON, John, 1849–1898, English electrical engineer; worked with brother engineer Edward Hopkinson [1859–1922], alternating current theory, magnetic circuits (1866), patented three-wire electric distribution system (1882), professor at King's College London (1890–1898), 40 inventions (E-64 MWBD WWW WWWIS)

HOPPER, Grace Murray, 1906–1992, American admiral, mathematician, and engineering scientist; data processing, computation, computer programing, co-inventor of COBOL (1959), design and operation of large-scale high-speed digital calculation machines, approximate methods of solving partial differential equations, professor at Vassar College (1931–1944), Harvard Univ. (1946–1971), consultant to computer companies Digital Electronic Co. and Sperry Rand Corp. (1959–1971), with US Navy, becoming rear admiral (1983), NAE (1973) (AIE AMOS AWIS AWIT Bridge LDOS MADOI MT6 MWBD PA RHW TBDOS(Po) TEIH WIS(2) WWIE WWWIA)

HORLICK, William, 1846–1936, English American industrialist; food manufacture, food for infants, malted milk (1887), with brother James Horlick [1844–1921] organized J & W Horlick Co. (1873), which became Horlick's Malted Milk Co. (1906–1921), of which he was president (MWBD NC27 WWWIA)

HORNBECK, John Austin, 1918–1987, American physicist and electronics engineer; physical processes in inert gas, glow discharges, scattering of high-voltage X-rays, semiconductor devices, transistors, research and management at Bell Labs. (1946–1962), president of Bellcomm, Inc. (1962–1966), vice president of Western Electric Co., and president of Sandia (1966–1972), vice president of Bell Telephone Labs. (1972), NAE (1975) (AMS MT4 WWIE)

HORNBLOWER, Jonathan Carter, 1717–1789, English engineer; first used double cylinder or compound steam engine design (1781), later used by Arthur Woolf [1766–1837] in the Woolf engine, worked for James Watt [1736–1819] (MMAH MWBD WWWIS TEIH)

HORNBLOWER, Jonathan Carter, 1753–1815, English engineer; son of Jonathan Carter Hornblower [1717–1789], double cylinder or two-stage expansion steam engine design (1781), worked for James Watt [1736–1819] (EIH EE MWBD TEIH WWWIS)

HORNBLOWER, Josiah, 1729–1809, English American engineer; built pumping plant for copper mines (NJ) (1753), served on several government committees and commissions to settle disagreements (DAB MEIA NC6 WWWIA WWWIS)

HORNE, William Dodge, 1865–1960, American chemical engineer; devoted life to sugar chemistry and technology, with Oxnard-Fulton Sugar Refinery (NYC) (1886), chief of chemistry at National Sugar Refining Co. (NY) (1893ff), for which he worked for 27 yr. (ACCE NYTObit)

HORTON, Robert Elmer, 1875–1945, American hydraulic engineer and inventor; natural drainage networks, water level gauge, Horton laws representing drainage networks, consulting engineer (1911ff) (MWBD WWWIA)

HOTCHKISS, Benjamin Berkeley, 1826–1885, American inventor and manufacturer; machinist, ordnance improvements, machine gun (1872), magazine rifle (1875), improvement in percussion and time fuses, punch projectile for use against ironclads, improved rifling, formed Hotchkiss & Co. (1882), several patents (DAB MEIA MWBD WA)

HOTELL, Hoyt Clarke, 1903–1998, American chemical engineer; radiant heat transmission, optical methods for temperature measurement, flame propagation, pyrometry, fuels technology, heterogeneous combustion, solar energy utilization, fire research and suppression, faculty at Massachusetts Inst. of Technology (1928–1968), cofounder of Combustion Inst., NAS (1963), NAE (1974) (Bridge HOCE MT10 SAI WWIA WWIE WWWIA)

HOUDRY, Eugène Jules, 1892–1962, French American mechanical engineer and inventor; French army in WWI, to US (1930), founder and president of Houdry Process Corp. (1931–1948), patented catalytic refining (1937), founded Oxy-Catalyst Co. (1940), catalytic cracking of petroleum, basic catalytic converter used in automotive mufflers to reduce exhaust emissions, over 100 patents (AIE ANB CEOS&T I&T20 NYTObit NYT-A WWWIS)

HOUGEN, Olaf A., 1893–1986, Swedish American chemical engineer; grain dust explosion, gas absorption, heat transmission, drying, diffusion, thermodynamics, chemical kinetics, faculty at Univ. of Wisconsin (1918–1963), numerous university and government assignments, NAE (1974) (AMOS MT3 WWIE)

HOUGH, Richard Ralston, 1917–1992, American electrical engineer; communications, telephone network analysis and design, missile systems development, Bell Labs. and AT&T (1940–1959), vice president of Ohio Bell Co. (1959ff), NAE (1983) (AMOS Bridge MT6 WWIA)

HOUSE, Henry Alfonzo, 1840–1930, American inventor and manufacturer; self-operating farm gate (1860), with brother James House built sewing machine to make buttonholes (1862), they built horseless carriages (1866), consulting, over 300 patents (DAB MEIA NC23)

HOUSE, Joe Estes, 1923–1998, American chemist and engineer; mining chemicals, reagents, flotation process for separation, taught high school and college chemistry (1947–1956), research and development with General Mills Chemical Co. (1956–1976), vice president of minerals division of Henkel Corp. (1977–1986), consultant (1986ff), NAE (1995) (MT5)

HOUSE, Royal Earl, 1814–1895, American inventor; submerged water wheel, printing telegraph, telegraph devices, machine to make barrel staves (1839), used stranded wire for telegraph (1840–1844) (1846), patented phonetic telegraph (1868), designed postal telegraph for US Post Office (AIE ANB ASAI BITAS DAB MEIA MWBD NC1 WWWIA)

HOUSTON, Edwin James, 1847–1914, English American electrical engineer; to US (1858), educator who organized and chaired civil engineering and geography department at Central High School, Philadelphia (1867–1894), wrote textbooks, developed laboratory, formed partnership with Elihu Thomson [1853–1937] to test and develop electrical equipment, developed arc lighting system and company to sell devices that became a part of General Electric Co., with Arthur E. Kennelly [1861–1939] established consulting engineering firm (1894ff) (ANB BITAS DAB WWWIA)

HOVEY, Otis Ellis, 1864–1941, American civil engineer; engineer for railroad system from Massachusetts to Vermont, with Union Bridge Co. and its successor, the American Bridge Co. (1896–1931), consulting firm (1931ff), recognized for movable bridge design (ANB)

HOVGAARD, William, 1857–1950, Danish American naval architect; to US (1901), naval design and construction, faculty at Massachusetts Inst. of Technology (1901–1933), consulting architect for US Navy, NAS (1929) (BM36 MWBD NC41 PS WWWIA)

HOWARD, Henry, 1868–1951, American chemical engineer; with Merrimac Chemical Co., which became a subsidiary of Monsanto Co. (1889–1920), becoming vice president (1902–1920), leader in improving sulfuric acid production, developed Howard dust chamber, in charge of research and development for Grasselli Chemical Co. (OH) (1920–1927), was a leading yachtsman in New England, 89 patents (DAB)

HOWE, Elias, 1819–1867, American inventor and engineer; machinist, textile machinery, workable sewing machine (Howe) that was widely used in the home (1846) and in the shoe industry, made improvements to the sewing machine of Isaac Merritt Singer [1811–1875] (1851), organized Howe Machine Co., (1865), the Remington sewing machine followed (1870) (AIE ANB ASAI BEST EAI EI EOWB GI IAI IIA I&T17 MEIA MMAH MWBD NC4,9 PS RHW TGS WWWIA WWWIS)

HOWE, Frederick Webster, 1822–1891, American machine tool builder and inventor; machinist, mechanical drafting, superintendent of machine shop (1848), profiling machine used for gun manufacturing (1848), first commercially successful universal milling machine (1850), established armory (1856), assisted Elias Howe [1819–1867] in sewing machine development (1865), later worked for and was administrator of Brown & Sharpe Manufacturing Co. (1868ff), developed turret lathe (DAB MEIA)

HOWE, Harrison Estell, 1881–1942, American chemist and chemical engineer; chemist for Sanilac Sugar Refinery Co. (MI) (1902), Bausch & Lomb Optical Co. (NY) (1904), with Arthur D. Little Inc. in Boston and Montreal as chemical engineer, army ordnance during WWI, National Research Council (1919–1921), editor of *Industrial and Chemical Engineering* (1921–1942) (ACCE)

HOWE, John Ireland, 1793–1876, American inventor and manufacturer; initially a physician, developed and then abandoned rubber products plant, invented pin-making machine, organized Howe Manufacturing Co. (1833), several patents on machines to make pins (DAB)

HOWE, William, 1893–1852, American civil engineer and inventor; built bridges, improved truss bridge particularly for wooden bridges, patented improved Howe truss design (1840), trusses used for railroads, bridges, and roofs, improved on Long type of truss (1840) developed by Stephen H. Long [1784–1864] (1930) (BDOACE DAB EIH MWBD NC7)

HOWELL, John Adams, 1840–1918, American naval officer and inventor; astronomy and navigation, hydrographic survey, inspection of steel ordnance, commanded ships, in charge of navy yards in Washington (DC) and Philadelphia, promoted to rear admiral, patents dealing with gyroscope, high-explosive shells, torpedo launching apparatus (1885–1892), and disappearing gun carriage (1896) (DAB MEIA)

HOWEY, Walter Crawford, 1882–1954, American journalist and inventor; journalist and editor of several midwestern newspapers, invented and patented an automatic photoelectric engraving machine (1931), sound photo method of transmitting halftone pictures over telephone lines (1935), obtained 17 patents (ANB)

HOXIE, Charles A., c. 1867–1941, American inventor; with General Electric Co. (1912–1932), father of sound pictures, invented process for changing sound into light and recording on film, message transmission and recording, developed photophone (WWWIS)

HOXIE, William Dixie, 1866–1925, American mechanical engineer and inventor; marine work, Babcock & Wilcox Co. (1919–1924), adaptation of boilers to marine uses, promulgated water-tube boilers, advocate of high-pressure superheating (DAB NC24)

HUBBARD, Henry Griswold, 1814–1891, American inventor and manufacturer; machine production of elastic webbing (1841), purchased control of Russell Manufacturing Co. (1850) to make webbing of various types, spinning wheels (DAB MEIA)

HUBER, Maksymilian Tytus, 1872–1950, Polish mechanical engineer; strength of materials, theory of orthotropic plates, Huber-Mises-Honcky theory, measure of hardness, concrete plates, professor at Lvov Inst. of Technology (1899–1906) becoming rector, at Warsaw Inst. of Technology (1928–1938), work interrupted by WWI and WWII, after WWII at Gdansk Inst. of Technology (1945ff) (DOSB PS)

HUBERT, Conrad, 1855–1928, Russian American inventor; invented electric device for lighting gas (1900), electric alarm (1902), electric battery (1902), small electric lamp (1902), organized and president of American Ever Ready Co., sold company to National Carbon Co. (OH), formed Yale Electric Corp. (DAB NC24 WWWIA)

HUDSON, Donald Ellis, 1916–1999, American mechanical engineer; shock and vibration, dynamic instrumentation, earthquake engineering and seismological studies, General Petroleum Corp. (1937ff), faculty at California Inst. of Technology (1941ff), NAE (1973) (AMOS Bridge WWIE)

HUDSON, Herbert E., Jr., 1910–1983, American sanitary engineer; water supply, filtration, water resources, hydrology, worked with Chicago Water Department, research at Univ. of Illinois, Illinois State Water Survey,

Army Corps of Engineers in WWII, associate and partner in firm of Hazen & Sawyer (1955–1971), Univ. of Florida (1971ff), NAE (1978) (MT3 WWIE)

HUDSON, William Smith, 1810–1881, British American mechanical engineer and inventor; machinist, locomotive manufacturing in England, to US (1835), worked with several companies as locomotive engineer, construction of facilities, superintendent of locomotive works, made many improvements to locomotives mainly through simplification, about 15 patents relating to boilers and locomotives (DAB MEIA NC24 PS)

HUEBNER, George Lee, Jr., 1918–1996, American environmental engineer; development of passenger car and light truck gas turbine engines, administrator of environmental research institute, engineer with Chrysler Corp. (1931–1975), becoming executive engineer and director of research, NAE (1975) (Bridge MT9 WWWIA)

HUGGINS, Leroy Gale, 1900–1971, American electrical engineer and inventor; Westinghouse Electric Co. (1923–1949), working on transmission engineering and switchboard engineering, air conditioning departments in Westinghouse factories, Horix Manufacturing Co. (1949–1950), sales engineer (1951–1962) for air conditioning systems, consulting engineer, 8 patents (NC56 PS)

HUGGINS, Sir William, 1824–1910, English astronomer and inventor; with William A. Miller [1817–1870] invented stellar spectroscope (c. 1860), fNAS (1904) (MWBD WWWIS)

HUGHES, David Edward, 1831–1900, English American inventor; printing telegraph or early telex (1855), carbon microphone (1878), induction balance (1879), electromagnets (CEOS&T DAB MWBD NYPL RHW)

HUGHES, George Wurtz, 1806–1870, Irish American civil engineer; topographical engineering, surveys of District of Columbia for US government, did mapping in Mexico in advance of army, surveyed railroad route across Panama, consulting engineer (DAB NC4 WWWIA)

HUGHES, Hector James, 1871–1930, American civil engineer; town engineer, municipal and sanitary engineering work, hydraulics, after government service and industry experience was on faculty and dean of engineering at Harvard Univ. (1902–1914) (1918ff), faculty and administrator at Massachusetts Inst. of Technology (1914–1918) (WWIA WWIE DAB)

HUGHES, Howard Robard, 1869–1924, American inventor and manufacturer; mining, drilling business, patented rotary drills for penetrating hard rock for oil well drilling (1903) (1909), organized Sharp-Hughes Tool Co. (1909), which became Hughes Tool Co. (1912ff), developed gate valve (AIE ANB DAB WWWIA)

HUGHES, Howard Robard, 1905–1976, American industrialist, aviator, and inventor; airplane design, speed records with airplanes, wooden flying boat, took over Hughes Tool Co. (1923ff), formed Hughes Aircraft Co., controlled RKO Studios (1948–1955), had controlling interest in Trans World Airlines (TWA) (ante 1966) (AIE IAI MWBD)

HUKILL, William V., 1901–1987, American agricultural engineer; preservation of farm products, refrigeration of fresh fruits in storage and in transit, principles of aeration and drying, grain respiration, duct airflow, joint USDA and faculty at Iowa State Univ. (1943–1967), 5 patents (AE49 WWIE)

HULBERT, Edwin James, 1829–1910, American mining engineer; surveyor and engineer for several copper mining companies, discovered copper deposits for mines, Calumet & Hecla Co. mines in northern Michigan (BITAS DAB)

HULETT, George H., 1847–1923, American mechanical engineer and inventor; developed series of patents for hoisting, conveying, loading and unloading (ante 1921), important invention was Hulett automatic ore unloading machine (1890), worked for several companies, vice president of Hulett Engineering Co. (OH) (1918ff) (MEIA)

HULL, Albert Wallace, 1880–1966, American physicist, electrical engineer, and inventor; instructor (1909–1912), physicist at General Electric Co., becoming assistant director (1914–1950), said to have developed more types of radio tubes than any other person, including such devices as dynatron, magnetron (1921), screw-grid tube, and thyratron (1936), communications during WWII, consultant (1950ff), NAS (1939), 91 patents (BM41 DOSB MWBD NC53 PS TFE&I WWWIA)

HULLS, Jonathan, 1699–1758, English inventor; tugboat powered by Newcomen (Thomas Newcomen [1663–1729]) steam engine (1736), steam-driven vessel (EIH MWBD)

HUMANN, Carl, 1839–1896, German railway engineer and archeologist; construction of railways, archaeology of Asia Minor, discovered sculpture and art works in excavations (1878–1896) (MWBD)

HUME-ROTHERY, William, 1899–1968, English metallurgist and chemist; alloys and intermetallic compounds, size of component atoms in alloys, construction of equilibrium diagrams for many alloys, at Oxford Univ. (1926–1968) (CEOS&T MWBD RHW WWWIS)

HUMPHREY, Arthur Luther, 1860–1939, American mechanical engineer; worked for railroad, mining, machinery, and gold mining, after work with Washington Iron Works (WA) became locomotive engineer for Puget Sound Shore Railroad, was superintendent of motive power for several railroads, with Westinghouse Air Brake Co. (1903–1938), becoming director and chairman of executive committee, except for service in ordnance in WWI (MEIA)

HUMPHREYS, Alexander Crombie, 1851–1927, Scottish American mechanical engineer; chief engineer of lighting company building oil-gas plants, Humphreys & Glasgow Co. (Arthur B. Glasgow [1865–1955]) (1892) and later Humphreys and Miller, Inc. (1910), built water-gas plants throughout the world, research in illumination and photometry, consulting engineer, president of Stevens Inst. of Technology (1902ff) (DAB MEIA NC22 WWWIA)

HUMPHREYS, Andrew Atkinson, 1810–1883, American hydraulics engineer; topographical engineer in US artillery, harbor and bridge projects, coastal survey, Mississippi River Delta survey, railroad surveys, NAS (1863) (ANB BDOAS(El) BITAS BM2 DAB NC7 PS)

HUND, August, 1887–1952, German American electrical engineer and physicist; to US (1912), testing of electric equipment, design engineer, research engineer for General Electric Co. (1912–1914), faculty at Univ. of Southern California (1915–1917), consulting research engineer (1918–1922), physicist with National Bureau of Standards (NBS) (1929–1936), consulting engineer, author of numerous articles covering a wide range of interests in telephone, radio, and electronics, at least 45 US patents (NC41 PS)

HUNLEY, Horace Lawson, 1823–1863, American inventor; submarine vessels, served in Confederate Navy (1861ff), vessel named Hunley submarine (1863) (ASAI MWBD)

HUNSAKER, Jerome Clarke, 1886–1984, American aeronautical engineer and naval architect; airplane design, aerodynamics, wind tunnel development (1916), naval attaché, faculty at Massachusetts Inst. of Technology (1912–1916) and head of department (1933–1947), US Navy (1916–1926), assistant vice president of Bell Telephone Labs. (1926–1928), president of Zeppelin Corp. (1928–1933), numerous advisory committees and councils, NAS (1935), NAE (1967) (AMOS ANB BM78 IWAP MT3 NYTObit WWIA WWIE)

HUNT, Alfred Ephraim, 1855–1899, American metallurgist and engineer; chemist and assistant manager of open-hearth plant of Bay State Steel Co. (1876–1877), worked in iron ore fields of northern Michigan and Wisconsin, superintendent of Nashua Iron & Steel Co. (1877–1881), with various companies as metallurgist and consulting engineer, helped establish Pittsburgh Production Co. for aluminum manufacture (ANB BITAS DAB MEIA NC25)

HUNT, Charles Wallace, 1841–1911, American mechanical engineer and manufacturer; formed C. W. Hunt Co. (1871), patented coal handling system (1872), built coal handling systems and storage plants, designed bucket conveyor systems for handling coal and ashes, developed flexible steel rope drives, was president of ASME (DAB MEIA NC13)

HUNT, Robert Woolston, 1838–1923, American mechanical and civil engineer; worked on development of Bessemer process and installation of plants in Pennsylvania and Michigan, superintendent of Troy Iron & Steel Co. (1875–1888), with others invented automatic rail mills, rebuilt blast furnaces, consulting engineer (1888ff) (MEIA NC19)

HUNT, Walter, 1796–1859, American inventor; sewing machine, double-lock stitch (1834), Glove stove, safety pin (1849), paper collar (1854), repeating rifle, revolver, metal cartridges, alarm gong, machine to make nails and rivets (ASAI GI IIA MEIA MWBD NC19 1000Y TEOUT WA)

HUNTINGTON, Collis Potter, 1821–1900, American railroad builder; transcontinental railroad (1861ff) joining Central Pacific Railroad and Union Pacific Railroad (1869), Southern Pacific Railroad (1884), becoming president (1890ff), also involved in steamship activities (MWBD NC15 WWWIA)

HUNTINGTON, Henry Edwards, 1850–1927, American railway executive; nephew of Collis Potter Huntington [1821–1900], developed interurban transit systems, railroad interests, executive with Huntington Railroads (1881–1890) (MWBD NC15 WWWIA)

HUNTSMAN, Benjamin, 1704–1776, English inventor; clockmaker, instrument maker, spring steel, invented cast or crucible steel manufacture (1740ff) (EIH MMAH MWBD PS WBE)

HURLBUT, Lloyd Wendling, 1908–1965, American agricultural engineer; faculty at Univ. of Nebraska (1934–1943), US Navy (1943–1946) working on underwater sound detection, becoming lieutenant commander, returned to Univ. of Nebraska (1947–1965) as professor and chairman of department, pioneer in development and use of tires for farm tractors, use of combine for harvesting corn, corn drying, minimum tillage, several patents (AE46)

HUSSEY, Obed, 1792–1860, American inventor; worked on models of reaper at agricultural implement factory of Richard B. Chenoweth (MD), invented and manufactured Hussey reaper (1833) with reciprocating knives against stationary guards, steam plow (1855) (a steam plow pulled by cable was invented by E. C. Bellinger (SC) (1833), invented machine for making hooks and eyes, grinding mill for corn, horse-powered husking machine, sugar cane crusher, ice-making machine, Englishman John Fowler [1826–1864] developed steam plow (1854) (DAB IAI MEIA MWBD NC11 PS RHW TAT TBDOS TCBE TTOT WWWIA)

HUTCHINS, Thomas, 1730–1789, American military engineer; US geographer, boundary surveys, fortifications, map of Virginia, cartographer (BDOACE BDOAS(El) BITAS DAB MWBD)

HUTTON, Frederick Remsen, 1853–1918, American mechanical engineer; educated as a mining engineer, faculty and administration at Columbia Univ. (1877–1907), including dean (1899–1905), author of several textbooks, edited magazines, consulting engineer (BITAS DAB MEIA)

HUTTON, William Rich, 1826–1901, American civil engineer; surveyor, chief engineer of Washington viaduct (1862–1863), chief engineer of Annapolis water works (1865), Chesapeake and Ohio Canal (1869–1874), Western Maryland Railroad (1871–1874), consulting engineer (1871–1881), chief engineer of Hudson River tunnel (1889–1891), chief engineer of Washington Bridge over Harlem River (NYC) (BDOACE)

HUYGENS, Christiaan, 1629–1695, Dutch scientist and inventor; improved accuracy of pendulum clock particularly for marine use (1649ff), gunpowder-fired engine, new method of grinding lenses for telescopes (1655), impact of elastic bodies (1669), spiral balance spring watch (1675), wave theory of light (1678), probability theory (AmSci84 BEST EIH EOWB GEAPIT GI MMAH MWBD NYPL 1000Y SAI TEIH TGS WA WWWIS)

HYATT, John Wesley, 1837–1920, American inventor; developed composition billiard ball (a substitute for ivory), celluloid—the first synthetic plastic (1869), water filter, roller bearings (1891–1892), lock-stitch sewing machine, sugar cane mill, machine for cold rolling steel shafting, established Hyatt Pure Water Co. (1887), established Hyatt Roller Bearing Co. (1892), over 200 patents (AIE ANB BEST BITAS DAB EB GI IAI MWBD NYPL PS RHW TBDOS WA)

HYATT, Thaddeus, 1816–1901, American civil engineer and inventor; developed sidewalk grating to permit light and fresh air to basements (c. 1845), floor slabs of reinforced concrete, patented reinforced concrete (1878) (BDOACE)

HYDE, (J.) Franklin, 1903–1999, American inventor and chemist; process for making fused silica glass, silicon products for lubricants, hydraulic fluids, water repellents, work led to optical fibers, work led to founding of Corning Glass Co. (1943) (ET22 NYPLSciRef)

HYPATIA, c. 370–415, Greek woman natural philosopher; designed planisphere, an astrolabe, a hydrometer to measure specific gravity (TBDOS(Po))

I

IBUKA, Masaru, 1908–1997, Japanese electrical engineer; president, co-founder with Morita Akio [1921–1999] and chief advisor of Japanese Sony electronics company (1954ff), leader in development of transistor radio, television receiver, videotape recorder, fNAE (1976) (Bridge IAI WWWIA)

ICTINUS, 5th century, Greek architect and builder; Parthenon in Athens (447–438 BC), construction of Acropolis, designer of other temples and public buildings (EB MWBD)

IDE, Albert L., 1841–1897, American mechanical engineer; during Civil War was major in infantry, built railroad line in Springfield (IL), steam heating business (1870), electric lighting at Edison Lab. (NJ) (1880), designed and built Ideal engine with governing system and self-oiling system (1886) (MEIA)

IL'YUSHIN, Aleksei Antonovich, 1911–??, Soviet mechanical engineer; professor at Moscow Univ. (1938–1943), USSR Academy of Sciences (1943ff) becoming director of Inst. of Mechanics (1953), strength of materials, viscous-plastic flow and its stability for metals (1936–1938), pile driver (1937), theories of elasticity and plasticity, theory of plates and shells beyond limits of elasticity, preparing metals by pressure, postulate on isotropy (1953), gaseous dynamics (SMOS WWWIS)

ILYUSHIN, Sergei Vladimirovich, 1894–1977, Soviet aeronautical engineer and aircraft designer; military aircraft for bombing (1936) and for attack (1941), after WWII concentrated on four engine airliners identified by IL-, returned to Air Force Engineering Academy as professor (1948ff) (BDOTHOT MWBD)

IMHOFF, Karl, 1876–1965, German engineer and inventor; developed the Imhoff tank, which was originally a two-level tank with flow based on gravity with one level for sedimentation and digestion of sludge collected at a lower level, invented Imhoff cone, which was a graduated container to measure settleable solids, consulting engineer in Essen (Germany), wrote several books on water treatment (EB NUC)

IMHOTEP, fl. c.2980–c.2950 BC, Egyptian engineer and scholar; master of stone construction, pyramids, first engineer and architect to be known (recorded) by name, chief of works for King Zoser (or Djoser), father of stone masonry (BEST E-64 EIH GEAPIT MWBD PS TAE TEIH TGS TSOE WWWIS)

INGENHOUSZ, Jan, 1730–1799, Dutch physician, physiologist and inventor; explained photosynthesis, plant respiration, invented improved apparatus for generating large amounts of electricity, explained role of oxygen in plants and animals, invented hydrogen-fueled lighter, use of air-ether mixture to propel bullet in electrically fired pistol (BEST RHW WWWIS)

INGERSLEV, Fritz Halfdan Bent, 1912–1994, Danish electrical engineer; acoustics, noise-control engineering, faculty at Technical Univ. of Denmark (1936–1981), fNAE (1982) (Bridge MT8 WWWIA)

INGERSOLL, Robert Hawley, 1859–1928, American industrialist; manufacturer of inexpensive watch (then for $1), developed mail order business (MWBD WWWIA)

INGERSOLL, Simon, 1818–1894, American inventor; mechanical experimenter, rotating shaft for steam engine (1858), steam wagon, friction clutch, gate latch, spring scale, compressed air drilling machine (1871), his works became a part of Ingersoll, Betts & Co., used vegetable farming as source of income, 20 patents dealing with rock drilling (DAB MEIA WA WWWIA)

INGLIS, Claude Cavendish, Sir, 1883–1974, British civil engineer; irrigation research and development, served in India, hydraulic engineering research, control of rivers and canals (PS RS21)

INOUYE, Goro, 1899–1981, Japanese electrical engineer; power engineering, high-voltage transmission, pump storage plant, research and design of fast breeder reactor, nuclear power plants, Chubbu Electric Power Co. (1923ff) becoming president (1951) and chairman of the board (1961–1967), fNAE (1977) (MT2)

INSLEE, William Harvey, 1841–1898, American mechanical engineer; machinist, foreman, draftsman at Hewes & Phillips Iron Works (NJ), designed special tools for Singer Co. with development of patentable devices (1869ff), to Scotland to improve Singer plant in Glasgow, later superintendent and general manager of Kilgore Works (MEIA)

INSULL, Samuel, 1859–1938, English American industrialist; to US (1881), built large turbines for General Electric (1903), vice president of Edison General Electric Co. (1889), president of Chicago Edison Co. (1892), companies merged to form Commonwealth Edison (1898), companies went into receivership (1932) (I&T20 MWBD TMA)

IPATIEFF, Vladimir Nikolayevich, 1867–1952, Russian American chemist and inventor; hydrocarbons, explosives, polymerization of ethylene in reduction process (1913), to US (1930), high-temperature catalysis of petroleum products to provide a higher octane rating (1930), faculty at Northwestern Univ. (1930ff) and Universal Oil Products (1930ff) including director of Catalytic High Pressure Lab., over 200 patents, consultant, NAS (1939) (BDOS(Wi) BEST BM47 CBD DOSB NCE PS RHW WA WWWIS)

IPPEN, Arthur Thomas, 1907–1974, American civil engineer; water resources, hydrodynamics, free surface flow, open channel flow, sediment transport, waves, coastal engineering, faculty at California Inst. of Technology (1934–1938), Lehigh Univ. (1938–1945), Massachusetts Inst. of Technology (1945ff), NAE (1967) (AMOS MT1 WWIE)

IPPONMATSU, Tamaki, 1901–1985, Japanese electrical engineer; nuclear engineering, development of electric power in Japan, authority on the nuclear fuel cycle, managing director of Kansai Electric Co. (1947–1957), with Japan Atomic Power Co. (1957–1985) becoming president (1962) and chairman (1970–1977), fNAE (1978) (MT3)

IRVIN, George Rankin, 1907–1998, American physicist and mechanical engineer; pioneer in development of modern fracture mechanics, penetration of projectile on target, engineering design, research and administration at US Naval Research Lab. (NRL) (1937–1967), faculty at Lehigh Univ. (1967–1972), Univ. of Maryland (1972ff), NAE (1977) (MT10 WWIA WWIE)

IRVING, Roland Duer, 1847–1888, American mining engineer; superintendent of smelting works, Ohio Geological Survey, Wisconsin Geological Survey, USGS studies of Lake Superior region, studied copper ore and iron ore of the region (BDOAS(El) BITAS DAB PS)

ISAACS, John Dove, 1913–1980, American oceanographer and consulting engineer; waves, motions of organisms, instruments for deep sea work, current meters, communication and signaling devices, seaplane base construction, chief engineer for Austin Co., faculty at Scripps Inst. of Oceanography (1948–1971), director of California Inst. of Marine Resources (1971ff), NAS (1974), NAE (1977) (BM57 MT2)

ISHERWOOD, Benjamin Franklin, 1822–1915, American naval and civil engineer; mechanical department of Utica and Schenectady Railroad, construction of Croton aqueducts, construction of lighthouses for US Treasury, designed new type of lens, US Navy chief engineer who improved use of steam engines, Bureau of Steam Engineering (1862–1870), research on increasing the speed of ships, presiding official for special navy boards, retired as rear admiral (1915) (ANB BDOACE BITAS DAB MEIA MWBD NC12 WWWIA)

ISHLINSKII, Aleksandr Yulevich, 1913–??, Soviet mechanics specialist; Moscow Univ. (1935–1948) becoming professor (1945), director of mathematics at Ukrainian SSR Academy and professor at Kiev Univ., general mechanics, elasticity, oscillation, gyroscopic device, academician at USSR Academy of Sciences in Moscow (1960) (SMOS WWWIS)

ISRAELSON, Orson Winso, 1887–1968, American irrigation and drainage engineer; water rights, design of earth dams, control of water, erosion control, seepage from irrigation ditches, technology of soil-water relationships, faculty at Utah State Univ. (1915–1953) including acting dean of engineering (1928–1929), collaborator with the US Department of Agriculture (USDA), consulting engineer (AE49 AMOS WWIE(I))

ISSIGONIS, Alexander Arnold Constantine, 1906–1988, Turkish British engineer; automobile designer for Morris Minor (1948) and Morris Mini-Minor (1959), first worked on suspension systems (1936) then later was in charge of separate approaches to various automotive activities including styling, interior, body engineering, and chassis for steering and suspension, credited with use of term mini in English language (EAI RHW)

IVES, Frederick Eugene, 1856–1937, American inventor; photoengraving process, halftone process of printing (1886), three-color printing (1892), optical screen process, binocular microscope (ANB MWBD NC15 RHW WA WWWIA WWWIS)

IVES, Frederick Eugene, 1884–1924, American agricultural engineer; engineering drafting, at Ohio State Univ. (1909–1924) in charge of farm structures then head of department (1920–1924), former campus building named in his honor, consulting engineer (AE5)

IVES, Herbert Eugene, 1882–1953, American physicist and inventor; son of Frederick Eugene Ives [1856–1937], research with Bell Labs. (1919ff), illuminating engineering, development of television and transmission of pictures by wire, NAS (1933) (ANB BM29 EIH MWBD)

IVES, Joseph Christmas, 1828–1868, American topographic engineer; survey of 35th parallel (1853–1854), explored Colorado River (1857–1858) including hydrographic survey, engineer and consultant for Washington monument (DC) (1859–1860), survey of boundary of California, served in Confederate Army during Civil War (DAB WWWIA)

IWASAKI, Yataro, 1835–1895, Japanese industrialist; manager of finances for Tosa Co., developed shipping line (1868), founded Mitsubishi Co. (1873) as one of Japan's largest industrial and financial businesses in automobiles and machinery (MWBD 1000Y WWWIS)

J

JABLOCHKOFF, Paul, 1847–1894, Russian electrical engineer and inventor; lighted boulevard in Paris with arc lights (1877) (E-78 EIH PS WBE)

JACKLING, Daniel Cowan, 1869–1956, American mining engineer and metallurgist; mining and processing low grade copper ore, work on metallurgy (1894ff) led to establishment of Utah Copper Co. (1905) (MWBD NCD)

JACK OF NEWBERY, 16th century, English industrialist; built large-scale weaving factory (MMAH)

JACKSON, Willis (aka Baron Jackson of Burnley), 1904–1970, British electrical engineer; electrical insulating materials, dipole rotation in solids, dielectric properties, wave-guides, lecturer at Bradford Technical College (1926–1929), Metropolitan-Vickers Co. (1929–1930) then director of research (1954ff), faculty of Manchester College of Technology (1930–1933), research at Oxford Univ. (1933–1936), faculty at Imperial College

of Science and Technology (1946ff) becoming administrator (1967ff), FRS (1953) (DNB PS RS17 WWWIS)

JACOBS, Arthur Irving, 1858–1918, American mechanical engineer and inventor; with Knowles Loom Works (1880ff), worked on improved methods of making harness chain for looms, patented book sewing machine (1887), then with Smyth Manufacturing Co. where he patented book cover machines, method of making needles, cloth cutting machine, patented drill chuck (1902), established Jacobs Manufacturing Co. of which he was president (1903) (MEIA NC16)

JACOBSEN, Lydik Siegumfeldt, 1907–1976, Danish American mechanical engineer; engineering seismology, vibration of mechanical systems, structural dynamics, earthquakes, Westinghouse Electric & Manufacturing Co. (PA) (1921–1924), faculty at Stanford Univ. (1924ff) including department head (1944), US Navy in WWII, NAE (1975) (AMOS MT1)

JACOBY, Henry Sylvester, 1857–1955, American civil engineer; surveying (rodman, transitman, draftsman) for US engineering office, second geological survey of Pennsylvania, Corps of Engineers (1878–1885), faculty at Lehigh Univ. (1886–1890), faculty at Cornell Univ. for civil engineering, bridge engineering, graphics (1890–1922), active in numerous professional education societies, numerous publications (BDOACE NCD WWWIS)

JACQUARD, Joseph-Marie-Charles, 1752–1834, French engineer and inventor; textile manufacturer, automatic loom called Jacquard loom for figured and patterned weaving particularly for carpets (1804), mechanical silk loom (1805), invented punch cards (1790) called Jacquard cards to partially control weaving later used for computing, punch cards used by American Herman Hollerith [1860–1929] in mechanical tabulator (EAI CEOS&T EAI GEAPIT GI MMAH MWBD NYPL 1000Y PS RHW SAI WBE)

JACQUET, Pierre Armand, 1906–1967, French chemical engineer; electrochemical research at Le Matériel Telephoniques Society (1921), developed micrographs and solved metallurgical problems while with the French navy (1945–1966), nondestructive metallurgy (CDOSB DOSB)

JACUZZI, Candido, 1903–1986, Italian American inventor; made commercially available health and recreational products (mid-1950s), pumps for agriculture wells, Jacuzzi for whirlpool and hot tubs named after him, issued 31 US patents in his name with 21 other patents issued to eight different family members (AIE ANB RHW)

JACUZZI, Rachele, 1886–1937, Italian American inventor; brother of Candido Jacuzzi [1903–1986], invented laminated wooden metal tipped airplane propeller called toothpick, built J-7 high winged monoplane called the Mosquito, crop defroster called Frostifuga, small steam jet, jet pump, high pressure car wash pump, several patents (ANB)

JADWIN, Edgar, 1865–1931, American soldier and engineer; military education (1890), Corps of Engineers with duty on river and harbor improvements (1890–1897), assistant chief of engineers (1897–1898), engineering projects on Pacific coast and Galveston (TX), assisted in Panama Canal construction (1907–1911), brigadier general in Army Expeditionary Force (WWI), numerous US army projects retiring as lieutenant general (1920), consulting engineer (ANB DAB NC27,A)

JAFFEE, Robert Isaac, 1917–1991, American metallurgist; metallurgy of titanium, molybdenum, chromium, copper, nickel, germanium, platinum, tungsten, niobium, nickel-manganese alloys, order-disorder transformation in nickel-manganese alloys, research and chief of materials science at Battelle Memorial Inst. (1943–1975), head of materials research at Electric Power Research Inst. (EPRI) (1975ff), 40 US patents, NAE (1969) (AMOS Bridge MT6 WWIE)

JAHAN, Shah, 1592–1666, Indian builder; built the Taj Mahal—mausoleum to emperor's wife—that took 22 yr. to complete (1000Y)

JAMES, Charles Tillinghast, 1805–1862, American engineer; construction of textile machinery, superintendent of Slater's steam powered cotton mills, proponent of steam cotton mills in east coast area, started 23 steam cotton mills of which 16 were in New England, senator in US Congress, worked on improvement of firearms (DAB NC3)

JAMES, Richard, ??–1974, American marine engineer and inventor; invented inexpensive novel toy called slinky (1943) introduced (1945), employed at Philadelphia Cramp Shipyard (PA), moved to Bolivia where he was involved with a religious cult (WOI)

JANCKE, Gunnar, 1913–1996, Swedish electrical engineer; research and development of extra-high voltage power systems, series capacitors, 400 kV electric transmission, direct current transmission, consultant with SwedPower AB, fNAE (1977) (Bridge WWIE)

JANG SZ-HSUN, 10th century, Chinese mechanician; built grand clock (976 AD) (SACIC)

JANIN, Louis, 1837–1914, American mining engineer; western metal mining, methods of extracting silver, in charge of Enriquetta quicksilver mine in Ecuador, gold, silver and copper, Herbert C. Hoover [1874–1964] attracted to his work, consulting engineer (DAB NC18)

JANNEY, Eli Hamilton, 1831–1912, American inventor; served in Confederate Army in quartermaster rising to rank of major, improved method of automatically coupling railroad cars (1868), formed Janney Car Coupling Co., coupling adopted by Pennsylvania Railroad

Co. (PRR) (1874ff), several improvements made on original patent (ANB DAB MEIA WBE)

JANSKY, Karl Guthe, 1905–1950, American electrical engineer; radio communications, static in atmosphere that interfered with radio reception, extraterrestrial origin of some radio waves (1932), birth of radio astronomy, at Bell Telephone Labs. (1928–1950) (ANB BEST GI IAI MWBD NYPL TBDOS PA RHW WBE WWWIA)

JANSSEN, (Pierre) Jules César, 1824–1907, French astronomer, physicist and inventor; developed spectrohelioscope (1868), faculty at École Spéciale d'Architecture (1865ff), studied solar spectrum, director of Meudon Observatory in France (1876ff), established observatory on Mont Blanc (1893) in the Alps, fNAS (1901) (CDOSB DOSB GLOSB RHW MWBD MWGD WWWIS)

JANSSEN (or JANSEN), Zacharias, 1580–c.1638, Dutch inventor; spectacles, invented compound microscope (1590), with Hans Lippershey [c.1570–c.1619] made first telescope (1608) (ASAI HIW MWBD WWWIS)

JAQUES, William Henry, 1848–1916, American mechanical engineer; served on US Coast Survey (1870), ordnance inspector (1881), invented improvements in manufacturing heavy ordnance and armor, served in US Navy (1867–1887), consulting engineer (1895ff), president of Holland Submarine Boat Co. (1904ff), president of Progress Manufacturing Co. (1913–1916) (BITAS MEIA WWWIA)

JARVES, Deming, 1790–1869, American chemist and inventor; leader in the glass industry, became co-owner of New England Glass Co. manufacturing flint and crown glass, innovated use of various minerals in glass to provide color, appearance and texture, made special glass door knobs, containers, bottles, tableware and other glassware (DAB WWWIS)

JATHO, Karl, 1873–1933, German aviator and manufacturer; early flight in mechanical airplane (1903), airplane manufacturing plant (1913), aviation school (MWBD)

JEFFERSON, Thomas, 1743–1826, American politician and inventor; using mathematics developed best shape for plow moldboard—not patented, invented spherical sun dial (c.1810), recognized the value of interchangeable parts for military arms, farmer, writer, educator, surveyor, architect, lawyer, politician at state and national level, president of US (1801–1809) (ANB ASAI BEOS BEST BITAS CDOSB DAB DOSB EB EIH HBW IIA MMAH MWBD NC2,3 1000Y WBE WWWIA WWWIS)

JEFFREY, Max Leroy, 1887–1971, American inventor and manufacturer; US Army in WWI, built stationary gas engine, first model of a steering gear, Gemmer Steering Gear Co., Haynes Automobile Co., Warner Gear Co, that became Borg Warner Co., four cylinder automobile engine, equipment for oil well industry, brushes for Bissel carpet sweepers, clutches and differentials, with White Motor Truck Co., Chevrolet Motor Co. (MI) (1915), Wright Aeronautical Co. (NJ) (1929), 20 patents often licensed to other companies (DAB NC56 PS)

JEFFRIES, John, 1745–1819, American physician and aeronaut; two balloon ascensions—London (1784) and Dover across channel with Frenchman Jean Pierre François Blanchard [1753–1809] (1785), carried first air mail letter from London to Paris on flight (1785), meteorologist, air sampling, altitude relationships (ASAI BDOAS(El) I&T15 MWBD NC24 WWWIS)

JEFFRIES, Zay, 1888–1965, American metallurgist and mining engineer; ore analysis, mine superintendent, faculty of Case School of Applied Science (1911–1917), X-ray diffraction, Jeffries method (1916) for grain structure identification of metals, strength of tungsten, developed slip theory of metal hardening (1921), executive of Carboloy Co. (OH) (1932) that became a part of General Electric Co., Manhattan project in WWII, vice president of General Electric Co. (1945), consulting engineer, NAS (1939) (ANB NCA WWWIS)

JENKIN, Henry Charles Fleeming, 1833–1885, British (Welsh) electrician, mechanical and electrical engineer and inventor; marine telegraphy with Liddell & Gordon, resistance and insulation of electric wires, submarine cables (1858–1873), professor at Univ. College London (1865–1868), Edinburgh Univ. (1868ff), invented telpherage (1882), electrical standards and units, 35 British patents, FRS (1865) (DNB DOT MWBD PS WWWIS)

JENKINS, Charles Francis, 1867–1934, American physicist, engineer and inventor; radio photography, inventor of projector and first commercial movie presentation (1894), radio photography, president of Jenkins Lab., contributed to development of television, vice president of Jenkins Television Corp., 400 US and foreign patents (AMIH EIH GI IAI PS TBOF WA WWWIA WWWIS)

JENKINS, Lawrence Eugene, 1933–1996, American chemical engineer; Lockheed Missiles & Space Co. (1957ff) becoming manager of space programs (1963–1971) and vice president (1981), NAE (1984) (MT9 WWWIA)

JENKINS, Nathaniel, 1812–1872, American inventor and manufacturer; coppersmith, activities culminated in firm of Jenkins & Co. (1853), silversmith and clock maker, invention and production of water faucets (1866) with rubber compound packing, steam globe valve (1867), manufacture of valves by his sons with Jenkins Brothers Co. (DAB)

JENKINS, Ronald, 1907–1975, British civil engineer; reinforced concrete shells, plates, slabs, worked for Ove Arup & Partners (1943ff) becoming partner (1949), retired (1973) (TGE:B)

JENKS, Joseph, 1602–1683, Welsh American inventor; casting utensils and tools, sawmill, cut dies for first coins in America (1652), probably first fire engine (1654), manufacture of scythe to cut grass (1655) (ANB DAB IIA)

JENKS, Stephen Moore, 1901–1974, American mechanical engineer; research and administration with US Steel Corp. (1927–1967) becoming executive vice president, consultant, NAE (1968) (MT1 WWIE)

JENKS, William Hampden, 1852–1938, American mechanical engineer; with Neafie & Levy Shipbuilding Co. (PA), became partner of Brown, Roth & Jenks (1877), built a company that he and his father purchased, patented inertia governors (1889) (MEIA)

JENNER, Edward, 1749–1823, English physician and inventor; discovery and development of vaccination (1798), developed procedure for vaccination against smallpox (1796) that was announced (1798) (BEST MWBD SAI-MP TBOF WWWIS)

JENNINGS, Burgess Hill, 1903–1996, American mechanical engineer; air conditioning, refrigeration, ventilation, thermodynamics, steam power, gas turbines, research and management in industry, faculty at Lehigh Univ. (1926–1940), faculty and administration at Northwestern Univ. (1940–1972) except for 3 yr. as director of ASHRACE, consulting engineer, NAE (1977) (AMOS Bridge MT10 WWIE)

JENNINGS, Isaiah, 1782–1862, American inventor; began as blacksmith, invented machine for making thimbles and eyelet holes, started business in England then returned to US, invented cigar-shaped boat, invented threshing machine (1810), invented steam boiler for pressure up to 500 psi, invented a pump, repeating gun (1822), a liquid fuel for lamps (I&T15)

JENNINGS, James Hennen, 1854–1920, American mining engineer; hydraulic gold mine in California (1877), developed small gravel-gold mine, manager of gold mine in Venezuela, consulting engineer for mines in South Africa, worked in England, contributed to US Bureau of Mines (USBM) development and history (DAB)

JENNINGS, Thomas L., 1791–1859, Black American engineer; thought to be first black person to receive a US patent (1812), developed a turpentine-based fluid used to clean clothing, worked to improve educational opportunities for blacks (c.1825ff) (AAI BI)

JENNY, William Le Baron, 1832–1907, American civil engineer; engineer for US railroad company in Mexico (1856–1859), chief engineer of 15th Army Corps in charge of offices, manufacturer of maps (1861–1865), introduced pressed brick in construction of office buildings, built numerous commercial buildings in Chicago and New York City, used Bessemer steel skeleton construction, known as father of skyscrapers (BDOACE DAB EIH)

JEROME, Chauncey, 1793–1868, American inventor; clocks, design and manufacture and sale of clocks, started own shop (1817), shop for making wooden cases for clocks (1822), bronze looking glass clock (1825), formed Jerome Manufacturing Co. (1850–1856) (DAB MEIA MWBD NC7)

JERVIS, John Bloomfield, 1795–1885, American engineer and inventor; canal, railway, inventor of railroad track, swivel trunk to negotiate curves (1831), Croton aqueduct (NYC), water supply systems, engineer in charge of building section of Erie canal, worked on development and management of other canals, became chief engineer of Mohawk & Hudson Railroad (1830), consulting engineering (ANB ASAI BDOACE BITAS DAB EIH MWBD NC9 TEIH)

JESSOP, William, 1746–1814, English civil engineer; built first canal in England dependent on reservoirs (1793–1797), designed aqueduct 1000 ft. long over the Dee River (Scotland), designed forerunner of the iron rail later universally adopted, navigation in England, served as chief engineer of the Grand Union Canal (1793–1805), construction of wet dock area, consulting engineer (DNB EAI GEAPIT LOTE MWGD RHW SFLOE)

JEWETT, Frank Baldwin, 1879–1949, American electrical engineer; telecommunications, development of transcontinental telephone line, research and management of research at AT&T (1904ff) and then president of Bell Telephone Labs. (1925–1944), electron microscope, NAS (1918), president of NAS (1937–1947) (ASAI BM27 DOSB TFE&I NCC PS WWWIS)

JOCONDE, Jehan, 1433–??, Italian builder and drainage specialist; recognized for restoring old bridges in Verona, invited to France to restore bridges, returned to Italy where he planned and built drainage canal near Venice (TSOE)

JOHNSON, Bruce Gilbert, 1905–1989, American civil engineer; structural engineer, metal column behavior, eccentric loading, studied vibration and shock, faculty at Lehigh Univ. (1932–1950) including director of Fritz Engineering Lab. (1938–1950), during WWII with US Navy Bureau of Yards and Docks, with John Hopkins Univ. Lab. of Applied Physics, with Univ. of Michigan (1950–1968), consulting engineer (1968ff) (MT4)

JOHNSON, Clarence Leonard 'Kelly,' 1910–1990, American aeronautical engineer; flight test engineer, stress analysis, tool designer, aerodynamicist, wind tunnel, numerous planes designed, at Lockheed Aircraft Corp. (1933ff) becoming chief engineer and senior vice president (1969), NAE (1965), NAS (1965) (BM67 MT6 PS TEIH WWIA WWIE WWWIA WWWIS)

JOHNSON, Edwin Ferry, 1803–1872, American civil engineer; professor of mathematics and civil engineering at Literary, Scientific and Military Academy (VT, CT) (1823–1828), in charge of land surveys for canals and railroads, promoter of Northern Pacific Railroad (1854ff), chief engineer for 14 railroads and 4 canals (1833–1861), bridge construction, waterworks and sewage systems, consulting engineer (DAB NC17)

JOHNSON, Eldridge Reeves, 1867–1945, American inventor; machinist and shop foreman, spring-motor for gramophone (1896), disc-recording process, organized Victor Talking Machine Co. and its successor (1901–1927), Victrola (1906), worldwide marketing, 70 patents on talking machines (DAB NC30)

JOHNSON, Herbert H., 1931–1989, American materials scientist and engineer; fracture, hydrogen embrittlement of metallic alloys, environmental cracking of materials, fatigue, faculty at Lehigh Univ. (1957–1960), Cornell Univ. (1960–1973) including director of material science and engineering, NAE (1987) (MT4)

JOHNSON, John Butler, 1850–1902, American civil engineer; US Great Lakes survey (1878–1881), Mississippi River Commission (1881–1883), flood control, faculty of Washington Univ. (1883–1899), dean of engineering at Univ. of Wisconsin (1899–1902), wrote several text and reference books (DAB NC11)

JOHNSON, Reynold Benjamin, 1906–1998, American electrical engineer; public school teacher (1929–1934), senior engineer, data processing, business machines and computer research and development at International Business Machines (IBM) (1934ff), invented hard discs, research manager (1952–1959), advanced systems management (1959ff), over 90 patents, NAE (1981) (AMOS FAL MT9 NYTObit)

JOHNSON, Thomas, fl early 1800s, English inventor; added improvements to the power loom (1803–1805) (MMAH)

JOHNSON, Tom Loftin, 1854–1911, American inventor; street railroad, invented first fare box for coins, rehabilitated, built and operated street railroad systems in Indianapolis, Cleveland and Detroit, established steel mill, invented and manufactured so-called Trilby rail, successful in politics, was mayor of Cleveland (OH) (DAB)

JOHNSON, Warren S., 1847–1911, American mechanical engineer and inventor; taught school, became superintendent, later taught at normal school (1876), with William Planfinton [1844–1905] formed company (1883) later called Johnson Electric Service Co. (1885), invented Johnson temperature control regulator, elevator and controls, pipe coupling, electric junction box, heating system, pneumatic clock, at least 18 patents (MEIA NC3)

JOHNSON, Wendell Eugene, 1910–1970, American civil engineer; water resources, military construction, dam construction, canal locks, Corps of Engineers (1945–1970), consulting engineer, NAE (1970) (MT2 WWIE WWWIA)

JOHNSON, Wilfrid Estill, 1905–1985, English American mechanical engineer and inventor; refrigeration, aircraft engines, atomic energy, manager of several related projects for General Electric Co. (1930ff), Hanford Atomic Products Operation (WA) (1948ff) becoming general manager (1952), Atomic Energy Commission (1960–1972), 12 patents, NAE (1968) (MT4 WWIE WWWIA)

JOHNSTON, Bruce Gilbert, 1905–1989, American civil engineer; dam construction, structural research and management of laboratory, plastic design and steel columns, inelastic stability, response of structures to shock loads, with construction firms (1927–1931), faculty at Lehigh Univ. (1938–1942), (1945–1950), Columbia Univ. (1934–1938), Johns Hopkins Univ. (1942–1945), Univ. of Michigan (1950ff), NAE (1979) (AMOS Bridge MT4 WWIE WWWIA)

JOHNSTON, John, 1791–1980, Scottish American drainage engineer and farmer; to US (1821), considered as father of tile drainage in US (c.1835ff), installed tile in western New York beginning with open bottom horseshoe-shaped tile imported from Scotland (AE16)

JOHNSTON, Samuel, 1835–1911, American inventor and manufacturer; improvement of farm machinery, patented a corn and bean planter (1855), self-rake (1856), corn husker (1858), reaping machine particularly for rake (1863) (1865), combined rake and reel (1865), established Johnston, Huntley & Co. (1868) then Johnston Harvester Co. (NY) (1871), manufacturer of his inventions including rotary and disc harrows and a fuel burner (CDOAB DAB MEIA NC24 WWWIA WWWIA-S&T)

JOLLY, Philipp Johann Gustav von, 1809–1884, German physicist and inventor; invented air thermometer, invented a spring balance known as a Jolly balance for determining specific gravity, determined coefficient of expansion of air and other gases, osmosis, professor at Heidelberg Univ. (1839ff) (MWBD WWWIS)

JOLY, John, 1857–1933, Irish physicist and inventor; devised a photometer, method of color photography, invented steam calorimeter, Dublin method of radiotherapy with Walter C. Stevenson [1877–1931], engineering faculty at Trinity College (1897–1933) (GLOSB MWBD WWW WWWIS)

JONES, Amanda Theodosia, 1835–1914, American woman inventor and author; wrote and edited poems, articles and magazines, developed preservation of food in vacuum with and without cooking, desiccated foods in vacuum (patented 1873), fruit jar for preserving fruit (1873), apparatus for exhausting container filled with fruit juices at 100–120 deg. F. (1873), perfected liquid fuel burner (1880) for glass furnaces and steam boilers, devised several types of valves, a can opener, improvements to increase safety of oil drilling, six patents issued (AIE ANB AWIT DAB IAI MOI NC7 WI WWWIA)

JONES, Sir Bennett Melvill, 1887–1975, English aeronautical and mechanical engineer and inventor; during WWI with Royal Aircraft Foundry, then with Air Armament Experiment Station, worked on gunnery and armaments and instruments to fly in clouds, faculty at Cambridge Univ. (1919–1952), aerial surveying, study of airplane stalls, studies to reduce drag, consultant (1952ff) (DNB ITCOF PS RS23 WWWIS)

JONES, David P., 1840–1903, American mechanical engineer; principal examiner for Utah Public Power Survey (1858–1859), assistant engineer with US Navy (1862), US Naval Academy (1874–1879), became chief engineer of Bureau of Steam Engineering (1889), consulting engineer (BITAS MWBD)

JONES, Evan William, 1852–1908, Welsh American mechanical engineer and inventor; to US (1854), with iron and machine works, inventor of underfed stoker, manager then president of Union Iron Works (OR), interested in burning Oregon fir (1888), developed method then designed system for bituminous fuel (1897), Jones Stoker manufactured by Under-Feed Stoker Co., port engineer at Northwest Commercial Co. (Canada) (DAB MEIA NC25)

JONES, Fred R., 1893–1983, American agricultural engineer; farm power and machinery authority, faculty at Texas A & M Univ. (1921–1958) including department head (1940–1958), author and editor of several textbooks (AE64, AEsupp WWIE)

JONES, Frederick McKinley, 1892–1961, Black American mechanical engineer and inventor; served in WWI, in charge of machinery and tractors (1912–1929) for a 30,000 acre farm for James J. Hill (MN), improved microphone, designed and drove racing cars (1925), developed air conditioning and refrigeration equipment for food (1935) for trucking (1938) and for trains (1948), with Joseph Numero formed Thermo King Corp. (1935), ticket-dispensing machine (1939), co-founded US Thomas Control Co., preservation of blood and plasma, during WWII provided refrigeration for field hospitals and blood banks, granted 61 patents, consultant (AAI AAISMAI ANB BCST BI DAAS NC50 PS SBPY WWWIA)

JONES, Horace Kimball, 1837–1925, American mechanical engineer and inventor; with several companies including John Hasting (1857), Ball Machine Shop (1858), Hart Manufacturing Co. (1863), with Samuel Hotchkiss established Colt Northeast Armory later bought by Russell & Erwin Co. (1875), developed and patented machine for making carpenters' squares, manufacture of wood screws and chucks (MEIA)

JONES, John Luther (aka Casey Jones), 1864–1900, American railroad engineer; engineer for high speed train, ballad named after him (Casey Jones), engineer for Illinois Central Railroad (1890–1900) (MWBD)

JONES, Mack Marquis, 1896–1979, American agricultural engineer; served in WWI, on faculty at Univ. of Missouri (47 yr.) including chairman of department (1948–1962), tools, manual arts, author (AE60 WWIE)

JONES, Robert T., 1910–1999, American aeronautical engineer; aerodynamics, fluid mechanics, theory of sweep-back wings for high speed aircraft, aircraft stability, research and staff scientist with NASA (1934–1962), Avco Everett Research Labs. (1965–1970), senior scientist at Ames Lab. (NASA) (1970–1982), consulting professor at Stanford Univ. (1982–1997), NAE (1973), NAS (1981) (Bridge MT9 WWIE)

JONES, Thomas Franklin, Jr., 1916–1981, American electrical engineer; radio, acoustics, instrumentation, electronics and measurements, guided missiles, machine computation, Naval Research Lab. (1941–1947), faculty at Massachusetts Inst. of Technology (1947–1958) (1975ff), department head at Purdue Univ. (1958–1962), president of the Univ. of South Carolina (1962–1974), NAE (1969) (AMOS MT2 WWIA WWIE WWWIA)

JONES, Webster Newton, 1887–1962, American chemical engineer; taught chemistry at Universities of Harvard, Maine, Missouri, Radcliffe, and Montana, during WWI an officer in Chemical Warfare Service, with B. F. Goodrich Co. (1919–1932) where in addition to research on rubber oxidation was manager of processing division and directed technical training of new employees, at Carnegie Inst. of Technology (PA) (1932–1956) as director of college of engineering and science and was vice president of industrial and government relations (ACCE WWIE(I) NCF)

JONES, William Richard, 1839–1889, American engineer and inventor; in Civil War as captain, machinist, lumber man, master mechanic (1859), erected steel plant and rolling mills at Bessemer then became superintendent of Edgar Thomson Street Co. (PA) (1875), consulting engineer for Carnegie Co. (1888), patented method of operating ladles, improvements in hose connections, designs for Bessemer converters, washers for ingot molds, hot beds for rolling mills, feeding apparatus for rolling mills, Jones mixer apparatus for handling, setting and removing rolls (DAB NC15)

JONSSON, John Erik, 1901–1995, American mechanical engineer; numerous management positions relating to research, development and sales for Alcoa Co. (1922–1927), founder of Texas Instruments, Inc. (1930ff) including president (1951–1958) and chairman of the board (1958–1966), numerous civic activities and elected offices, philanthropist, NAE (1971) (Bridge CB WWIA WWIE)

JORDAN, Edward Conrad, 1910–1991, Canadian American electrical engineer; radio station operator, antennas, radio direction finding, electromagnetic waves and radiation, faculty at Worcester Polytechnic Inst. (MA) (1940–1941), Ohio State Univ. (1941–1945), Univ. of Illinois (1945ff) including department head (1954), NAE (1967) (AMOS Bridge MT6 WWIA WWIE)

JOUKOWSKI, Nikolai Jegorowitch, 1847–1921, Russian applied mathematician and aerodynamicist; airfoil principles and design, Kutta-Joukowski (1910) (German Wilhelm M. Kutta [1867–1944]) airfoil and theory (CEOS&T EDOP L&M)

JOULE, James Prescott, 1818–1889, English physicist and inventor; measurement of heat, Joule law (1842), mechanical equivalence of heat (1850), Joule-Thomson effect (1852) (William Thomson [1824–1907]), Joule cycle or constant pressure engine (1884), unit of energy or work named after him, RS (1850), fNAS (1887) (BEST EE FINE L&M MWBD SAI S:TLAW WWWIS)

JOY, D., 1825–1903, English inventor and mechanical engineer; locomotive valve gear (1880) (EE)

JOY, James Frederick, 1810–1896, American railroad builder; officer of several railroads in Michigan and Midwest, built canal at Sault St. Marie (MI) (DAB)

JUDAH, Theodore Dehone, 1826–1863, American civil engineer; employed by railroad companies, constructed railroad bridges, research engineer for part of Erie Canal, engineer for railroad through gorge of Niagara River (1847–1854), sought to build transcontinental railroad (1859) (ANB BDOACE DAB NC15)

JUDSON, Egbert Putnam, 1812–1893, American civil engineer, inventor and manufacturer; explosives, use of dynamite in US (1867), formed and served as president of Judson Powder Co. and patented Giant Powder (1873), an especially designed explosive for railroad development for moving banks of dirt, rock and gravel (1876) known as Judson's RRP (railroad powder) (DAB NC24)

JUDSON, Whitcomb L., ??–1909, American mechanical engineer and inventor; automatic closing device with sliding mechanism called clasp locker (1893), with Lewis A. Walker [1855–1938] formed Universal Fastener Co. (1894), company reorganized (1904) as Hook and Eye Co. with improvements to zipper due to Swedish engineer Gideon Sunback [1880–1954], first practical device (1913), name zipper provided by B. F. Goodrich Co. when applied to rubber galoshes (1923), a similar device was patented in Europe (1912) by Katherine Kuhn-Moos, 30 patents (BDOTHOT EOET HFP GI IAI IB NC28 WOI)

JULIAN, Percy Lavon, 1899–1975, Black American chemist and inventor; methods of manufacturing processes for drugs, vitamins, hormones, flame retardants, paint and paper, synthesis of cortisone (1954) and other hormones, on faculties at West Virginia State College, Howard Univ. and DePauw Univ. (IN) (ante 1935), director of research at Glidden Co. (18 yr.), founded Julian Labs., Inc. (1953–1964) that merged with E. I. Lilly Corp. (1961), NAS (1973), over 130 patents, honored by USPS commemorative stamp (1993) (AIE BDOBA BEOS BI BM52 BPOSAT CB CBB CDOAB Chem-Herit23 DAB LOTCS NC28 NOAIMA NYPL NYTO-bit S:TLAW TFM TGLOSB WWWIA)

JULLIOT, Henri, 1856–1923, French American aeronautical engineer; aeronautics, directed aeronautical work in France during WWI, became general manager of aircraft division of Goodrich Co. (1915), developed balloons and small nonrigid airships (ITCOF PS)

JUNKERS, Hugo, 1859–1935, German airplane engineer and builder; developed all-metal airplane (1915), established airplane factory (1919), established motor works (1924), established mail and passenger routes, airplane named after him (MWBD WWWIS)

JUNKINS, Jerry R., 1937–1996, American electrical engineer and manufacturer; management of industrial processes particularly advanced technologies for high-volume production, defense industry, chairman, director, president and CEO (1985) of Texas Instruments, Inc. (1959–1996), NAE (1988) (Bridge MT9 WWWIA)

JUSTINIAN, 6th century, Roman builder and last Roman emperor; fortifications, developed overflow channel, church construction (TAE)

JUVE, Arthur Edgar, 1901–1965, American chemical engineer; rubber products technology, rubber-coated products—rolls, blankets, typewriter platens, jar rings—testing and development of natural and synthetic rubber, research and administration at B. F. Goodrich Co. (1925–1965) becoming director of technical services (1954–1961) followed by position of director of corporate rubber technology (ACCE PS)

K

KABLE, George Wallace, 1888–1950, American agricultural engineer; Alaska boundary survey and US Geological Survey (1907–1910), numerous positions including inspector and foreman, draftsman for Port of Seattle (WA) (1913–1914), with faculty in irrigation at New Mexico State Univ. (1915–1916), faculty in land clearing and drainage at Oregon State Univ. (1917–1928), range manager, director of National Rural Electric Project (REA) (1928ff), ASAE award named after him (AE31 WWIE(I))

KAFER, John Christian, 1842–1906, American engineer; US Navy engineer (1864–1868), instructor in steam engineering at US Naval Academy (1868–1874) (1876–1882), Bureau of Steam Engineering, general manager and vice president of Morgan Iron Works (NYC) (1888ff), later vice president of Quintard Iron Works (NYC), consulting engineer (DAB MEIA)

KAHLES, John F., 1914–1993, American chemical engineer; correlation of effects of processing on surfaces, improved productivity by dissemination of data on metal removal, machinability, failure analysis, Metcut Associates, Inc. (OH) (1951–1990) including vice president of group (1978), NAE (1984) (Bridge MT7 WWIE)

KAHN, Albert, 1869–1942, German American industrial architect and builder; to US (c.1880), began private practice (1902), creator of modern factory design, developed huge single-story steel framed structures, all assembly processes on one floor, architect for America's car manufacturers (30 yr.), designed over a thousand projects for Henry Ford [1863–1947] including the River Rouge plant (MI) (1917) (CB CDOAB NC31,C 1000Y)

KAISER, Henry John, 1882–1967, American industrialist; highway construction worldwide (1914–1930), dam building, shipbuilding—particularly Liberty cargo ships used widely in WWII (1940ff)—expanded business (1945ff) to steel and aluminum, built automobile called Henry J, established Kaiser Foundation for health services (1942) (AIE MWBD NCG)

KALMUS, Herbert Thomas, 1881–1963, American engineer, scientist and inventor; teacher and researcher in Canada and US, with wife Natalie Kalmus given credit for development of motion pictures, discovered abrasives and formed Exolon Co. of which he was officer (1915–1925), formed consulting firm, involved in color film motion pictures with Technicolor, Inc. (DAB WWWIA)

KANE, William, 1849–1922, Irish American mechanical engineer and inventor; patented first automatic dry-joint fire extinguisher (1886), patents dealing with steam generator (1901), circulating gas hot water heater (1904) and established manufacturing company, patented other devices for heating and gas regulation, Frenchman François Carlier invented chemical fire extinguisher (1866), Russian Alexander Laurent developed extinguisher using sodium bicarbonate and aluminum sulfide (1905) (MEIA TTOT)

KAPELYUSHNIKOV, Matvei Alkunovich, 1886–??, Soviet petroleum engineer; turbodrilling of oil wells (1912), director of turbodrilling and cracking at Baku (1914–1937), with V. G. Shukhov built first Soviet cracking plant (1924), pneumatic control of drilling rig (1933) with S. D. Zalkin, effect of dissolving oil in gas under pressure (1952), USSR Academy of Sciences (SMOS)

KAPITSA, Pyotr Leonidovich, 1894–1984, Soviet physicist and inventor; practical method of liquefying air (1939), superfluidity of liquid helium (1941), low-temperature physics, administrator of research institutes, invented turbine for producing liquid air, equipment for handling liquid oxygen, high-energy physics, fNAS (1946), Nobel prize (1978) (AHOI BEOS MMAH MWBD TBDOS WWWIS)

KAPLAN, H. C. Viktor, 1876–1934, Austrian inventor; variable propeller turbine with blades adjusted by governor (1913) named after him (1920), Kaplan water turbine with vertical shaft, diesel engines (EE TBDOS(Po) WBE MWBD)

KARANDEEV, Konstantin Borisovich, 1907–??, Soviet electrical engineer; professor, signal engineering, metrology, machine studies and automation, USSR Academy of Sciences (1958), worked with semiconductor rectifiers, telemetry, computers, geophysical apparatus (PS PSOCE SMOS WWWIS)

KÁRMÁN, Theodor von, 1881–1963, Hungarian American physicist and aeronautical engineer; director of Aeronautical Inst. at Aachen (1912–1930), to US (1930), fluid flow, turbulence, vortices named after him, rocket technology, professor and academic administrator at California Inst. of Technology (1930ff), NAS (1938), cofounder of Jet Propulsion Lab. (JPL) (CA) (1944) (BM38 CDOS EOWB FNIE IWAP MWBD PS TFE&I WWWIA WWWIS)

KATTE, Walter, 1830–1917, English American civil engineer; from rodman to division engineer for railroad (1851–1854), then to Pennsylvania Railroad (PRR) and several regional railroads, during Civil War in US Army as colonel in charge of bridge and railroad construction around Washington (DC), with Keystone Bridge Co. (1865–1875), and St. Louis and New York for railroad and bridge construction (DAB NC16)

KATZ, Alexander E., 1887–1957, Russian American chemical engineer; specialized in saccharin and byproducts, in charge of Ritter & Co. (1913ff), helped develop standards for Federal Drug Administration (FDA), during WWII was consultant on war gases, worked on essential oils for California Department of Agriculture and California Polytechnical School (1945–1947), cofounder of Florsynth Lab (ACCE)

KATZ, Donald L., 1907–1989, American chemical engineer; technology involving gases, petroleum and nuclear energy, at Phillips Petroleum Co. (OK), faculty at the Univ. of Michigan (1936–1977) including department chairman (1951–1962), consulting engineer, NAE (1968) (AMOS Bridge MT4 NCK PS WWIE)

KAVANAUGH, Thomas C., 1912–1978, American civil engineer; planning, design and supervision of building heavy structures and manor public projects, faculty at New York Univ. (1948–1952), department chairman at Pennsylvania State Univ. (1952–1954), adjunct professor at Columbia Univ. (1955–1969), senior vice president of Madigan-Praeger, Inc. (1955ff), NAE (1964) (MT1 PS WWIE WWWIA)

KAY, Edgar Boyd, 1860–1931, American civil engineer and inventor; contractor and consulting practice

(1885–1895), on faculties at Rensselaer Polytechnic Inst., Union Univ. and Cornell Univ. (1896–1903), Univ. of Alabama (1903–1912) including dean of engineering, consultant for railroad construction in Alabama (1903–1915) and other works—water, sewer, lighting, steam, electric—invented US Standard Incinerator, testing and setting of concrete, consulting in sanitary engineering (DAB NC22 WWWIA)

KAY, John, 1704–c.1780, English inventor; invented improved textile machinery, invented flying shuttle for weaving textiles (1733), improvement of the ribbon loom (GI MMAH MWBD NYPL RHW TBDOS(Po) TSOE WBE)

KEELY, John Ernst Worrell, 1827–1898, American inventor; worked on mechanical devices that were proven not to be what was claimed (DAB NC9)

KEEN, Morris Longstreth, 1820–1883, American inventor; pattern and iron factory business, with his brother Joseph Keen specialized in making flat irons, pulp and paper manufacture, boiler for making paper pulp from poplar wood (1859), formed American Wood Paper Co. (1863), developed several patents as a result of his work at The Experimental Mill that he established (DAB NC11 WWWIA)

KEENAN, Joseph Henry, 1900–1977, American mechanical engineer; thermodynamics, turbine design, properties of steam, steam tables, compressible fluid mechanics, conversion of heat into work, steam turbine engineer with General Electric Co. (1928–1934), faculty at Stevens Inst. of Technology (1934–1939), faculty at Massachusetts Inst. of Technology (1939–1966) including department head, consultant, NAE (1976) (AMOS MT1 WWWIA)

KEITH, Percival Cleveland, Jr., 1900–1976, American chemical engineer; research and administration for several oil companies including Cross Engineering Co. as vice president (1925–1927), chief engineer of M. W. Kellogg Co. (1927–1929) (1932ff), petroleum refining, catalytic cracking, reforming, gaseous diffusion for separating uranium (WWII), formed Hydrocarbon Research, Inc. (1945–1964) serving as president, 40 patents on refining, NAE (1968) (MT2 NC59 PS)

KEITHLEY, Joseph F., 1915–1999, American electrical engineer; underwater mine firing devices, electronic test, measurement and instrumentation, with Bell Labs., founded Keithley Instruments Co. (1964ff), NAE (1992) (Bridge WWIE)

KELDYSH, Mstislav Vsevolodovich, 1911–??, Soviet mathematician and mechanics specialist; aerohydrodynamics, flutter engineering, oscillators, aerodynamics, theory of waves, hydrodynamics, rocket propulsion (SMOS WWWIS)

KELLNER, Carl, 1851–1905, Austrian chemical engineer and inventor; manufacture of caustic soda by electrolysis of brine (1894), Castner-Kellner process (1892) (American Hamilton Y. Castner [1858–1898]), set up Alkali Co. in England (1892) (BEOS GLOSB PS TBDOS(Ab) UEOIC)

KELLOGG, Will Keith, 1860–1951, American manufacturer and businessman; with his brother physician John Henry Kellogg [1852–1943] organized W. K. Kellogg Co. (1906–1946) for making cereals for food, served as president and board chairman, developed processing machinery, established W. K. Kellogg Foundation (1930) (AIE MWBD WBE WWWIA)

KELLY, Clarence Francis, 1907–1976, American agricultural engineer; engineer with Gleaner Co. manufacturing combines, livestock environment particularly in hot climates, grain storage, use of energy in agriculture, government services with US Department of Agriculture and Cooperative State Research Service, US Navy in WWII (1942–1945), faculty at Univ. of California including director of California Agricultural Experiment Station (1950ff), NAE (1968) (AE57 AMOS MT1 WWIA WWIE WWWIA)

KELLY, Mervin Joseph, 1894–1971, American mining engineer; research at Western Electric Co. (1918–1925), Bell Telephone Labs. working on vacuum tubes, transmission systems and administration (1925–1959) including president, chairman and director, consultant to International Business Machines (IBM) (1959–1965), NAS (1945) (ANB BM46 NC56 WWWIA)

KELLY, William, 1811–1888, American metallurgist and inventor; manufactured sugar kettles, iron and steel making, iron converter (parallel to Bessemer process) (1850), later using forced air or airblast (1857), axe handle manufacturing (AIE ANB DAB IIA MEIA MMAH MWBD NC27 TBDOS(Po) WBE WWWIA)

KELVIN, Lord (aka William Thomson 1st Baron), 1824–1907, Irish physicist, electrical engineer and inventor; contributed to understanding of converting heat to mechanical energy, thermodynamics, proposed absolute temperature scale called Kelvin (1848), laid first successful transatlantic cable (1857), gas filled thermometer, mirror galvanometer (1858), theory of elasticity, heat pump (1882), over 70 inventions, fNAS (1883) (BEST EE EIH EOWB FNIE IAI MWBD SAI-MP TGE WBE WWWIS)

KEMBLE, Gouverneur, 1786–1875, American manufacturer; established West Point Foundry (1818), cannon manufacturer (MWBD TNAS WWWIA)

KEMP, James Furman, 1859–1926, American geologist and mining engineer; faculty first at Cornell Univ. then Columbia Univ. (1891–1926), petrographic research, economic problems related to deposition and concentration of lead and zinc ores, NAS (1911) (BM16 DAB TNAS WWWIA)

KENEDI, Robert M., 1921–1998, Hungarian Scottish civil engineer; British citizen (1947), structural engineer, light structures, stress analysis, vibrations, worked with plastic surgeon, faculty at Univ. of Strathclyde that became Royal Technical College in Glasgow, fNAE (1976) (MT9)

KENNEDY, Alexander Blackie William, 1847–1928, English mechanical and electrical engineer; kinematics of mechanisms, testing of materials and machines, machine design, electrical power systems (1899ff), use of compressed air for power, professor at Univ. College London (1874–1889) including laboratory for students, consulting engineer (1889ff), member of several boards and commissions (1900ff), FRS (1887) (CDOSB DNB DOSB PS)

KENNEDY, John F., 1933–1991, American civil engineer; hydraulics, ice engineering, cooling tower technology, density of stratified flows, Massachusetts Inst. of Technology (MIT) (1961–1966), Univ. of Iowa (1966ff) as professor and director of Iowa Inst. of Hydraulic Research, NAE (1973) (Bridge MT6 WWIE WWWIA)

KENNEDY, Julian, 1852–1932, American mechanical engineer; superintendent of blast furnaces of Brier Hill Steel Co. (OH) (1876), then Struthers Iron Co. (1877), Morse Bridge Works, then Carnegie Steel Works (PA), Lucy Furnace (PA), general superintendent of Carnegie, Phipps & Co. (PA), patented hot blast furnace oven (1881), engines, stoves, mill for steel, chief engineer of Latrobe Steel Works (PA), consulting engineer (MEIA WWWIA)

KENNELLY, Arthur Edwin, 1861–1939, British Indian American electrical engineer; electronics, alternating current circuits, ionosphere described as Kennell-Heaviside layer (1902) (announced by Oliver Heaviside [1850–1925]), worked with Thomas Edison [1847–1931] (1887–1894), faculty at Harvard Univ. (1902–1913), Massachusetts Inst. of Technology (MIT) (1913ff), consulting engineer, NAS (1921) (BEST BM22 CEOS&T DOSB MWBD TFE&I WWWIA)

KENT, Arthur Atwater, 1873–1949, American inventor and manufacturer; formed Atwater Kent Manufacturing business (1902), made variety of electrical devices—voltmeters, telephones, automatic ignition systems, as well as many of his own inventions such as jump-spark ignition, radio sets—philanthropist (DAB NC38 WWWIA)

KENT, William, 1851–1918, American mechanical engineer and inventor; properties of copper-tin and copper-zinc alloys, established private engineering consulting business, superintendent of open hearth production, manager of Pittsburgh office of Babcock & Wilcox as manufacturer of water tube boilers, with William F. Zimmerman founded Pittsburgh Testing Lab. (1882ff), with Babcock & Wilcox Co. (1882–1887), combustion of fuel and design of steam boilers, invented wing-wall furnace and a gas producer, was dean at L. C. Smith College in Syracuse (NY) (1903–1908), editor of engineering journals (ANB BITAS DAB MEIA WWWIA WWWIS)

KENYON, David Ecclestone, 1914–1970, American electrical engineer and inventor; flash tube construction, assistant at Massachusetts Inst. of Technology (MIT) Vacuum Tube Lab., electronics, instrument landing program with Federal Aviation Authority (FAA), Sperry Gyroscope Co. that became Sperry Rand where he became head of section (1960–1969), radar development, vice president of engineering for Ken Lab., Inc. (NC55 PS)

KEPLER, Johannes, 1571–1630, German astronomer and inventor; professor at Graz (1594), at observatory near Prague (1600), mathematician in Austria (1612), Ulm (1626), Silesia (1628), Kepler laws of planetary action (1609, 1619) based on observations of Tycho Brahe [1545–1601], his work led to invention of calculus, astronomical telescope (1611), double convex microscope (1611) (DOT L&M MWBD NYPL SAI SAI-MP TGS WA WWWIS)

KERENSKY, Oleg Alexander, 1904–1984, British civil engineer; structural engineer, joined firm of Freeman Fox & Partners (1946) becoming partner (1956) (TGE:B)

KERR, Walter Craig, 1858–1910, American mechanical engineer; faculty at Cornell Univ. (1879–1882), engineering practice (1882ff), large complex plants for power development with backing of Henry H. Westinghouse [1863–1933], organized Westinghouse, Church, Kerr & Co., involved in South Station in Boston (MA), director of Merchant's Association of New York City (1907–1910) (BITAS DAB MEIA WWWIA)

KESSLER, George W., 1908–1983, American mechanical engineer and inventor; various engineering positions from sales to chief engineer and vice president at Babcock & Wilcox Co. (1930–1973), steam engineering, marine design, several patents, NAE (1969) (MT3 WWIE)

KESTIN, Joseph, 1913–1993, American mechanical engineer; measurement of transport properties, thermodynamics, free-stream turbulence effects on heat transfer, stagnation-line stability, worked with several universities in Europe and US particularly Brown Univ. (1953ff), NAE (1982) (Bridge MT7 WWIE WWWIA)

KETCHLEDGE, Raymond W., 1919–1987, American electrical engineer; management of research at Bell Telephone Labs. (1942ff), submarine cable system, broadband coaxial carrier systems, gas discharge devices, switching devices, granted 58 patents, NAE (1970) (MT5 WWIE WWWIA)

KETCHUM, Milo S., 1910–1999, American civil engineer; folded-plate and thin-shell structures, innovative and creative and aesthetic designs, founded consulting firm of Ketchum Konkel et al. (CT) (1945), structures and structural design, faculty at Univ. of Connecticut (1967–1978), consultant (1967ff), NAE (1982) (Bridge WWIE WWWIA)

KETTERING, Charles Franklin, 1876–1958, American electrical and mechanical engineer and inventor; first electric cash register, with Edward A. Deeds [1874–1960] founded Delco (Dayton Electric Co.) (OH) (1909), farm lighting systems, automobile self-starter (1911), automotive electric systems, organized Dayton-Wright Airplane Co. (1914), research director and officer of General Motors Corp. (1920–1947), high speed two-cycle diesel engine (1951), antiknock fuels, variable speed transmissions, 185 patents, supported numerous philanthropic and education activities, NAS (1928) (AIE ANB BM34 CDOAB HOCS&T IAI MWBD NC48,E NYTObit PS TEIH T7-AP TTOS WBE WWWIA WWWIS)

KEULEGAN, Garbis H., 1890–1989, Armenian American civil engineer; hydraulics, fluid mechanics, waterways experiment station, erosion, open-channel flow, wave action, National Bureau of Standards (NBS) (1921–1932), National Hydraulics Lab. (1933–1963), consultant, NAE (1979) (Bridge MT5 WWIE WWWIA)

KEYES, Charles Rollin, 1864–1942, American geologist and mining engineer; chief geologist of Missouri Geological Survey (1890–1892), Iowa Geological Survey (1892–1894), president of New Mexico School of Mines (1902–1906), president and general manager of several mining companies, traveled extensively and initiated professional journals (ANB NC31 WWWIA)

KHAN, Fazlur Rahman, 1929–1982, American civil and structural engineer; structural systems for tall buildings, reinforced concrete structures including Plaza Building in Houston (TX), John Hancock Center and Sears Tower in Chicago (IL), NAE (1973) (ANB MT2 WWIE WWWIA)

KHARKEVICH, Aleksandr Aleksandrovich, 1904–??, Soviet radio engineer; theory, design, and construction of electro-acoustical apparatus, communications, information transmission, professor and department head, USSR Academy of Sciences (1960) (SMOS WWWIS)

KHITRIN, Levee Nikolaevich, 1907–??, Soviet heat engineer; physics of burning processes, heterogeneous burning, utilizing fuels, development of new intensive-heating processes, All Union Heat Engineering Inst. (1931–1941), professor at Moscow Univ. (1936ff) (SMOS WWWIS)

KHRENOV, Konstantin Konstaninovich, 1894–??, Soviet electric welding engineer; originated methods of electric welding and cutting under water, teacher at several institutes and academies in Moscow (1928–1947), taught electric welding at Ukrainian Academy of Science (1945–1948), Kiev Polytechnical Inst. (1947), Inst. of Electrotechnics (1952ff) (SMOS)

KHRISTIANOVICH, Sergei Alekseevich, 1908–??, Soviet mechanical engineer; mechanics of solids and gases, hydrotechnical structures aerodynamics, flow of gas at supersonic speeds, worked with hydrological institute (1930–1937), Central Aerohydrodynamics Inst. (1937–1953), Inst. of Chemical Physics (1956ff) (SMOS WWWIS)

KHUFU-ONEKH (CHEOPS), 3969?–3908? BC, Egyptian builder; master builder, directed building of the great pyramid (EB GEAPIT TSOE)

KIDDER, Wellington Parker, 1853–1924, American automotive engineer; patented rotary steam engine (1868), several patents relating to printing press (1874ff), organized the Kidder Press Manufacturing Co. (1880), invented Franklin typewriter (1887), Wellington typewriter (1891), noiseless typewriter (1895), formed the Parker Machine Co. and later Noiseless Typewriter Co., steam motor car (1902) (MEIA NC26 WWWIA WWWIS)

KIELY, John R., 1906–1996, American civil engineer; design and construction of industrial facilities and public works projects, nuclear power plants, industry executive, NAE (1967) (Bridge WWIE)

KILDALL, Gary, 1943–1994, American computer scientist; at Naval Postgraduate School (CA) (1972–1976), formed Digital Research Co. (1976–1980), developed CP/M control program for micro-computers, developed early programs for microcomputer (Apple) (1983), worked with International Business Machines (IBM), worked on CD-ROMs (FAL GLOSB I&T20 TMA)

KILLIAN, James R., Jr., 1904–1988, American mechanical engineer; proponent of theme that engineers are important in public service, editor of Technology Review (MIT) (1927–1939), faculty and officer of Massachusetts Inst. of Technology (1927ff) as president and chairman (1948–1971) during which time he served on several presidential commissions and as special assistant to US president for science and technology, NAE (1967), NAS (1988) (Bridge MT5 PS WWIA WWIE)

KILROY, James Jackson, Jr., 1925–1987, American ship technologist; shipyard worker, legislator, inspector of ships, his signature became basis of WWII fame expression—Kilroy was here (AOFP NatVetNet1999)

KIMBALL, Dexter Simpson, 1865–1952, American mechanical and industrial engineer; machinist, machine tools, standardization, production engineering, Pope & Talbot(WA)(1881–1887), Union Iron Works(CA)(1887–1898), designed hydraulic press and hoisting equipment at Anaconda Mining Co., faculty and administration at Cornell Univ. (1898–1901) (1905–1936) including dean of engineering (1920–1936), works manager of com-

pany making electric generators (1901–1904), consulting engineer (1936ff) (ANB NC42,D WWWIA)

KIMBARK, Edward Wilson, 1928–1982, American electrical engineer; transient power stability, power system analysis, switching, over-voltages, direct current power systems, faculty at Massachusetts Inst. of Technology (MIT) (1939–1950), Inst. Technologico de Aeronautica (Brazil) (1950–1955), dean of engineering at Seattle Univ. (1955–1962), Bonneville Power Administrator (1963ff), NAE (1979) (MT3 WWIE WWWIA)

KING, Clarence, 1842–1901, American geologist and mining engineer; explorer, geological survey of 40th parallel, studied glaciers in US, first head of US Geological Survey (1878–1880), mining engineer (1881ff), NAS (1876) (BM6 DAB MWBD NC13 WA WWWIA)

KING, Franklin Hiram, 1848–1911, American agricultural scientist and inventor; inventor of above-ground vertical cylindrical or tower silo (1882, 1889), contributed to knowledge of storing forages, barn ventilation, River Falls State Normal School (1878–1888), taught agricultural physics at Univ. of Wisconsin (1888–1901), US Department of Agriculture (1901–1904), wooden square silos above ground for storage of forages credited to Fred L. Hatch (IL) (1873), Chinese stored grain in cylindrical airtight containers (c.50 AD) (DAB NC19 TTOT WBE WWWIA)

KINGSBURY, Albert, 1862–1943, American mechanical engineer and inventor; developed bearings adopted by US Navy (1917), lubrication was specialty, faculty at Univ. of New Hampshire (1891–1899), Worcester Polytechnic Inst. (1899–1903) with industry mixed in, Westinghouse Electric & Manufacturing Co. (1903ff), friction of screws and nuts, air lubrication, tilting segmented oil-thrust bearings, bearings used in large hydroelectric stations and pumping stations, established Kingsbury Machine Works, Inc. (PA) (1921) (ANB NC32 NYTObit)

KINGSFORD, Thomas, 1799–1869, English American inventor and manufacturer; became president of starch factory, developed method of separating starch from corn with lye, started Kingsford Factory for starch manufacture (1846), later founded Oswego Starch Factory (NY), developed food–quality cornstarch (1850) (DAB NC5 WWWIA)

KINNERSLEY, Ebenezer, 1711–1778, English American inventor; investigations on positive and negative properties of electricity, demonstrated that heat could be produced by electricity, invented electrical air thermometer, a contemporary of Benjamin Franklin [1706–1790] (DAB NC1)

KINNICUTT, Leonard Parker, 1854–1911, American chemist, sanitary engineer and inventor; studied in Germany, made refinements to gas analysis, papers on phenylglycine acid with German Richard Anschütz

[1852–1937], faculty at Worcester Inst. of Technology (1882ff) as department director, sewage disposal and sanitary problems (1885ff), authority on sanitation of air, water, gas, water and watersheds, contamination of rivers and ponds, methods of analysis, consultant (DAB NC25 WWWIA)

KINZEL, Augustus Braun, 1900–1987, American metallurgical engineer and inventor; research at General Electric Co., research at Union Carbide Co. (1926–1965) becoming vice president (1955), president of Salk Inst. (CA) (1965–1967), 40 patents, NAS (1960), NAE (1964) (ANB Bridge MT6 WWIE WWWIA)

KIPP, Petrus Jacobus, 1808–1864, Dutch chemist and inventor; began as pharmacist and developed business in chemicals and apparatus, invented Kipp device to generate hydrogen sulfide (1833), with sons continued business under firm of P. J. Kipp and Sons that produced selenium cells, electrodynamometer, and microtone (ChemHerit20/21)

KIRCHHOFF (KIRCHOFF), Gustav Robert, 1824–1887, German physicist/chemist and inventor; Kirchhoff laws (1845), with Robert W. Bunsen [1811–1899] developed method of spectrum analysis (1859), radiation laws of emission and absorption, professor at Heidelberg Univ. (1854–1875), Berlin Univ. (1875–1886), fNAS (1883) (BEST FNIE MWBD SAI WA WBE WWWIS)

KIRCHMAYER, Leon Kenneth, 1924–1995, American electrical engineer; electric power system control, utility planning, economic simulation, planning and dispatch, manager of advanced systems technology, with General Electric Co. (1958ff) as manager of operations, NAE (1979) (Bridge IWWIE WWIE WWWIA)

KIRKALDY, David, 1820–1897, Scottish metallurgist and mechanical engineer; recognized for his drawings, known for testing of iron and steel, manufacture of testing machinery, consultant (DOSB PS)

KIRKBRIDE, Chalmer Gatlin, 1906–1998, American chemical engineer; petroleum engineering, development of catalytic cracking, research in petroleum processes, president of Houdry Process Corp. (1952–1956), worked for several petroleum companies, advisor to government on energy, NAE (1967), consultant (Bridge WWWIA)

KIRKPATRICK, Sidney Dale, 1894–1973, American chemical engineer; served in WWI, leader in editing and writing about engineering, editor of McGraw-Hill Chemical Engineering magazine (1928–1959), and with Chemical and Metallurgical Engineering (1928ff), during WWII was active with several government agencies (ACCE NYTObit WWWIA)

KIRKWOOD, James Pugh, 1807–1877, Scottish American civil engineer; sanitary engineering, waterworks engineering, first work in US was in railroad engineering, officer with two railroad companies (1840s–1853), public water supplies (1855ff) including reservoirs, pumping systems, pipes, water filtration, first as chief engineer and later as consulting engineer (ANB DAB EIH NC9)

KISHIDA, Yoshikuni, 1905–1986, Japanese agricultural engineer; founder and president of Farm Machinery Industrial Research Corp. Shin-Norinsha Co., Ltd. (1940ff) becoming president (1943) and chairman (1968), initiated Agricultural Machinery News (1933) and Farming Mechanization (1940), publisher and editor of Agricultural Mechanization of Asia, Africa and South America (1971ff), ASAE Kishida International Award named for him (AE[supp.] CIGR)

KISHKIN, Sergei Timofeevich, 1906–??, Soviet metallurgist; metal sciences working primarily on diffusion of metals, radiology, effect of radiation on metals, USSR Academy of Sciences (SMOS)

KJELDAHL, Johan Gustav Christoffer Thorsager, 1849–1900, Danish chemist and inventor; headed chemistry research at Carlsberg Lab. (1876–1900), developed rapid method of determining nitrogen (1883) using sulfuric acid followed by amount of ammonia produced—named after him—worked on hydrolysis of starch and estimation of reducing sugars (CDOSB BDOA(Wi) MWBD WWWIS)

KLEBANOFF, Philip Samuel, 1918–1992, American aeronautical engineer; fluid mechanics including boundary layer flow, transition, turbulence, hot wire anemometry, research with National Bureau of Standards (NBS) now National Institute of Standards and Technology (NIST) (1941ff), NAE (1981) (AMOS Bridge MT6 WWWIA)

KLEIN, John S., 1849–1903, German American mechanical engineer and inventor; employed in oil fields (PA), then with shops of Smith & Crombie (PA) and was in charge of repair shop, several patents (1886ff) dealing with steam pumps, piping and coupling, rod-packing (1901) and water cooled rod and piston (MEIA)

KLEIN, Joseph Frédéric, 1849–1918, French American mechanical engineer; engineering studies at Yale Univ. (1868–1871), draftsman to chief engineer at Colt Co., returned to Yale Univ. as faculty member (1877–1881) in thermodynamics, established course in mechanical engineering at Lehigh Univ. (1881–1918) including thermosciences, machine design and kinematics and was university administrator (BITAS DAB MEIA NC18 WWWIA)

KLEINSCHMIDT, Edward Ernst, 1875–1940, German American electrical engineer and inventor; owner and operator of electric shop (1898–1913), president of Kleinschmidt Electric Co. (1913–1925), vice president of Morkrum-Kleinschmidt Corp. that made signaling apparatus, high speed teletype-typewriter (1914), teletype (1928) (NYPL TTOS WWWIA)

KLINE, Stephen Jay, 1922–1997, American mechanical engineer; rocket propulsion, thermosciences, mechanics and fluids, diffusivity turbulence, separated flows, modeling, US Army in WWII (1944–1946), North American Aviation Co. (1946–1948), faculty including department chairman at Stanford Univ. (1948ff), consultant, NAE (1981) (AMOS WWIE WWWIA)

KLOPSTEG, Paul Ernest, 1889–1991, American applied scientist and inventor; electrical measuring instruments, prosthetic devices, research and development with Central Scientific Co. (1921–1955) during which he served as manager of manufacturing (1922), director (1922) and president (1930–1944), during WWII in Office of Scientific Research and Development (OSRD), faculty and administration at Northwestern Univ. (1944–1954), government research administration at National Science Foundation (NSF) (1951–1957), issued approximately 50 patents (CB59 CB91 TP WWIE WPObit WWWIA)

KNEASS, Samuel Honeyman, 1806–1858, American civil engineer; brother of Strickland Kneass [1821–1884], design of triumphal arch in Philadelphia to honor Gen. Marie J.-P.-Y.-R.-G. Lafayette [1757–1834] (1824), survey of Chesapeake and Delaware Canal, study of public works in England (1825), in Corps of Engineers for Pennsylvania state canal, chief engineer of several railroads, city surveyor, canal and railroad builder (ANB DAB MWBD NC25 WWWIA)

KNEASS, Strickland, 1821–1884, American civil engineer; assistant engineer and topographer on Pennsylvania state survey (1839), Bureau of Engineering of US Navy, maps for Northeast boundary between US and Canada (1842), Pennsylvania Railroad surveys and construction of railroads over mountainous route, worked as chief engineer for several railroad companies, surveyor of Philadelphia (10 yr.), drainage system for city, officer of several regional railroad companies (ANB DAB NC39 WWWIA)

KNIGHT, Edward Collings, 1813–1892, American industrialist and inventor; formed E. C. Knight & Co. as agents for sugar refining, ship owners, held patent for railroad sleeper car (1859) known as Knight cars, sold interests to Pullman Co. (George M. Pullman [1831–1897]) (1868), president of American Line steamships, east coast regional railroad construction and executive (DAB NC6 WWWIA)

KNIGHT, Jonathan, 1787–1858, American civil engineer; surveyor in Pennsylvania, county commissioner, Pennsylvania legislature (1822–1828), federal commissioner for road development in west through Ohio, Indiana and Illinois, chief engineer of Baltimore & Ohio Railroad (1829–1842) in charge of planning structures, machinery and location, consulting engineer (EIH DAB NC11 TEIH WWWIA)

KNIGHT, Margaret E., 1838–1914, American woman artist and inventor; designed devices to increase safety of workers in cotton textile mills (1870), machine to make square bottom paper bags (1868), protective lining for dresses, clothing clasp, shoe cutting machine (1890–1894), window frame, devices for steam engine improvements, sleeve valve engine, issued 27 patents (AIE ANB AWIT BDOTHOT I&T15 MEIA MOI NAW WI WIS(1))

KNOLL, Max, 1897–1969, German engineer and inventor; with German engineer Ernst A. E. Ruska [1906–1988] developed electron microscope (1929) (1931) (IAI MWBD NYPL S&TF WA)

KNOWLES, Chester L., 1894–1972, American chemical engineer; research chemist advancing to manager of division at DuPont Co. (1917–1923), sales engineer advancing to sales director with Dorr-Oliver Co. (NYC) (1923–1944), technical director of process equipment design of General American Transportation Corp. (NYC) (1944–1947), with others formed Knowles Assoc. consulting engineering firm (1947–1959) (NC57)

KNOWLES, Hugh Shaler, 1904–1988, American electrical engineer and inventor; radio, electro-acoustics, loud speakers, receivers and microphones, with several industrial companies including chief engineer and vice president of Jensen Manufacturing Co. (IL), formed Knowles Electronics Co. (1954ff), officer of several other companies, consulting engineer, NAE (1969) (MT4 WWIA)

KNOWLES, Lucius James, 1819–1884, American inventor and manufacturer; cotton-spinning, spool-threading equipment, formed partnership to manufacture cotton warp (1849) then expanded to include woolen mill (1853), patents credited to Henry R. Worthington [1817–1880], invented improvement to looms (1856), boiler feed water regulator (1862), worked on steam valves, legislator (DAB MEIA NC5 WWWIA)

KNOX, Thomas Wallace, 1835–1896, American inventor; wrote of his extensive worldwide travels, patented device for transmitting plans of battlefields by telegraph, invented device for telegraphing by S. F. B. Morse [1791–1872] signals the spot where a bullet struck a target (DAB NC7 WWWIA)

KNUDSEN, William S., 1879–1948, Danish American industrialist; to US (1900), officer of several automotive companies including Ford Motor Co. (1911–1921), vice president then president of Chevrolet Motor Co. (1922–1937), president of General Motors Corp. (1937–1940), government service during WWII as director of production with rank of lieutenant general (1942–1945) (MWBD NCF WWWIA)

KOBAYASHI, Koji, 1907–1996, Japanese electrical engineer; NEC Corp., formerly Nippon Electric Co., (1929–1988) becoming president (1964) and chairman emeritus (1988), manufacturer of communications and power equipment in Japan, helped build business oriented activities of the company after WWII, promoted concept of computers and communications, fNAE (1977) (MT9)

KOBAYASHI, Shiro, 1924–1995, Japanese American mechanical engineer; understanding of metal manufacturing processes including forging, extrusion, rolling and sheet-metal forming and cutting, taught high school in Japan (7 yr.) and at Doshisha Univ. (1953–1956), to US (1956), faculty at the Univ. of California (1961–1991), NAE (1980) (Bridge MT8 WWIE)

KOBZAREV, Yurii B., 1905–??, Soviet electrical engineer; radio engineering, developed frequency stabilization using quartz crystals (1926–1931), development of radar, worked at universities and institutes of USSR Academy of Sciences, Physico-Technical Inst. of USSR (1926–1943), professor at Moscow Inst. of Energetics (1944–1955), Inst. of Radio Engineering & Electricity (1955ff) (PS SMOS WWWIS)

KOCH, Robert, 1843–1910, German physician, scientist and inventor; professor at Univ. of Berlin (1885ff), identified several bacteria and treatments, invented pure culture technique of growing microbes using agar, hanging drop technique of observing microbes with a microscope, developed photomicrographs, fNAS (1903), Nobel prize (1905) (EIH MWBD WBE SAI SAI-MP WWWIS)

KOCHINA, Pelageya Yakovlevna, 1899–??, Woman Soviet hydrodynamicist; theory of filtration, movement of oil and ground water in soil, dynamics, theory of tides in basins, meteorology, geophysical observatory (1919ff) and taught at Leningrad Univ. becoming professor (1934), mathematics and mechanics institute (1935ff) (SMOS WWWIS)

KOITER, Warner T., 1914–1997, Dutch mechanical engineer; stiffness, strength and stability of structures, aircraft safety, Aeronautical Research Inst. (1936–1938), Department of Civil Aviation (1939–1949), faculty at Delft Univ. (1949–1979), fNAE (1977) (Bridge WWIE)

KOLBE, Adolph Wilhelm Hermann, 1818–1884, German chemist and inventor; discovered Kolbe reaction, produced salicylic acid based on earlier work of Karl Gerhardt [1833–1902] (1853) that led to the manufacture of aspirin (acetylsalicylic acid) (1859), was assistant to Robert Bunsen [1811–1899] (1842–1845), working in London on analysis of mine gases to prevent explosions (1845–1851), professor at Marburg (1851–1865), professor at Leipzig (1865ff), Felix Hoffmann [1868–1946] developed method of making the compound in quantities that the German firm Bayer introduced trademarked aspirin (1899), introduced synthesis into chemistry (B-ACGA EB EOS&T GLOSB IAOSTA MWBD S&FT TBOF WWWIS)

KOLSTER, Frederick August, 1883–1950, Swiss American engineer and inventor; Stone Telegraph & Telephone Co. (1902–1909), DeForest Radio Co. (1909–1911), National Bureau of Standards (NBS) (1912ff), inventor of Kolster decremeter, directional systems for radio communication, developed radio compass (1913) (NC41,E TTOS WWWIA)

KOMAROV, Vladimir M., 1927–1967, Soviet cosmonaut; first reported space flight death (1967) when parachute for spacecraft failed (MWBD WBE)

KOMPFNER, Rudolph, 1909–1977, Austrian American electrical engineer and inventor; British Admiralty Research (1941–1944), Oxford Univ. (1944–1951), traveling-wire tube, satellite communications, optical fibers, acoustic microscopy, administration Bell Labs. (1951), microwave tubes, light wave communication, NAE (1966), NAS (1968) (BM54 MT1 WWIE WWWIA)

KÖNIG (KOENIG), Friedrich, 1774–1883, German English printer and inventor; invented steam printing press (1810), first flatbed cylinder press (1806), used rollers to make impression in a power-driven press (1811–1814), co-founded König and Bauer Co. (Andreas Friedrich Bauer [1783–1860]) (1817) that manufactured steam presses (MMAH MWBD PS SAI WBE WWWIS)

KORN, Arthur, 1870–1945, German physicist and inventor; workable system of radio picture telegraphy (1907) (MMAH S&TF)

KOROLYEV, Sergei Pavlovich, 1906–1966, Soviet aeronautical engineer; design of rockets, space flight, spacecraft (1945ff), director of Soviet space program with first manmade satellite in space (Sputnik) (1957), launched first dog (Laika) into space, launched first man Yuri A. Gagarin [1934–1968] into space (1961), first woman Valentina N. Tereshkova [b. 1937] into space, probes to Venus and Mars (1961–1962) (BDOTHOT BEOS EB ITCOF MWBD 1000Y PS TBDOS(Po) TBOF TGS)

KOSONOCKY, Walter F., 1931–1996, Polish American computer and electrical engineer and inventor; to US (1949), optoelectronics and solid state circuits, development and application of charge-coupled and optical-switching devices and circuits, New Jersey Inst. of Technology (9 yr.), Radio Corp. of America Labs. (RCA) (1957–1987), inventor of the year award (1987), NAE (1992) (Bridge MT9)

KOSTENKO, Mikhail Polievktovich, 1889–??, Soviet electrical engineer; theory of electrical machines and experimental studies, polyphase commutating machines, electrodynamic modeling of energy systems, design of hydroelectric plants, taught and was chair of Electric Machines, worked for several institutes related to electric power (PS SMOS WWWIS)

KOTELNIKOV, Vladimir Aleksandrovich, 1908–??, Soviet electrical engineer; radio engineering and electronics becoming institute director (1954ff), communication effectiveness introducing error stability as related to transmission (PS SMOS WWWIS)

KOUWENHOVEN, William Bennett, 1886–1975, American electrical engineer; developed heart machines, first practical heart defibrillator (1930s), external heart massage (1959), faculty at John Hopkins Univ. (1914–1975) (BEOS MSAE MWBD WWWIA WWWIS)

KOVALENKOV, Valentin Ivanovich, 1884–??, Soviet electrical engineer and inventor; theory of wire transmission and communications, processing signals, several inventions in electrotechnics and sound movies, worked for electrical-technical Inst. of Automation and Remote Control (1940–1948) and director of laboratory at USSR Academy of Sciences (1946–1956) (PS SMOS)

KOVALEV, Nikolai Nikolaevich, 1908–??, Soviet electrical engineer; design of hydroturbines and hydroelectric power plant, chief of hydroturbine construction (1945–1959), at Leningrad Metal Plant (1933ff), taught at Leningrad Polytechnical Inst. (PS SMOS WWWIS)

KOYL, Charles Herschel, 1855–1931, Canadian American civil engineer; taught physics and electrical engineering (1887ff), patented a parabolic semaphore for railroad signaling, practice of engineering (1890ff), authority on disposal of municipal water, pioneer in treatment of industrial water supplies, schemes for softening water, worked for railroad related to water use, supply and conservation (DAB NC27 WWWIA)

KRAFT, James Lewis, 1874–1957, American food manufacturer and inventor; established Kraft Food Corp. (1909) that became part of the National Dairy Products Corp., patented cheese-making process (1917), developed mechanical wrapping individual slices of cheese (AIE ANB WBE WWWIA)

KRANICK, Frank N. G., 1881–1954, American agricultural engineer; started with J. I. Case Co. (10 yr.) and returned as assistant to the president (1928–1947), chief engineer at Bell City Manufacturing Co. (1907–1913) designing a line of threshers, with Rumley Co. (1913–1917), Hyatt Roller Bearing Co. (1917–1920), then Timken Roller Bearing Co. (1920), often in charge of experiments for companies for which he worked (AE36 WWIE(I))

KREBS, Arthur C., 1847–1935, French engineer and inventor; with Charles Renard [1847–1905] built airship La France with electric motor propulsion (1884), built multi-pole electric motor, electric motor for submarine (1888), invented carburetor (EB MWBD TTOT WBE WWWIS)

KREUGER, Ivan, 1880–1932, Swedish industrialist; founded match company that he directed (1913ff), during WWI placed company along with other companies under Swedish Match Co. (MWBD)

KRISHNA, Jai, 1912–1999, Indian civil engineer; earthquake engineering, soil mechanics, design standards, State Public Works Department (1935–1939), faculty at Roorkee Univ. (1939ff) becoming vice chancellor (1971), engineering consultant, fNAE (1979) (Bridge MT9)

KROEGER, William John, 1906–1966, American electrical engineer and inventor; field engineer at Westinghouse Co. while obtaining advanced degrees, ballistics for Federal Bureau of Investigation (FBI) (1938–1940), engineer at Frankford Arsenal (PA) (1940ff) becoming chief scientist (1963–1966), recoilless rifle (WWII), ejection seats for pilots, at least 18 patents (NC51)

KROLL, William J., 1889–1973, Luxembourg American metallurgist and inventor; developed method to produce metallic titanium (1932) called Kroll process, with US government laboratory, developed process to produce zirconium (ET22 NYPL WWWIA)

KRUESI, John, 1843–1899, Swiss American mechanical expert and inventor; to US (1870), first worked with Singer Sewing Machine Co., then for Thomas Edison [1847–1931] responsible for mechanical inventions of his many ideas such as first phonograph (1877), perfected incandescent lamp, developed machines to manufacture electrical equipment and lighting systems, helped develop underground distribution systems for electricity (1882–1887), became manager and chief engineer for Edison Machine Works (1866ff), 10 patents (ANB DAB NC26 WWWIA)

KRUPP, Alfred, 1812–1887, German metallurgist and manufacturer; son of Friedrich Krupp [1787–1826] who succeeded his father and perfected process of making cast iron guns, welded steel tire for railroad vehicles, manufacture of ordnance (c.1847), supplied arms internationally (EIH MWBD 1000Y TBDOS(Po) WWWIS)

KRUPP, Bertha, 1886–1957, German manufacturer; daughter of Friedrich Alfred Krupp [1854–1902] whose husband took the name Gustav Krupp [1870–1950] who assumed management of Krupp Works after WWI and whose son Alfred Krupp [1907–1967] assumed control (c.1939) and took over properties confiscated by Germany in WWII, imprisoned at Nuremberg (1946) and granted amnesty (1951) and rebuilt the company (MWBD)

KRUPP, Friedrich, 1787–1826, German ironmaster; founder of Krupp Works (1811), initiated process for making cast steel (MWBD)

KRUPP, Friedrich Alfred, 1854–1902, German manufacturer; son of Alfred Krupp [1812–1887], succeeded father in management of Krupp Works, manufacturer of iron and steel machinery (MWBD)

KRUZHILIN, Georgii Nikitich, 1911–??, Soviet heat engineer; distribution of heat emission from surface of a body, condensation of steam, removal of moisture in steam from boilers, with several institutes including Central Boiler Turbine Inst. (1933–1948) becoming director at Inst. of Energetics, USSR Academy of Sciences (1960) (SMOS WWWIS)

KRYLOV, Aleksandr Petrovich, 1904–??, Soviet petroleum engineer; mining and drilling, rational methods of exploring oil deposits, taught at Moscow Petroleum Inst. (1933), then to USSR Academy of Sciences where he became administrator (PS SMOS WWWIS)

KTESIBIOS (also written as CTESIBIUS), fl. 285–247 BC, Greek physicist and inventor; force pump, water organ, hydraulic pipe organ, musical keyboard, metal spring, improvement to water clock (BEST MWBD TAE TEIH)

KUHRT, Wesley Amos, 1917–1988, American aerospace engineer; research and development, executive at United Aircraft (1941–1967), executive at Sikorski Helicopter Co. (1968ff), NAE (1980) (MT4 WWWIA)

KULEBAKIN, Viktor Sergeevich, 1891–??, Soviet electrical engineer; electronic computers, automatic regulation, design of regulators, electrical ignition of aircraft engines, taught at Moscow Higher Technical Inst. (1917–1923), Air Force Engineering Academy (1923ff), All-Union Electrotechnical Inst. and Inst. of Automation (PS SMOS)

KUMMER, Fred Alfred, 1906–1980, German American agricultural engineer; soil physics as related to tillage implement design, mechanization, consultant to US Army ordnance in WWII, faculty at Auburn Univ. (1935ff) including department head (1948ff) (AE61 AMOS WWWIA)

KUMMER, Joseph T., d. 1997, age 77, American engineer; industry research on ionic conductors, sodium-sulfur battery, catalysis, combustion, fuel cells, NAE (1986) (Bridge)

KURDIUMOV, Georgy Vyacheslavovich, 1902–1996, Russian engineer and metallurgist; processes relating to hardening and tempering of steel, phase transformations, diffusion in metals, structural changes in alloys, at Dnepropetrovsk Research Inst. and Univ. (1932–1944), director of metallurgy and physics in Moscow (1944ff), USSR Academy of Sciences (1946), fNAE (1977) (Bridge SMOS WWWIS)

KUZNETSOV, Vladimir I., 1913–??, Soviet mechanical engineer; applied mechanics, worked for several research and construction organizations (1938ff), USSR Academy of Sciences (1958) (SMOS)

KYLE, John Montgomery, Jr., 1904–1970, American civil engineer; worked in construction (1925–1943), US Army Corps of Engineers (1943–1946), chief engineer of Port Authority (NY and NJ) (1946ff), terminal and transportation facility, tunnel, airport terminal and marine facilities (NY) (NJ), NAE (1967) (MT1)

L

LABELYE, Charles, 18th century, Swiss French engineer; builder of Westminster bridge over Thames (1738), first to use caisson in major construction, construction of piers and use of stone arches (1746, 1750) (TSOE)

LACER, Caius Julius, 2nd century AD, Spanish engineer and builder; built bridge at Alcantara in Spain dedicated to Roman emperor Trajan [53–117] (TSOE MWBD)

LAËNNEC, René-Théophile-Hyacinthe, 1781–1826, French inventor and medical doctor; invented stethoscope (1816, 1819), specialist in treating tuberculosis and emphysema (BEST EB GI NYPL 1000Y WA WBE WWWIS)

LAGRANGE, Joseph-Louis, 1736–1813, Italian French geometer and applied scientist; applied his mathematical skills to a systemization of mechanics, calculus of variations (1755), headed commission in France to reform weights and measures (1793) from which the metric system developed, professor at Royal School in Turin (1755–1766), director of Berlin Academy of Sciences (1766–1781), professor at École Polytechnique in Paris (1795–1799) (BEST FNIE MWBD TGS WWWIS)

LAIDLAW, Walter, 1849–1914, Scottish American mechanical engineer; worked with Caird & Co. as shipbuilder and engineer in Scotland and later with Trinity House, to US (1881), employed by Lane and Bodley (OH), construction engineer for Procter & Gamble Co. (OH) (1883), organized Laidlaw & Dunn Co (OH) (1897) manufacturing steam pumping and hydraulic equipment, plant manager for International Steam Pump Co. (1899) then member of executive committee (1909) (MEIA WWWIA)

LAKE, Simon, 1866–1945, American naval architect and inventor; steering high-wheeled bicycles (1887), submarine pioneer, inventor of even-keel submarine, built Argonaut (1897), torpedo boat design, invented apparatus for recovering sunken vessels, over 200 patents (ANB MWBD NC15 PS)

LALOR, Peter, 1823–1889, Irish Australian engineer and politician; to Australia (1852), gold miner and involved in associated worker concerns (1854ff), politician (DNB MWBD)

LAMB, Isaac Wixom, 1840–1906, American inventor and clergyman; machine to braid whip lashes (1859), knitting machine (1863), Lamb Knitting Machine Manufacturing Co. (1864), improved windmill (1883), Perry Glove and Machine Co. (1895), 15 patents for knitting machines plus other patents (DAB MEIA NC7 WWWIA)

LAMÉ, Gabriel, 1795–1870, French mathematician, physicist and engineer; professor at École Polytechnique

School (1832–1844) and Univ. of Paris (1851–1862), railroad engineer, chief engineer of mining, applied mathematics, elasticity, thermodynamics, introduced curvilinear coordinates (CEOS&T FNIE MWBD WWWIS)

LAMM, A. Uno, 1904–1989, Swedish American electrical engineer and inventor; with firm Vasteras Sweden (1928–1969), to US (1965), high-voltage direct current (HVDC) transmission, mercury arc rectifier, switch gear, electrotechnical director of ASEA Ab. a joint venture with General Electric Co. (1965–1970), consultant (1969ff), about 150 patents, fNAE (1976) (Bridge MT5 WWIE)

LAMME, Benjamin Garner, 1864–1924, American mechanical and electrical engineer and inventor; mechanics, flow of natural gas through long pipes, testing of electrical machinery at Westinghouse Corp. (1889ff), reduction-gear motor (1889), direct-current railway developments, issued 162 patents (DAB NC28 WWWIA)

LANCHESTER, Frederick William, 1868–1946, English engineer and manufacturer; built first automobile in Great Britain (1895), worm gear (1897), founded Forward Gas-Engine Co. (1893), Lanchester Motor Car Co. (1899), built first English automobile (1901) with novel features, pioneer in aeronautics, glider, disc brakes (1902), Lanchester Prandtl theory (Ludwig Prandtl [1875–1953]) (1894?), powered flight, published Aerodynamics (1907) and Aerodonetics (1908), FRS (1922), Lanchester Lab. Ltd. (1925ff), consulting engineer (DNB DOSB E-64 ITCOF MWBD NYPL PS RHW RSObit TBDOS (Po) TGS)

LAND, Edwin Herbert, 1909–1991, American physicist, inventor and manufacturer; founded Land-Wheelwright Lab. (1932), devised unique light-polarizing apparatus (1932), founded Polaroid Corp. (1937), introduced Polaroid Land camera (1948) and black-and-white film (1950) and color film (1963), three-dimensional viewing, NAE (1965), NAS (1953) (AIE ANB BEST BM77 Bridge DOT EOWB IAI MT7 MWBD NCH PA SAI S:TLAW WBE WWWIA WWWIS)

LANDAUER, Rolf W., 1927–1999, American physicist and electronics engineer; US Navy in WWII, NACA (2 yr.), research at International Business Machines Co. (IBM) (1952ff), injection laser, large-scale integration (LSI) in electronics, physics of computing devices, NAE (1978), NAS (1988) (Bridge MT9 WWIE WWWIS)

LANDER, Frederick West, 1821–1862, American civil engineer and explorer; survey work for Western railways, survey for crossings of Mississippi River, chief engineer for overland wagon route, survey for railroad from Mississippi River to Salt Lake City (ANB DAB NC8 WWWIA)

LANDIS, James Noble, 1899–1989, American mechanical engineer; design and development of energy systems, design of fossil and nuclear steam power plants, with various units of Edison Co. (1923–1948), Bechtel Corp. (1948–1964), consultant (1969–1977), NAE (1964) (Bridge MT4 WWIE)

LANDIS, Walter Savage, 1881–1944, American metallurgist and engineer; thermochemistry of metallurgical operations, nitrogen fixation, fertilizers, cyanide products, nitric acid, faculty at Lehigh Univ. (1903–1912), American Cyanide Corp. (1912–1944) becoming vice president, numerous patents on processes (ANB)

LANDRIANI, Marseillio, c.1751–c.1816, Italian Austrian instrument developer; developed instrument to measure properties of air called audiometer, used by Joseph Priestley [1733–1804] in his studies (DOSB PS)

LANDSBERG, Helmut E., 1906–1985, German American meteorologist; to US (1934), pollution and environmental problems, climatology and microclimatology as applied to engineering, effects of urbanization, biometeorology, government employee, faculty positions at Pennsylvania State Univ. (1934–1941), Univ. of Chicago (1941–1946), and Univ. of Maryland (1964ff), NAE (1966) (AMOS MT5 WWIA WWIE WWWIA)

LANDWEBER, Louis, 1912–1998, American mechanical engineer; hydrodynamics, David Taylor Model Basin (DC) (1932–1954), ship hydrodynamics, faculty at Univ. of Illinois (1954ff), contributed to modern architecture and marine engineering, NAE (1980) (Bridge WWIE)

LANE, Henry Marcus, 1854–1920, American mechanical engineer; designed power equipment for Elm Street Inclined Plane Railway, supervised reconstruction for Inclined Plane Railway (1879), built cable railways (OH, CO, RI), became president of Lane & Bodley (1890ff), consulting mechanical engineer (MEIA WWWIA)

LANE, John, 19th century, American blacksmith; invented first steel (of strips of steel) moldboard plow (c.1833) (MWBD)

LANE, John, 1824–1897, American manufacturer and inventor; son of John Lane [19th century], manufacturer of steel plows, invented improvements on steel plows (MWBD)

LANGBEIN, Walter B., 1907–1982, American civil engineer and hydrologist; worked for construction company for many years, US Geological Survey (USGS) (1935–1969) followed by 12 yr. consulting, water utilization, flood investigations including social and economic risks, contributed to developments such as rain gauges, field conductivity bridge, flow distribution chart, overland flooding, use of tritium for determining age of groundwater, use of trees for reference data, NAS (1970) (HOWRD)

LANGEN, Eugen, 1833–1895, German engineer and inventor; internal combustion engine developed with Nikolaus A. Otto [1832–1891] (1867), promoted idea of overhead suspension monorail (EIH MWBD WBE WWWIS)

LANGEVIN, Paul, 1872–1946, French physicist and inventor; invented sonar system for submarine detection (1916), molecular structure of gases, theory of magnetism, faculty at College of France (1902–1909), professor at the Sorbonne (1909ff), fRS (1934) (BEST BITAS CBD DOSB GI IAI MWBD NYPL RSObit TBOS(Po) TCBE TP TTOS WBE)

LANGLEY, Samuel Pierpont, 1834–1906, American astronomer, engineer and inventor; pioneer in astrophysics at Univ. of Pittsburgh (1867–1887), studied solar eclipses of 1869, 1870, 1878, invented bolometer (1878), solar radiation at various wavelengths, solar activity and weather (1887ff), director of Smithsonian Inst. (1887ff), aerodynamics (1891), studied flight of heavier than air machines, mechanically propelled flight attempted (1903), NAS (1876) (ANB ASAI BEST BM7 CEOS&T DAB EIH EOWB IIA ITCOF MWBD WBE WWWIS)

LANGLOIS, Claude, c.1700–c.1756, French instrument maker; official instrument maker for government, made instruments primarily for astronomers, researchers, laboratories, and schools (DOSB PS)

LANGMUIR, Irving, 1881–1957, American chemist, engineer and inventor; metallurgical engineering (1903), with General Electric Co. Research Lab. (1909–1950), invented vacuum gauges (1913), invented gas-filled light bulbs (1914), mercury vacuum pump (1916), Langmuir isotherm (1916), Lewis-Langmuir atomic theory (1919) (Gilbert N. Lewis [1875–1946]), atomic hydrogen welding (1924), Langmuir-Blodgett film (1934) (Katherine B. Blodgett [1898–1979]), NAS (1918), Nobel prize (1932) (ANB ASAI BEST BM45 DOT EE EOWB LOTCS MWBD NCC TFE&I TNAS WBE WWWIA WWWIA-S&T WWWIS)

LANGREN, Michael F. van, c.1600–1675, Belgian engineer; cartography, selenography, maps including a map of moon (DOSB PS)

LANHAM, Frank Bristol, 1913–1978, American agricultural engineer; faculty at Univ. of Georgia (1936–1941), US Army in WWII (1942–1946), wholesale hardware jobbing to general manager for Bower Co. (1946–1950), secretary of ASAE (1951–1955), faculty and department chairman at Univ. of Illinois (1955ff) (AE59 WWIA WWIE)

LANSTON, Tolbert, 1844–1913, American inventor and lawyer; served in Civil War, US Patent Office (DC) (1865–1887), invented padlock (1870), hydraulic dumbwaiter (1871), railroad car coupler (1871), typesetting machine (1887), Lanston Type Machine Co. (DC) (1887), monotype (1897), sewing machines, adding machine (1899) (ANB DAB MEIA MWBD NC13 NYPL WA WBE WWWIS)

LAPIDUS, Leon, 1910–1977, American chemical and computer engineer; computing machinery, lattice theory, high-speed digital computation, optimum control of chemical processes, numerical solutions, at Princeton Univ. (NJ) (1951–1977) including department chairman (1968–1977), consultant (1977ff), NAE (1976) (AMOS MT1 WWWIA)

LAPLACE, Pierre-Simon de Marquis, 1749–1827, French mathematician and applied scientist; early work on specific heat, heat production by a process (1780), latent heat, tides, capillary action, probability, turned attention to astronomy, known for his monumental work in five volumes on celestial mechanics, equations named after him used extensively by engineers and others, professor at École Militaire (1767ff), French senate (1790) and its vice president (1803) (BEST DOSB FNIE HOFL TGS MWBD WWWIS)

LAQUE, Francis Lawrence, 1904–1988, Canadian chemical and metallurgical engineer; research, development and administration including director of corrosion engineering (1940–1945) with International Nickel Co. (Inco) (42 yr.) becoming assistant to the president (1952–1969), fNAE (1985) (MT4)

LARIONOV, Andrei Nikolaevich, 1889–??, Soviet electrical engineer; theory, design, and construction of electric machines and electric drive, at All Union Electric Technical Inst. (1921–1941), worked on automation at Academy of Sciences (1953ff) (PS SMOS)

LARNED, Joseph Gay Eaton, 1819–1870, American inventor and industrialist; patent lawyer, perfected steam fire engine (1855), Wellington Lee and Joseph Larned formed Novelty Iron Works (NYC) (1855–1863) to develop, manufacture and sell fire engines (DAB MEIA)

LARSON, Clarence E., 1909–1999, American chemist and engineer; radiochemistry, nuclear energy, electromagnetic separation for purification of uranium, Union Carbide Co. (1943–1950), nuclear power plant design, Oak Ridge (TN) (1950–1969), commissioner of Atomic Energy Commission (AEC) (1969), consultant, NAE (1973) (Bridge MT9 WWIE)

LARSON, John Augustus, 1892–1965, Canadian American psychiatrist and inventor; forensics, police work, criminology, Larson Fingerprint System (1923), lie detector (1928), taught at several universities, medical schools and hospitals, consultant to justice system (ABES NC42 NYPL TBOF WBE WOI WWWIA)

LARSON, Maurice Allen, 1927–1999, American chemical engineer; research and development at Dow Corning Corp. (1951–1954), process dynamics and control, fertilizer technology, faculty and college administration at Iowa State Univ. (1954ff) (AMOS WWIE)

LARSON, Thurston E., 1910–1984, American chemical engineer; corrosion, water chemistry, water treatment, water standards, Illinois State Water Survey (1932ff), NAE (1978) (MT3 WWIE)

LATÉCOÈRE, Pierre, 1883–1943, French aircraft manufacturer; established aircraft building firm (1917), commercial air flights (1918), company taken over by Air France (1932) (MWBD)

LATIMER, Lewis Howard, 1848–1928, Black American inventor and electrical engineer; Union Navy during Civil War, with W. C. Brown developed water closet for railroad cars (1874), expert in drafting particularly for patents, incandescent electric lighting system (1880ff), process for manufacturing carbon for filaments (1882), friend and colleague of Alexander Graham Bell [1847–1922] (1876) and Thomas A. Edison [1847–1931] (1884–1911), General Electric Co. and predecessors, consultant (1911ff) (AAI ANB BDOBA BI BPOSAI CBB LTM SAI-MP TBDOS(Po) WBE)

LA TOUR, Le Blond de, ??–1723, French engineer; draftsman (1702), engineer in Louisiana for France (1703), engineer-in-chief for Province of Louisiana (1720–1723) (DAB WWWIS)

LATROBE, Benjamin Henry, Sr., 1764–1820, English American architect and engineer; to US (1796), building design, city water supply systems (1799), surveyor, construction of public buildings in Washington (DC), led work to reconstruct the capitol after it was burned by British (1915), with Robert Fulton [1765–1815] built steamboats (1813–1815) (ANB BITAS DAB EIH FEL IIA MWBD NC9 TEIH TGS WWWIA WWWIS)

LATROBE, Benjamin Henry, 1806–1878, American civil engineer; son of Benjamin Henry Latrobe [1764–1820], survey between Baltimore and Washington (DC) (1832), known for stone arch railroad bridge southwest of Baltimore (MD), engineer for developing railroads from Baltimore to Havre de Grace (MD) (1835), several railroad construction projects including tunnels and bridges at Pittsburgh (PA) and Wheeling (WV) (DAB EIH FNIE TEIH TGS)

LATROBE, Charles Hazlehurst, 1832–1902, American civil engineer; railroad construction including bridges, Baltimore and Ohio Railroad, construction of railroads in Florida, engineer with Confederate Army, Baltimore Bridge Co. (1866–1877), city engineer for Baltimore (MD) (1875–1889), several commissions for Peruvian government, consulting engineer (1899ff) (DAB NC9 WWWIA)

LATROBE, John Hazlehurst Boneval, 1803–1891, American inventor and lawyer; son of Benjamin H. Latrobe [1764–1820], helped develop charter for Baltimore & Ohio Railroad (1827), helped obtain railroad right-of-ways, Latrobe stove for house heating, lawyer for railroad patents (ANB DAB NC9 WWWIA)

LATTA, Alexander Bonner, 1821–1865, American inventor and manufacturer; machinist, foreman in machine shop, built locomotives (1847–1865), Latta steam engines for fire department apparatus (1852) (DAB MEIA NC13 WWWIA)

LAUDISE, Robert A., 1930–1998, American chemist, materials engineer and inventor; growing single crystals, discovery of electronic materials, physical and inorganic chemical research, with AT&T Bell Labs. (1956ff) including several research management positions, 14 US patents, NAE (1980), NAS (1991) (MT10 WWIE WWWIA)

LAUFER, John, 1921–1983, American aerospace engineer; turbulence and airflow stability, jet noise abatement, pressure radiation field from a turbulent boundary layer, California Inst. of Technology (1944–1949) and Jet Propulsion Lab. (JPL) (1952), National Bureau of Standards (NBS) (1949–1952), NAE (1977) (AMOS MT2 WWIE)

LAURIE, James, 1811–1875, Scottish American civil engineer; maker of engineering instruments (1832), railroad construction as associate then chief engineer, superintendent and inspector of railroads in New York, Nova Scotia, and Massachusetts (1855ff) (DAB)

LAVAL, Carl Gustaf Patrik de—see DE LAVAL

LAVOISIER, Antoine-Laurent, 1743–1794, French chemist and inventor; principles of combustion, conservation of matter, human respiration, doubled gunpowder production, public service, devised metric system adopted in France (1795), devised plan for lighting city streets (1766), guillotined by government (BEST 1000Y MWBD SAI SAI-MP S:TLAW TEIH TGS WBE WWWIS)

LAW, Harold Bell, 1911–1984, American physicist and electrical engineer; researcher with RCA Labs. (1941ff), television pickup tubes, color display tubes, tube fabrication techniques, NAE (1979) (Bridge MT3 WWIE)

LAWLER, Joseph Christopher, 1920–1983, American civil and sanitary engineer; known for complex engineering projects, applied hydraulics with Camp Dresser & McKee, Inc. (1947ff) from project engineer to partner to president (1970–1978) to chairman and CEO (1978–1983), NAE (1973) (MT2 WWIE WWWIA)

LAWRANCE, Charles Lanier, 1882–1950, American aeronautical engineer; radial air cooled engines (1921), engines for airplanes while working for Wright Aeronautical Co. (1923–1930), founded Lawrance Engineering and Research Corp. (1930) serving as its president and chairman (1930–1950) (MWBD)

LAWRENCE, Ernest Orlando, 1901–1958, American physicist and inventor; radiation laboratory director, with associates invented cyclotron (1929), radioisotopes, biology and medicine, medical cyclotron (1939), faculty at Univ. of California (1930ff) including founder and director of Radiation Lab. (1936ff), NAS (1934), Nobel prize (1939) (ACS AG ASAI BEST BM41 IAI

MWBD NYPL PharmCent SAI-MP TFE&I TGS TP WWWIS)

LAWRENCE, Richard Smith, 1817–1892, American inventor; gunsmith, tool manufacturer, devised barrel-drilling and rifling machine, furnished machinery to English Enfield Armory, superintendent of Sharp Rifle Co. (1851–1872), worked for city of Hartford (CT) (1872ff) (EB DAB MEIA)

LAWROWSKI, Stephen, 1914–1997, American chemical engineer; Fischer-Tropsch gas synthesis, nuclear chemical engineering, separation processes for spent reactor fuels, feed materials processes, fast-reactor technology, at Standard Oil Development Co. (1943–1944) (1946–1947), Argonne National Lab. (1947ff) including head of chemical engineering (15 yr.), NAE (1969) (Bridge WWIE NC24)

LAY, John Louis, 1832–1899, American engineer and inventor; perfected torpedo, assistant engineer for US Navy during Civil War, developed mines in Peru (1865–1867), developed Lay Moveable Torpedo submarine (1867), 10 patents (ANB DAB NC7 NYT-A WWWIA)

LEA, Matthew Carey, 1823–1897, American chemist and mechanics; considered father of mechanochemistry—the study of chemical changes as a result of pressure—shear, impact, friction, photography, photochemistry, chemistry of coal, founded Chemical Testing Lab. (PA), NAS (1892) (ACCE BM5 ChemHerit20)

LEAR, William Powell, 1902–1978, American electrical engineer and manufacturer; formed and headed Lear Avia Corp. (1934) for manufacturing navigational equipment, company followed by Lear, Inc. (1939–1962), then by Lear Jet Corp. (1962ff) to produce small aircraft, attempted to build steam-powered automobiles and buses with Lear Motors Corp. (1967) (AIE ANB MWBD)

LEATHERS, Joel Franklin Monroe, 1920–1987, American chemist and chemical engineer; industry project leader, assistant superintendent, director of research and development and executive vice president of Dow Chemical Co. (1942–1974), Office of Technology Assessment (OTA) (DC) (1974ff), OTA existed (1972–1995), NAE (1978) (MT4 WWWIA)

LEAVITT, Erasmus Darwin, Jr., 1836–1916, American mechanical engineer and inventor; engine construction for ships, pumping engine (1867ff) used principally for waterworks and mines, Calumet and Hecla Mining Co. (1874–1904), consulting engineer (BITAS DAB MEIA)

LEAVITT, Frank McDowell, 1856–1928, American mechanical engineer and inventor; steam-steering apparatus for US Navy, sheet metal working machinery, power presses, torpedo development and manufacture, issued over 300 patents (DAB MEIA NC15 WWWIA WWWIS)

LEBAUDY, Paul, 1858–1937, French industrialist; with brother Pierre Lebaudy [1861–1924] built semirigid dirigibles, military dirigible (1902), roundtrip across channel (1910) (BDOTHOT MWBD)

LEBEDEV, Sergei Alekseevich, 1902–??, Soviet electrical engineer; radio engineering, stability of power stations, high-speed computers, at various institutes as researcher, teacher and administrator at All Union Electrical Institute (1928–1945), and several other technical positions (PS SMOS WWWIS)

LEBLANC, Maurice, 1857–1923, French electrical engineer and inventor; several developments named after him—dynamic balancing machine, phase advancer, damper winding, induction machine and exciter (EE MWBD WWWIS)

LEBLANC, Nicolas, 1742–1806, French physician and inventor; physician (1780–1793), developed procedure for producing sodium carbonate quickly and cheaply that is used in thousands of everyday products, such as aluminum, soap, detergents, and paper, devised soda process (1787), manufacture of soda from salt (1789) (BEST GEAPIT MWBD 1000Y TGS TodChem9 TSOE WBE WWWIS)

LEBON, Philippe, 1767–1804, French civil engineer, chemist and inventor; highway engineer, artificial gas made from wood, gas for illumination (1799), Lebon engine for using coal gas (1799) (EE EIH GEAPIT MWBD PS TBDOS(Po) WWWIS)

Le BRUN, Albert, 1871–1950, French mining engineer and politician; after engineering work, minister of colonies and minister of war (1911–1914), senator (1920–1932), president of France (1932–1940) (EB MWBD)

LE CHÂTELIER, Henry-Louis, 1850–1936, French chemist, mining engineer and metallurgist; cement, combustion, equilibria, alloys, iron and steel, professor at the School of Mines (1877ff), École Polytechnique, Le Châtelier principle (1888), fRS (1913) (BEST CDOSB CEOS&T DOSB MWBD WWWIS)

LECLANCHÉ, Georges, 1839–1882, French electrical engineer and inventor; galvanic cell, Leclanché battery (1866)—the first dry cell, electric cells (1867, 1873), railway engineer (CEOS&T DOSB MWBD TBDOS WBE WWWIS)

LE CORBUSIER (aka Charles-Édouard Jeanneret), 1887–1965, Swiss architect and industrialist; city planning, adopted Henry Ford system of mass production, consulting architect for firm of Auguste Perret [1874–1954] and Gustave Perret in Paris (1916), settled in Paris (1917), established Societe d'Études Industrielles et Techniques (1917), built private houses, apartment complexes, public buildings, emigrated to Great Britain and then to US (1937) where he continued his projects and influence, identified round concrete silos as a uniquely US building (AG MWBD)

LECORNU, Leon Francois Alfred, 1854–1940, French mechanical engineer; mechanics, Corps of Mines (1874–1924), chief engineer in charge of railroads, inspector general, faculty at various schools (1896–1934) including École Polytechnique (1896), School of Mines (1900ff) and National Polytechnic Superior School of Aeronautics (1927–1934) (CDOSB DOSB PS WWWIS)

LEE, James Paris, 1831–1888, American Canadian Scottish inventor; emigrated to Canada (1836), invented Lee-Enfield rifle used by British army (1902) and other magazine rifles, used in WWI modified and called the American Enfield rifle (1917), manufactured his rifle in Milwaukee (WI) and then Remington (NY) (WWWIA WWWIS)

LEE, Robert Edward, 1808–1870, American army officer and engineer; US Army corps of engineers (1829–1846), superintendent of US Military Academy at West Point (NY) (1842), served as general, later commander of Confederate Army troops (1861–1865), president of what is now Washington and Lee Univ. (1865–1870) (DAB EB MWBD NYPL TEIH WBE)

LEE, William, c.1550–1610, English inventor and clergyman; first hosiery knitting machine (1589) established in France, silk knitting machine (1599) (MMAH MWBD NYPL WBE WWWIS)

LEE, William S., 1929–1996, American civil engineer; development of large multipurpose electric power projects, environmentally sensitive to designs, Duke Power Co. (1955ff) becoming chairman (1982), NAE (1978) (Bridge WWIE WWWIA)

LEEDS, Pulaski, 1945–1903, American mechanical engineer; worked primarily with several railroads as locomotive fireman, locomotive engineer, superintendent of motive power (1877), patented turntable (1881), master mechanic (1882), invented labor-saving devices (MEIA)

LEEGHWATER, Jan Adriansz, 1575–c.1650, Dutch engineer; mill maker, engineer on Beemster drainage project (1608–1612), windmills for drainage (EIH GEAPIT PS SAI TSOE)

LEES, Lester, 1920–1986, American aeronautical engineer; subsonic and supersonic gas flows, boundary-layer studies, gas dynamic aspects of combustion, faculty of California Inst. of Technology (1942–1944) (1953ff), government research and development, NAE (1971) (AMOS MT5 WWWIA)

LEEUWENHOEK, Antoni van, 1632–1723, Dutch scientist and inventor; surveyor in Holland (1669), ground lenses and made microscopes, first observed bacteria (1674), red blood cells (1674), blood capillaries (1683), protozoa, spermatozoa (1677), yeast (BEST CDOSB DOSB IAI MWBD SAI TGS WBE WWWIS)

LEFFEL, James, 1806–1866, American inventor; built and operated water-powered sawmills, iron foundry (1840ff), waterwheel development (1845), level jack (1850), cookstove (1852), perfected double-turbine water-ter wheel (1862) (ANB DAB WWWIA)

LEFFERTS, Marshall, 1821–1876, American engineer; built telegraph lines, developed galvanizing of iron, became executive officer of different telegraph companies, consulting engineer (DAB IIA NC10 WWWIA)

LEGENDRE, Adrien-Marie, 1752–1833, French mathematician; theory of elliptic equations, Legendre functions published (1825), method of least squares, number theory, Legendre polynomials (1784), Legendre theorem (1785), Legendre transformation (1786), weights and measures for French government (1794ff) and professor at Inst. de Marat, chief of Bureau of Longitudes (1813ff) (BDOS(Wi) CDOSB DOM DOSB MWBD RHW WWWIS)

LEGGET, Robert F., 1904–1994, English Canadian civil engineer; heavy building construction, building research, geological engineering, to Canada (1929), Queen's Univ. and Univ. of Toronto (11 yr.), consultant, National Research Council of Canada, fNAE (1988) (Bridge MT8 WWIE)

LEHAN, Frank W., 1923–1997, American electrical engineer; frequency-modulated telemetry, secure communication systems, development of electronic systems for missiles, jet propulsion, space programs, US Army Signal Corps (1942–1944), California Inst. of Technology Jet Propulsion Lab. (1944–1954), space electronics for TRW (1954–1967), assistant secretary of transportation (1964–1977), NAE (1970) (Bridge WWIE)

LEHMANN, Ernst August, 1886–1937, German aeronautical engineer; manufacture and piloting Zeppelins (1913ff), commanded several Zeppelins including the Hindenburg, which exploded (1937) (MWBD)

LEIBNIZ (LEIBNITZ), Gottfried Wilhelm, 1646–1716, German mathematician and inventor; calculating machine capable of performing multiplication and division (1671), dynamic theory of motion (1676), provided foundation for calculus (1676), which he published (1684), diplomatic missions to Paris (1672–1676) and London (1673) (BEST EB EIH GI HOFL MWBD NYPL TGS WWWIS)

LELAND, Henry Martyn, 1843–1932, American automobile manufacturer; with son built transmission used by Ransom Eli Olds (REO) [1864–1950], organized company (1893) that merged with Cadillac Motor Co. (1904) and then became part of General Motors Corp. (GMC) of which he was president (1909–1917), founded Lincoln Motor Co. (1917) (ASAI MWBD NC40 WWWIA)

LEMAIRE, Jacques, 18th century (fl. 1720–1740), French instrument maker; sundials, front-view reflecting telescope that was later improved by English astronomer Sir William Herschel [1738–1822] (DOSB MWBD PS)

LEMAIRE, Pierre, 18th century (fl. 1733–1760), French instrument maker; son of Jacques LeMaire [fl. 1720–1740], Lemaire workshop, sundials, compass maker, telescope, developed device that permitted several screws to be advanced at once while tightening a single screw (BDOS DOSB PS)

LEMEHAUTE, Bernard, 1927–1997, French American ocean engineer; development and application of analytical techniques and hydraulic modeling in coastal processes, harbor design, coastal engineering, to US (1961) and citizen (1966), faculty at École Polytechnic (Montreal) and Queen's Univ. (Kingston) and Univ. of Miami (FL), National Engineering Science Co. (1961–1966), founded and operated Tetratech (CA) (1966ff), NAE (1991) (Bridge WWIE WWWIS)

LEMELSON (LEMUELSON), Jerome H., 1923–1997, American inventor and industrial engineer; independent inventor, industrial automation, bar-code reader, cordless telephone, cassette player that became a part of Sony Walkman, camcorder, elements in ATM machine, fax machine, toys such as crying dolls and water pistols, Smithsonian Inst. Lemelson center named after him, obtained 554 patents (ranks 5th among individual patent holders in US) (FAL NYTObit WPost2001 WWWIA)

L'ENFANT, Pierre Charles, 1754–1825, French soldier and engineer; fought for independence of US, converted old buildings in New York City to temporary US government use, city planner, surveyed site and developed plans for the new Federal City of Washington (DC) (ANB CEOS&T DAB IIA TEIH)

LENOIR, Jean-Joseph-Étienne, 1822–1900, Belgian French engineer and inventor; to France (1838), invented first effective internal combustion engine (1859) using illuminating gas, also called coal gas, and air as fuel, (Frenchman Alphonse-Eugène Beau de Rochas [1815–1893] developed rules for four-stroke engine (1863) later developed by German Nikolaus August Otto [1832–1891]) (1867), electric brake for trains (1855), electric motor (1856), electric spark ignition system presented (1857) by Barsanti and Mattuci Co., internal combustion using illuminating gas (1859), automobile with internal combustion engine (1862), boat with engine (1866), made two-crank, four-cylinder engine (1883) (ASAI BDOS(Wi) BEST E-63 EE EIH FNIE GI IB MWBD NYPL PS SAI TBDOS(Po) TGS WBE WWWIS)

LEONARD, Harry Ward, 1861–1915, American electrical engineer and inventor; superintendent of light company, head of firm to engage in central-station and electric railway construction that became part of Edison General Electric Co. (1881), electric lighting systems (1889), motor control (1891), numerous inventions related to electric control of trains (DAB)

LEONARDO DA VINCI, 1452–1519, Italian (Florentine) artist, sculptor, military engineer and inventor; technical advisor on architecture and engineering, canal lock gates (1495), military engineer (1502–1503), best known for painting Mona Lisa (1503–1506), catapults, numerous writings on a wide variety of subjects including guns, hydraulics, hydrology, meteorology, mechanics, friction, machines and vehicles, gears, link chain, weaponry, fortifications (BDOTHOT BEST DOT E-64 EIH EOWB FNIE GEAPIT IAI MMAH MWBD 1000Y SAI SAI-MP TBDOS(Po) TEE TEIH TSOE WWWIS)

LEONARDS, Gerald A., 1921–1997, Canadian American civil engineer; soil mechanics, geotechnical engineering, worked for Canadian government, faculty at Purdue Unv. (1946–1976) including department head (1965–1968), consulting geotechnical engineer (1972–1976), NAE (1988) (Bridge MT9 WWIE)

LEONHARDT, Fritz, 1909–1999, German civil engineer; long-span concrete bridges, high-rise towers, bridges over Rhine River, Columbia River, and autobahn, telecommunication towers, with construction firm (ante 1938), consulting engineer (1938ff), faculty at Univ. of Stuttgart (1957–1974) including rector (1967–1969), fNAE (1983) (Bridge MT9 WWIE)

LE PLAY, Pierre-Guilliame-Frédéric, 1806–1882, French mining engineer and sociologist; professor (1840) and inspector (1848), at French School of Mines, then went into sociology (c.1855) (MWBD WWWIS)

LERY, Joseph Gaspard Chaussegros de, 1721–1797, Canadian (New France) engineer; fortifications for Quebec and Montreal, involved in Louisiana conflict with Indians, erected forts along Allegheny River (DAB)

LESSEPS, Ferdinand-Marie Vicomte de, 1805–1894, French civil engineer; designer and builder of Suez Canal connecting Mediterranean and Red Seas completed (1869), construction projects, attempted Panama Canal construction (EIH EOWB MWBD 1000Y TBDOS(Po))

LESSON, Martin, 1920–1999, American mechanical engineer; aerospace, energetics, hydrodynamic stability, turbulence, thermoplastic shock, US Navy (1940–1946), NACA (1947–1949), faculties at Pennsylvania State Univ. (1949–1953) and Univ. of Pennsylvania (1953–1960) and Univ. of Rochester (1960–1970), consulting engineer (Prism WWIE WWWIA WWWIS)

LeTOURNEAU, Robert Gilmore, 1888–1969, American inventor; had own machine and welding shop, invented and built earth-moving and hauling machines, leveling equipment, electric wheel (motor in wheel), president of R. G. LeTourneau, Inc. (1929–1969) and associated companies, 187 US patents (CB DAB NCF T7-AP WWWIA)

LEUPOLD, Jakob, 1674–1727, German engineer and inventor; operated mechanical shop, machines, statics, hydrostatics, built improved air pump, built multiplication machine, built waterwheel driven by force of river current to drive a system of pumps to provide water supply (1724), Leupold high-pressure steam engine (1725) (BDOS(Wi) EB EE PS WWWIS)

LEVASSOR, Émile, c.1844–1897, French automotive engineer; with French engineer René Panhard [1841–1908] built a vehicle with an internal combustion engine mounted in front (1891–1892) (MWBD WWWIS)

LEVICH, Benjamin Gregory, 1917–1987, Soviet electrochemist and engineer; used concentration polarization as a research tool, rotating disc electrode, turbulence, research, teaching, and administration at Soviet Inst. for Electrochemistry (ante 1978), to Israel at Univ. of Tel Aviv (1978), to City College of New York (NYC) (1979), theoretical turbulence, fNAE (1982) (MT5 WWWIA WWWIS)

LEVISON, Wallace Goold, 1846–1924, American chemist, chemical engineer and inventor; chemical and electrical research, directed Chemical Lab. at Cooper Union (1865?–1884), invented spectropolariscope (1881), automatic single lens camera (1887), an early arc lamp (1886), founded and head of department of mineralogy, active in forming several scientific organizations (NC19 PS)

LEVITT, William, 1907–1994, American housing specialist; assembly line techniques used to build low-cost or affordable housing, divided building a house into 27 basic steps, contributed to urban sprawl, several communities named after him as Levittown (1000Y WWWIA)

LEVY, Louis Edward, 1846–1919, American inventor; optics, microscope, surveying, photography, photoengraving called linotype (1875), with brothers established companies and developed half-tone screen, many publishing activities (DAB NC13 PS WWWIA)

LEVY, Max, 1857–1926, American inventor and manufacturer; brother of Louis Edward Levy [1846–1919], photoengraver, design and construction of machines to make screen for half-tone reproduction (1881), photographic plate holders, eyepieces for range finders, invented hemocytometer (DAB NC13 WWWIA)

LEWIS, Isaac Newton, 1858–1931, American inventor and army officer; artillery position finder (1891), Lewis machine gun (1911), fire-control system for artillery, gas-propelled torpedo, after retirement moved to Europe and manufactured machine gun in Belgium and England (ANB DAB MEIA MWBD NC16 WWWIA)

LEWIS, Warren Kendall, 1882–1975, American chemical engineer; faculty and administration at Massachusetts Inst. of Technology (1916–1947) except for work with US government during WWI and WWII, with William H. Walker [1869–1934] and William Henry McAdams [1892–1975] developed classic book on principles of chemical engineering, worked with textiles, leather, rubber, refining and production of petroleum, with Edwin K. Gilliland [1909–1973] invented

fluidized-bed cracking of petroleum, NAS (1938), NAE (1966) (BM70 HOCE MT1 NUC PS WWIA WWIE WWWIA)

LEWIS, Wilfred, 1854–1929, American mechanical engineer and inventor; mechanic, draftsman, designer, assistant engineer and director for William Sellers & Co. (PA) (1875ff), invented gears and machinery for building and testing gears, more than 50 patents (MEIA WWWIA)

LEWIS, Wilfred Bennett, 1908–1987, English Canadian engineer; administrator and work with US government (1939–1945), radioactivity, guidance and development of CANDU reactor system, officer of Atomic Energy Canada, Ltd., reactor economics, fNAE (1976) (AMOS Bridge MT5 WWWIA WWWIS)

LEWIS, William Deming, 1915–1989, American mathematician and electrical engineer; radar, communications, Apollo program, useful use of satellites, research at Bell Labs. (1941–1962), Bellcomm, Inc. (1962–1964), president at Lehigh Univ. (1964–1982), 33 patents, NAE (1967) (Bridge MT4)

LEWIS, William Gaston, 1835–1901, American engineer; surveyor for government, assistant engineer for railroad, ensign in Confederate Navy in Civil War, state engineer, chief engineer of Albany and Raleigh Railroad, chief of engineers in North Carolina National Guard (DAB NC24)

LEY, Willy, 1906–1969, German American engineer; rocketry, pioneered work with liquid propellants (1931ff), space exploration, first rockets in US (1936), popular writer and lecturer, consulted with several US government agencies (ANB BEST WWWIA WWWIS)

LEYBENZON, Leonid Samuilovich, 1879–1951, Russian mechanical engineer and geologist; built wind tunnel, tested propellers, developed methods of aerodynamic and structural design of airplanes, Aerodynamics Inst. (1904), professor of mechanics at Yurev Univ. (1916–1919), Tbilisi Polytechnical Inst. and Univ. (1919–1929), Polytechnical Inst. at Baku (1921–1922), Moscow Univ. (1922), research at Central Aerohydrodynamics Inst. (1933–1936), USSR Academy of Sciences (1933) (CDOSB DOSB PS)

LEYBOURN, William, 1629–c.1700, English surveyor; layout of streets, surveying in London (1666ff), early book on surveying (1650) (MWBD TSOE WWWIS)

LEYNER, John George, 1860–1920, American inventor and manufacturer; engineer for mining company, then for milling company, machine shop and foundry (1886ff), designed several improvements to mining machinery, compressed-air rock drilling (1893) (1899), jackhammer called Water Leyner Rock Drill (c.1902), drill-sharpening equipment (1909) (DAB MEIA NC25)

LIBBY, Willard Frank, 1908–1980, American chemist and inventor; atomic clock for carbon dating materials (1947), Univ. of California (1934–1941), Manhattan project (1941–1945) working on atomic bomb, Univ. Chicago (1945–1959), Univ. of California Los Angeles (UCLA) (1959–1980), NAS (1950), Nobel prize (1960) (BEST MWBD NCI NYPL 1000Y SAI-MP TGS WWWIS)

LIEB, John William, 1860–1929, American electrical and mechanical engineer; draftsman, with Edison Electric Illuminating Co. (NY) (1881–1900) developed central lighting plants, several positions at Milan Italian Edison Station (Italy) (12 yr.), long-distance transmission of high-voltage alternating current underground, president of Electrical Testing Co. (NYC) (1900), officer of New York Edison Co. becoming senior vice president (1901ff) (ANB DAB MEIA NC13 WWWIA WWWIS)

LIEBERMAN, Gerald J., 1925–1999, American statistician and mechanical engineer; control and reliability, operations research, quality control, engineering statistics, US Army (1943–1944), National Bureau of Standards (NBS) (1949–1950), faculty at Stanford Univ. (1953ff) and consultant at Stanford Research Inst. (SRI) (1959ff), NAE (1987) (AMOS Bridge WWWIA)

LIGHTHILL, Michael James, 1924–1998, British aerodynamicist and mathematician; aerodynamics division of National Physical Lab. (1943–1945), faculty at Cambridge Univ. (1945–1949) (1969–1978), Univ. of Manchester (1946–1959), research professor at Imperial College (1964–1969), provost of Univ. College London (1979–1989), consulting (1989ff), FRS (1953), fNAS (1976), fNAE (1977) (Bridge WWW WWWIS)

LILIENTHAL, David Eli, 1899–1981, American lawyer in charge of engineering; management of related projects at Tennessee Valley Authority (TVA), chief engineer was Arthur E. Morgan [1878–1975], dam construction, conservation of natural resources, telephone rate reduction, flood control, electric power production, NAS (1951) (ANB FEL WWIE WWWIA)

LILIENTHAL, Otto, 1848–1896, German aeronautical engineer; gliders (1891), studied air flow over wings, studied flight of birds, wrote basic books on air flow (ASAI BEST EIH HIW IIA ITCOF MWBD TBDOS(Po) WWWIS)

LILJEDAHL, John Bruce, 1919–1996, American agricultural engineer; served in WWII, faculty at Purdue Univ. (1954ff) specializing in power and equipment for agriculture, after similar positions at Iowa State Univ. (1946–1949) and Univ. of Tennessee (1949–1952), consultant serving in overseas countries, expert witness on product liability and product safety (Res3 WWIE)

LINCOLN, Abraham, 1809–1865, American president and inventor; first and only patent issued to a person who became president of US, patent was for a device to help lift a vessel over shoals in water (1849), other presidents George Washington [1732–1799] invented a seeding plow and a wine coaster and Thomas Jefferson [1743–1826] invented a swivel chair, folding buggy top, plow share and a writing desk but neither applied for a patent (EB HFP MWBD TEIH WOI WWWIA)

LINDBERGH, Charles Augustus, 1902–1974, American aviator and inventor; studied mechanical engineering in college, made first solo nonstop flight across Atlantic Ocean (1927) in his airplane the *Spirit of St. Louis,* test pilot, technical advisor to several airlines, consultant to Ford Motor Co., sterilizable glass pump for circulating culture fluid through an excised body organ (ANB BEST DAB EOWB GI IAI MWBD NYPL 1000Y WWWIA)

LINDE, Karl Paul Gottfried von, 1842–1934, German engineer; professor at Technische Hochschule Munich (1868–1879), methyl ether refrigerator (1874), vapor compression refrigerating machine using ammonia as a refrigerant (1876), process of liquefying air (1895), method of separating oxygen from liquid air, with Charles Brush [1849–1929] founded Linda Air Products Co. (1905) (AAI BEST EE MWBD NYPL PS WWWIS)

LINDENTHAL, Gustav, 1850–1935, Austrian American civil engineer; to US (1874), consulting engineer in bridge and railroad construction (1877–1890), commissioner of bridges in New York City, consulting engineer for Pennsylvania Railroad (PRR) tunnels under rivers in NYC, Hell Gate steel arch bridge (NYC) (1917) (ANB EIH MWBD WWWIA)

LINDER, Clarence Hugo, 1903–1994, American electrical engineer; manufacturing and management in fields including turbines, searchlights, and consumer appliances, to electric utilities equipment with General Electric Co. (1924–1963), vice president of engineering (1953–1960), retired as vice president (1960–1963), founding member of NAE (1964) (Bridge MT8)

LINDSEY, William, 1808–1900, English civil engineer; chief engineer of Hamburg-Bergedorf Railway (1838–1860), consulting engineer for several cities (1865–1869), constructed sewage system for Frankfurt-am-Main (MWBD)

LINDVALL, Frederick C., 1903–1989, American mechanical engineer; railway equipment, underwater ordnance, rocket applications, faculty at California Inst. of Technology (1930–1969), vice president of Deere & Co. (1969–1972), consulting, NAE (1967) (Bridge MT4 WWIE WWWIS)

LINGO, Donald Percy, 1912–1981, American mathematician and engineer; space technology, real-time control systems, systems engineering, large-system design, guidance techniques, missile systems, engineer on government committees and panels, with Bell Labs. (1945–1958), NAE (1967) (AMOS MT2)

LINING, John, 1708–1760, Scottish American physician and inventor; investigated electricity and lightning, importance of grounding and insulation, meteorological observations (1738ff) (ANB DAB NC25 WWWIA WWWIS)

LINK, Edwin Albert, Jr., 1904–1981, American inventor, engineer and industrialist; with brother George Link invented Link flight simulator for training pilots (1929), founder and president of Link Aviation (1935–1959), General Precision Corp. (1958–1959), equipment for oceanographic use, NAE (1965) (ANB CB ITCOF MT2 MWBD WWWIA)

LINNAEUS, Carolus (aka Carl von Linné), 1707–1778, Swedish botanist and inventor; invented system of plant classification (1735), explored Scandinavian area for plants (1738), professor at Uppsala (1741ff), introduced binomial system of nomenclature (1753) (CDOSB DOSB MWBD NYPL SAI SAI-MP TGS WBE WWWIS)

LINSLEY, Ray Keyes, 1917–1990, American civil engineer; engineer with Tennessee Valley Authority (TVA) (1937–1940), hydrologist with US Weather Bureau (USWB) (1940–1950), surface water hydrology, faculty at Stanford Univ. (1950–1975), chairman of Hydrocomp, Inc. (1966–1978), resource planning, professional practice (1979ff), NAE (1976) (AMOS Bridge MT5 WWIE WWWIA)

LINTLAER, Jean, C. 1600s, Flemish engineer; worked in France, built waterwheel under arch of Pont Neuf (1602), waterwheel adjusted for water height, pumping stations, water systems for cities (EIH TAH TEIH TSOE)

LINVILL, William Kirby, 1919–1980, American electrical engineer; servomechanisms, feedback control, interactive control systems, engineering economic systems, faculty at Massachusetts Inst. of Technology (MIT) (1949–1958), Rand Corp. (1958–1960), faculty at Stanford Univ. (1960ff), NAE (1979) (AMOS MT2 WWWIA)

LIPE, Charles E., 1851–1895, American engineer and inventor; worked at Bradley Co. (NY), Remington Co. where he worked on Spooner water meter and foundry molding machines, numerous patents on such items as horse hayrake (1875), tool post for lathes (1882), heads for milling machines and universal milling machines (1884), plus several other mechanical devices (MEIA)

LIPPERSHEY, Hans, c.1570–c.1619, Dutch optician and inventor; made eyeglasses, credited for invention of telescope (1608) (BEST GI HIW MWBD NYPL RHW SAI TBOF TGS WA WWWIS)

LIPPISCH, Alexander Martin, 1894–1976, German American aeronautical engineer; design of tailless and delta-winged planes (1920s, 1930s), rocket-propelled airplane (1928), chief designer of Messerschmitt 163 (Me 163) (1944) fighter plane, liquid-fuel rocket aircraft (1944), to US (1945), established Lippisch Research Corp. (IA) (1965) (MWBD)

LIST, Hans, 1896–1996, Austrian engineer and inventor; art, science, and technology for design and manufacture of vehicular engines, designer at diesel engine factory in Graz (1926–1945) including several years with Tonji Univ. (China), founder and chairman of engineering office of AVL (1948ff) and professor at Technical Hochschule Dresden and Technical Univ. Graz, awarded 364 patents, fNAE (1989) (Bridge MT10 WWIE WWWIA WWWIS)

LISTER, Joseph, 1827–1912, English surgeon and inventor; received medical degree (1852), professor at Glasgow Univ. (1859–1869), Edinburgh Univ. (1869–1877), King's College London (1877–1893), introduced carbolic acid as an antiseptic, improved methods of surgery, founder of antiseptic surgery, baron (1897), fNAS (1898) (BEST MWBD SAI SAI-MP TGS WWWIS)

LISTER, Samuel Cunliffe (aka 1st Baron Masham), 1815–1906, English inventor; with brother John Lister [1827–1912] established worsted milling business (1838), invented wool-combing machine (1845), silk-combing machine (c.1865), velvet loom machine (c.1878), cloth-making devices, baron (1891) (DOSB MWBD PS WWWIS)

LITTLE, Arthur Dehon, 1863–1935, American chemical engineer and inventor; chemist and superintendent of mill making sulfite wood pulp (1845), organized chemical consulting firm (1886) reorganized as Arthur D. Little Co. (1909) for industrial research, several inventions (AIE MWBD NC15 WWWIA)

LLOYD, Marshall Burns, 1858–1927, American inventor and manufacturer; weighing scale for farmers, machine for weaving wire mattresses, machine to make wire wheels for baby buggies, manufacture of tubing made from strips of steel, a loom for manufacture of wicker, consultant (ANB DAB NC25 WWWIA)

LLOYD, Morris H., 1902–1988, American agricultural engineer; rural electrification in New York, Niagara Mohawk Power Corp. (1929–1963), followed by special assignments (until 1967), safety in use of electricity (AE WWIE)

LOCKE, John, 1792–1856, American scientist and inventor; surgeon in US Navy, taught in and supervised female academies, lectured at Mechanics Inst., thermoscopic galvanometer, terrestrial magnetism, level, portable compass, electromagnetic chronograph, US Geological Survey (USGS) (1844–1848) (ANB BITAS DAB NC15 WWWIA WWWIS)

LOCKE, Joseph, 1805–1860, British civil and railway engineer; surveyor, railroad engineering, construction, politician as member of parliament (MP), had an aversion to tunnels (TBDOS(Po) TEIH TGE:B)

LOCKE, Sylvanus Dyer, 1833–1896, American mechanical engineer and inventor; joined with Walter A Wood [1849–1934] to develop and sell wire-binders (1871, 1873), Locke wire tie baler with automatic self-binding (1875ff), obtained 43 patents on harvesting devices (1865–1879) (ASAI MEIA NC6 TGH WBE)

LOCKHEED, Allan Haines, 1889–1969, American aeronautical engineer; worked in automotive manufacturing then aircraft industry in mid-1910s, aviator, with brother Malcolm Lockheed [1887–1958] designed seaplane (1912), Loughead (original spelling of name) Aircraft Manufacturing Co. formed (1916), hired John K. Northrop [1895–1981] as chief engineer, numerous planes built, company purchased by Detroit Aircraft Co. (1929) (AIE ANB WWWIA)

LOCKHEED, Malcolm, 1887–1958, American aircraft engineer and inventor; worked with White Steam Car Factory, developed four-wheel hydraulic brake system (1904–1917), introduced several planes along with brother Allan H. Lockheed [1889–1969] to design and build aircraft, early best known plane was Lockheed Vega (1930–1935) (AIE ANB)

LODGE, Sir Oliver Joseph, 1851–1940, English applied physicist and inventor; lightning, electromagnetic waves, wireless telegraphy, perfected coherer for wireless telegraphy, professor at Univ. College Liverpool (1881–1900), author of several books (EIH MWBD TFE&I WWWIS)

LODGE, William, 1848–1917, English American mechanical engineer and inventor; to US (1869), built machinery for folding paper (1869–1872), machinist and foreman with Steptoe, McFarlan, Mottingham & Co. (1872–1880), founded Lodge and Banker (1880), Ohio Machine Tool Co. (1892), and Lodge & Shipley Tool Co., specialized in manufacture of engine lathe (MEIA)

LOENING, Grover Cleveland, 1888–1976, American aircraft manufacturer and inventor; formed two companies named after him for aircraft production (1917–1938), Loening Aeronautical Engineering Corp. (1917–1928) and Grover Loening Aircraft Co. (1928–1938), invented strut-braced monoplane, amphibian plane, numerous aeronautical patents (MWBD NCB WWWIA)

LOEWY, Raymond Fernand, 1893–1986, French American industrial designer; to US (1919), founded design firm of Raymond Loewy Associates (1927), which he headed (1930–1961), becoming Raymond Loewy International (1961–1975), designed wide variety of products (MWBD WWWIA)

LOGAN, John Alexander, 1908–1987, Canadian American civil engineer; with Saskatchewan Highway Department (1929–1932), to US (1934), planning, design, operation of large environmental control systems,

applications of system analysis to environmental work, served as chief engineer, with several faculties including Iowa State Univ. (1935), Univ. of Missouri (1937), Northwestern Univ. (1954), and president of Rose-Hulman Inst. of Technology (1962ff), NAE (1968) (Bridge MT5 WWIA WWIE WWWIA)

LOMAX, Harvard, 1922–1999, American mechanical and aerospace engineer; in US Navy during WWII, research at NACA (1945ff) that became NASA (1958) and with its national center for computational fluid dynamics, supersonic aerodynamics, derived the supersonic area rule, was chief of Computational Fluid Mechanics Branch at NASA, was father of Aerodynamic Simulator Program of NASA, NAE (1987) (Bridge MT10 WWWIA)

LONG, James Dewey, 1899–1996, American agricultural engineer; machinist with Hart-Parr Co, (1918), extension and research at universities, field tests on wood preservation, structural characteristics of building materials, design of structures for animals and produce, grain storage (Res3 WWIE)

LONG, Stephen Harriman, 1784–1864, American army explorer and engineer; with Army Corps of Engineers, topographical engineer, explored rivers in western US, an authority on railroad engineering particularly relating to grades and curves of rails, surveys, bridge construction, consulting engineer (ANB BITAS DAB NC11)

LONGSTREET, William, 1759–1814, American inventor; operated steam engine (1788), breast roller for cotton gins, steam-operated cotton gins (1801), operated steamboat on Savannah River (GA) (1806), portable steam sawmill (ante 1812) (DAB MEIA NC9 WWWIA WWWIS)

LORING, Charles Harding, 1828–1907, American engineer; naval and marine engineer, engineering service with the US Navy (1851–1890) and chief engineer during the Civil War, inspector of iron-clad steamers, tested water-tube boilers, consulting engineer, US & Brazil Steamship Co. (MEIA NC12 WWWIA)

LOVELACE, (Augusta) Lady Ada (aka Countess of Byron, Countess of Lovelace), 1815–1852, British mathematician and computer expert; worked with Charles Babbage [1792–1871] analytical engine (1834ff), calculating machine (1842), developed instructions for operating the device, first computer programmer, designed punch card system, conceptualized GIGO (garbage in, garbage out), Ada computer language named after her (BDOTHOT DNB EB GI GLOSB LDOS LDS MADOI MOI 1000Y S:TLAW TBDOS(Po) TEIH WIS(1))

LOVELL, Mansfield, 1822–1884, American civil engineer; artillery officer, superintendent of street improvements in New York City (1858), served in Confederate Army, surveying (ANB DAB NC4 WWWIA)

LOW, Frederick Rollins, 1860–1936, American mechanical engineer; telegraphy, stenography, secretary to editor then became editor of magazine Power (1888–1930), Clark & Low Machine Co., with F. M. Clark invented Clark vertical boiler flue cleaner and then steam engine devices (MEIA NCL WWWIA)

LOW, George Michael, 1926–1984, Austrian American aeronautical engineer; to US (1940), NACA (1949), to headquarters of NASA (1958) then held administrative positions in manned space flight and Apollo spacecraft programs (1967) and deputy administrator of NASA (1969–1976), president of Rensselaer Polytechnic Inst. (1976–1984), NAE (1970) (Bridge MT3 WWIE WWWIA WWWIS)

LOW, John Routh, Jr., 1909–1988, American metallurgist; research and development with various steel companies and General Electric Co. (1948ff), physical and mechanical metallurgy, fracture of metals, NAE (1978) (AMOS MT4)

LOWE, Thaddeus Sobieski Coulincourt, 1832–1913, American topographical engineer, aeronaut and inventor; ballooning authority (1856ff), studied upper air currents, US Army chief of aeronautic section (1861–1865), commander of balloon corps, mobile hydrogen gas generator, invented compression ice machine (1865), airborne use of camera, apparatus for producing water gas (1873–1875), New Lowe Coke Oven system for producing coke (1897) (ASAI BITAS DAB MWBD NC9 PS WWWIA WWWIS)

LOWELL, Francis Cabot, 1775–1817, American industrialist; mercantile business (1793–1810), cotton spinning and weaving mill, with Paul Moodey [1779–1831] built Massachusetts town of Lowell named after him (BITAS IIA MEIA MWBD NC7 PS WWWIA)

LUBINSKI, Arthur, 1910–1996, Belgian American civil and mechanical engineer; with French Forces in WWII, to US (1947), industry development with Amoco Production Co. (1950–1975), applied mechanics in drilling for oil and gas, offshore and arctic operations, seven patents, NAE (1986) (AMOS MT9 WWIE WWWIA)

LUCAS, Anthony Francis, 1855–1921, Australian American mechanical and mining engineer; study and discovery of salt domes, prospected with diamond drill, discovered oil in Beaumont (TX) area called Lucas Gusher, consulting engineer (DAB NC29 WWWIA)

LUCAS, Jonathan, 1775–1832, English American millwright and inventor; patented a new type of machine for removing husks from rice (1808) using vertical conical cylinders, returned to England with a British patent and developed a major business (DAB WWWIA)

LUCKIESH, Matthew, 1883–1967, American physicist and electrical engineer; illuminating engineer, most of career with National Electric Lamp Assoc. (NELA) and was director of Lighting Research Lab. (1924–

1949), NELA became a part of General Electric Co., wrote 25 books, after retirement continued promoting proper lighting (ANB NCF WWWIA)

LUDLOW, William, 1843–1901, American engineer and soldier; chief engineer in army in Civil War, in several roles as chief engineer with military, surveys of Yellowstone National Park (1873ff), river and harbor works in Philadelphia (1876–1882), chief engineer of Philadelphia Water Department (1883–1886), served in war with Spain, became brigadier general (DAB NC9 WWWIA)

LUDWIG, John H., 1913–1995, American civil engineer; meteorologist, structures and hydraulics, US Army Air Corps in WWII (1943–1946), water pollution and control (1934–1949) in Burlington (VT), USPH and EPA (1951–1972), organizer and director of technological resources to meet rapidly changing environmental challenges, consultant, NAE (1971) (Bridge MT8 WWIE WWWIA)

LUHR, Otto, 1860–1932, German American engineer and inventor; designer with Keystone Bridge Works (PA), chief engineer with Chicago Brewer & Malting Co. (IL), consulting engineer in refrigeration, inventions related to lubrication of engine cylinders (1903), dehydrating gaseous fluids (1923), manufacturing ice (1924), Frigicar (1927) and others (MEIA)

LUKIN, Lionel, 1742–1834, English inventor and coach builder; unsubmergible boat (1785), raft for rescuing people under water, rain gauge, hospital bed (MWBD)

LULL, Ramon, c.1232–1316, Spanish philosopher and inventor; Catalan cyclopedist, forerunner of modern symbolic logic and computer science, used manually operated concentric rotating rings to make determinations (calculations), combinatory methods (DOSB HOFL MWBD)

LUMIÈRE, Auguste-Marie-Louis-Nicolas, 1862–1954, and brother LUMIÈRE, Louis-Jean, 1864–1948, French inventors; invented cinematograph (1895), first to capture on film pictures that moved (1895), motion picture industry, color photography (1904), developed improved dry photographic plate (1880) that was financially successful (GI IAI 1000Y SAI TBDOS(Po) TGS WWWIS)

LUN, Ts'ai, fl. c.105 AD, Chinese inventor; one of persons credited with invention of paper (T-100)

LUNDIE, John, 1857–1931, Scottish American engineer and inventor; railroad work, harbor engineering, drainage, canal, water supply, designed electric hoist, Lundie Ventilated Rheostat (1901), water power development, formed Lundie Tieplate Co. (1913), consultant (DAB NC24 WWWIA)

LURIE, Anatolii Isakovich, 1901–??, Soviet mechanics specialist; faculty of Leningrad Polytechnical Inst. (1925ff) becoming chairman (1935), theoretical mechanics, machine strength and dynamics, automatic control systems, elasticity, USSR Academy of Sciences (1960) (SMOS WWWIS)

LUX, John Herbert, 1918–1996, American chemical engineer and manufacturer; held various positions at Union Carbide Corp. (WV) (1942ff), research and development at Neville Co. (PA), was with Atomic Basic Chemical Co. (PA), Witco Chemical Co. (IL), General Electric Co. (MA), Shea Chemical Co. (MD), Hercules, Inc. (DE), president of Haveg Industries, Inc. (1955–1964), president, CEO and director of AMETEK (NYC) (1966ff) a manufacturer of scientific and specialty equipment, 13 patents (NCL WWWIE)

LYALL, James Broadwood, 1836–1901, Scottish American inventor and manufacturer; to US (1839), made Jacquard looms, served in Civil War, invented enameling cloth (1863), with brother formed J. & W. Lyall Co. (1865), invented Lyall positive motion loom (1868), established and headed plant for manufacturing cloth mixing cotton and jute, established several cotton mills in northeast, manufacturer of jute binder twine (1888), invented woven fabric for pneumatic tires and fire hoses (1893–1896), at least eight patents (1880s) (DAB MEIA NC7 WWW WWWIA)

LYANG LINGDZAN, 8th century, Chinese inventor; built first clock with an escapement (724) (TAE)

LYDDAN, Robert H., 1910–1990, American civil engineer; after several engineering jobs joined US Geological Survey (USGS) (1933ff), topographical mapping of areas from Puerto Rico to Alaska, to Washington (DC) as assistant director (1956–1968), retired with the topographic division (HOWRD)

LYLE, Samuel P., 1892–1987, American agricultural engineer; head of agricultural engineering at Arkansas A & M College (1922–1924), professor and head of department at Univ. of Georgia (1924–1930), federal extension service of US Department of Agriculture (USDA) (1930–1962) as engineer subject-matter specialist related to agriculture (AE(supp) WWIE)

LYMAN, Benjamin Smith, 1835–1920, American geological and military engineer; topographical and geographical survey work, coal lands in Nova Scotia, gold mining in California, consulting mining engineer, consultant to governments of India, Japan, China and Philippines (BITAS DAB NC9 WWWIA)

LYON, G. Albert, 1881–1961, American inventor and manufacturer; plant superintendent of E. F. Houghton Co. (1900–1903), with Texas Co., Sun Oil Co., worked independently, worked for Metal Stamping Co., established experimental shop (1924), founded Lyon Cover Co. (1927), founded Lyon Co. (1930ff) that he operated for the remainder of his life, inventions primarily in the automotive field with approximately 1,200 patented devices (NC49 PS)

LYON, Howard, 1860–1926, American scientist and inventor; taught science at normal schools (ante 1909), was high school principal, with Welsback Co. (NJ) as investigator and inventor, experimented with X-rays, invented Kinetic Burner and CEZ gas light, photometric standards (NC20 PS)

LYSE, Inge Martin, 1898–1990, Norwegian American engineer; reinforced concrete, related engineering research to practice, research engineer with Portland Cement Association (1927–1931), director of Fritz Lab. (structures) at Lehigh Univ. (1931–1938), faculty at Univ. of Norway (1938–1968), fNAE (1981) (Bridge MT7 WWIE WWWIS)

LYULKA, Arkhip Mikhailovich, 1908–??, Soviet aeronautical engineer; utilization of gas turbine as an aviation engineer, worked for several aviation factories and institutes in the Kharkov area (1931–1939), followed by work at the Leningrad Central Boiler-Turbine Inst. (1939–1941) (SMOS)

M

MACADAM, John—see MCADAM, John Louden (as in DNB)

MACDONALD, Charles, 1837–1928, Canadian American civil engineer; Grand Trunk railroads in Canada and US (1852–1853, 1857–1858), Philadelphia and Reading Railroad (1863), served in Civil War and was taken prisoner by Confederates in battle of Gettysburg, chief engineer of railroads (1867), iron bridge construction (1868ff), organized and president of Union Bridge Co. (1884–1900), bridge construction in US and Australia (BDOACE)

MACDONALD, Sir James Ronald Leslie, 1862–1927, British soldier and engineer; British soldier (1883–1913), chief engineer for Uganda Railroad, surveying and mapping in Africa (MWBD)

MacDONALD, Thomas Harris, 1891–1957, American civil engineer; study on highway improvements, becoming state of Iowa Highway engineer followed by position as chief engineer (1907–1919), chief of US Public Roads (1919–1953) (ANB BDOACE)

MacDONNELL, Wilfred Donald, 1911–1999, American metallurgist; methods of increasing steel production, Bethlehem Steel Co. (1937–1957), National Steel Co. (1957) vice president then president of Great Lakes Steel Co. (1957–1962), chairman and CEO of Kelsey-Hayes Corp. (1962–1976), NAE (1969) (Bridge WWIA WWIE)

MACFARLANE, Charles William, 1850–1931, American engineer; builder, authored several works on economics (CDOAB DAB)

MACH, Ernst, 1838–1916, Austrian physicist and applied scientist; university teacher and researcher (1867–1901) including Univ. of Vienna (1860ff), Mach principle (1883), studied flight of projectiles, experiments in supersonic airflow (1887) that resulted in Mach number (1887) (BEST DOT FNIE IWAP L&M MWBD NYPL 1000Y SAI WWWIS)

MACHIN, Thomas, 1744–1816, British American military engineer; surveyor, large canal for coal transport in England, then to West Indies and US where he served in the Continental Army, built fortifications, built chain bridge over Hudson River (ANB)

MA CHUN, fl. 220–260 AD, Chinese mechanical and hydraulic engineer; contributed to design of chain pump, theater of puppets powered by water wheel, nonmagnetic direction pointer (to the south), a loom for weaving silk, rotary ballista (an engine that threw rocks) (GEAPIT)

MACINTOSH, Charles, 1766–1843, Scottish chemist and inventor; invented waterproof fabrics, impregnation of cloth with naptha-rubber solution (1823), opened factory (1834) with Thomas Hancock [1786–1865] (AHOI GEAPIT PS WWWIS)

MACKELLAR, Patrick, 1717–1778, Scottish military engineer; sent to US to build fortifications and roads, made maps of areas, involved in British war with French and Indians, wounded several times (ANB CDOSB DAB)

MACLAURIN, Colin, 1698–1746, Scottish mathematician; faculty at Marischal College (Scotland) (1717) and at Edinburgh (1725), developed theory of maxima and minima, MacLaurin expansion series (a special case of the Taylor series) (DOT GLOSB MWBD WWWIS)

MACPHERSON, Herbert G., 1911–1993, American nuclear engineer; development of reactor graphite, molten salt reactors, Union Carbide Co. (1937–1956), Oak Ridge National Lab. (1956–1970) including deputy director of laboratory, faculty of Univ. of Tennessee (1970–1976), consulting engineer (1976ff), NAE (1978) (Bridge MT7 WWIE)

MADDOCK, Thomas, Jr., 1907–1991, American civil engineer; with several firms on construction of pipelines and canals (1928–1935), with Soil Conservation Service (SCS) (NM) (1930s) in charge of rainfall and run-off studies in desert areas, during WWII worked in Central America in agriculture developing a farm marketing system to assure source of fresh fruits, USGS water resource staff and consultant, retired (1974) (HOWRD)

MAGNUS, Heinrich Gustav, 1802–1870, German chemist and physicist; professor of technology and physics at Univ. of Berlin (1828ff), green salt of Magnus (1828), elucidated the Magnus effect now well known by mechanical engineers (1853), fluid mechanics and aerodynamics, manufacture of sulfuric acid (CDOSB EOAP FNIE LDOS WWW? WWWIS)

MAHAN, Dennis Hart, 1802–1871, American military and civil engineer; professor and dean of faculty at US Military Academy, wrote many books, NAS (1863), son was Alfred Thayer Mahan [1840–1914], graduated from US Naval Academy and served extensively with US Navy (ANB BITAS BM2 PS)

MAHONE, William, 1826–1895, American civil engineer; engineer, construction of Norfolk and Petersburg Railroad (1851–1861) becoming president and superintendent, Confederate Army (1861–1864) becoming major general, Atlantic, Mississippi, & Ohio Railroad as president (1867–1873), senator (1880–1887) (BDOACE)

MAILLART, Robert, 1872–1940, Swiss civil engineer; reinforced concrete for masonry arch bridge design and construction, several bridges in the Alps, constructed factories in Russia (1912–1919) (EIH MWBD TEIH)

MAIN, Charles Thomas, 1856–1943, American mechanical engineer; maritime and steam engineering, constructed mills, water, steam power plants, invented receiver-pressure regulator for compound engines, formed consulting firm with Francis Winthrop Dean [1852–1940] (1893–1907), organized firm of Charles T. Main (1907, incorporated 1926), active in professional organizations (ANB BDOACE CDOAB DAB MEIA NC33 WWWIS)

MAKAREVSKII, Aleksandr Ivanovich, 1904–??, Soviet aeronautical engineer; professor at Aero-Hydrodynamic Inst. (1927ff) becoming director (1950), investigated external loads acting on aircraft in flight, establishment of domestic standards for aircraft (SMOS WWWIS)

MALINA, Frank, 1912–1981, American mechanical engineer; aeronautical engineering, with Theodor von Kármán [1881–1963] founded Aerojet General Corp. (1942), rocket-assisted take-off of large aircraft (1943), served in several government roles in aerospace activities and consultant, wind erosion (AMOS ANB CEOS&T ITCOF)

MALLET, Alain Manesson, 1630–1706, French military engineer; fortifications (book, 1671), combined telescopic sight and level bubble called the French-spirit level (1702, 1707) or dumpy level (AHOEAT TSOE)

MALLET, Robert, 1810–1881, Irish seismologist, engineer and inventor; earthquake design, study of speed of velocities in earth by exploding gunpowder (1846ff), ordnance, hoisting and moving devices, bridges, lighthouses, construction, coal mining, corrosion of materials, coined word seismology, FRS (1884) (CDOSB PS WBE WWWIS)

MALOOF, Alfreda Ward, 1911–1998, and husband Saul Maloof [b.1916], American inventors and designers; designed and built world-famous Maloof joint (1974), made slotted-rabbet joint, introduced ergonomic flat-spindle chairs (1971) (from display at Renwick Museum (DC))

MALOZEMOFF, Plato, 1909–1997, Russian American metallurgical and mining engineer; research at Montana

School of Mines (1931–1934), with various mining enterprises and then with Newmont Mining Corp. (1945ff) advancing to president and chairman (1966), NAE (1969) (WWIE)

MALUS, Étienne-Louis, 1775–1812, French engineer and physicist; military engineer (1796ff), research in optics, light rays, polarization of light by reflection (1808), double refraction of light (CDOSB CEOS&T MWBD WWWIS)

MANCE, Sir Henry Christopher, 1840–1926, English electrical engineer; with telegraph company in India (1863–1885), invented heliograph and method of locating faults in submarine cables (MWBD)

MANDEL, Heinrich, 1919–1979, Czech German mechanical engineer; electric power demand and total energy requirements, with RWE organization (1948ff), developed nuclear power plant technology (1955ff), built a nuclear plant (1958), developed power plants burning brown coal, concerned with developing nations emphasizing energy at a reasonable cost, fNAE (1976) (MT2)

MANDROKLES of Samos, 512 BC, Greek (Samian) bridge builder; successfully bridged the Bosporus (half mile), floating bridges, credited as being first foot bridge in history (GEAPIT TAE)

MANGIN, Joseph François, 1764–c.1818, French American military engineer and architect; mapped northern part of Santo Domingo (1785–1793), fortification of New York City, city surveyor (1796–1804) (1810–1818), invented stone-cutting machine (ANB CDOAB DAB)

MANKOVSKII, Grigorii Ilich, 1897–??, Soviet mining engineer; chief of mine construction (1924–1932), building of Moscow subway (1932–1939), USSR Academy of Sciences (1939–1954), professor and researcher at Skochinski Mining Inst. (1954ff) (SMOS)

MANLY, Charles Matthews, 1876–1977, American mechanical engineer and inventor; design and construction of radial five-cylinder water-cooled gasoline engine, Manly drive—a hydraulic device to transmit power at variable speed from a constant speed engine (1902), Manly oil hydraulic transmission of piston type (1905) (CDOAB DAB EE EIH)

MANN, Sir Donald D., 1853–1934, Canadian railroad builder; with partner William Mackenzie [1849–1923] formed contracting firm (1886ff), important role in building and organizing Canadian Northern Railroad (1895ff) later merged into Canadian National Railroad System (MWBD)

MANN, Jacob J., 19th century, American inventor; with son Henry Mann invented a reaper which lifted cut crop into a receptacle (1849), several harvesting machines developed and built based on Mann reaper (TGH)

MANNES, Leopold Damrosch, 1899–1964, American inventor; taught music and was administrator at Mannes School of Music (1953ff), research in color photography at Eastman Kodak Lab. (1930–1941), with Leopold Godowsky [1870–1938] invented Kodachrome process (1935), soundtrack for motion pictures (1941) (CDOAB MWBD NC52 WWWIA)

MANNESMAN, Reinhard, 1856–1922, German inventor; worked in iron foundry, invented process of making seamless stainless steel tube (EAI TBDOS)

MANNING, Robert, 1816–1897, Irish civil engineer; made surveys, developed hydraulics and hydrology in context of others as a newly appointed district engineer, engineer in Irish Office of Public Works (33 yr.), river engineering, water supply, flow-velocity formula related to Chézy equation (Antoine de Chézy [1718–1798]) for open-channel flow that carries his name (1888), managed estates in Ireland (CFR MWBD)

MANNING, Vannoy Hartrog, 1861–1932, American mining engineer; topography, US Geological Survey (USGS) to Bureau of Mines (USBM) as director (1885–1920), during WWI did research on chemical weapons (CDOAB DAB NCA WWWIA)

MANNLICHER, Ferdinand Ritter von, 1848–1904, Austrian engineer and inventor; with Austrian Arms Co. (1866ff), created 150 types of repeating guns and automatic rifles, credited with development of cartridge clip (1885) (MWBD 1000Y)

MANNY, John H., 1825–1856, American inventor; combined reaper and mower (pat. 1851), inventions of improved reaper and mower with Walter Abbott Wood [1815–1891] (1852ff) (ASAI MEIA NC11 TGH)

MANSFIELD, Joseph King Fenno, 1803–1862, American military engineer; army chief engineer, inspector general of army (1853–1861), mortally wounded in Civil War where he commanded Union army's XII Corps (CDOAB DAB)

MANSON, Philip W., 1905–1983, American agricultural engineer; faculty at Univ. of Minnesota (1926–1970), worked on soil and water conservation, known for improvement of concrete pipe and tile exposed to acid and sulfate water, hydraulic behavior of liquids flowing through joints (AE64 WWIE)

MAO, Yi-Sheng T. E., 1895–1989, Chinese civil engineer; transportation, design of major bridges in China, general manager of China Bridge Engineering Co., designer with McClintock-Marshall Co. (Pittsburgh, PA), chief structural consultant for Great Hall of Peoples, guiding role in engineering education, in China professor at five universities including president of four, fNAE (1982) (Bridge MT5)

MARCHANT, Guillaume, 1531–1605, French engineer; master mason, chief engineer for bridge building in Paris (TSOE)

MARCIUS REX, Quintus, fl. 145 BC, Roman administrator of engineering works; repair and expansion of city water supply (Rome), restoring and protecting old aqueducts, construction of a new aqueduct named after him called Aqua Marcia (145 BC) (GEAPIT)

MARCONI, (Marchese) Guglielmo, 1874–1937, Italian physicist, electrical engineer and inventor; successful transmission of wireless telegraphy first in Italy (1895) then in England (1896), formed Marconi Wireless Telegraph Co. Ltd. (1897), established communication across English channel between England and France (1899), formed American Marconi Co. (1900), sent signals across the Atlantic (1901), invented several communication devices, VHF electromagnetic waves (1917), magnetic detector (1902), Russian engineer Alexander Stepanovich Popov [1859–1905] (1895) also claimed invention of radio, Nobel prize (1909), fNAS (1932) (AHOI ASAI BEST CEOS&T DOSB EAI EIH GI IAI MWBD NYPL 1000Y SAI SAI-MP S:TLAW TBDOS TGE:B TGS WA WWWIS)

MARCUS, Siegfried, 1831–1898, German inventor; telegraphic relay system (c.1850), established engineering laboratory (1860), built one-cylinder automobile (1864), built automobile with an electric system (1874), electric lamp (1877) and other electrical devices, carburetor (ASAI MWBD)

MARK, Herman Francis, 1895–1992, Austrian American chemist; various teaching positions in Vienna and in Germany, Kaiser Wilhelm Inst. Dahlen (1922–1926), I. G. Farben (1927–1932), to US (1940), Polytechnic Inst. of Brooklyn (1940ff) becoming director of the Polymeric Research Lab. (1946), reactions and properties, materials, elasticity, micromolecules, polymers, rubber, consulting with numerous government and research organizations, NAS (1961) (AMOS BM68 HOCS&T WWWIA WWWIS)

MARKS, Amasa Abraham, 1825–1905, American inventor; inventor and manufacturer of artificial body limbs (CDOAB DAB)

MARLOWE, Donald E., 1916–1999, American civil and mechanical engineer; experimental stress analysis, research on undersea warfare during WWII, US Naval Ordnance Lab. on underwater warfare becoming associate director (1950–1955), faculty and administration including dean of engineering and vice president at Catholic Univ. (DC) (1955–1975), ASEE executive director (1975–1981) (Prism WWIE)

MARSEL, Charles Joseph, 1921–1964, American chemical engineer; petrochemicals, rocket propellants, ram jet fuels, waxes, Colgate-Palmolive Peet Company (1945), faculty at Columbia Univ. (1946ff) (AMOS NC51 PS)

MARSH, Charles Wesley, 1834–1918, Canadian American farmer and inventor; patented binder attached to reaper (1858), built factory for manufacturing binder (1863) with Lewis Steward then sold major interest to Gammon & Deering Co. (1865), established Marsh Harvester Manufacturing Co. (1884) that became a part of International Harvester Co. (IHC), editor of Farm Implement News (1885) (AIE ANB CDOAB DAB MEIA)

MARSH, Sylvester, 1803–1884, American inventor; invented devices for mechanical handling of grain and made improvements in grain dryers and for making kiln dried meal called Marsh's Caloric Dry Meal (1855–1865), built inclined railway (1866–1869), designed engines for ascending grades, cog rail (1867), atmospheric brake (1870) (CDOAB DAB MEIA MWBD WWWIA WWWIS)

MARSH, William Wallace, 1836–1918, Canadian American farmer and inventor; with brother Charles Wesley Marsh [1834–1918] patented harvester for binding grain on Mann reaper (1858), superintendent of Steward and Marsh Co. (1864), designed cultivator, corn harvester, corn husker, wire stretcher, and windmill, reorganized stove and lumber company (1892), retired (1895) (ASAI CDOAB DAB MEIA)

MARSHALL, William Louis, 1846–1920, American civil engineer and inventor; graduated from and was instructor at US Military Academy, served in Corps of Engineers in Civil War, explored Rocky Mountain area (1872–1876), river improvements in southeast US (1876–1881), invented automatic dam or sluiceway gate for canal (1890), various channel, harbor, coastal defenses, was chief engineer retiring as brigadier general (1910), consulting engineer (BDOACE CDOAB DAB NC11 WWWIA)

MARSHALL, William Robert, 1916–1988, Canadian American chemical engineer; at DuPont (1941–1947), faculty and administration including dean of engineering at Univ. of Wisconsin (1947ff), recognized authority particularly for work in spray drying, leadership in professional organizations, NAE (1967) (AMOS Bridge MT4 WWIE(I))

MARTENS, Adolf, 1850–1914, German mechanical engineer and inventor; Martens hardness tester using ball and a diamond, built railway bridges for the Prussian State Railroad (1871–1883), director of Mechanical Technical Research Center (1884ff) (CDOSB EE)

MARTIN, Benjamin, 1704–1782, English instrument maker; invented and made optical instruments, pocket reflecting microscope, made spectacles, business under the name of Hadley's Quadrant and Visual Glasses, published several books (DNB DOSB PS WWWIS)

MARTIN, Glenn Luther, 1886–1955, American airplane manufacturer; established airplane factory (1909), formed Glenn L. Martin Co. (1917), built airplanes for US army in WWI and bombers in WWII, built clipper ships for distance flying (1935ff) (MWBD NCF WWWIA)

MARTIN, James, 1893–1981, British aeronautical engineer; design and manufacture of ejection seats for aircraft, small engines, three-wheeled car (EAI TBDOS)

MARTIN, James W., 1908–1998, American agricultural engineer; faculty including department head at Univ. of Idaho (1946–1973), previously on faculties of Kansas State Univ., Iowa State Univ., and Univ. of Illinois, WWII naval officer (1943–1945), engineer with John Deere Plow Co. (1935–1938), consulting engineer on farm mechanization and safety (AE54 Res6 WWIE)

MARTIN, Pierre Émile, 1824–1915, French metallurgist and inventor; regeneration principle of waste heat recovery, Siemens-Martin open-hearth process (1863), melted pig iron and wrought iron to produce steel (BDOS(Wi) BDOTHOT EIH PS WWWIS)

MARTINI, Francesco di Giorgio, 1439–1502, Italian engineer; municipal engineer for Siena, painter, sculptor, architect, military engineer (DOSB PS TSOE)

MASAMUNE (aka Okazaki Goro Nuido), fl. 1300, Japanese sword maker; superb craftsman, chief swordsmith of court (1287–1297), used multiple pieces or layers of wrought iron and steel forged together to make superior swords (GEAPIT)

MASON, Arthur John, 1857–1933, Australian American engineer and inventor; to US (1881), designed and invented excavating and ore-handling machinery, mechanized systems for agriculture to decrease soil erosion (1910ff) (CDOAB NC29)

MASON, Claibourne Rice, 1800–1885, American civil engineer; carpenter, contractor with the construction of the Midlothian Railroad (VA) (1829), captain of Confederate Army volunteers, after the war was a contractor with several railroad companies (BDOACE CDOAB DAB)

MASON, John, 1832–1902, American inventor; glass jars, food preservation, patented Mason jar (1858), which had a wide mouth with a screw top, Ball Brothers (NY) (IN) of which Frank Ball was president and got into the business when the Mason patent expired (1875) and Mason Ball term used for which Frank Ball [1857–1943] invented a semi-automatic glass-blowing machine (1898), Mason patented a life raft, baby bottle, cigar case, brush roller, etc. (CDOAB I&T9 OED)

MASON, Stanley George, 1914–1987, Canadian chemist and chemical engineer; worked on munitions at Suffield Experimental Station (1941–1945), after service with the National Research Council (Canada), on faculty at McGill Univ. (1946–1987), developed field of microrheology, fNAE (1980) (AMOS MT4 PS)

MASON, William, 1808–1883, American farmer and inventor; cotton factory mechanic (1824), machine shop, invented power loom to make diaper cloth, built loom to make tablecloths (1828), numerous designs and patents for cotton machines (1840ff), manufactured Springfield rifles during Civil War, reorganized company as Mason Machine Works (1873) (ANB CDOAB DAB MEIA)

MASON, William Horatio, 1877–1940, American engineer and inventor; US Navy during Spanish-American war, worked for Portland Edison Co. (1902–1915), worked for Thomas Edison Lab. and General Motors Corp., developed use of wood chips bonded into hardboard (1924) forming Masonite, established Mason Fiber Co. (MS) (1925) that became Masonite Corp. (1929) to manufacture a durable building material, received 34 patents dealing with wood and wood-fiber processes (OED NC37 PA)

MASON, William Pitt, 1853–1937, American civil engineer; also received a medical degree, studied in Europe as well as US, studied water supply systems and sewage farms, studied plankton and other organisms in water, was an expert on sanitary engineering and water supply, active in several organizations interested in health and water and sewage systems (BDOACE NC27 WWWIA)

MASSIAH, Frederick McDonald, 1884–??, West Indies Black American architect and civil engineer; established construction business, used combination of steel and concrete for special building design, outstanding beam and girder design, elliptical dome structure, many well-known buildings (BIS)

MAST, Phineas Price, 1825–1898, American farmer, teacher and inventor; invented cider mill (1856), grain drill and corn plow, with John H. Thomas [1829–1897] patented seed planter (1858), patented improved seeding machines, cultivators, fertilizer distributors and made improvements (1872–1880), organized Mast, Foos & Co. (1880) to manufacture lawn mowers, windmills, and other small equipment (CDOAB DAB MEIA)

MASTERS, Sybilla (Isabelle) Righton, ??–1720, American woman inventor; invented method for stamping instead of grinding corn including a trough for drying (1715), which produced a mash, also patents for a method of processing and weaving palmetto leaves, chips and straw for decorating hats and bonnets, considered first woman inventor in US (ANB AWIT BDOWIS NAW(2) NWS)

MASURY, John Wesley, 1820–1895, American inventor and manufacturer; devised metal paint containers holding ready-mixed paints, invented mill for fine grinding of colors (1870) (CDOAB DAB NC5 WWWIA)

MATHEWSON, Edward Payson, 1864–1948, Canadian American mining engineer; management of corporate smelting and refining of lead, copper, and nickel, mineral industries, faculty at Univ. of Arizona (1926–1942) (CDOAB DAB)

MATOBA, Sachio, 1899–1987, Japanese metallurgist; physical chemical reactions in steelmaking, faculty member and dean of engineering at Tohoku Univ.

(1924–1962), vice president and director of Fuji Iron and Steel later named Nippon Steel Co. (1962–1987), fNAE (1988) (MT5)

MATSUSHITA, Konosuke, 1894–1989, Japanese industrialist; founded plant to make small electric fixtures (1918), formed Matsushita Electrical Industrial Co. based on a previous plant and became one of largest manufacturers of consumer electric appliances (1935ff) (MWBD)

MATTEUCCI, Felice, 1803–1887, Italian hydraulic engineer; land reclamation, worked with Eugenio Barsanti [1821–1864] (1851ff) to design and produce a gas engine with the two people forming Barsanti and Matteucci Co. (1860) to produce engines, engineering consulting (1864ff) (BDOTHOT)

MATTHES, Gerard Hendrik, 1874–1959, Dutch American hydraulic engineer; with US Geological Survey (USGS), then US Bureau of Reclamation Service (USBRS), then with private industry, with Army engineer's office (1929ff) where he worked on several river and flood control projects (CDOAB DAB NC43,H WWWIA WWWIS)

MATTHEWS, John, 1808–1870, English American inventor; machinist, opened machine repair shop, manufactured soda water and built manufacturing machinery, invented soda fountains made of cast iron lined with tin, manufactured generators made of cast iron lined with lead, developed water purification method (CDOAB DAB MEIA WWWIA)

MATTICE, Asa Martines, 1853–1925, American mechanical engineer; graduated from US Naval Academy and served in various capacities as engineer, Bureau of Steam Engineering as assistant chief engineer then engineer and chief designer (1885–1889), assistant to Erasmus D. Leavitt, Jr. [1836–1916] (1889) where he designed machinery for mining, steel and tube manufacturing, chief engineer Westinghouse Machine Co. (1901), Allis-Chalmers Co. (1904) and works manager of Walworth Manufacturing Co. (1906), during WWI advisory engineer to Remington Arms Co. (CDOAB DAB MEIA)

MATZELIGER, Jan Ernest, 1852–1889, Black Surinam American inventor; machinist (1872) to Harnery Brothers shoe factory (1877ff), invented shoemaking machine (1883), patented several devices for shoemaking industry, regularly used by US Shoe Machinery Co., honored by US Postal Service with issue of a commemorative stamp (1991) (AAI AIE ANB BDOBA BI BPOSAI CDOAB DAB MWBD PS WA)

MAUCH, Hans Adolph, 1906–1984, German American mechanical and electrical engineer and inventor; control of pneumatic conveyors (in Berlin) (1930–1935), engineer with German Air Ministry (1935–1939), consulting engineer (1939–1946), established Mauch Labs.

(1959ff), which specialized in rehabilitation devices such as artificial knees and joints, Henschke-Mauch model of knee joint operation, over 80 patents, NAE (1973) (MT3 WWIE WWWIA)

MAUCHLY, John William, 1907–1980, American physicist, electrical engineer and inventor; with John P. Eckert [1919–1995] invented first electronic computer (ENIAC) (1946), later model was UNIVAC (1951), formed corporation to build computers that was sold to Remington Rand (1950), he continued with Rand as director of special projects, NAE (1967) (AHOI AIE EAI IAI MT2 MWBD NCG PA TBDOS TTOS WWWIS)

MAUDSLAY, Henry, 1771–1831, English mechanical engineer and inventor; toolmaker, made machines that made machines, worked for Joseph Bramah [1748–1814] (1789–1798), established own business (1798ff), patents for printing calico (1805), invented industrial metal lathe (1790s), improved lathe design, cutting threads with a lathe (1810), planer, perfected measuring machine and other precision instruments, micrometer, considered father of machine tool industry (AHOI ASAI DNB GEAPIT GI MMAH MWBD 1000Y PA PS TBDOS TEIH TGE:B TSOE WWWIS)

MAUDSLAY, Joseph, 1801–1861, English engineer and industrialist; son of Henry Maudslay [1771–1831], patented and built marine engines, screw steamer, feathering screw propeller, direct-acting annular cylinder screw engine (DNB MWBD)

MAUDSLAY, Thomas Henry, 1792–1864, English engineer; son of Henry Maudslay [1771–1831], built father's business, constructed ships for British navy for 25 yr. (DNB MWBD)

MAURY, Matthew Fontaine, 1806–1873, American naval officer and oceanographer; prepared topographical chart of the Atlantic Ocean that preceded Atlantic cable, laid out lanes of travel across ocean (1855), superintendent of Department of Depot and Charts (1842), US Naval observatory, torpedo defense and electric mines for Confederate forces (ASAI CDOAB DAB NC6 PA WWWIS)

MAUSER, Peter Paul, 1838–1914, German inventor; with brother Wilhelm Mauser [1834–1882] invented weapons, breech-loading gun known as Mauser model-1871, purchased arsenal to manufacture guns, invented pistol, revolver and repeating rifle, Mauser magazine repeating rifle (1897) (EB MWBD)

MAUVE, Anton, 1838–1888, Dutch painter and inventor; leader in painting particularly landscapes using oils and water colors, invented color with a strong purple or violet named after him made from synthetic aniline dye, used for painting and dyeing silk (EB MWBD)

MAXIM, Hiram Percy, 1869–1936, American inventor; son of Sir Hiram Stevens Maxim [1840–1916], invented silencer for firearms, adapted principle to mufflers, gun silencer (1908), safety valves, blowers, designed Columbia electric automobile (c.1895ff) for Electric Vehicle Co. (ANB BDOAB EAI DAB MWBD NC15 PA PS WWWIA WWWIS)

MAXIM, Sir Hiram Stevens, 1840–1916, American English inventor and engineer (born in US); engineering works in US (1865), chief engineer of US Lighting Co. (1878), organized Maxim Gun Co. (England) (1884) that later merged with other companies, invented Maxim automatic recoil machine gun (1884) and several other devices such as electric pressure indicator, curling iron, carbon filaments for light bulbs, automatic steam-powered water pump, vacuum pumps, and engine governors, gasoline- and electric-driven carriage (1899), over 100 patents (ANB ASAI BEST BITAS CDOAB DAB EOWB GI MEIA MWBD NC6 NYPL PA PS TBDOS TEE TEIH TTOS WWWIA WWWIS)

MAXIM, Hudson, 1853–1927, American inventor; brother of Sir Hiram Stevens [1840–1916], explosive expert, organized Maxim Powder Co. (1893), sold to E. I. DuPont de Nemours & Co. (1897), consultant to company (1897–1927), invented high explosives, smokeless powder, torpedo propellant, process for making calcium carbide (CDOAB DAB MWBD NC13 WWWIA WWWIS)

MAXWELL, James Clerk, 1831–1879, Scottish physicist, mathematician and discoverer; developed electromagnetic theory of light, electromagnetic radiation (1873), postulated the existence of radio waves later discovered by Heinrich R. Hertz [1857–1894], study of heat and kinetic theory of gases, viscoelasticity, Cambridge Univ. (1871ff) (AHOI ASAI EIH FNIE IAI MWBD NYPL PA SAI TGS WBE WWWIS)

MAYBACH, Wilhelm, 1846–1929, German engineer and inventor; Maybach carburetor—constant level float-feed induced carburetor (1893), produced one of first gasoline engines (1883), with Gottlieb W. Daimler [1834–1900] formed Daimler Motor Corp. (1890), was its technical director (1895ff), constructed first Mercedes automobile (1900–1901), invented spray-type carburetor, internal expanding brakes, honeycomb radiator, change-speed gear, made automobiles (1922–1939) (EE PS MWBD MMAH PA PS TBDOS WWWIS)

MAYNARD, Edward, 1813–1891, American dentist and inventor; discovered dental fibrils (1836), began filling nerve cavities with gold (1838), invented dental tools, invented several advances relating to firearms including percussion caps (1845), converting muzzle to breech-loading rifles (1860), method of joining rifle barrels (1868) (DAB MEIA NC11 WWWIA WWWIS)

MAYNARD, George William, 1839–1913, American mining engineer; devised the process for the treatment of pyritic ores in Ireland (1863–1864), established an engineering assay office in Colorado (1864–1867), consulting engineer in England (6 yr.), then consulting engineer in US (BDOACE BITAS CDOAB DAB WWWIA)

McADAM, John Loudon, 1756–1836, Scottish civil engineer and inventor; surveyor of roads (1815ff), paving commissioner (1806), crushed stone roads, macadamized roads (named after him) (1815), road administration (AHOI BEST DNB EAI EIH GEAPIT LOTE MWBD NYPL OED PS SAI SFLOE TBDOS WWWIS)

McADAMS, William Henry, 1892–1975, American chemical engineer; heat transfer, distillation, flow of viscous liquids, absorption, Goodyear Tire and Rubber Co. (1917), faculty at Massachusetts Inst. of Technology (1919–1957), lecturer at Harvard Univ. (1925–1952), federal government and National Research Council assignments, during WWII with NACA (AMOS PS SAI)

McAFEE, Jerry, 1916–1995, American chemical engineer; research and development of hydrogenation and petroleum processes, Gulf Oil Corp. (1945ff) becoming chairman (1967), NAE (1967) (Bridge WWIE)

McALLISTER, Charles Albert, 1867–1932, American marine engineer; engineer with Revenue-Cutter Service (1902–1916), US Coast Guard (USCG) (1916–1919), vice president and president of American Bureau of Shipping (1919ff) (CDOAB DAB NC25 WWWIA)

McALPINE, William Jarvis, 1812–1890, American civil engineer; chief engineer of eastern division of Erie Canal (1836ff), chief engineer of government dry dock (Brooklyn), engineer in charge and consultant to several railroads, planned water supply for several cities, in charge of several bridge projects (CDOAB DAB MWBD NC10 WWWIA)

McCABE, Warren Lee, 1899–1982, American chemical engineer; McCabe-Thiele (c.1927) (Ernest W. Thiele [1895–1993]) graphical method for design of distillation column, enunciated law of crystal growth, faculty at Massachusetts Inst. of Technology (MIT) (1923–1925), Univ. of Michigan (1925–1936), Carnegie Inst. of Technology (1936–1947), including dean at Polytechnic Inst. of Brooklyn (1953–1970), except for period as director of research at Flintkote Co. (1947–1953) and Office of Scientific Research and Development (1940–1946), NAE (1977) (AMOS MT2 PS WWWIS)

McCALLUM, Daniel Craig, 1815–1878, Scottish American civil and mining engineer; patented arched truss for bridge, construction engineer of High Bridge over the Harlem River (1852), general superintendent of NY & Erie Railroad (1855–1856), consulting engineer, military director and superintendent of all railroads in US (1862–1864) during Civil War reaching rank of major general (1866), inspector of the Union Pacific Railroad (ANB BDOACE CDOAB DAB NC7 WWWIA)

McCARROLL, James, 1814–1892, Irish Canadian journalist and inventor; wrote for newspapers in Canada, patents on improved elevators, invented fireproof wire gauge (CDOAB DAB)

McCARTHY, Gerald T., 1909–1990, American civil engineer; water resource engineering in US and abroad, large dams, US Army Corps of Engineers (1930–1938), joined Parsons, et al. consulting (1938), then consulting with TAMS, Inc. (1947ff), NAE (1973) (CDOAS MT6)

McCARTHY, John Francis, Jr., 1925–1986, American aeronautical engineer; Aeroelastic & Structures Laboratory at Massachusetts Inst. of Technology (MIT) (1951–1955), supersonic flutter, Mach number supersonic tests, Strategic Air Command (1955ff), vice president North American Rockwell (1961ff), professor and researcher at MIT (1971–1978), to Northrop Co. (1982ff), NAE (1981) (MT4)

McCLELLAN, Carswell, 1835–1892, American civil engineer; participated in Civil War in which he was twice wounded and taken prisoner (1862–1864), engineer in charge of location and construction of several railroads (1867–1881), US assistant civil engineer (1881–1892) (BDOACE)

McCONNELL, Ira Welch, 1871–1933, American civil engineer; engineer with C. E. Cornell (1897), US Reclamation Service (1903–1909), irrigation and hydraulics expert with several companies, Dwight P. Robinson & Co. and United Engineers and Constructors (1918ff) (CDOAB DAB)

McCORMICK, Cyrus Hall, 1809–1884, American inventor and industrialist; hillside plow (1831), invented successful reaping machine at Walnut Grove Farm (VA) called Daisy (1831, pat. 1834), opened factory in Chicago (1847), formed and was president of the McCormick Harvesting Machine Co. (1879–1884), succeeded by son Cyrus Hall McCormick [1859–1936] who became president of the International Harvester Co. (IHC) (1902–1919) (AIE ANB ASAI CDOAB DAB EAI EOWB IAI IIA I&T16 LAI MEIA MWBD NC21 NYPL 1000Y PS Res14 SAI TBDOS TEIH TGS WA WWWIA WWWIS)

McCORMICK, John Buchanan, 1834–1924, American engineer-mechanic; developed improved water turbine (early 1870s), entered into partnership with James S. Brown [1802–1879] (1872), with Holyoke Machine Co. (1877) to produce Hercules turbine, and his Achilles turbine, large turbines installed at Niagara Falls (1901) and Sault Ste. Marie (1902) (DAB MEIA)

McCORMICK, Robert, 1780–1846, American inventor; father of Cyrus Hall McCormick [1809–1884], developed several agricultural implements including hemp brake and threshing machine, manufactured reaping machine developed by son (CDOAB DAB)

McCORMICK, Stephen, 1784–1875, American inventor; improved shape of millstone, invented cast iron plow (1816, pat. 1819), cast iron moldboard with adjustable wrought iron point, designed 12 types of plows

(1826), worked with Jethro Wood [1774–1834] in introducing cast iron plow (CDOAB DAB MEIA)

McCOY, Caldwell, Jr., 1933–1990, Black American electrical engineer; US Air Force (1956–1959), naval research on antisubmarine warfare, US Department of Energy (1976–1983) where he managed the National Magnetic Fusion Energy Computer Network, NASA (1983–1990) where he managed large information systems (BCST NBAS)

McCOY, Elijah J., 1843–1929, Black Canadian American inventor and mechanical engineer; worked on railroad, developed automatic lubrication systems, the term 'real McCoy' based on his reliable product, drip cup (1872), McCoy lubricator (1882), patented 58 designs, Elijah McCoy Manufacturing Co. (1872), consultant to railroad companies (AAI AIE ANB BDOBA BI BPO-SAI CBB CDOAB DAB NBAS 1000Y PS SAI-MP S: TLAW TFM)

McCRORY, Samuel Henry, 1879–1949, American agricultural engineer; private practice (1904–1906), drainage engineer doing research with Office of Experiment Station of USDA (1907–1912), chief of investigations (1913–1921), chief of Bureau of Agricultural Engineering (1931–1939) followed by chief of Bureau of Agricultural Chemistry and Engineering (1939–1942), director of hemp division of CCC (1942–1946), consulting engineer (1946ff) (AE30 WWWIA)

McCULLOCH, Robert Paxton, 1911–1977, American engineer and inventor; founded McCulloch Engineering Co. (1936), McCulloch Aviation (1943) and McCulloch Corp. (1945), McCulloch Oil Co. (1958), developed lightweight engine-powered portable chain saws, etc., involved in land development, sold McCulloch Corp. to Black and Decker Manufacturing Co. (1973) (CDOAB)

McCULLOUGH, Ernest, 1867–1931, American civil engineer; structural engineer, specialized in use of reinforced concrete and structural steel, edited periodicals (CDOAB DAB)

McCUNE, Francis Kimber, 1906–??, American electrical engineer; atomic energy instruments and apparatus at General Electric Co. (1928ff), was assistant general manager of Hanford Atomic Products (WA) (1946) becoming vice president (1954) then vice president for engineering (1960ff) (CB61 PS)

McDONNELL, James Smith, Jr., 1899–1980, American aeronautical engineer and manufacturer; engineer with several aircraft companies, all-metal monoplanes, pioneer aircraft that was forerunner of Ford Trimotor, started business (1928) after working with Glenn L. Martin Co., organized McDonnell Aircraft Co. (1939), major effort was supplying parts to Boeing and Douglas, developed a plane (FD-1 Phantom) (1946) for US Navy to fly off of carriers, primarily devoted to supplying airplanes for navy and air force, combined to form

McDonnell Douglas Aircraft Co. (1967), commercial aircraft and space contracts, NAE (1967) (ANB MT2)

McDOUGALL, Alexander, 1845–1923, Scottish American inventor; worked on Great Lakes fleet, patented 'whaleback' freight ship used for transporting iron ore, grain, and coal, organized American Barge Co. (1888), Collingwood Shipbuilding Co. (1899), St. Louis Steel Barge Co., inventions for ore and grain handling equipment, ship's equipment, mining machinery, peat fuel, canal boat (1914) (DAB MEIA WWWIA)

McFARLAND, Walter Martin, 1859–1935, American naval engineer; US Navy, Bureau of Steam Engineering, served on several ships in engineering capacity, vice president of Westinghouse Electric and Manufacturing Co. (1899), Babcock & Wilcox (1910), development and installation of oil burning marine boilers during WWI, president of Webb Inst. (1926ff) (MEIA WWWIA)

McGAUHEY, Percy H., 1904–1975, American civil and sanitary engineer; water resources, sanitary engineering, faculty at several universities including Virginia Polytechnic Inst., Univ. of Southern California, director of Sanitary Engineering Research Lab. at Univ. of California (1956–1969), consultant (1969ff), NAE (1973) (MT1 WWIE)

McGEORGE, John, 1852–1933, English American inventor and engineer; draftsman, built special machinery for making Mason fruit jar cap, chief engineer of Wellman-Seaver Engineering Co. (1896), developed hearth equipment, consulting engineer (1903) (MEIA)

McGUINNESS, Charles L. 1914–1971, American geologist and hydrologist; first with Soil Conservation Service (SCS), then to USGS (32 yr.), developed groundwater hydrology, with others developed Water Resources Research Institutes (1964), chief of Technical Reports Section of USGS (1946–1961) (HOWRD)

McHENRY, Keith W., Jr., 1928–1994, American chemical engineer; catalytic processes for utilization of heavy, sulfur-containing petroleum feedstocks, helped develop lead-free gasoline, synthetic oils, waste management, from research and development to corporation officer with Amoco Co. (1955–1993) and its affiliates where he advanced to vice president (1989), NAE (1982) (Bridge MT8)

McINTIRE, Ray, d. 1996, age 77, American chemical engineer and inventor; invented Styrofoam, engineer for Dow Chemical Co. (1940–1981), retired (1981) as director of technology (NYTObit)

McINTOSH, Robert E., Jr., 1940–1998, American chemical engineer; microwave and millimeter wave radar remote sensing and applications, first as microwave engineer at Bell Labs. (1962–1965), then professor at Univ. of Massachusetts and director of Microwave Research Sensory Lab. (1967ff), NAE (1997) (Bridge MT9)

McKAY, Donald, 1810–1880, Canadian American shipbuilder; US navy carpenter, first with shipbuilder Isaac Webb Co., then with William Currier (1841), formed McKay and Pickett Corp. (MA) (1843) to build clipper ships, with brother Lauchlan McKay [1811–1895] formed partnership to build ships (until 1875) (ASAI BDOTHOT DAB GEAPIT NC2)

McKAY, Gordon, 1821–1903, American mechanical engineer and inventor; worked with engineering corps, railroad and canal companies, established machine shop (1845), treasurer and general manager of Lawrence Machine Shop (1852), shoemaking machinery (1860), organized McKay Associates to manufacture shoes, with Blake received eight patents for improving machines to make shoes, sold company to Goodyear Co. (1895), established McKay Inst. for education of blacks, over 40 patents (ANB CDOAB DAB MEIA NC10,15)

McKAY, Lauchlan, 1811–1895, Canadian American shipbuilder; brother of Donald McKay [1810–1880], wrote first treatise on shipbuilding (1839), invention of waterline models, formed partnership with Donald McKay and Pickett (1843), builder of great clipper ships (ASAI)

McKEE, Jack Edward, 1914–1979, American civil and sanitary engineer; teaching and research in environmental and health engineering first at California Inst. of Technology and later (1949ff), US Army Corps of Engineers (1941–1946), partner in Camp, Dresser, and McKee—sanitary engineering consultants, professor at Massachusetts Inst. of Technology (1946–1949), NAE (1969) (MT2)

McKEEN, John E., 1903–1978, American chemical engineer; mass production of antibiotics, with Charles Pfizer Co. (1926–1968), progressed through several positions becoming president and chairman of the board, NAE (1965) (MT1)

McKENNA, Charles Francis, 1861–1930, American chemical engineer; chemist for Johnstown Steel Co. (1886–1889), followed by work at several companies, materials of construction and explosives, consulting chemist and chemical engineer (25 yr.) (1897ff) (ACCE PS WWWIA)

McKIBBEN, Eugene George, 1895–1974, American agricultural engineer; faculty at Univ. of California (1922–1927), Iowa State Univ. (1928–1942) including administrator of Work Progress Administration (1936–1937), head of department at Michigan State Univ. (1942–1945), head of Pineapple Research Inst. (HI) (1945–1950), chief of agricultural engineering research branch at USDA (1950–1956), ASAE McCormick Gold Medal (1949), mechanization of agriculture, tillage, soil conservation (AE55 WWWIA)

McLEAN, William Burdette, 1914–1976, American physicist, engineer and inventor; missile systems engineering, original concept of the sidewinder air-to-air missile, research with National Bureau of Standards (NBS) (1941–1945), Naval Weapons Center (1945–1967) as researcher and director (13 yr.), technical director of Naval Undersea Weapons Center (1967–1974), 49 patents, NAE (1965), NAS (1973) (BM55 MT1 PS)

McLENNAN, Ian, d.1998, age 88, Australian engineer; introduced advanced technologies into iron and steel industries, chairman of Broken Hill Proprietary Co., fNAE (1978) (Bridge)

McLEOD, Herbert, 1841–1923, English chemist and inventor; School of Mines (1860–1871), faculty of Royal Indian Engineering College (1971–1901), liquid level McLeod vacuum gauge (1874), FRS (EB EE RS11 WWW)

McMASTER, Robert Charles, 1913–1986, American electrical engineer; pioneer in nondestructive testing (NDT), welding and X-ray radiography, Battelle Memorial Inst. (1945–1954), faculty at Ohio State Univ. on nondestructive testing, 19 patents, NAE (1970) (MT3)

McMATH, Robert Emmet, 1833–1918, American civil engineer; surveys, designs, and construction to improve the Mississippi River (1858–1859), deputy county surveyor of St. Louis (1859–1862), assistant engineer of US Army Corps of Engineers (1865–1883), devised McMath formula for determining the size of storm sewers, consultant (BDOACE CDOAB DAB NC26 WWWIA)

McMILLAN, Charles, 1841–1927, Russian American civil engineer; assistant engineer, draftsman, waterworks, reservoir construction, faculty at Rensselaer Polytechnic Inst. (RPI) and consulting (1865–1871), Lehigh Univ. (1871–1875), Princeton Univ. (1875–1914), consulting engineer (BDOACE WWWIA)

McNAIR, Ronald Erwin, 1950–1986, Black American engineering physicist; astronaut, joined NASA space program (1978), in first journey away from earth he orbited 122 times (1984), perished in Challenger explosion (1986) (BI CBB WA)

McNAUGHT, William, 1813–1881, Scottish mechanical engineer and inventor; invented compound steam engine (1845)—first high pressure then exhausted low pressure steam to force pistons, established firm of J. & W. McNaught for manufacture of small steam engines (until 1914), his ideas led to many improvements in the steam engine (BDSE CB62 EAI NYTObit PS TBDOS)

McNEELY, Eugene Johnson, 1900–1974, American electrical engineer; engineer at Bell System (1922ff), became president of AT&T Co. (1961–1964) during which time the commercial communications satellite Telstar launched (1962), consulting (CB62 PS)

McNEILL, James McFadyen, 1892–1964, British naval architect and engineer; designer and builder of a variety of ships, Royal Naval Volunteers and then with Royal Field Artillery in WWI, collaborated in production of Queen Mary (1936) and Queen Elizabeth (1940), when Clydebank Works became a separate company he was appointed managing director (1953–1962) (DNB PS RS11)

McNEILL, William Gibbs, 1801–1853, American civil engineer; graduate of US Military Academy (1817), worked with George Whittle on plans for many new railroads in eastern US (1828–1837) (CDOAB DAB TEIH NC9 WWWIA)

McNOWN, John S., 1916–1998, American civil engineer; hydraulics, taught at several universities, was dean of engineering and architecture at Univ. of Kansas (1957–1965), consultant with numerous agencies in US and internationally principally in Africa and Asia, NAE (1987) (WWIE)

McPHERSON, James Birdseye, 1828–1864, American military engineer; graduated from US Military Academy (1853), instructor in engineering (1854–1857), Boston harbor fortifications (1861), in charge of railroads for a time, rose to colonel then battlefield commander becoming major general, was a casualty in the Civil War (ANB NC4 WWWIA)

McTAMMANY, John, 1845–1915, Scottish American inventor; Union solder, invented mechanical player piano (conceived in 1863, pat. in 1881), pneumatic registering voting machine (1892) (ANB CDOAB DAB MWBD PS WWWIA WWWIS)

MEAD, Elwood, 1858–1936, American civil engineer; hydraulics, irrigated agriculture (1882), professor of irrigation engineering at Colorado State Univ. (1886–1889) and worked for USDA (1889ff) in Wyoming, developed water use plans in Australia (8 yr.), was professor of rural institutions at Univ. of California (1915ff), US Commissioner of Reclamation (1924ff) (ANB CDOAB)

MEADE, George Gordon, 1815–1872, American military engineer; served in US Army after graduation from military academy in numerous activities, including ordnance, survey, boundary studies, lighthouses, served in topographical engineers, reconnaissance (Mexico), geodetic survey, numerous commands in Civil War, promoted to major general in regular army, commissioner of Fairmont Park (Philadelphia) (1866–1872) (ANB BDOACE EB)

MEADE, Richard Kidder, 1874–1930, American chemist, chemical engineer and inventor; several positions as chemist, after 1902 devoted most of career to cement industry, Meade Testing Labs. (1911ff), invented multitubular dryer for cement industry, founded and edited the Chemical Engineer (1904–1910) (ACCE NC10 PS)

MECHAIN, Pierre François André, 1744–1804, French surveyor and astronomer; mapmaker, hydrographer, made network of maps for large military areas (CDOSB DOSB PS WWWIS

MEDINA, Bartolomé de, ??–1580, Spanish metallurgist; to Mexico (1554), invented patio process in Mexico for extracting silver from ore (1557) (MWBD)

MEGE MOURIES, Hippolyte, 1817–1880, French chemist, inventor and manufacturer; invented margarine (1869), first made of tallow then improved process by addition of emulsified milk (AHOI PS WA)

MEIER, Edward Daniel, 1841–1911, American mechanical engineer and inventor; partially educated in Germany (1862), served in Civil War in engineer corps and cavalry, industry as draftsman and superintendent of machinery of railroad companies (1867), chief engineer of Illinois Patent Coke Co. (1870), built blast furnaces and was secretary of Meier Iron Co. (1870), designed machinery for compressing cotton, organized Heine Safety Boiler Co. (1884ff) of which he was chief engineer and president, president of ASME (BITAS MEIA NC23)

MEIGS, Montgomery Cunningham, 1816–1892, American military engineer; US Military Academy graduate first assigned to artillery then transferred to engineers, several construction projects including Fort Mifflin (PA), navigational improvements of Mississippi River (16 yr.), Washington Aqueduct Project (1853–1861), was quartermaster general during Civil War and promoted to brigadier general, spent 46 yr. in military service, NAS (1865) (ANB BITAS BM3 CDOAB DAB NC4 PS WWWIA)

MEIKLE, Andrew, 1719–1811, British (Scottish) mechanical and agricultural engineer and inventor; millwright, invented fantail for windmill that would adjust to the direction of the wind (1750), first rotary-drum-type thresher (1786), spring sail for windmill (1772), threshing machine using drum and concave design (1788) (DNB GEAPIT LOTE MEIA MMAH MWBD SFLOE TEIH TSOE)

MEISSNER, Alexander, 1883–1958, Austrian German radio engineer; insulators, oscillators, invented radio transmitter with vacuum tubes (1913) (CEOS&T TTOS WWWIS)

MELCHER, James R., 1936–1991, American electrical engineer and physicist; faculty at Massachusetts Inst. of Technology (MIT) (1962ff) and director of Lab. for Electromagnetic and Electrical Systems, development of continuum electromechanical principles, electrohydrodynamics, magnetohydrodynamics, electrofluidized beds, nine patents, NAE (1982) (Bridge MT6 WWIE)

MELENTEV, Lev Aleksandrovich, 1908–??, Soviet energetics engineer; power plants, teacher to professor in Leningrad Energetics (1929–1942), in various power plants, director of Siberian Branch of Inst. of Energetics (1960ff) (PS SMOS)

MELNIKOV, Nicholai Vasilevich, 1909–??, Soviet mining engineer; utilization of mines, investigation of new systems of open pit mining, director of Inst. of Mining of USSR Academy of Sciences (1961ff), professor at Academy of Coal Industry (1950–1956) (SMOS)

MELVILLE, David, 1773–1856, American inventor; maker of pewter, patent for making coal gas (1813) (CDOAB DAB WWWIA WWWIS)

MELVILLE, George Wallace, 1841–1912, American mechanical and military engineer; naval officer and engineer with US Navy (1861ff), arctic explorer, installation of water tube boilers on ships, became engineer-in-chief and commodore (1887), chief of Bureau of Steam Engineering (1887–1903), designed triple-screw vessels, invented steam turbine reduction gear with John Henry Macalpine [1859–1927], retired as rear admiral (1903), consulting engineer (1903ff) (ANB BDOACE BITAS BDOTHOT CDOAB MEIA MWBD NC20 WWWIA)

MENABREA, Luigi Federico, 1809–1896, French structural and military engineer; developed energy methods in the theory of elasticity and structures, professor of mechanics and construction (Turin), major general and commander-in-chief of Army Engineers (1859), became lieutenant general (1860), involved in political and diplomatic activities (DOSB PS)

MENDELEYEV (MENDELEEF), Dimitri Ivanovich, 1834–1907, Russian chemist and inventor; faculty at Technical Inst. of St. Petersburg (1864–1890), Bureau of Weights and Measures (1890ff), invented periodic table of elements (1869), suggested nonbiological source of hydrocarbons (1877), fNAS (1903) (BEST MWBD 1000Y PA RHW SAI SAI-MP S:TLAW TGS WBE WWWIS)

MENDELSOHN, Samuel, 1895–1966, American inventor and manufacturer; invented three-cell dry battery, powered flashgear for photographers, Mendelsohn Speedgun Co. (1932–1951), devices for aircraft and target controls, developed microwave components, coaxial connections, held approximately 30 patents (CDOAB DAB NC53 WWWIA)

MENEELY, Andrew, 1802–1851, American bell foundryman; built scientific instruments, bells, and clocks, established foundry (1826), known internationally (CDOAB DAB WWWIA)

MENG T'IEN, fl. 220 BC, Chinese general; although not an engineer, he was commissioned to build the first section of what later became known as the Great Wall of China, the ancient wall was 3,080 mi. long (modern wall is 1,700 mi.) (CDOAB GEAPIT)

MENOCAL, Aniceto Garcia, 1836–1908, Cuban American civil engineer; engineering officer in US Navy (1874–1898), engaged in surveys for interoceanic canal in Panama and Nicaragua (CDOAB DAB)

MENTZER, William Cyrus, 1907–1971, American aeronautical engineer; engineer (1934) to senior vice president (1958ff) at United Air Lines (UAL), leader in development of Douglas DC8 and Boeing 707s into one company, transition of UAL from piston to jet planes, NAE (1968) (MT1 NC56 WWWIA)

MENZIES, Michael, ??–1766, Scottish inventor; invented water-driven thresher (1732), flail threshing machine (1734), coal cutter (1761), use of steam engine for raising coal (1750ff) from mines (BDOTHOT DNB MMAH)

MERCALLI, Giuseppe, 1850–1914, Italian seismologist; formulated scale to represent earthquake intensity (1902) with a scale of 1–12 on modified Mercalli scale (IAI WBE)

MERCATOR, Gerardus (aka Gerhard Kremer), 1512–1594, Flemish instrument maker and cartographer; directed his own Institute of Geography (1534–1592), published first map (1537), professor at Duisberg (1592ff), made terrestrial globe (1549), made navigation instruments, produced numerous maps, made map of the world (1569) using a projection named after him, introduced word atlas as a collection of maps (1570) (AHOI BEST GI MWBD PA SAI TGS WBE WWWIS)

MERCER, Henry Chapman, 1856–1930, American archaeologist and inventor; lawyer, archaeologist, studied drift gravels and flint quarries (US and abroad), collected tools and built museum, developed a new method of making mural tiles (1899) and process for mosaics (1902), invented process for printing large designs on fabrics and paper (1904) (CDOAB DAB NC21 WWWIA WWWIS)

MERGENTHALER, Ottmar, 1854–1899, German American inventor; instrument maker, first Linotype typesetting machine (1884) with improvements (1885), including automatic justification, business partner with August Hahl (1880), constructed clocks, bells, weather devices, and patent models, patented over 50 improvements to linotype, formed company that became Mergenthaler Printing Co. (1885) that became Mergenthaler Linotype Co. (1891), awarded at least 60 patents (AIE ANB CDOAB DAB EAI EOWB IAI IIA LAI MEIA MMAH MWBD NYPL PA PS SAI SATF TBDOS TGS WWWIS)

MERICA, Paul Dyer, 1889–1957, American metallurgist; mechanical and magnetic properties of metals under elastic and plastic deformation, faculty at Univ. of Illinois, then researcher with the National Bureau of Standards (NBS), expert on alloys, developed concept of precipitation hardening, heat treatment of duralumin, International Nickel Co. (1919–1954) moved through several positions becoming president of company (1952), consultant, (1952ff), 21 US patents, NAS (1942) (ANB BM33 NCG,46 WWWIA WWWIS)

MERRILL, Frank Dow, 1903–1955, American civil engineer; roads and other public works, graduate of US Military Academy (1924), active in Asian theater of op-

erations in WWII until 1948, commissioner of public works and highways (NH) (1949ff) (CDOAB DAB)

MERRILL, William Emery, 1837–1891, American civil engineer; graduated from US Military Academy and taught there, assigned to Corps of Engineers, military engineer and chief engineer of various army units during Civil War (1861–1865), captured, chief engineer of Division of Missouri (1867–1870), improvement of Ohio River for transportation (CDOAB BDOACE DAB)

MERRITT, Israel John, 1829–1911, American inventor; marine salvage business, agent of Marine Underwriters (1854), agent of Coast Wrecking Co. (1860), patented pontoon for raising sunken vessels (1865), with son organized Merritt Wrecking Co. (1880) (CDOAB DAB MEIA NC5)

MERSENNE, Marin, 1588–1648, French mathematician and philosopher; became Catholic priest (1611), taught at convent at Place Royale, important role in communicating with people in the physical sciences, optics and acoustics, proposed Mersenne numbers (1644) that are prime numbers generated by formula $2^p - 1$ (CDOSB DOT EOM GLOSB RHW TGS WOM WWWIS)

MESEREAU, Theodore T., 1860–1937, American engineer and inventor; machinist with J. S. Graham & Co. (NY) (1880) becoming chief engineer, engineer and inspector for various marine and power companies, instructor for marine and stationary engineers (1908), inspector of boilers and engines (1914ff), consulting engineer, retired (1933) (MEIA)

MESSEL, Rudolph, 1848–1920, German chemist and inventor; with W. S. Squire patented a process for the improved manufacture of sulfuric acid (1875) (AHOI WWWIS)

MESSERSCHMITT, Wilhelm (Willy) Emil, 1898–1978, German aircraft designer and manufacturer; first plane built (1916) and founded company under his name (1923), all metal airplane (1926), Me 109 fighter (1935) used in WWII and the Me 262 the first jet flown in combat (1944), transport aircraft, after WWII worked with Spanish manufacturer Hispano, developed vertical/short takeoff and landing (V/STOL) aircraft (EAI MWBD PA TBDOS)

MESSINGER, Robert P., 1888–1964, English American agricultural engineer; to US (1907), first with Adriance, Platt and Co. (1907–1910), with International Harvester Co. (IHC) (1910–1956) in several capacities overseas and in US, headed engineering and patent works, vice president (1947) and executive vice president (1950) (AE45)

MESTRAL, Georges de, 1904–1990, Swiss American engineer and inventor; toy airplanes, asparagus peeler, fabric and fastener called Velcro (1948, 1951), established Velcotrex Factory in Switzerland (1957), son of agricultural engineer Albert-Georges-Constantin de

Mestral [1878–1966] (EOET GI IAI ITTC PA S&TF WA WOI WPI)

METAGENES of Knossos, fl. 550 BC, Greek engineer; son of Chersiphron (6th century BC) who was also an engineer, designed temple of Artemis, constructed temple with foundation of quarried marble on soft soil (GEAPIT MWBD)

METCALF, John, 1717–1810, British civil engineer and road builder; distinguished road builder in England, as engineer contracted for 3 miles of new turnpike, built road over bog, later built 180 miles of road (DNB GEAPIT LOTE TEIH SFLOE)

METEZEAU, Clément, 1581–1652, French architect and engineer; worked on French harbors, his grandfather Clément I was a builder, his father Thebault advised on Pont Neuf (TSOE)

METFORD, William Ellis, 1824–1899, English inventor; invented explosive rifle bullet (1863), which was outlawed (1869), breech-loading rifle (1871) combined in the Lee-Metford rifle by American James Paris Lee [1831–1904] with bolt action and detachable magazine (DNB MWBD WWWIS)

METIUS, Adriaen Anthony, c.1543–1620, Dutch military engineer; cartography, built fortifications, sons involved in technology activities, drew charts of cities and military works, wrote publications on sundials (BDOAS BDOSB DOSB PS)

MEUSNIER DE LA PLACE, Jean-Baptiste-Marie-Claude, 1754–1793, French general, mathematician and engineer; Army engineer (1775ff), theorem on curvature of a point on a surface (1776), dynamics and equilibrium (1783), designed dirigible balloon (1784), field marshall (1792) (GEAPIT MWBD WWWIS)

MEYER, Henry Coddington, 1844–1935, German American sanitary engineer; Union soldier, dealer in plumbing fixtures, pioneer in sanitary engineering, founded magazine (1870) that became Engineering Record (1890) (CDOAB DAB NC25 WWWIA)

MEYER, J. J., 1804–1877, British inventor; Meyer cutoff gear for steam engine (1843) (EE)

MEYER, Viktor, 1848–1897, German chemist and inventor; faculty at Stuttgart Polytechnic (1870–1871), Zurich (1872–1885), Göttingen (1885–1889), Heidelberg (1889–1897), invented apparatus for measuring vapor density (1871), leader in organic chemistry (CDOS MWBD RHW WWWIS)

MICHAEL, Harold L., 1920–1999, American civil engineer; faculty and department administration at Purdue Univ., practice was in highway traffic engineering, planning and safety, pioneered use of statistical methods on problems of highway engineering, NAE (1975) (Bridge WWIE)

MICHELANGELO, Buonarroti, 1475–1564, Italian architect, artist and engineer; fortifications, domes, decorated ceilings of Sistine Chapel (1508–1512), numerous well-known paintings and sculptures, some consider him the creator of the Renaissance (GEAPIT MWBD TAE)

MICHELIN, André, 1853–1931, French industrialist; with brother Édouard Michelin [1859–1940] partners in Michelin & Cie. founded by father Jules Michelin (1831) and formed firm (1888) to manufacture rubber tires for bicycles and pneumatic tires for automobiles (1895), created the Michelin Guide (1900) for travel (MWBD WWWIS)

MICHELL, Anthony George Maldon, 1870–1959, Australian civil and mining engineer and inventor; first with B. A. Smith (1899) in Melbourne, then began an independent engineering practice centered on hydraulic engineering, hydroelectric installations in Australia, machinery for Murray Valley Irrigation Works (1919), Michell tilting pad thrust bearing (1905), Michell crankless engine (1922) (DNB EE PS RS8)

MICHELSON, Albert Abraham, 1852–1931, German Polish American physicist and inventor; faculty of Case School of Applied Science (OH) (1882–1889), Clark Univ. (1889–1892), Univ. of Chicago (1892–1929), instruments for measuring the speed of light in air (1878) and in vacuum (1928), invented Michelson inferometer (1880), Michelson-Morley experiment (1887) with Edward W. Morley [1838–1923], stellar inferometer (1890), NAS (1888), first American to receive Nobel prize (1907) (ASAI BM19 RHW SAI MWBD WWWIS)

MICHIE, Peter Smith, 1839–1901, Scottish American physicist and civil engineer; US Military Academy assigned to Corps of Engineer (1863–1901), including assignment to Academy (1867–1871) and as professor (1871–1901) (BDOAS(El) BITAS DAB PS)

MIDGLEY, Thomas, Jr., 1889–1944, American industrial chemist, engineer and inventor; Dayton Engineering Lab. (Delco) (1916–1923) that became a part of GMC, discovered use of tetraethyl lead in gasoline for antiknock properties (1921), vice president Ethyl Corp. (1923–1930), freon refrigerant (1930), vice president of Kinetic Chemicals and director of Ethyl-Dow Chemical Co. (1930ff), research on hydrocarbon cracking, synthetic rubber, vice president of Ohio State Univ. Research Foundation (1940ff), NAS (1942) (BM24 CDOAB IAI I&T17 MWBD NC34 PA RHW TBDOS WWWIA WWWIS)

MIKHEEV, Mikhail Aleksandrovich, 1902–??, Soviet power engineer; Physico-Technical Inst. of USSR (1925–1934) plus the Moscow Energy Inst. (1936ff), processes of heat transfer of various heat carriers under free and forced convection (SMOS WWWIS)

MIKOYAN, Artem Ivanovich, 1905–1970, Soviet aeronautical engineer; military air academy, military aircraft design, major general in the engineering technical service, with M. L. Guervich designed MIG fighter plane (1939–1940) and modified to MIG 3 used in front line in WWII (1941–1945), pioneers in jet aviation and demonstrated first turbojet (1946) (BDOTHOT SMOS WWWIS)

MIKULIN, Aleksandr Aleksandrovich, 1895–??, Soviet aeronautical engineer; designer in Scientific Automotive Inst. (1923ff) engine (AM-34) used in aircraft (1931), plane with this engine flew several distance routes, developed first Soviet turbocompressor and variable pitch propeller, directed development of jet engines (1945ff), served as major general in engineer-technical service (SMOS WWWIS)

MILLER, Arthur Merkel, 1894–1959, American chemical engineer; from draftsman to engineer in charge at Eastman Corp. (1920–1927), operations director General Chemical Co. (1927–1932), administration of chemical engineering for Tennessee Valley Authority (TVA) (1933–1945), assistant to president of Rohm & Haas Co. (1944–1955), managing director of Charles Lennig Ltd. of England (1955–1958), advisor and consultant, including analysis of industrial projects (AMOS NC47 PS)

MILLER, Ezra, 1812–1885, American civil engineer and inventor; topographical, mechanical, civil, and hydraulic engineering, active in artillery (1833–1842) reaching the rank of colonel, engineering practice (1842–1848), survey of public lands, patented improved railroad car coupling (1863), railroad car platform, coupler and buffer (1865) (BDOACE CDOAB DAB NC7 WWWIA)

MILLER, Ferdinand von, 1813–1887, German bronze foundryman; director of foundry (1844ff), cast bronze door of the capitol in Washington (DC) (MWBD)

MILLER, Fred J., 1857–1939, American mechanical engineer; foreman of Wander, Bushnell & Blessner Co. (OH), developed devices for machine tool industry with one device of note for gear cutting, assistant editor (1887–1897) for American Machinist and editor-in-chief for Hill Publishing Co. (1897–1907), manager of plants for Union Typewriter Co. (1909), served with engineers in WWI (1917), engineering management dealing with labor issues (1918), consultant (ante 1925) (MEIA NC30 WWWIA)

MILLER, Kempster Blanchard, 1870–1933, American mechanical engineer; telephone design, construction, and operation, designed and built hydroelectric plants, designed fire alarm system for New York City (CDOAB DAB NC41 WWWIA)

MILLER, Lewis, 1829–1899, American mechanical engineer and inventor; C. Aultman & Co. (OH) (previously called Ball Brothers) as member of firm (1852ff), manufactured mowing machines and reapers, designed double-jointed cutting bar of mowing machine, low-down binder, twin binder incorporated in Buckeye Machine, managed plant (1863) (CDOAB DAB MEIA)

MILLER, Oskar von, 1855–1933, German electrical engineer; co-founder and director of German Edison Co. from which developed Allgemeine Elektrizitats-Gesellschaft (AEG) or General Electric Co. and Berlin Electrical Works (1883–1890), contributed to high voltage electric power transmission, organized industrial exhibitions (MWBD)

MILLER, Otto Neil, 1909–1988, American chemical engineer; with Chevron (formerly Standard Oil of California) (1934–1974) serving as president (1961–1970) and chairman of the board and CEO (1966–1974), authority on catalytic cracking, planned and constructed refineries, involved in development and utilization of natural gas, NAE (1968) (MT4 WWIE)

MILLER, Patrick, 1731–1815, Scottish inventor; experimented with steamboat and credited by some for its invention (1788–1790) (EIH MWBD)

MILLER, Stewart Edward, 1918–1990, American electrical engineer; research with AT&T Bell (1941–1983), lightwave telecommunications, microwave radar, development of waveguide communications systems, consultant to Bellcore (1983ff), 80 patents, NAE (1973) (Bridge MT5 WWIE)

MILLER, William Hallowes, 1801–1880, Welsh crystallographer and mineralogist; St. John's College of Cambridge Univ. (1826ff), known for development of Millerian indices (1839) representing shapes and surfaces of crystals (CDOSB L&M RHW WWWIS)

MILLHOLLAND, James, 1812–1875, American engineer and inventor; machinist, mechanic (1832), master mechanic (1838) with Baltimore and Susquehanna Railroad (1838), remodeled locomotives and introduced cast iron crank axle, railway car springs (1843), iron plate railroad bridge (1847), patented several improvements in boilers for locomotives while working for the Philadelphia and Reading Railroad (1848ff), retired (1866) (MEIA)

MILLIGAN, Robert Wiley, 1843–1909, American naval and civil engineer; navy during Civil War, instructor US Naval Academy (1879–1882, 1885–1889), Board of Inspectors and Survey (1893–1896), chief engineer of battleship Oregon, contributed greatly to its success and performance during Spanish American war and retired as rear admiral (ANB CDOAB WWWIA)

MILLIKAN, Clark Blanchard, 1903–1966, American physicist, mathematician, and aeronautical engineer; faculty at California Inst. of Technology (1928–1966) becoming professor (1940), advised and consulted on aeronautics and ballistics in WWII for army, navy, and air force, NAE (1964), NAS (1964) (ANB BM40 CDOAB DAB MT1 PS WWWIA WWWIS)

MILLIKAN, Robert Andrews, 1868–1953, American physicist and inventor; recognized for incorporating research results in teaching, first at Univ. of Chicago (1896–1921), worked on submarine listening devices during WWI, then at California Inst. of Technology (1921ff) becoming CEO (24 yr.), invented cosmic ray ionization chamber, active in developing national science policy, NAS (1915), Nobel prize (1923) (ANB ASAI BM33 DAB IAI MWBD NC42 PS WWWIA WWWIS)

MILLIKEN, Frank R., 1914–1991, American mining engineer; with Kennecott Copper Co. (1952–1979) serving as president and CEO (1961) and chairman, following various positions with companies associated with mining, development of latent mineral resources, NAE (1975) (Bridge MT6 WWIE)

MILLINGTON, John, 1779–1868, English American civil engineer; studied law and was a patent agent, engineer of an English water works, to US for an English company to serve as chief engineer of mines and superintendent of a mint (1829ff) and remained in US, opened a scientific shop to supply instruments, tools, and materials, in academic work at Univ. of Mississippi and Memphis Medical School (TN) (1848–1860) (BDOACE BITAS CDOAB DAB MEIA)

MILLIONSHCHIKOV, Mikhail Dmitrievich, 1913–??, Soviet mechanical engineer and physicist; taught at institutes—Groznyi Oil, Moscow Aviation, and Moscow Engineering Physics and Mechanics (1932–1949), professor (1949), major work was in theory of turbulence, theory of filtration, and applied gas dynamics including methods of exploiting oil wells, studied gas ejects and their use (SMOS)

MILLS, Anson, 1834–1924, American military officer and inventor; Union army officer, surveyor in Texas (1857–1961), patented and manufactured cartridge belts for US Army (CDOAB DAB)

MILLS, Hiram Francis, 1836–1921, American sanitary and hydraulic engineer; chief engineer at cities in Massachusetts, worked on sewage, water purification, and work in hydraulics that advanced turbine development (CDOAB DAB EIH)

MILLS, John, 1880–1948, American electrical engineer, physicist and inventor; faculty of several universities (1901–1911), inventor of several methods for wire and radio telephony, signals, AT&T (1911–1915), Western Electric Co. (1915–1924), Bell Telephone Labs. (BTL) (1925–1945), California Inst. of Technology (1946–1948) (ANB NYTObit WWWIA)

MILLS, Robert, 1781–1855, American architect and civil engineer; considered by some as first professional architect in US, architect for public buildings in US principally in Washington (DC) (1836–1851) for treasury building, general post office, Washington monument (ANB CDOAB DAB MWBD)

MILLS, Victor, 1897–1997, American chemical engineer and inventor; worked for Procter and Gamble Co. (OH), known as father of disposable diapers, independently along with Marion O. Donovan [1917–1998] who developed a new tape that made disposables practical (FAL WWWIA)

MILLS, Sir William, 1856–1932, English marine engineer and inventor; research on alloys, established first British aluminum foundry (1885), Mills hand grenade (1915) (DNB MWBD WWWIS)

MILNE, John, 1850–1913, English mining engineer and seismologist; engineer in Canada, professor of geology, geologist at Tokyo Imperial College of Engineering (1874–1894), invented seismograph (1880), established earthquake stations throughout the world after returning to England (1894ff) (DNB MWBD WWW WWWIS)

MILNER, John Turner, 1826–1898, American civil engineer; assistant in construction and engineer in construction of railroads, joined gold rush (1842) to west and became city surveyor of San Jose, returned east (1852) and worked on railroad and navigation to transport mineral resources for munitions to supply the Confederate Army, formed the Elyton Land Co. (1870) that founded the city of Birmingham (AL), established the Milner Coal and Railroad Co. (1879–1896) (BDOACE CDOAB DAB NC19 WWWIA)

MINDLIN, Raymond D., 1906–1987, American civil engineer; solid mechanics, protection of electronic equipment from shock and vibrations, faculty at Columbia Univ. (1932–1975), NAE (1966), NAS (1973) (Bridge)

MINIÉ, Claude-Étienne, 1804–1879, French military officer; army captain who invented conical-tip cylindrical bullet (1849) called 'Minié Ball' that was particularly used in musket rifles and for rifled barrels, Minié rifle (1851), was widely used in Civil War (MWBD OED S&TF SAIAD)

MINTS, Aleksandr Lvovich, 1895–??, Soviet communications engineer; with radio technical scientific units of USSR Army (1920–1928) with work that dealt mainly with radio-telephone modulating high-power broadcasting systems, directional antennas, radio engineering, director of Radio Engineering Inst. (1946ff) (PS SMOS WWWIS)

MISHIN, Vasillii Pavlovich, 1917–??, Soviet mechanical engineer; design and research following graduation from Moscow Aviation Inst., devoted to various problems of applied mechanics (SMOS)

MITCHELL, Alexander, 1780–1868, Irish civil engineer and inventor; brickmaking and building (until 1832), invented Mitchell screw-pile and mooring (1842), established company in Belfast to make his invention (DNB)

MITCHELL, Henry, 1830–1902, American surveyor and engineer; hydrography, US Coast Guard Survey

(USCGS) (1849–1888), also professor at Massachusetts Inst. of Technology (MIT) (1869–1876), dynamics of tides on shorelines and channels, several studies for government, NAS (1885) (BDOAS BITAS BM20 CDOAB DAB PS)

MITCHELL, Nolan, 1915–1984, American agricultural engineer; engineer then head of agricultural engineering section at Tennessee Valley Authority (TVA), early work on forage and grain drying, tobacco curing, dehydration of fruits and vegetables, supplemental irrigation, Aerovent Fan & Equipment Co. (MI) (1947ff) first as vice president and chief engineer then as president (1966ff), application of electricity to agriculture (AE(supp) WWIE)

MITCHELL, Reginald Joseph, 1895–1937, English aeronautical engineer; aircraft designer and engineer for Supermarine Aircraft Works (1916–1937), chief engineer (1919ff), recognized for design of flying boats, high-speed seaplanes, work led to Spitfire fighter plane (1936) (EAI MWBD TBDOS)

MNESICLES, 5th century BC, Greek architect and builder; strengthened beams by putting iron bar on underside, architect and builder of Propylaeo of the Acropolis (c.437 BC) in Athens (EB MWBD TSOE)

MODJESKI, Ralph, 1861–1940, Polish American civil engineer; consulting bridge engineer in Chicago (1892ff), chief engineer on building several bridges in US—including McKinley (St. Louis, MO), Broadway (Portland, OR), Benjamin Franklin (Philadelphia, PA), Huey Long (New Orleans, LA), and worked on bridge reconstruction, NAS (1925) (ANB BM23 CDOAB MWBD NC15 PS WWWIA)

MOHR, Christian Otto, 1835–1918, German civil engineer; engineer for state railroads, built notable bridges, professor of civil engineering and mechanics (1867–1900) first at Stuttgart then Dresden, failure of materials, stress, strain, Mohr stress circle (1882) (CEOS&T DOSB DOT FNIE L&M PS)

MOHS, Friedrich, 1773–1839, German mineralogist; relative scratch hardness of materials, Mohs scale (1–10) for classifying hardness of materials (1812), which was generally adopted by 1820, professor of mineralogy in Graz (1812), later he taught at Freiberg and Vienna (1826ff) (CDOS CDOSB DOSB DOT EB RHW TBDOS WBE WWWIS)

MOISSEIFF, Leon Solomon, 1872–1943, Latvian American civil engineer; worked with engineering improvements to Erie Canal (1896), joined NYC Department of Bridges (1898) spending about 20 years with this department, consulting (1915ff) in design of many bridges mostly suspension bridges, including the Tacoma Narrows Bridge (ANB CDOAB WWWIA)

MOLARD, François Emmanuel, 1744–1829, French mechanical engineer; mechanization of French industry,

served in military (1793–1795), developed a technique for sawmills to cut curved pieces, developed braking system for wagon wheels, set up a shop to build iron plows, threshers, windrowers and machines for chopping straw and roots for animal feed (GEAPIT WWWIS)

MOLL, Friedrich Rudolf Heinrich Carl, 1882–1951, German naval engineer and wood technologist; preservation of wood (1907ff), built plants to treat wood for which he was internationally known (1911ff), lectured at technical schools (1922–1936) (DOSB PS)

MOLLIER, Richard, 1863–1935, Italian German physicist and engineer; faculty at Dresden Technische Hochschule (1897ff) including director of Machines Lab., developed thermodynamic charts and diagrams, enthalpy-entropy relationships, Mollier chart (c.1932) CEOS&T DOSB EE FNIE.

MOLNAR, Julius Paul, 1916–1973, American physicist, engineer and inventor; research and development at Bell Labs. and Bell System (1945–1957) (1960–1973), M-band optical absorption band named for him, Hornbeck-Molnar (John A. Hornbeck [1918–1987]) effect on ionization process, president of Sandia Corp. (1958–1960), NAE (1969) (MT1)

MONASH, Sir John, 1865–1931, Australian civil engineer and soldier; brigade commander in defense of Suez Canal (1915), commanded division and army corps in France in WWI (1916–1918), retired as a general (1930), specialist in reinforced concrete (DNB MWBD WWW)

MONCRIEFF, Sir Alexander, 1829–1906, British army engineer; originated Moncrieff system of disappearing artillery gun mountings (c.1868), designed hydropneumatic carriage (1869) (DNB MWBD)

MONIER, Joseph, 1823–1906, French gardener and inventor; used reinforcement for concrete planters for trees, work led to use of reinforced concrete in construction (1867) (EIH 1000Y)

MONROE, Jay Randolph, 1883–1937, American inventor; Western Electric Co. (1906–1910) (IL), machine for automatic multiplication and addition, calculating machine (1911ff) based on developments of Frank S. Baldwin [1838–1925], Monroe Calculating Machine Co. (NJ) (1912ff) including president, president of Defiance Manufacturing Co. and Calculator Equipment Corp. (DAB NC27 NYPL WWWIA)

MONTALEMBERT, Marc-René de Marquis, 1714–1800, French military engineer; served in several mid-European wars, developed simplified polygonal fortifications adopted by military (MWBD WWWIS)

MONTEAGLE, Robert Charles, 1859–1932, Scottish American engineer and inventor; chief draftsman of Fulton Iron Works (1882), Honolulu Iron Works (1889), Charles Hillman Ship and Engine Building Co. (1892),

Burke Dry Dock Co. (1895), then chief engineer of several engine, boiler, and power companies, with patents on packing, piston rings, valves, and water tube boiler (EAI MEIA)

MONTEITH, Alexander Crawford, 1902–1979, Canadian American electrical engineer; Westinghouse Electric Corp. (1924–1967) becoming senior vice president, nuclear power plant development, authority in power plants, auxiliary power transmission and distribution, lightning protection of transmission lines, NAE (1965) (MT2 WWIE)

MONTGOLFIER, Jacques Étienne, 1745–1799, and brother Joseph Michel, 1740–1810, French inventors; pioneers in flight, hot air ballooning for prolonged flights (1783), self-acting hydraulic ram for pumping (1797)   (AHOI BEST CDOSB EAI EE EIH EOWB FIOTEC GEAPIT GI MMAH NYPL PS SAI TGS)

MONTGOMERY, Benjamin Thornton, 1819–1877, Black American inventor and merchant; surveyor, construction of levees, plantation buildings, trade on the Mississippi River, invented a boat propeller (AAI ANB)

MONTGOMERY, John J., 1858–1911, American scientist and inventor; aeronautics leading to successful gliding machines, continued studies leading to devices to control aircraft, died during experiment with one of his planes (BITAS NC15 PS)

MONTRESOR, James Gabriel, 1702–1776, British (Scottish) military engineer; known for design and construction of military works particularly in North America, chief engineer in North America (1754ff), surveyed Lake Champlain area (NY and VT) (1756) and for fortifications in New York and Nova Scotia, also for defense of Gibraltar (1731) (ANB CDOAB DAB MWGD)

MONTRESOR, John, 1736–1788, British military engineer; son of James Gabriel Montresor [1702–1776], served as an engineer and infantry field officer in Pennsylvania and Quebec and surrounding area, made engineering survey of the St. Lawrence River, was involved in the Revolutionary War as a British officer (ANB CDOAB DAB DNB WWWIA)

MOODY, Lewis Ferry, 1880–1953, American hydraulic engineer; development and design of pumps and turbines, Moody turbine for low-head installation (pub. 1921) (CDOAB CEOS&T EE NC47 WWWIA WWWIS)

MOODY, Paul, 1779–1831, American engineer and inventor; operated cotton mill (1800), established Boston Manufacturing Co. (1814) with Francis Cabot Lowell [1775–1817], repaired and manufactured machinery, patented numerous devices for making cotton rope, supervised cotton machinery manufacturing with Lowell Machine Works (1825) (DAB GEAPIT MEIA NC25 WWWIA WWWIS)

MOON, William, 1818–1894, English inventor; blind from 1840 on, he devised embossed Moon type (1845), claimed to be easier to read than Braille but required more space (DNB MWBD)

MOORE, Daniel McFarlan, 1869–1936, American electrical engineer; radio, X-ray, lighting, neon discharge lamp (1917), over 100 patents issued (MWBD)

MOORE, Hiram, 1801–1874, American farmer and inventor; patented a machine that combined harvesting, threshing, and cleaning of grain known as Moore-Hascall harvester (1834) [John Hascall], after pressure from Cyrus Hall McCormick [1809–1884] and Obed Hussey [1792–1860] he turned his efforts to farming (BDOTHOT HIW NYPL TGH)

MOORE, Ian David, 1951–1993, Australian agricultural engineer; erosion mechanics, salination, water quality (HOCS&T)

MOORE, Philip North, 1849–1930, American mining engineer; work that stimulated production of strategic minerals for the US government (1917) (CDOAB DAB)

MOORE, Robert, 1838–1922, American civil engineer; chief engineer and assistant engineer in building railroads in the Ohio, Indiana, Illinois area (1868–1877), sewer commissioner (1877–1881), consulting engineer (1881–1920) (BDOACE)

MOORE, Walter Parker, Jr., 1937–1998, American civil engineer; improvement of quality of structures, creative design of structures, effects of earthquakes on concrete buildings, engineer to chairman for Walter P. Moore & Associates (1966ff), visiting professor and lecturer at Rice Univ. (TX), Univ. of Illinois, and Univ. of Texas, architecture and civil engineering at Texas A&M Univ. (1994–1998), NAE (1991) (Bridge WWWIA)

MOORE, William James Perry, 1858–1930, American engineer and inventor; first with Buckeye Engine Co. (OH), then Worthington Pump & Machinery Co. from salesman to European representative to vice president (1883–1902), general sales manager for Alberger Pump & Condenser Co. (1902ff), numerous patents relating to internal combustion engines, tires, and automobiles (1918–1927) (MEIA)

MOORHEAD, James Kennedy, 1806–1884, American electrical engineer; canal builder, pioneer in telegraphy, president of associated companies (Monongahela Navigation Co., Atlantic and Ohio Telephone Co.), congressman from Pennsylvania (1859–1869) (CDOAB DAB)

MORAN, Daniel Edward, 1863–1937, American civil engineer; known for designing difficult foundations for bridges and other structures, worked on foundations for lighthouses, piers for bridges, foundations for buildings working under Charles Sooysmith [1856–1916], started own consulting firm and in partnership with others

(1896–1937) recognized for his innovative approach, designed Moran air lock (ANB CDOAB DAB WWIE(I))

MORDECAI, Alfred, 1804–1887, American military engineer; graduated (1821) and taught at US Military Academy (1821–1825) as assistant professor of philosophy and engineering, moved to Department of Ordnance (1883) after serving in engineers corps, commanded arsenals at several locations, weapons testing, served in ordnance (29 yr.), resigned commission during Civil War (ANB BITAS CDOAB DAB)

MOREELL, Ben, 1892–1978, American naval and civil engineer; in US Navy (1917–1946) retiring as an admiral, created Seabees for US Navy (1941), was chief of civil engineers (1937), chief of Bureau of Yards and Docks (1937), CEO of Jones and Laughlin Steel Corp. (1947–1958), NAE (1976) (CDOAB DAB MT1 WWIE)

MOREHEAD, John Motley, 1870–1965, American chemical engineer and inventor; devised first large-scale means for producing calcium carbide (1892), with Union Carbide Co. in charge of research and development for new processes (1897–1925), consulting engineer (1933–1965) (CDOAB DAB)

MORELL, George Webb, 1815–1883, American civil engineer; graduate of US Military Academy and assigned to Corps of Engineers, served in improvement of Lake Erie harbors, on Ohio and Michigan boundary survey (until 1837), studied and practiced law, appointed colonel and served as quartermaster and chief of staff during Civil War and participated in several battles, including the defense of Washington (DC) (1861–1864) (BDOACE CDOAB DAB NC4)

MOREY, Samuel, 1762–1843, American mechanical and civil engineer and inventor; lumber and sawmill business, consulting engineer for construction of locks for canal, steamboat construction (1797), Morey American Water Burner (1817), treble pipe steam boiler (1818), gasoline powered internal combustion engine (1826), 20 patents (BITAS CDOAB DAB GEAPIT NC11 PS WWWIS)

MORGAN, Arthur Ernest, 1878–1975, American civil engineer; wetlands preservation while at Univ. of Colorado (1902), then with private firm, drainage engineer with USDA (1907–1910), designed Miami Conservatory District (1913), president of Dayton Morgan Engineering Co., (1915–1930), president of Antioch College (OH) (1921–1933), chairman of TVA (1933–1938) (ANB CDOAB DAB NCE NYTObit WWIE)

MORGAN, Charles Hill, 1831–1911, American engineer and inventor; draftsman for Erastus B. Bigelow [1814–1879] (1853), where he designed and built cam curves for carpet looms, with brother founded paper bag manufacturing company (1860), superintendent of I. Washburn & Moen Co. (MA) a wire-making company (1864ff), numerous patents on rolling mill known as

Morgan mill and wire making, president of American Wire & Steel (1887), Morgan Construction Co. (1891), consulting engineer (CDOAB DAB MEIA NC23)

MORGAN, Garrett Augustus, 1877–1963, Black American inventor; improved sewing machine part (belt fastener), gas mask called inhalator (1912), Morgan safety hood (1914), automatic three-way electric traffic signal (1923), electric curling iron (1960) (AAI AI AIE BDOBA BI BPOS CBB TBDOS TFM PS)

MORGAN, Julia, 1872–1957, American woman civil engineer and architect; reported to be first woman graduate in civil engineering in US at Univ. of California (1894), became licensed architect (1904), design and construction, made reputation for quality of construction particularly following the devastating 1906 earthquake, had an extensive consulting practice (AWIT AWOA)

MORGAN, Paul W., 1911–1992, American chemist and inventor; corporate research at DuPont (1941ff), polymer synthesis and technology, layed out the synthetic foundation for development of numerous materials, interfacial method of low-temperature polymerization, 38 US patents, NAE (1977) (Bridge MT7)

MORISON, George Shattuck, 1842–1903, American civil engineer; after being a lawyer became an engineer, worked on big bridges over Missouri River at Kansas City, assistant engineer to Octave Chanute [1832–1910] building wooden and steel bridges, devoted 25 yr. to building bridges between Mississippi River and the Rocky Mountains (ANB BITAS CDOAB DAB MEIA)

MORITA, Akio, 1921–1999, Japanese physicist, engineer, and inventor; electronics and communications, invented CD (compact disc) system that along with Philips Co. became an international standard (1981ff), with Masaru Ibuka [1908–1997] founded Sony Corp. (1954), originally named Tokyo Telecommunications Engineering, became president (1971), then board chairman (1976) (CBY IAI NYTObit)

MORRIS, Thomas Armstrong, 1811–1904, American civil engineer; after graduation from US Military Academy and commissioned in artillery, resigned and was in charge of construction of Central Canal in Indiana, chief engineer and superintendent of several railroad operations in Ohio and Indiana (1836–1861), commissioned as brigadier general and commanded troops in Civil War (1861–1862), chief engineer and president of railroads in Ohio and Indiana area (1863–1869), road construction in Indiana, president of the Indianapolis Water Co. (1888) (BDOACE CDOAB DAB)

MORRIS, William Richard (aka Viscount Nuffield), 1877–1963, English inventor, manufacturer and philanthropist; manufactured pedal cycles, motorcycles (1902), Morris Oxford auto (1912) (1919), organized Morris Motors Limited (1919), also established Morris Garage (MG) Car Co., transfer mechanism for making cylinder block for MG (1924), Wolseley Co., SU Combustion Co., and Morris Commercial Cars (BDOTHOT EIH MMAH)

MORRISON, Charles Samuel, 1919–1967, American agricultural engineer; Huber Manufacturing Co. (1941), US Navy (USN) in WWII (1942–1946), farmer (1947), from research and development to manager of research for Deere & Co. (1947ff), president of ASAE (1965–1966) (AE48 WWWIA)

MORRISON, Ivan G., 1893–1957, American agricultural engineer; served in France in WWI (1917ff), farmed and taught vocational agriculture (8 yr.), taught agricultural engineering at Cornell Univ. (1934–1936) and at Purdue Univ. (1936ff), wrote widely used reference, pioneer in preventive maintenance, repair and care of farm machinery, during WWII was consultant to US Office of Education (AE38)

MORSE, Charles Adelbert, 1859–??, American civil engineer; division engineer, resident engineer, chief engineer of railroads in Iowa, Kansas, and Colorado, chief engineer of Rock Island Railroad Lines (1913–1918) (1919–1929), US railroad commission (DC) (1918–1919) (BDOACE WWWIA WWWIA-S&T WWWIS)

MORSE, Edward Kirtland, 1865–1942, American civil engineer; Morse Bridge Co. draftsman and western agent (until 1887), went to Australia to build largest bridge in southern hemisphere at that time, general contracting work (1889–1893), consulting engineer for Jones & Laughlin (12 yr.) and Allegheny Co. (3 yr.), transit commissioner, private consulting practice (20 yr.) (sBDOACE)

MORSE, Philip McCord, 1903–1985, American physicist and engineer; faculty and administration primarily devoted to behavior of electron flow through gases and computation center at Massachusetts Inst. of Technology (MIT) (1931–1946) (1949–1969), identified with Morse potential, Associated Universities, Inc. (1946–1948), during WWII consultant to MIT Radiation Lab. Antisubmarine Warfare Operations, operations research, NAS (1955), NAE (1985) (AMOS ANB BM65 MT4 WWWIS)

MORSE, Richard Stetson, 1911–1988, American physicist and engineer; Eastman Kodak Co., (1935–1940), technology of high vacuums, founded National Research Corp. of Cambridge (1940–1959), pioneered in use of vacuum in manufacturing processes for food, drugs, vitamins, optical lenses, director of research and development for US Army (1959–1961), involved in entrepreneurship and managing innovation with MIT, US Department of Commerce (USDC), and consulting (1961ff), NAE (1976) (AMOS MT5 WWWIA)

MORSE, Samuel Finley Breese, 1791–1872, American artist and inventor; founder and first president of National Academy of Design (1826–1845, 1861), professor at New York Univ. (1832ff), experimental associate was Leonard D. Gale [1800–1883], magnetic telegraph (1832ff), developed Morse code (1837), sent first message between Washington (DC) and Baltimore (MD) (1844), honored by 3-cent US postage stamp (1944) (AHOI AIE ANB ASAI BEST BITAS CDOAB CEOS&T DAB EAI EIH GEAPIT IAI IIA I&T2000 LAI MWBD NC4 NYPL 1000Y SAI TBDOS TGS WA WBE WWWIS)

MORSE, Sidney Edwards, 1794–1871, American inventor and journalist; brother of Samuel F. B. Morse [1791–1872], started newspaper and author (until 1858), with Henry A. Munson developed new method of making maps with letterpress (1840ff), patented device with his brother for raising water and other fluids (1866), invented a bathometer to explore sea (1866) (CDOAB DAB NC13 PS WWWIS)

MORTON, George Luton, 1858–1937, American mechanical engineer and patent attorney; taught mechanical drawing (1884), draftsman and machinist Cleveland Rubber Co. (1885), work devoted to patenting activities (1886ff) (MEIA)

MORTON, Jack Andrew, 1913–1971, American electrical engineer; engineer to vice president of electrical technologies at Bell Labs. (1936–1971), microwave technology, vacuum tube development, transistor and solid state devices, NAE (1967) (MT1)

MORTON, James St. Clair, 1829–1864, American military engineer; graduate of US Military Academy, expert in fortifications, chief engineer of army of Ohio and army of Cumberland in Civil War, killed while serving as chief engineer of IX Army Corps (CDOAB DAB)

MORTON, Thomas, 1781–1832, Scottish inventor and shipbuilder; slip for docking vessels (1819), carriage submerged on railroad tracks used as dry dock (DNB MWBD)

MOSELEY, Henry Gwyn Jeffreys, 1887–1915, English scientist and inventor; Univ. of Manchester (1910–1914), evolved idea of atomic numbers while studying X-rays, X-ray spectroscopy (1913), proposed that elements can have only integral numbers, Moseley law, killed in WWI (1915) (BEST DNB L&M MWBD PA RHW WWWIS)

MOSHKIN, Panteleimon Afanasevich, 1891–??, Soviet chemical engineer; efforts devoted to industrial methods for chemical synthesis involved with tar of humus coal and plasticizers, became professor at Moscow Chemical Technological Inst., chief of Lab. of Scientific Research Inst. of Plastics (1943) (SMOS WWWIS)

MOSS, Sanford Alexander, 1872–1946, American mechanical engineer; worked with General Electric Co. (1903–1938), development of turbosupercharger for aircraft engines (1939), consultant (CDOAB DAB NCF)

MOUTARD, Théodore Florentin, 1827–1901, French engineer; engineering corps (1849ff), became professor of mechanics at School of Mines (1875ff), worked on theory of algebraic surfaces (CDOSB DOSB PS)

MOUTON, Gabriel, 1618–1694, French mathematician and priest; as a result of his studies the metric system developed as the international system of weights and measures, similar suggestions were made by Frenchman Jean Picard [1620–1682] (1671), Paris Academy of Sciences made recommendations based on suggestions to the French assembly (1791) (DOSB EB WPost2005 WWWIS)

MOWBRAY, George Mordey, 1814–1891, English American industrialist; produced first refined oil in Pennsylvania field (c.1859), manufactured nitroglycerin for blasting (1866ff), improved explosives, smokeless gunpowder (1886) (ANB DAB CDOAB MWBD NC23 WWWIA)

MUDGE, Thomas, 1715–1794, English inventor; clock and watchmaker, invented lever escapement, originally with Clockmakers' Co. (until 1738), formed partnership with William Dutton making clocks (1755ff), marine chronometers, clockmaker for George III (1776) (BDOTHOT PS WWWIS)

MUELLER (or MÜLLER), Erwin Wilhelm, 1911–1977, German American physicist and inventor; to US (1952), invented field emission microscope (1936) and field ion microscope (1951) and atom probe field ion microscope (1967), following service at Fritz Haber Inst. (Berlin) (ante 1952), was on faculty at Pennsylvania State Univ. (1952ff), first to visualize the atom (1955), NAE (1975), NAS (1975) (BM82 CDOS GLOSB MT1 PA RHW TPA WWWIA WWWIS)

MUENCH, Carl Gebhard, 1887–1971, American inventor; first as a draftsman then assistant foreman of Flaxlinum Co. (1909ff), made rigid insulation board from wood fibers called Insulite and established a new company by that name, then with associates formed and was vice president of Celotex Corp. (1927–1952), consulted internationally (NC58 PS)

MUESER, William Henry, 1900–1985, American civil engineer; authority on problems of foundation engineering, his first job was with a consulting firm that was a predecessor of his own firm of design and construction of drydocks in WWII, became senior partner of consulting firm renamed Mueser Rutledge Consulting Engineers (1923–1975), specialty was design of concrete bridges, partner in firm in Venezuela (1944–1951), consulting, NAE (1978) (Bridge MT4)

MULHOLLAND, William, 1855–1935, Irish American civil engineer; sanitary water supply, Los Angeles City Water Supply Co. (1878ff) becoming superintendent (1886), chief engineer of Water Works and Supply, aqueduct to supply water to Los Angeles (1907–1913),

first American to use hydraulic sluicing to build dams, internationally recognized expert on water resources development (AmSci91 ANB CDOAB ET22 WWWIA)

MULLAN, John, 1830–1909, American military and civil engineer; a graduate of US Military Academy commissioned in artillery (1852), railway survey, road building in Pacific Northwest, transferred to topographical engineers, explored crossing of Rocky Mountains, built roads in state of Washington (1858–1863), built Mullan Road from Walla Walla (WA) to Coeur d'Alene (ID) (1863), studied and practiced land law (1867ff) (ANB)

MULLER, Gertrude Agnes, 1887–1954, American woman inventor; manufacture of child care equipment, child safety expert, health, nutrition and human development, first worked with General Electric Co. (1904–1910), then at Van Arnam Manufacturing Co. where she became assistant manager, established Juvenile Wood Products Co. (1924) that was renamed Toidey Co. (1944ff) (AIE AWIT MADOI NAW PS)

MULLER-BRESLAU, Heinrich, 1851–1925, German civil engineer; design of steel structures primarily bridges, also design of cathedral in Berlin, construction of Zeppelin airship, faculty at the Polytechnic Inst. of Hannover (1883–1888), chair of structural engineering at Berlin-Charlottenburg Inst. (1888ff) (CDOSB DOSB PS)

MULLIGAN, James H., Jr., 1920–1996, American electrical and computer engineer; served in WWII, worked at Bell Telephone Labs., Naval Research Lab., DuMont Lab., faculty at New York Univ. (1949–1968) and Univ. of California-Irvine (1974–1991) including chair of department, officer of NAE (1968–1974), electrical network theory, system theory, system applications, NAE (1974) (Bridge WWIE)

MUMMA, Albert G., 1906–1997, American military engineer; US Navy (1922–1959) becoming rear admiral, chief of Bureau of Ships (1955–1959), Worthington Corp. (1959–1971) becoming chairman (1962), design and construction of naval ships, NAE (1976) (Bridge MT10 NCK WWIE WWWIA)

MUNCASTER, Walter James, 1850–1934, American engineer and inventor; machinist (1867), machinist at several companies including US Navy Yard, several patents relating to steam engine, water tube steam boiler, screw propellers, metal bending and straightening machine, boring cylinders and turning pulleys, grinding machines, measuring instruments, crane for loading boxcars (MEIA NC28)

MUNGER, Robert Sylvester, 1854–1923, American inventor; operated cotton gin plant (1873), patented saw cleaner (1878) and saw-sharpening tool (1882), established manufacturing firm (1885), formed and was president of Munger Improved Cotton Machine Co. (1888ff), several patents issued relating to pressing, handling, cleaning, pneumatic handling, and baling of cotton (CDOAB DAB MEIA NC25)

MUNK, Max Michael, 1890–1986, German American physicist, mechanical and aeronautical engineer; aerodynamics and gas dynamics, research engineer for German Navy (1918) and Zeppelin Co. (1919), NACA (1920–1926), Naval Ordnance Lab. (NOL) (1945–1958), faculty at Catholic Univ. (1958–1960) (AMOS ITCOF WOI)

MURDEN, William R., Jr., d. 1997, age 74, American marine expert; development of dredging techniques, dredge plant and associated equipment, president of Murden Marine, Ltd., NAE (1979) (Bridge)

MURDOCK, William, 1754–1839, Scottish engineer and inventor; with firm of Boulton and Watt (1777), distilled coal gas for home lighting (1792), invented improvements on steam engine, worked on road locomotive, invented D-slide valve (1799) and apparatus for using compressed air, steam gun (1803), introduced gas lighting, production of coal gas (AHOI BEST EAI EE EIH GEAPIT MMAH MWBD NYPL PA PS (SAI TBDOS TGS WWWIS)

MURPHREE, Eger Vaughan, 1898–1962, American chemical engineer; studied the mechanism of the physical and chemical interactions at the interface separating solids, liquids, and gases, effect of agitation on reactions, petroleum processing, problems of distillation particularly multicomponent systems, Murphree plate efficiencies, faculty at Massachusetts Inst. of Technology (MIT), Solvay Process Co. (1924–1930), Standard Oil Co. (1930ff) to president of Standard Oil Development Co. (1947ff), leader in development of rubber in WWII, NAS (1950) (BM40 PS WWWIA)

MURPHY, John Wilson, 1828–1874, American civil engineer; surveyor, with Squire Whipple [1804–1888] introduced pin connections in bridges, built levees on Alabama River (1851–1852), with George Washington Plympton [1827–1907] developed testing machine, built bridges in Pennsylvania, built Union Hall in Philadelphia (PA) (1864), built pipe aqueduct over valley (1869), built Broad Street bridge in Philadelphia (BDOACE CDOAB DAB WWWIAH-S&T WWWIA)

MURRAY, Charles, 1864–1941, Scottish poet and engineer; secretary for public works (1910), South Africa director of Defense (1917), author of several poems (MWBD)

MURRAY, Jerome, d. 1998, age 85, American engineer and inventor; heart-surgery pump, electric knife, founded Murwood Labs. after WWII, 75 patents (FAL NYTObit)

MURRAY, Matthew, 1765–1826, English mechanical engineer and inventor; developed new applications for steam power, new techniques for forging iron, new machines for textile industry, new machine for spinning flax yarn (1790), introduced wet spin process, made many improvements in steam engines, developed D-slide valve, steam power use for railroads, ships, founded factory (GEAPIT)

MURRAY, Thomas Edward, 1891–1961, American mechanical engineer and inventor; partner in business established by father to develop and market inventions and to design power plants of which he became chief executive (1929ff), new designs for manufacture of mortar shells, member of Atomic Energy Commission (AEC) (7 yr.), over 200 patents (ANB CDOAB DAB NC49 WWWIA)

MUSHET, David, 1772–1847, British (Scottish) mechanical engineer; production of iron and cast steel, made cast steel using wrought iron with charcoal (pat. 1800), discovery of beneficial effects of manganese on iron and steel, improved quality of wrought iron (GEAPIT)

MUSHET, Robert Forester, 1811–1891, English metallurgist and inventor; improved Bessemer process (1856), developed manganese-tungsten process for making self-hardening steel (1868), steel alloys for tools (EIH MMAH MWBD PS WBE)

MUSKAT, Morris, 1906–1998, American physicist and inventor; flow of fluids in the earth, established field of reservoir engineering, Gulf Research and Development Co. (1929–1950), Gulf Oil Corp. (1950–1971), 15 US patents, NAE (1983) (Bridge WWIE WWWIS)

MUSSCHENBROEK, Pieter (Petrus) van, 1692–1761, Dutch instrument maker; experimentalist, with Ewald Georg von Kleist [1700–1748] developed what became known as Leyden jar (c.1746) also called Kleistian jar, taught at Duisberg (1719–1723) and Utrecht (1723–1740) and Leiden (1740–1761) (CDOSB DOSB EB EIH GEAPIT MWBD NYPL WWWIS)

MUSSERT, Anton Adriaan, 1894–1946, Dutch engineer and political leader; leader of Nazi party in Netherlands (1931ff), commissar leader named by Germans (1942), executed for treason (MWBD)

MUSTEL, Victor, 1815–1890, French manufacturer and inventor; musical instruments, established factory in Paris (1853), invented typophone, metaphone, and other instruments and improvements (MWBD)

MUTO, Kiyoshi, 1903–1989, Japanese civil engineer; earthquake engineer, structural design, developed D-method of resistance to seismic moments, worldwide construction engineering, faculty at Univ. of Tokyo (1925–1963), vice president of Kajima Corp. (1963–1977), fNAE (1978) (Bridge MT6)

MUYBRIDGE, Eadweard (sic) (Edward) (aka Edward James Muggeridge), 1830–1904, English American photographer and inventor; use of photograph to study gait and running of horse showing a galloping horse has all four feet off the ground (1872) (1878), successive positioning of human being in motion, inventor of first theatrical motion picture projector, had a commission with US Coast Guard and Geodesic Survey making photographic surveys of Pacific Northwest, invented zoopraxiscope, experiments supported by Univ. of Pennsylvania (1884–1887) (ANB ASAI GI HIW MWBD NC19 PA WWWIS)

MYDDELTON, Hugh, 1556–1631, British (Welsh) financier for pumping works; goldsmith, clothmaker, promoted engineering and mining enterprises, built canal and reservoir system to supply water to London (1613) (DNB EIH GEAPIT LOTE MWBD SFLOE TEIH TSOE)

MYLNE, Robert, 1733–1811, Scottish engineer and architect; noted architect who also became engineer, master mason (1755), used elliptical arches for bridges, built bridges in London, engineer for New River Co. (1770–1791) (GEAPIT)

MYLNE, William Chadwell, 1781–1863, Scottish engineer; son of Robert Mylne [1733–1811] who developed water supply for London using cast iron instead of bored log pipes (TSOE)

N

NABOPOLASSAR, ??–605 BC, Mesopotamian ruler and engineer; built bridge over Euphrates River, helped found Babylonian empire and was king (625–605 BC) (MWBD TAE)

NACE, Raymond L., 1907–1987, American hydrologist; in water resources division of USGS (1941–1977), WWII service (1942–1946), groundwater studies in Nebraska and Idaho (1946–1949), assigned to water supply and waste disposal with AEC (1949–1956), several special assignments including development of the International Hydrological Decade (IHD), after retirement (1977) continued working until (1983) (HOWRD WWWIS)

NAGEL, Theodore J., 1913–1986, American electrical engineer; American Electric Power (AEP) Service Corp. (1939–1982) (43 yr.) rising from assistant engineer to assistant to chairman, US Navy in WWII (1942–1946), power supply planning and reliability, NAE (1973) (MT3)

NAGHDI, Paul M., 1924–1994, Iranian American mechanical engineer; continuum mechanics, shell theory, inelastic behavior of materials, faculty at Univ. of Michigan (1951), then on faculty including administration at Univ. of California (1958ff), NAE (1984) (Bridge MT8 WWWIA)

NAISMITH, James, 1808–1890, British engineer and inventor; established business to manufacture steam locomotives and machine tools (1834), slotting, shaping, planing, and milling machines, developed Naismith (Naysmyth) hammer in which block is raised by piston rod using steam and released to fall under gravity (1839), hydraulic punching machine (AHOI AIE ASAI CDOSB DOSB EAI EE GEAPIT PS SAI TBDOS TEIH TGS MWBD)

NAKAGAWA, Ryoichi, 1913–1998, Japanese mechanical and automotive engineer and inventor; aircraft engine designer with Nakajima Aircraft Co. (1936–1945), was executive managing director of Prince Motor Co. when it merged with Nissan Motor Co. (1964), engineer and corporate officer of Nissan Motor Co. (1966–1979), development of high-performance aircraft and automotive engines (1936–1945), chairman of Japanese Electronic Control System (1977–1984), automobile emission control systems, electronics, fNAE (1990) (Bridge MT10 WWIE)

NAPIER, John, 1550–1617, Scottish mathematician and inventor; invented logarithms (1614), systems of mathematical calculation with mechanical aids called Napier's bones or rods, designed instruments of war, John Napier and Henry Briggs [1561–1630] published the first tables of logarithms to the base 10 (1624), Joost Burgi [1552–1632] invented antilogarithms, William Oughtred [1574–1660] invented the first slide rule (1621) based on logarithmic scale (AEOTHOT AHOI AHOT DNB DOSB EIH GLOSB MWBD NYPL 1000Y PS TTOS WWWIS)

NAPIER, Robert, 1791–1876, Scottish marine engineer; built engines for steam ships (1840), built early ironclad warship (1860) (DNB MWBD)

NAPIER, Robert Cornelis, 1810–1890, British army officer and engineer; engineer for public works (1826–1845), built roads, canals and defenses, involved in war against Sikhs (1845, 1848), commander in chief of forces in India (1870–1876), general (1874), governor of Gibraltar (1876–1882), field marshal (1883) (DNB MWBD)

NARUTOWICZ, Gabriel, 1865–1922, Polish engineer and politician; minister of public works (1920–1922), foreign minister (1922), president of Poland (1922) shortly before he was assassinated (EB MWBD)

NASMYTH, James, 1808–1890, Scottish mechanical engineer; son of Alexander Nasmyth [1758–1840], built steam carriages (1827–1828), worked for Henry Maudslay [1771–1831] (1829–1832), manufacturer of machine tools (1834), developed Bridgewater Foundry, invented steam hammer (1839), developed planing machines, nut-shaping machines, steam pile driver, hydraulic punching machine (DNB EIH MWBD TEIH WWWIS)

NASON, Carleton Walworth, 1849–1906, American mechanical engineer and inventor; invented pneumatic lift, sheet iron diaphragm radiator used by US Navy, wrought iron boiler for house heating, managed business (1884ff) (MEIA)

NASON, Joseph, 1815–1872, American inventor and manufacturer; established Walworth & Nason (1841) specializing in heating and ventilating equipment, established first steam heating system in America (1842), valves, piping and fittings, steam traps and heating accessories patented by Nason, also developed a hydraulic elevator, screw attachment for lathes, and Argand gas burner, upon splitting of the company (1852) the Nason Manufacturing Co. continued to develop and produce steam heating equipment (MEIA NC24)

NATTA, Giulio, 1903–1979, Italian chemist and inventor; faculty and director (1938ff) of Milan Polytechnic Inst. (1938–1974), developed a catalytic process for manufacture of polypropylene, polymerized propylene to make a high-strength plastic (1954), Nobel prize (1963) (AHOI CB64 MWBD PS TGS WWWIS)

NAUMOV, Aleksei Aleksandrovich, 1916–??, Soviet physicist and engineer; Scientific Research Inst. (1941–1945), Inst. of Atomic Energy (1945–1959) becoming associate director of Inst. of Nuclear Physics, radio engineering, accelerators of elementary particles (PSOCE)

NAVIER, Claude-Louis-Marie-Henri, 1785–1836, French civil engineer; with French Engineer Corps (1806ff), pioneer in application of mathematics to engineering, equilibrium and vibration of elastic solids (1821), strength of materials and structural analysis (1821), with George Gabriel Stokes [1819–1903] used momentum of fluids to derive the Navier-Stokes equation still widely used (DOSB GEAPIT L&M MWBD PS WWWIS)

NEILSON, James Beaumont, 1792–1865, Scottish inventor; worked in industry with Glasgow Gasworks (1814–1847), patented use of hot blast in iron manufacturing, improved methods of manufacturing gas (GEAPIT MWBD)

NEILSON, William George, 1842–1906, American mining engineer; locomotive builder, pioneer in bauxite industry, helped establish Adirondack Reserve (NY) (1880) (CDOAB DAB MWGD WWWIA)

NERI, Antonio, 1575–c.1614, Italian chemical technologist; a priest, developed glassmaking procedures including coloring of glass, made lead glass of high refractive index, made enamel glass (CDOSB DOSB PS)

NERNST, Walther Hermann, 1864–1941, Polish German physical chemist; chemical equilibrium, ions, solutions, galvanic cell, enunciated what is now called third law of thermodynamics (1905) also called heat theorem, professor at Göttingen (1891–1904), Berlin (1905ff) including director (1924–1933), Nobel prize (1920) (BEST L&M MWBD PA TGS WWWIS)

NERO Claudius Caesar (aka Lucius Domitius Ahenobarbus), 37–68 AD, Roman builder and emperor; became emperor (54 AD), rebuilt city of Rome (64 AD) after fire, called natural gifted engineer of the day (AIn EB)

NESMITH, John, 1793–1869, American inventor and merchant; textile equipment and manufacture of textiles, politician in Massachusetts (CDOAB DAB WWWIA)

NESSLER, Karl Ludwig—see NESTLE, Charles

NESTLE, Charles (born as Karl Ludwig Nessler), 1872–1951, British German inventor; Nestle-wave for hair treatment after experiencing the Marcel wave in Paris, work included study of different hair, heating methods, and chemical treatments, to US (1915), produced permanent wave, founded Nestle Co. that manufactured and distributed hair dressing equipment and supplies, developed new health-care products (ANB CDOAB DNB)

NETTLETON, Edwin, 1831–1901, American civil engineer; county surveyor (1865), surveyor and engineer for Greeley (CO), built largest irrigation system in Colorado at that time, invented weir for company, surveyed sites for several towns in Colorado, state engineer of Colorado, consulting engineer as irrigation expert including invention of instruments for gauging water and worked for US Department of Agriculture (USDA) and in the states of Idaho, Wyoming, and Colorado (BDOACE BITAS CDOAB DAB WWWIA)

NEUMANN, Gerhard, 1917–1997, German American aeronautical engineer; Office of Strategic Services (OSS) in WWII, variable stator compressors and high-bypass turbofan engines for aircraft, designer, vice president and group leader at General Electric Co. (1948ff), NAE (1970) (Bridge WWIE WWWIA)

NEUMANN, John Louis von, 1903–1957, Hungarian American mathematician, engineer and inventor; Institute of Advanced Studies at Princeton (1930–1954), theory, design, and construction of computers, one of first computers of which was called the maniac (mathematical analyzer, numerical integrator, and computer), several capacities in WWII particularly in ordnance, missiles, antimines, and antisubmarine, Atomic Energy Commission (AEC) (1954), NAS (1937) (ANB BM32 HOFL IAI IWAP MWBD NC46 RHW 1000Y PS TGS WWWIA)

NEUMANN, Leonid Robertovich, 1902–??, Soviet electrical engineer; director of high-voltage center of Electric Physical Inst. (1931–1935), Inst. of Energetics (1946–1960), then Inst. of Electromechanisms, investigation of nonlinear electric circuits, skin effect on ferromagnetic bodies, direct current transmission of electricity (PSOCE)

NEUSTADTER, Arnold, d. 1996, age 85, American inventor; invented Rolodex (1940s), invented spring-mounted phone directory named Autodex, numerous work organizers, his company sold to Insilco Corp. (OH), owner of Zephyr American Corp. (ante 1961) (FAL NYTObit)

NEWBOLD, Charles, 1780–??, American inventor; first US patent for cast iron plow (1797) but could not sell it because some farmers thought the steel would poison the soil, developments preceded in England by first English patent (1720) by Joseph Foliambe and Scott James Smith (1767) and by Robert Ransome (1785) for a cast iron plow whose surface was hardened by chilling, Jethro Wood (NY) [1774–1834] introduced a plow with changeable parts (1819) (ASAI HFP)

NEWCOMB, Charles Leonard, 1854–1930, American mechanical engineer and inventor; worked in Murless Iron Foundry, was a machinist, master mechanic, superintendent of American Electric Lighting Co. (CT), superintendent of Deane Steam Pump Co. (MA) (1881), patented rotary nozzle for fire fighting (1898), was president of company then general manager when absorbed by Worthington Pump & Machinery Co. (MA) (1914), consulting engineer (CDOAB DAB MEIA NC22 WWWIA)

NEWCOMEN, Thomas, 1663–1729, British engineer and inventor; first effective atmospheric steam engine (1712) called Newcomen engine, widely used in mines, worked in association with John Cawley (or Calley) [d. 1725] in development of steam engine, worked with Thomas Savery [1650?–1715] in developing steam engine to pump water from mines (1698, 1712) (AHOI BEST CDOSB DNB EAI EB EE EIH GEAPIT GI MWBD NYPL PS SAI TBDOS TEIH TGS WWWIS)

NEWELL, Allen, 1927–1992, American computer scientist and engineer; at Rand Corp. (1950–1961), faculty and administration at Carnegie Mellon Univ. (PA) (1961ff), developed list processing, computer structures, human/computer information processing systems, considered one of the founders of artificial intelligence, methods of programming computers (beginning in 1956), NAS (1972), NAE (1980) (BM71 Bridge MT7 WWIE WWWIA)

NEWELL, Frederick Haynes, 1862–1932, American mining and civil engineer; land surveys, railway surveys, hydraulic engineer with US Geological Survey (1888–1914), chief engineer of reclamation service, leader in conservation, irrigation, faculty at Univ. of Illinois (1915–1920), worked to unify engineering profession, consulting (ANB CDOAB DAB NC23 WWWIA)

NEWMARK, Nathan Mortimer, 1910–1981, American civil engineer; faculty and administration at the Univ. of Illinois (1930–1976) including department head (1956–1973) and chairman of Digital Computer Lab. (1947–1957), structural analysis, structural materials, structural dynamics, mechanics, earthquake design criteria, NAE (1964), NAS (1966) (BM60 MMOS MT2 PS WWWIA WWWIS)

NEWTON, Henry Jotham, 1823–1895, American inventor; piano manufacture, dry plate photography (BDOAS(El) CDOAB DAB NC7 WWWIA WWWIS)

NEWTON, Sir Isaac, 1642–1727, English physicist, mathematician and inventor; invented early form of differential and integral calculus (1665–1667) and binomial theorem, theory of gravitation widely used by engineers, invented early form of reflecting telescope (1668), at Cambridge Univ. (1669–1701), proposed cooling law (1701), wrote papers and books of lasting value (BEST DNB EIH FNIE GEAPIT HIW IAI MWBD NYPL SAI SAI-MP WBE WWWIS)

NEWTON, John, 1823–1895, American military and civil engineer; graduated from US Military Academy and commissioned into Corps of Engineers and was instructor there, served in engineering capacities in construction of forts, surveys of breakwaters, improved lighthouses, floating docks, harbor improvements, defenses of Washington (DC) (1861–1862), commanded troops in Civil War as major general (discharged 1866), removed obstructions from waterways, became chief of engineers and brigadier general (1884), commissioner of Public Works for NYC (1886–1888), NAS (1876) (ANB BDOACE BITAS BM4 CDOAB DAB NC4 PS WWWIA)

NICHOLS, Mark Lovel, 1888–1971, American agricultural engineer; faculty at Univ. of Delaware (1915–1917), Virginia Polytechnic Inst. (1917–1919), head of department at Alabama Polytechnic Univ. (1919–1936), with US Department of Agriculture (USDA) soil conservation service, known for Nichols terrace, head of tillage laboratory (1936ff), collaborator USDA (1958ff), McCormick medal (1934) (AE52 WWWIA)

NICHOLS, Othniel Foster, 1845–1908, American civil engineer; machinist, elevated railroad in NYC, tunnel construction in Peru (1871–1876), various engineering tasks such as park development, bridge works, water works, chief engineer and general manager of Brooklyn Elevated Railroad (1888–1895), chief engineer of department of bridges in New York (1904–1906) (BDOACE BITAS NC9 WWWIA)

NICHOLSON, William Thomas, 1834–1893, American manufacturer and inventor; machinist, shop manager (1847–1852), established shop with Isaac Brownell then bought him out (1859), manufactured egg-beater and spirit level (1860), manufactured small arms, file cutting machine, also invented taper pin locking device, machinist's vice, machine for cutting file teeth, organized Nicholson File Co. (RI) (1864) (CDOAB DAB MEIA NC8 WWWIA)

NIELSEN, Jack N., 1918–1990, Welsh American mechanical and aeronautical engineer; Corps of Engineers during WWII (6 yr.), research engineer with NACA (13 yr.), later with NASA, missile aerodynamics, co-founder of Vidya Co. and later president of Nielsen Engineering and Research, Inc., NAE (1985) (MT6 WWIE)

NIEPCE, Joseph-Nicéphore, 1765–1833, French inventor; brief services as professor, soldier, and civil administrator, working with brother Claude Niepce developed an internal combustion engine fueled with powder (1807), produced primitive photograph (1816, 1822, 1826), formed partnership with Louis Daguerre [1789–1851] (1829), with nephew Claude-Félix-Abel Niepce [1805–1870] devised first successful photographic process on glass (1847) (ASAI BEST CDOSB MWBD PS PopSci SAI TGS WA WWWIS)

NIEUPORT, Édouard, 1875–1911, French aviator and airplane builder; pioneer airplane builder, built biplanes as used in WWI (MWBD)

NIKOLAEV, Ivan Ivanovich, 1893–??, Soviet railroad engineer; Moscow Inst. of Communication and Line Engineering (1921–1938) becoming professor (1935), also with Moscow Technical College (1921–1938), Academy of Railroad Transport (1947–1951), then with USSR Academy of Sciences particularly dealing with complex transport problems (PSOCE)

NILES, Nathaniel, 1741–1828, American theologian and inventor; developed water-powered process to draw iron from bar iron, manufactured an improved wool-carding machine, political activities and government in state of Vermont (ANB CDOAB DAB NC5 WWWIA)

NIPKOW, Paul Gottlieb, 1860–1940, German engineer and inventor; developed railroad-signaling devices, scanning disk that led to mechanical television system (1884) (EIH MWBD PS SAI WWWIS)

NISHIYAMA, Zenji, 1901–1991, Japanese physicist and metallurgist; faculty at Tohoko Univ. (1927–1936) and Osako Univ. (1936–1965), Nippon Steel Corp. (1965–1978), crystal structure, phase transformations in metals, developed high-strength low-alloy steels (HSLA), fNAE (1982) (Bridge MT6)

NOBEL, Alfred Bernhard, 1833–1896, Swedish chemical engineer, inventor and philanthropist; manufacture and exploitation of nitroglycerine under name of dynamite (1867), discovery of nitroglycerine was by Italian Ascanio Sobrero [1812–1888], use of explosives for road building, canal digging, foundation laying, blasting gelatin, smokeless powder (1887), established fund for Nobel prizes (1900) first awarded (1901), over 100 patents (ASAI BEST CDOAB DAB DOSB GI MWBD 1000Y PS SAI-MP TGS WWWIS)

NOBERT, Friedrich Adolph, 1806–1881, German instrument maker and inventor; clockmaker, optical instruments, university mechanic (1835ff), invented technique for ruling parallel lines for microscopic micrometers (1845), made compound microscope and objective lens (CDOSB PS)

NOBILE, Umberto, 1885–1978, Italian aeronautical engineer; designed airships, commanded expedition in dirigible (1928), which was wrecked, although not judged responsible he resigned commission as general (MWBD WWWIA WWWIS)

NOBLE, Daniel Earl, 1901–1980, American electrical engineer; radio engineering developed first FM commercial broadcast stations in world, faculty at Univ. of Connecticut (1929–1942), became vice chairman of Motorola, helped Motorola develop Walkie-Talkie for armed forces, nine patents, NAE (1968) (MT2 WWWIA)

NOLAN, Thomas B., 1901–1992, American metallurgist and geologist; USGS (1924–1971) first as junior geologist, research in mineral resources, with increasing administrative duties becoming director (1956–1965), NAS (1951) (HOWRD TNAS WWWIA WWWIS)

NONIUS—see DATUS

NORCROSS, Orlando Whitney, 1839–1920, American civil engineer and inventor; contractor, Union soldier in Civil War, invented flat-slab reinforced concrete construction, overall construction work (CDOAB DAB NC25 WWWIS)

NORDBERG, Bruno Victor, 1857–1924, Finnish American mechanical engineer and inventor; to US (1879), detailed engine parts and designed blowing machines for E. P. Ellis Co. (WI) (1879), organized Bruno Nordberg Co. (WI) (1886) to manufacture poppet valve cut-off governor (1887, 1888), enlarged company (1890), served as chief engineer and president, developed Nordberg regenerative cycle used in turbine plants, hoists, compound steam stamps, gas compressors, vacuum pumps, received over 70 patents (ANB CDOAB DAB MEIA NC34 WWWIA WWWIAH-S&T)

NORDEN, Carl Lukas, 1880–1965, Dutch American inventor and consulting engineer; to US (1904), private consulting engineer for US Navy (1915ff) where he developed flight instruments, radio-controlled target plane, drone flying bomb, catapults for aircraft carriers, formed Carl L. Norden Co. (1918) and produced Norden bombsight (1921ff) and perfected bombsight with help of Thomas H. Barth and US Army captain Frederick I. Entwistle (1931) widely used in WWII (ANB CB CDOAB MWBD NC52 PS TP)

NORTH, Simeon, 1765–1852, American firearms manufacturer; manufactured firearms for U. S. government—pistols (1799ff) and rifles (1823ff), made a 10-round repeating rifle (1825), use of interchangeable parts in manufacturing (CDOAB DAB IIA MWBD NC7 WWWIA)

NORTHEN, William Ezra, 1819–1897, American civil engineer; specialist in water problems, dams, floating docks and pumping machinery, with George W. Whistler [1800–1849] formed an engineering company (1842–1849), vice president and engineer of the New York and New Haven Railroad (1854–1866), sanitary engineer with New York Metropolitan Board of Health (1866–1869), chief engineer of the Chicago Main Drainage Canal (1890–1891) (BDOACE WWWIA)

NORTHROP, John Knudsen, 1895–1981, American engineer, aircraft designer and manufacturer; helped found Lockheed Aircraft Co (1927), founded Avion Corp. (1928), president of Northrop Corp. (1931–1937), president of Northrop Aircraft (1939ff) whose company designed several airplanes used by US—Vega monoplane, A-17, P-61 Black Widow, F-89 Scorpion jet, C-125 Raider, B-49 bomber, NAE (1979) (ANB CDOAB DAB MT2 MWBD WWWIA WWWIS)

NORTON, Charles Hotchkiss, 1851–1942, American mechanical engineer and inventor; designer and builder of grinding machines, upgraded Brown & Sharpe Co. (RI) (1986–1890, 1896ff) universal grinding machine, with others formed LeLand, Faulconer & Norton Co. (MI) (1890) from which he withdrew to go into private practice (1895), founded Norton Grinding Co. (MA) that merged to form Norton Emery Wheel Co. (MA) that was very successful in grinding crankshafts and cylinders, he received over 100 patents (1904–1940) for improvements in production grinding machines (MEIA NC31 WWWIA)

NORTON, Lewis Mill, 1855–1893, American chemical engineer; chemist at Moskeag Manufacturing Co. (NH) (1879–1881), at Massachusetts Inst. of Technology (MIT) (1881ff) where he was first chair of chemical engineering after being professor of industrial chemistry (HOCE NC4)

NORTON, Thomas Herbert, 1851–1941, American chemist and chemical engineer; managed chemical plant in France (6 yr.) after degrees in US and study in Europe, professor at the Univ. of Cincinnati (OH) (1883–1900), prepared hydrated silica, in government activities as specialist in dyestuffs among other activities (1900–1916), E. I. du Pont (DE) (1917–1920) and later with American Cyanide Co. (1930ff), worked in many foreign assignments, edited magazine Chemical Engineer (ACCE NC13 PS WWWIA WWWIS)

NOSTRAND, Peter Elbert, 1856–??, American civil engineer; with father's company John Lott Nostrand & Son in design, location, and construction of Brooklyn Elevated Railroad (1876–1880), mining engineering work, hydraulics, chief engineer of Ramapo Water Co. (1887–1890), also served as city surveyor, county engineer and superintendent of highways, consulting engineer (BDOACE NC12 WWWIA)

NOTT, Eliphalet, 1773–1866, American clergyman, engineer and inventor; clergyman, president Union College (NY) (1804–1866) for 62 yr., inventor of anthracite coal burning stove, developed a range of home heating devices, received 30 patents (ANB CDOAB DAB I&T16 MEIA NC7 WWWIA)

NOVIKOV (NOVIKOFF), M. L., ??–1956, Soviet inventor; Novikov and Wildhaber (USA), also referred to as Novikov helical gear with circular arc teeth (1923) (EE)

NOVOZHILOV, Valentin Valentinovich, 1910–??, Soviet engineer; mechanics specialist primarily in elasticity, plasticity, theory of shells especially related to problems in ship building, worked at a series of establishments, taught at Leningrad Univ. (1946ff) becoming professor (1949) (PSOCE WWWIS)

NOYCE, Robert Norton, 1927–1990, American electrical engineer; Shockley Semiconductor Lab. (1957), co-founder of Fairchild Camera and Instrument Co. becoming general manager and vice president (1957–1968), co-inventor with engineer Jack S. Kilby [1923–2005] at Texas Instrument Co. of monolithic integrated circuit (1959, 1961), transistor, electronics, with Gordon Moore [b. 1929] founded Intel (1968) and was first president and CEO of Intel Corp. (1968ff), helped organize and was CEO (1988) of Semitech Corp., Draper prize (1990) (shared), NAE (1969), NAS (1980) (AIE ANB Bridge IAI I&T15 MT6 MWBD 1000Y S:TLAW TPA WWIE WWWIA)

NOYES, La Verne, 1849–1919, American engineer and inventor; manufacture of hay tools (1874), horse hay fork (1878), wire dictionary holder (1879), traction wheel and harvester reel (1885), sheaf carrier for self-binding harvesters (1888), cord knotter for grain binders (1889), established Aeromotor Co. (IL) (1889) to build windmills and steel towers, interested in use of wind power to produce electricity, obtained more than 100 patents (CDOAB DAB MEIA MWBD NC17 WWWIA)

NUSSELT, Ernst Kraft Wilhelm, 1882–1957, German mechanical engineer; heat transfer and thermodynamics, use of dimensional analysis to study and represent heat flux in a flowing fluid, Nusselt equation, Nusselt number for heat transfer (1915), various teaching and industry positions (1907–1925), Technische Hochschule Munich (1925–1952) (CDOSB CEOS&T DOSB FINE L&M PS)

O

OATLEY, Charles William, 1904–1996, English electrical engineer; research led to commercial production of a successful electron microscope (1948), WWII with Ministry of Supply (1939–1945), Cambridge Univ. (1945–1971) and was chair of department, previously worked on a scanning microscope with Max Knoll [1897–1969] (1935), with first true scanning microscope by Englishman M. von Ardenne (1938), FRS (1969), fNAE (1979) (MSAE WWW)

O'BRIEN, Brian, 1898–1992, American physicist and electrical engineer; optical and electro-optical engineering, analysis of vision, night vision devices, motion picture equipment, developed process and equipment for irradiating milk to produce vitamin D, defense applications, faculty at Univ. of Rochester (1930–1953) including development of Inst. of Optics, American Optical

Co. (1953–1958), consultant (1958ff), NAS (1954), NAE (1981) (Bridge MT7)

O'BRIEN, Morrough Parker, 1901–1988, American civil and mechanical engineer; applied fluid dynamics and control, river engineering, coastal engineering, worked with Hudson River Regulating Department, faculty at Purdue Univ., Royal College of Engineering (Sweden), then faculty and dean of engineering (1943–1959) at the Univ. of California-Berkeley (1928ff), consulting, NAE (1969) (MT4 WWIA WWWIS)

OCKERSON, John Augustus, 1848–1924, Swedish American civil engineer; Mississippi River Commission (1879ff), authority on river and harbor improvement, navigation, construction of levees to control flood waters of Colorado River (1910) (CDOAB DAB)

O'CONNOR, Donald J., 1911–1997, American civil engineer; faculty at Manhattan College, consultant to HydroQual, mathematical modeling to evaluate the effects of pollution and abatement measures, NAE (1978) (Bridge)

ODEH, Aziz S., 1925–1994, Palestinian American engineer; research and administration with Mobil R & D Corp. (1953–1989) becoming senior scientist (1980), petroleum reservoir simulation, transient test analysis, consultant (1989ff), NAE (1987) (Bridge MT8 WWIE)

ODELL, William, 1840–1926, American mechanical engineer; engineer with Baldwin & Flagg (NY), heated dye kettles with steam (1868), Remington Arms Co. (NY) (1870), superintendent of William Wright & Co. (NY), specialist in economic generation and use of steam, introduced electric lights for tunneling, patented steam boiler (1888), evaporating coil (1888), feed water circulator and purifier (1908) (MEIA NC29 WWWIA)

OERSTED (ØRSTED), Hans Christian, 1777–1851, Dutch physicist and inventor; demonstrated that the flow of electricity caused deflection of magnetized needles (1820), measured the compressibility of liquids, first to prepare metallic aluminum (1825), Univ. of Copenhagen (1806ff) from lecturer to rector, established Copenhagen Polytechnic Inst. (BEST EIH GEAPIT GI MWBD WWWIS)

OFFNER, Franklin F., 1911–1999, American biomedical engineer; developed instruments particularly for medical purposes, president of Offner Electronics, Inc. (1939–1963), faculty of Northwestern Univ. (1963ff), electronic technology and its application to instrumentation and control in biomedical engineering, EEG and EKG machines, approximately 60 patents, NAE (1990) (Bridge MT10 WWIE)

OGDEN, Francis Barber, 1783–1857, American mechanical engineer; entered US Army (1812), patented steam engine (1813), designed and built low-pressure condensing engine in England (1817), became US consul to England (1840) (CDOAB DAB MEIA NC11 WWWIA)

OGLESBY, Clarkson H., 1908–1992, American civil engineer; faculty at Stanford Univ. (1943–1974), distinguished in highway engineering, civil engineering construction, NRC Transportation Research Board (1950–1988) administration and research, NAE (1989) (Bridge MT7)

OHAIN, Hans J. P. von, 1911–1998, German aeronautical engineer; with Aerospace (Aeronautical) Research Lab. (Germany), patented and designed turbojet engine for plane that flew (1939), after WWII consulted with US Navy and on propulsion systems for USAAF (ante 1979), consultant to US Department of Energy (1979), shared NAE Draper prize (1991) with British engineer Frank Whittle [1907–1996], fNAE (1980) (MT10 1000Y WWIE WWWIA)

OHM, Georg Simon, 1789–1854, German physicist; Ohm law for flow of electricity (1826) that formed the basis of much engineering work particularly electrical resistance (1827) named Ohm law, professor at Polytechnische Schule of Nüremberg (1833–1849), then Univ. of Munich (1849–1854), studied reception of sound by the human ear using Fourier analysis (1843) (EIH FNIE MWBD NYPL SAI WWWIS)

OKOZAKI Goro Nuido (aka Masamune), fl. 1300, Japanese swordsmith; made swords entirely of steel, used layers of steel of different carbon content and forged and hardened to obtain toughness throughout (not only the edges) (GEAPIT)

OLCOTT, Eben Erskine, 1854–1929, American military and mining engineer; pioneered in copper mines in Peru, managed Hudson River Day Line (1895ff) (CDOAB DAB NC5 WWWIA)

OLDFIELD, Berna Eli (Barney), 1878–1946, American automobile racer; driver for Henry Ford's racing cars (1902ff), first to achieve mile-a-minute in races (1903), involved in tire manufacturing and automobile safety engineering (MWBD)

OLDHAM, John, 1779–1840, British engineer and inventor; machinery used in making and numbering notes, patented paddle wheels for steamers, introduced system for warming buildings, Oldham coupling permitted the misalignment of connected shafts (DNB EE)

OLDS, Ransom Eli, 1864–1950, American inventor and manufacturer; founded Olds Motor Works (1899), manufactured the Oldsmobile (1901ff), founder and president of the Reo Motor Car Co. (1904–1936) including chairman (1924–1936), company became part of General Motors Corp. (AIE ASAI CDOAB MWBD NC39,F WWWIA)

OLIN, Franklin Walter, 1860–1951, American engineer, inventor and manufacturer; his son Franklin W. Olin, Jr. (d. ante 1951), mechanic, designer of knitting mill machinery, repair man for harvesting machinery, president and director of Olin Industries (1892–1944) and later Olin Mathieson Chemical Corp., F. W. Olin Foundation (1938ff) for philanthropic support, foundation established Franklin W. Olin College of Engineering (MA) (1997) for innovative education of engineers (ET NC41 Prism12 WWWIA)

OLIVER, Bernard M., 1916–1995, American electrical engineer and inventor; research with Bell Telephone Labs. (1940–1952) where he contributed to radar and early television systems, research and vice president of Hewlett-Packard Co. (1952–1981), developed electronic equipment, NASA Ames Lab. (1983–1993), NAE (1966), NAS (1973) (Bridge MT8 PS)

OLIVER, James, 1823–1908, Scottish American industrialist and inventor; invented process for making hard-faced plows (1868, 1869), forerunner of Oliver farm machinery manufacturing company (ASAI CDOAB DAB MWBD NC12)

OLIVER, Paul Ambrose, 1830–1912, American soldier and inventor; invented dynamite and black powder explosives at about the same time but independently of Alfred B. Nobel [1833–1896] (CDOAB DAB MWBD NC5 WWWIA)

OLIVETTI, Adriano, 1901–1960, American entrepreneur; son of Camillo Olivetti [1868–1943] who founded Olivetti Co., studied methods of mass production, design and manufacture of typewriters and office equipment, president of company (1931–1960) (CBD TCBE WWWIA)

OLMSTED, Frederick Law, 1822–1903, American engineer and architect; topographer, with partner English American Calvert Vaux [1824–1895] designed Central Park in NYC (1857), Belle Isle Park in Detroit (MI), Prospect Park in Brooklyn (NYC), South Park in Chicago (IL), park system of Boston (MA) and the grounds of the US capitol in Washington (DC), leading role in establishing National Park Service (MWBD NC2 1000Y WWWIA)

OLNEY, Raymond, 1885–1980, American agricultural engineer; early work with Allis Chalmers Manufacturing Co., Iowa Agricultural Experiment Station, and M. Rumely Co., editor of Farm Engineering, editor of Thresher's Review and Power Farming and Farm Power (1915–1925), editor of Agricultural Engineering (1920ff), secretary and treasurer of American Society of Agricultural Engineers (1925–1959) (AE35,61)

OLTMAN, Roy E., 1911–1977, American civil engineer; hydrology, hydraulics, large-scale flood investigations, several leadership positions in USGS (1938–1972), organized and directed flow measurements of Amazon River (HOWRD)

ONO GOROEMON, fl. 1250, Japanese architect and sculptor; his sculptors of Buddha were both a work of art and engineering in which he used materials to

produce his work with 1/50th weight of known works (GEAPIT)

ONSAGER, Lars, 1903–1976, Norwegian American chemical engineer, chemist and inventor; to US (1928), US citizen (1945), chemical solutions, statistical mechanics, devised gaseous diffusion method for manufacture of uranium 235, Onsager law of dilute solutions (1931), Onsager reciprocal theorem (1930s), Onsager relations sometimes called the fourth law of thermodynamics, faculty at Brown Univ. (1929–1933), Yale Univ. (1934–1972), NAS (1947), Nobel prize (1968) (BEST BM60 CB58 CDOSB DOT L&M MWBD RS24 WWWIA WWWIS)

OPEL, Fritz von, 1899–1971, German industrialist; director of Adam Opel AG maker of automobiles, bicycles (1929ff), experimented with rocket propulsion for automobiles and aircraft and piloted second rocket airplane to fly (1929) (MWBD)

OPPENHEIMER, J. Robert, 1904–1967, American physicist; director of Manhattan project that developed the atomic bomb (1942–1945) that was detonated (1945), Univ. of California and California Inst. of Technology (1929–1947), director of Inst. for Advanced Study at Princeton Univ. (1947–1966), development of the hydrogen bomb (1952) credited to leadership of Edward Teller [1908–2003] who was an adversary of Oppenheimer, NAS (1941) (BEST BM71 IAI MWBD NCG NYPL SAI-MP TGS WWWIS)

OQRUQCI, fl. 1290, Mongol hydraulic engineer; excavation of land for Grand Canal in China including dams and a feeder canal (1283) (GEAPIT)

O'REILLY, Henry, 1806–1886, Irish American electrical engineer and editor; pioneer builder of telegraph lines, leading exponent of enlarging and rebuilding of Erie Canal, with others obtained financing and built telegraph line from Pennsylvania to St. Louis (MO) (CDOAB DAB)

ORR, Hugh, 1715–1798, Scottish American inventor; to US (1742), owner of shop and was toolmaker, made muskets (1748ff), invented machine to clean flaxseed (1753), made muskets for rebels (1776), with a Frenchman established a cannon factory (CDOAB DAB MEIA NC2 PS WWWIA WWWIS)

OSBORN, Elburt F., 1911–1998, American geologist and ceramist; advances in ceramics, slag, mineral and steel technologies, served at several universities of which the longest tenure was Pennsylvania State Univ. (1946–1970) including dean (1953–1959) and vice president (1959–1970), director of US Bureau of Mines (1970–1973), Carnegie Institution (DC) (1973–1977), NAE (1968) (Bridge WWIE WWWIS)

O'SHAUGHNESSY, Michael Maurice, 1864–1934, Irish American civil engineer; to US (1885), hydraulics and hydrology, city engineer of San Francisco (CA)

(1912–1932), built water and power supply for San Francisco, dams, aqueducts, tunnels, dam in California named in his honor (ANB CDOAB MWBD NC24 WWWIA)

OSROW, Harold, 1917–1997, American inventor and entrepreneur; served in WWII, built improved broom and automotive items, chairman of Osrow Products, with brother Leonard Osrow introduced Twin-Sweep broom (1947), chairman of Osrow Products, invented cordless iron, founder of PTHY Co. (products that help you), 30 patents (FAL NYTObit)

OSTWALD, Friedrich Wilhelm, 1853–1932, Latvian German physical chemist; professor at Univ. of Riga (1881–1887), professor at Univ. of Leipsig (1887–1906), Ostwald law of dilution of an electrolyte (1882), rate of reaction, invented process for making nitric acid, a co-developer of field of physical chemistry, Nobel prize (1909) (BEST L&M MWBD RHW)

OSWATITSCH, Klaus, 1910–1993, Austrian aeronautical engineer; Kaiser Wilhelm Inst. Göttingham, faculty at Kungliga Teknika Hogskolan in Stockholm (1949–1956), Inst. for Theoretical Gas Dynamics Aachen (1956–1972) then returned to Austria, gas dynamics, high-speed aerodynamics, transonic flow, interacting boundary layer, leadership in academic programs, fNAE (1982) (Bridge MT7)

OTHMER, Donald Frederick, 1904–1995, American chemical engineer; development engineer for Eastman Kodak (1927–1933), faculty of Polytechnic Inst. of New York (Brooklyn) (1932ff) including head of department (1937–1961) and distinguished professor (1961ff), co-editor with Raymond E. Kirk [1890–1957] of internationally recognized Encyclopedia of Chemical Technology in 26 main volumes, extractive distillation, heat transfer, correlation of engineering data, manufacturing of wood products, plastics, sugar, over 150 patents (AMOS WWIE WWWIA)

OTIS, Charles Rollin, 1835–1927, American mechanic, engineer and inventor; son of Elisha Graves Otis [1811–1861], machine shop, with brother Norton Prentiss Otis [1840–1905] carried on father's business with elevator construction, elevator brakes (1864), steam elevator engine (1865), improved valve for steam engine (1867), hoisting apparatus (1868, 1871), first practical escalator resulted from independent inventions (1892) by Jesse W. Reno [1861–1947] and Charles Seeberger, rights obtained by Otis Elevator Co., company built first escalator (1900), numerous patents (ANB CDOAB DAB MEIA NC11 NYPL PA WOI WWWIA)

OTIS, Elisha Graves, 1811–1861, American farmer, legislator and inventor; manufactured wagons and carriages, constructed turbine waterwheel (1848), built factory for manufacturing (1851), invented automatic safety control for elevators (1852), patented steam elevator (1861), provided the basis of the Otis elevator enterprise, safety brake for railroad car (1852), steam plow (1857), rotary

oven (1858) (AIE ANB BEST CDOAB DAB DOSB EIH GI MEIA MWBD NC11 1000Y PS TBDOS)

OTTO, Nikolaus August, 1832–1891, German engineer and inventor; built his first internal combustion engine (1861), formed Nickolaus August Otto and Co. with Eugen Langen [1833–1895] (1864), with Langen invented early internal combustion engine (1867), built first four-stroke cycle engine known as Otto cycle (1862) (1875) (ASAI BDOS BEST EE EIH FNIE HFP HOCS&T IAI MWBD NYPL PS SAI TBDOS(Po) TGS)

OUGHTRED, William, 1574–1660, English mathematician and inventor; invented first rectilinear slide rule (1621) and with his pupil Richard Delamain invented the circular slide rule (1630), invented trigonometric abbreviations, invented symbols for multiplication and proportion (1631), Edmund Gunter [1581–1626] is credited by some for inventing the first slide rule, which was a device used for navigation (1620) (BEST DOSB EIH GI HIW MWBD NYPL PS WA WOM WWWIS)

OVINGTON, Earle L., 1879–1936, American aeronautical engineer and inventor; with New York Telephone Co. (1899–1900), formed and was president of Ovington Manufacturing Co. (1905–1908) and Ovington Motor Car Co. (1908–1910) and several other companies, first US airmail pilot (1911), invented electrical appliances including Ovington high-frequency apparatus (MWBD WWWIA)

OWEN, William Leonard, 1897–1971, British engineer; served in WWI, worked as an engineer in a company that became one of the companies of Imperial Chemical Industries (ICI), in British ministry of supply (1940–1945), design and construction of atomic energy plants in England (one of which was Calder Hall), director and manager of UK Atomic Energy Authority (1954–1964) (DNB PS WWW)

OWENS, Michael Joseph, 1859–1923, American inventor and manufacturer; invented bottle-blowing machine (1895, 1904), organized Owens Bottle Machine Co. (1903) of which he was vice president (1915–1923), organized Libbey-Owens Sheet Glass Co. (WV) (1916) (ANB CDOAB DAB MEIA MWBD NC28 WWWIA)

P

PACINOTTI, Antonio, 1841–1912, Italian physicist and inventor; invented dynamo with ring winding (1860), Pacinotti armature independently discovered and used by Belgian electrical engineer Zenobe T. Gramme [1826–1901] (1869), faculty at Univ. of Pisa (CDOSB DOSB EIH MWBD WWWIS)

PACKARD, David, 1912–1996, American electrical engineer, inventor, and manufacturer; radio engineering, innovative audio oscillator, manufacturer of computing, communications, and measurement products and provided services, co-founder and president or chairman of Hewlett-Packard Co. (TX) (ante 1993), (William R. Hewlett [1913–2001]), US Department of Defense (1969–1971) working on procurement and management, numerous philanthropic activities, NAE (1971), NAS (1989) (Bridge I&T16 MT9 1000Y WWIE WWWIA)

PACKARD, James Ward, 1863–1928, American engineer and inventor; with brother William Doud Packard [1861–1923] founded Packard Electric Co. (1890), designed and built first Packard automobile (1899), president of Packard Motor Car Co. (ante 1915) (CDOAB DAB MWBD NC20 WWWIA)

PACKARD, Walter Eugene, 1884–1966, American agricultural engineer; special agent for Packard & Niece of Chicago in London (1905–1909), federal irrigation research in California, established desert experiment station in Imperial Valley for Univ. of California, serving as superintendent (1909–1917), in WWI in France, consulted on numerous irrigation projects including financing and family development, Rural Resettlement Administration (1935–1938), consulted with federal agencies and other countries, also worked in reclamation, drainage, flood control, and hydroelectric power development (NC53 WWWIA)

PACKHAM, Frank Russell, 1855–1915, American inventor; machinist, superintendent and experimenter of Baker Drill Co. (OH) (1878), manager and designer of turner's tools for Packham Crimper Co., patented crimper machine (1886), patented grain drills and other agricultural implements for Superior Drill Co. (1887), manager of experimental department of American Seeding Co., patented disc drill (1903) (MEIA)

PAGE, Charles Grafton, 1812–1868, American physician and inventor; invented electrical devices, invented an induction apparatus (1836), use of bundled wires instead of iron bars for coils, invented self-acting circuit breaker, examiner in US Patent Office (1861–1868) (ANB BDOS(El) CDOAB EIH NC5 PS WWWIA WWWIS)

PAGE, Frederick Handley, 1885–1962, English airplane manufacturer; founded first British airplane manufacturing company called Handley Page, Ltd. (1909), designed first twin engine bomber (1915) (DNB MWBD WWW)

PAGE, Logan Waller, 1870–1918, American civil engineer; highway testing engineer for state of Massachusetts (1893–1900), contributed to knowledge of materials for highways, Bureau of Roads in USDA that was renamed Office of Public Roads which he headed (1900–1918) (ANB WWWIA)

PAINE, Thomas, 1737–1809, English American essayist and inventor; to US (1774), after living in England returned to US (1802), a revolutionary in politics, made earliest American design for a metal (cast iron) bridge (1788) not built until after his death (DNB I&T15 LOTE MWBD NC5 SFLOE WWWIA WWWIS)

PAINE, Thomas Otten, 1921–1992, American engineer and metallurgist; submarine officer in WWII, several assignments in research and management with General Electric Co. (25 yr.), administrator of NASA (1968–1970), president and CEO of Northrop Corp. (1976–1982), chairman of Thomas Paine Assoc. (1982ff), space exploration programs, contributed to Apollo program and other space missions, NAE (1973) (Bridge CB70 HOCS&T MT6 PS TEIH WWWIA)

PAINTER, William, 1838–1906, American engineer and inventor; patented fare box and car seats (1858), counterfeit coin detector (1862), lamp burners (1863), foreman of machine shop that built pumping machinery (1865), automatic magneto-signal for telephones, seed sower (1870), pump valves (1875), formed Triumph Bottle Stopper (MD) (1885), first single-use bottle seal (1885), metal single-use cap called crown cork (1892), Bottle Seal Co. (1892) then reorganized company as Crown Cork & Seal Co. (MD) (1892–1903) and invented machinery for applying caps to bottles, received 85 patents (CDOAB DAB MEIA NC31)

PALFREY, John Carver, 1833–1906, American military engineer; US Military Academy (1857), Union soldier in Civil War specialized in military fortifications (CDOAB DAB NC17)

PALLADINO, Nunzio Joseph, 1916–1999, American nuclear and mechanical engineer; design and development of nuclear reactors for submarine and power, life cycle evaluation of equipment and systems, numerous positions with Westinghouse Electric Co. (1939–1959), government, faculty and administration as department chairman and dean of engineering (1966ff) at Pennsylvania State Univ. (1959–1981), NAE (1967) (Bridge WWIE WWWIS)

PALLADIO, Andrea di Pietro, 1518–1580, Italian architect and civil engineer; bridge builder, extended length of truss bridges, reintroduced truss to modern bridge building, best known for church at San Giorgia Maggiore in Venice (AHOEAT EIH GEAPIT)

PALMER, Timothy, 1751–1821, American civil engineer and bridge builder; surveyor for highways, pioneer builder of covered timber truss bridges in America, bridge over Schuylkill River in Philadelphia (PA) (1804), several bridges in east including one in Georgetown (DC) over the Potomac River (1796) (EIH GEAPIT MWBD PS)

PAMBOUR, François Marie Guyonneau de, 1795–??, French civil engineer; entered military in artillery and transferred to general staff, improved the theory and practice of steam engines and locomotives, particularly with respect to calculation of work (CDOSB DOSB PS)

P'AN Chi-Hsun, 1521–1595, Chinese hydraulic engineer; director of Yellow River Works and Grand Canal during 16th century (GEAPIT)

PANHARD, René, 1841–1908, French automotive engineer; with Émile Levassor [c.1844–1897] built first automobile with engine in front (1891), founder and director of firm of Panhard and Levassor, early automobile manufacturers (MWBD WWWIS)

PAPIN, Denis, 1647–c.1714, French physicist and inventor; constructed first engine in which condensation of steam was used to create a vacuum, developed pressure cooker (1679) or autoclave, invented pressure valve to prevent explosion, steam engine (1690), Hessian pump—a centrifugal pump (CDOSB EE EIH GEAPIT GI MWBD NYPL TEIH TGS WWWIS)

PARACELSUS (aka Theophrastus Bombastus von Hohenheim, his original name), 1493–1541, Swiss German chemist and physician; investigated mechanics of mining and disease of miners, traveled through Europe and Middle East (c.1510–1524), Univ. of Basel (1526ff), connected specific diseases with certain mineral elements (BEST EB MWBD SAI WWWIS)

PARDEE, Ario, 1810–1892, American civil engineer and surveyor; survey to locate Beaver Meadow Railroad (PA) (1882), chief engineer of Hazelton Railroad and Coal Co. (1836–1840), founded firm later called Pardee & Co. (1840) for coal development in Pennsylvania, benefactor of Lafayette College (PA) (CDOAB DAB NC11 WWWIA)

PARETO, Vilfredo, 1848–1923, French Italian engineer and economist; engineer in Rome (1870–1875), faculty at Univ. Lausanne (Switzerland) (1893–1923), Pareto efficiency (1906), formula of economic equilibrium, Pareto law, 80/20 rule, application of methods of physical sciences to sociology (DOSB DOT L&M TodChem6 MWBD WWWIS)

PARK, Robert H., 1902–1994, American electrical engineer and inventor; president of Fast Load Control, Inc., electric power system stability, originated Park equations related to electric power and machines (1929), first worked at General Electric Co., during WWII at Naval Ordnance Lab. on development of mines, manufacturer of plastic bottles inventing the machines to make them (in the 1950s and 1960s), NAE (1986) (Bridge MT8 WWIE)

PARK, William Robert, 1831–1921, American engineer and inventor; merchant (1850), superintendent, manager, and consulting engineer of Hancock Inspirator Co. (MA) (1876ff), founder of W. R. Park & Son (MA), about 15 patents relating to boiler design, parts, and operation (MEIA)

PARKER, Earl Randall, 1912–1998, American metallurgist and engineer; faculty and administration at Univ. of California-Berkeley (1944ff), research at General Electric Co. (1935–1944), dislocation theory, brittle fracture, phase transformations, improvement of properties of engineering alloys, NAE (1969) (AMOS Bridge MT9 WWIE WWWIA WWWIS)

PARKER, Ely Samuel, 1828–1895, American soldier and engineer; served as engineer to General U. S. Grant [1822–1885] in Civil War (1864–1865) reaching rank of brigadier general (ante 1869), Commission of Indian Affairs (1869–1871) (CDOAB DAB NC5 WWWIA)

PARKER, Herbert Myers, 1910–1984, English American physicist and inventor; radiology, medical physics, head of radiation at Oak Ridge (TN) and Hanford (WA) (1943–1956), director of Hanford Lab. (1956–1965), consultant to Battelle-NW (1965–1971), consultant HMP Assoc. (1971ff), NAE (1978) (MT3 WWIE WWWIA WWWIS)

PARKER, James Wentworth, 1886–1957, American mechanical engineer; apprenticeship, then boiler engineer where he advanced through several positions becoming president (1944–1951) of Detroit Edison Co. (MI) (1910–1951), consultant in design and construction of several power plants, operation of electric systems and central heating systems (NC44 WWWIA)

PARKER, John P., 1827–1900, Black American inventor and entrepreneur; skilled iron molder, moved from south to north, assisted in underground railroad, established foundry (1854) producing castings, established Parker Foundry & Machine Co. (OH) after Civil War where he manufactured slide valve engines and reapers, patented a tobacco press (1884) and a Parker pulverizer for smoothing the soil (a harrow for farming) (1895) (ANB BI)

PARKER, Joseph L., 1898–??, Black American civil engineer; helped design and build Holland Tunnel (NY), Queens Midtown Tunnel (NYC), served on Triborough Bridge Authority (NYC) (BIS)

PARKER, Joseph Lloyd, 1913–??, Black American chemical engineer; researcher and plant supervisor of DuPont Co. (1940–1946), chief operating engineer of Girdler Corp. (1945–1951), owner of J. L. Parker Co. (1951ff), president of Amulco Asphalt Co. (1951ff), organic arsenicals, production and treatment of gases, asphalt and paving (AMOS)

PARKER, Theodore Bissell, 1889–1944, American civil engineer; waterways and hydroelectric engineering, with Stone and Webster, served in Federal Emergency Administration of Public Works (1933), chief engineer of construction of Tennessee Valley Authority (TVA) (1935) and chief engineer of TVA (1938–1943), faculty at MIT (1943ff) (CDOAB WWWIA)

PARKES, Alexander, 1813–1890, English chemist and inventor; cold vulcanization process (1846), Parkes process of using zinc for desilverizing lead (1850), invented celluloid (1855) (BDOS MWBD PS WWWIS)

PARKHURST, Edward G., 1830–1901, American mechanical engineer and inventor; Savage Arms Co. (CT) during Civil war, assistant superintendent of Pratt & Whitney Co. (CT) (1869–1895), later with American Ordnance Co. and Bethlehem Steel Co., patented machinery for feeding wire to machines (1871), gauge for machine drills (1872), numerous patents for weapons and accessories (1880), milling machine and turret lathe, approximately 10 patents (BDOTHOT MEIA)

PARME, Alfred L., 1909–1987, French American civil engineer; structural engineer, theory and design of shells, arch dams, and pre-stressed concrete structures, was stress analyst for Republic Aviation, with Corps of Engineers, Portland Cement Association (1940–1968), consulting engineer, NAE (1974) (Bridge MT7)

PARR, Charles H., 1868–1941, American inventor; co-inventor with Charles W. Hart [1872–1937] of Hart-Parr farm tractor (1902), formed Hart-Parr Tractor Co. (1903), company later consolidated with Oliver Farm Equipment Co. (IA) (1929) and other manufacturing companies, later became White Co. (1974) (GTHP MISAT S&TF WOI)

PARR, Samuel Wilson, 1857–1931, American chemist and inventor; at Univ. of Illinois (1891–1926), developed calorimeters for determining heat value of coal and other solids, studied embrittlement of boiler plates (ACCE CDOAB DAB NCC PS)

PARROTT, Robert Parker, 1804–1877, American army officer and inventor; after graduation from US Military Academy (1824) served in artillery (1824–1836), foundry business (1836–1877), use of wrought iron hoops on breech to strengthen cast iron guns (1861), expanding projectile for rifled cannon (1861), Parrott guns used during Civil War built at West Point Foundry (NY) (ANB ASAI CDOAB DAB MEIA MWBD NC5 WWWIA)

PARSEVAL, August von, 1861–1942, German aeronautical engineer; designed medium-size nonrigid airship named after him having a single large gas bag from which a car was suspended (1901–1903) (MWBD)

PARSHALL, Ralph Leroy, 1881–1959, American civil engineer and inventor; irrigation engineering, invented a flow measuring device that came to be known worldwide as the Parshall flume (1923) originally called improved Venturi flume, made other devices for measurement of water and snow, faculty at Colorado State Univ. (CSU) (1904–1913), CSU hydraulics laboratory and USDA (1913–1948) serving as director of laboratory (1918ff) (AE69 NC49 WWIE(I))

PARSONS, Sir Charles Algernon, 1854–1931, British mechanical engineer and inventor; utilization of steam power, produced first practical steam turbine with condenser (1884), introduced compound steam turbine (c.1884ff), founded firm (1889), started Parsons Marine Steam Co. (1894), used turbine to power ships and to produce electricity, worked on devices to gear down the rapid rotation of turbines, over 300 patents, fNAS (1925) (BDOS BEST DNB DOSB EIH FNIE GI IAI MWBD NYPL PS SAI S&TF TBDOS(Po) TEIH TGE: B WA WWW WWWIS)

PARSONS, James A., Jr., 1900–??, Black American metallurgist and inventor; in charge of research for Duriron Company (OH), authority on corrosion resistant alloys, patented austenitic alloy steels resistant to sulfur and other industrial chemicals, cementation process for treating metals (1943), melting technologies (AAFIS&T AAI PS SBPY)

PARSONS, William, 1800–1867, Irish politician, engineer, and astronomer; built telescopes (1827ff), made large-diameter telescopes, polished metal mirrors, important astronomical observations (MWBD TBDOS(Po) WWWIS)

PARSONS, William Barclay, 1859–1932, American civil engineer; designed and built initial section of New York City subway system (1899–1904), built East River Tunnel (NYC), chief engineer Cape Cod Canal (1905–1914), Chicago Transit Commission chairman (1916ff) (CDOAB DAB MWBD NC14 WWWIA)

PARTRIDGE, Everett Percy, 1902–1969, American chemical engineer; studied formation of boiler scales, water conditioning, embrittlement of boiler steel, US Bureau of Mines (1931–1935), director of research and then director of Hall Labs. (1935ff) (AMOS NC25 PS WWWIA)

PASCAL, Blaise, 1623–1662, French mathematician, scientist and inventor; invented and patented a calculating machine (1642–1645) based on rotating discs, barometer (1648), hydraulic press (1648), syringe, helped develop probability relationships, hydrostatics (ASAI BEST DOSB EIH FNIE GEAPIT GI MWBD NYPL SAI-MP TEIH TGS WWWIS)

PASLEY, Sir Charles William, 1780–1861, British army officer and engineer; introduced course for non-commissioned officers in military engineering (1811), director of program (1812–1841), general (1860), wrote several treatises on military engineering (DNB MWBD WWWIS)

PASTEUR, Louis, 1822–1895, French chemist, microbiologist, and inventor; professor and administrator of physics, chemistry, and geology at several schools (1848–1868), developed process of pasteurization (1864), director of Pasteur Inst. (1888–1895), use of polarized light and crystals, worked on fermentation, winemaking processes, silk production using silkworms, developed vaccinations (1880ff) to protect sheep from anthrax and dogs from rabies, fNAS (1883) (BEST DOSB EIH GI MWBD NYPL PS SAI SAI:MP S:TLAW TGS WBE WWWIS)

PATON, Angus, d. 1999, age 93, English civil engineer; senior consultant for Alexander Gibb & Partners

(England), planning, design, and management of international civil engineering projects, fNAE (1979) (Bridge)

PATTERSON, Richard Cunningham, Jr., 1886–1966, American engineer; inspector in NYC (1913–1920), J. G. White Engineering (1920), E. I. Du Pont (1921–1927), NYC Commissioner of Corrections (1927–1932), official in National Broadcasting Co., then US Department of Commerce, chairman of the board of RKO (1939–1945), followed by government appointments (CDOAB DAB WWWIA)

PATTERSON, Rufus Lenoir, 1872–1943, American inventor; developed equipment for handling tobacco (1887), vice president of American Tobacco Co. (1901), president and founder of American Machine and Foundry (1901ff) becoming chairman of the board (1941–1943) (CDOAB DAB NC16 WWWIA)

PATTINSON, Hugh Lee, 1796–1858, English engineer and inventor; invented a silver-extraction process from lead ore (1829), assay master (1825–1831), manager of Wentworth Beaumont Lead Works, held patents for manufacturing of white lead and developed new process (1833), established partnership in a chemical works, developed new manufacturing process (1841) (BDOHOT BDOS(Wi) DNB PS WWW WWWIS)

PAUGER, Adrien de, ??–1726, French engineer; engineer for defining French colony of Louisiana (1707ff), captain in Navarre regiment (1720) and chief engineer, plotted town of New Orleans (LA) (1721) (DAB EIH)

PAUL, Henry Martyn, 1851–1931, American astronomer and engineer; surveying and triangulating for US Coast Survey, then with Naval Observatory (1875, 1883), served with US Navy, then Bureau of Yards and Docks, to US Naval Academy (1897–1912) retiring as a captain (1913) (CDOAB DAB NC10 PS)

PAUL, Lewis, 1730–1759, British mechanical engineer; developed spinning machine that was forerunner of that of Richard Arkwright [1732–1792] patented (1748) (DNB GEAPIT PS)

PAUL, Robert W., 1869–1943, British engineer; exhibited a theatrograph (1896), an early form of motion picture projector (MWBD WWW)

PAUL, William Darwin, 1900–1977, American medical doctor and inventor; at Univ. of Iowa (1930ff) in clinical medical practice, developed buffered aspirin (1940s), developed buffered antiacid tablets that became known as Rolaids, pioneer in sports medicine (CDOAB)

PAULING, Linus Carl, 1901–1994, American chemist, chemical engineer, and biologist; educated first as an engineer, molecular structures, crystal structure, hydrogen bonding, chemical bonding, quantum mechanics, faculty at California Inst. of Technology (1925ff) serving as chair of Division of Chemistry and Chemical Engineering (1936ff), NAS (1933), Nobel prizes (1954) (1962)

(BEST BM71 LOTCS SAI-MP TBDOS(Po) TGS WA WWWIA WWWIS)

PAVLOV, Igor M., 1900–??, Russian metallurgist; son of M. A. Pavlov [1863–1950] who was also a metallurgist, worked in metallurgical plants, rolling and pressing of metals, began teaching in 1928 and was in several locations before going to Inst. of Metallurgy of the USSR Academy of Sciences (1953) (PS SMOS WWWIS)

PEABODY, Raymond Avery, 1923–1978, American mechanical engineer; design engineer with Electric Boat Co. (1948–1961) that became a part of General Dynamics, organized Astra Corp. (1961ff), co-founder and president of Underseas Engineering (1966–1971), formed other companies dealing with hydrospace systems (NC61 PS)

PEARCE, (Standen) Leonard, 1873–1947, English electrical engineer; initially with Metropolitan Electric Supply Co. after apprenticeship, then joined British Thomson-Houston Co. (1899), heavily involved in power station operation and electricity supply, became consulting engineer of the Manchester Corp. (1901–1925), electricity commission in Sydney (1926ff) (DNB PS WWW)

PEARL, William Abiel, 1893–1975, American mechanical engineer; heat effect on welding, metallography, welding of high-strength steels, State College of Washington (1919–1925) and later director of Industrial Research and professor at Washington State Univ. (1945–1954), head of gas engineering and aviation at Benson Polytechnic School (1926–1935), Armour Inst. and professor at Illinois Inst. of Technology (1935–1945), administrator at Bonneville Power Administration (1954ff) (AMOS NC63 PS)

PEARSON, Fred Stark, 1861–1915, American civil engineer; faculty at Tufts College (MA) (1883–1886), served as chief engineer of Dominion Coal Co., electrification of West End Street Railroad (MA) (1889–1893), Metropolitan Street Railway Co. of NYC (1894–1899), interests turned to foreign countries (Brazil, Mexico, Canada) (1899ff), died in sinking of Lusitania (CDOAB DAB NYTObit WWIE(I) WWWIA)

PEARSON, Gerald L., 1905–1987, American electrical engineer; research at Bell Labs. (31 yr.), faculty at Stanford Univ. (1960ff), solid-state electronic devices, thermistors, solar cells, experimental work relating to the development of the transistor and its reduction to practice, co-inventor of the solar battery, 34 US patents, consultant, NAE (1968) (Bridge MT5 WWIE)

PEAUCELLIER, Charles-Nicolas, 1832–1913, French military engineer; army engineer, invented Peaucellier linkage to change circular motion into linear motion (1873), Peaucellier cell, inverse geometry, pantograph (DOSB EB MWBD OED WWWIS)

PECK, Ralph E., 1910–1982, Canadian chemical engineer; faculty in chemical engineering at Drexel Univ.

(1936–1939), at Illinois Inst. of Technology (1939–1982) from instructor to chairman of department (1950–1968), thermodynamics and heat transfer (AMOS WWIA WWIE WWWIA)

PECLET, Jean-Claude-Eugène, 1793–1857, French engineer and physicist; rate of heat transfer from water to water across a metal plate, expanded on the mechanism of heat transmission, a dimensionless group used in heat transfer by forced convection named for him (1921) (HOHT L&M WWWIS)

PECORA, William Thomas, 1913–1972, American geologist and engineer; USGS (1939–1971), studied strategic mineral deposits in South America (1948), chief of Geochemistry and Petrology (1957–1961), recognized eight new minerals, chief of Ecologic Division (1964–1965), director (1964ff), worked with NASA on development of Earth Resources Observation System (EROS) using satellite (1966), NAS (1965), participated in numerous federal and university committees and councils (BM47 HOWRD WWWIA)

PEEK, Frank William, 1881–1933, American electrical engineer; General Electric Co. (1909ff), specialist in high-voltage generation and transmission, lightning generator (1931), built 10-million-volt generator (CDOAB DAB NC24 WWWIA)

PELLINI, William S., 1917–1987, American metallurgist; research and administration of metallurgy at US Naval Research Lab. (NRL), methods of fracture-safe design, fracture-control plans used internationally, NAE (1974) (Bridge MT4 WWIE)

PELTIER, Jean-Charles-Athanase, 1785–1845, French physicist and meteorologist; current flow between two different metals and heat flow at junctions called Peltier effect (1834), electrostatic induction (1840) (DOSB L&M MWBD WWWIS)

PELTON, Lester Allen, 1829–1918, American mining engineer and inventor; designed and erected machinery in mines (1864ff), invented impulse water turbine known as Pelton wheel (1870, pat. 1880), took up mining engineering using his turbine for crushing, pumping, and hydroelectric power (FNIE MEIA MWBD NC13 PS TBDOS(Po) TWABOI)

PENDER, Harold, 1879–1959, American electrical engineer; proved existence of magnetic field around a moving charged body (1903), wrote several books relating to electrical engineering (MWBD NC48 WWWIA WWWIS)

PENNELL, Maynard L., 1910–1994, American aeronautical engineer; Navy Bureau of Aeronautics, contributed to design of DC-3, and jet transport studies leading to Boeing 707, with Boeing Co. (1940–1974), contributed to B-29 and held several executive positions, NAE (1968) (MT8 WWIE)

PENNINGTON, Mary Engle, 1872–1952, American woman chemist; specialist in food quality and preservation, director of chemistry and instrumentation at Women's Medical College (PA) (1898–1906), poultry and food cooling, refrigeration, transportation at USDA Food Research Lab. (1907–1919), independent consultant (1922ff) (MOI NWITPS NWS WCBE WIS(1)(2) WWWIA)

PERCY, John, 1817–1889, English metallurgist; discovered a process of extracting silver from ores, improved process of making Bessemer steel, wrote basic book on subject, research and teaching at Royal School of Mines (1851–1879) (CDOSB DNB MWBD)

PERIN, Charles Page, 1861–1937, American steel engineer; specialist in geology and steel, worked with Tata Iron and Steel (India) (1902ff) the first steel plant in India (CDOAB WWWIA)

PERKIN, William Henry, 1837–1907, English chemist and inventor; processing involving coal-tar derivatives, made synthetic dyes including mauve (aniline purple) (1856), red dye alizarin, established a factory to manufacture dyes, rotation of polarized light in a magnetic field (C&EN78 DNB MWBD SAI WWW)

PERKINS, Jacob, 1766–1849, American inventor; machine to cut and head nails in one operation (1790, 1795), steel check plate for printing bank notes (c.1808), horizontal steam engine (1827) operating at 1,400 psi, and later steam engine operating at 2,000 psi pressure (1836), received patent for refrigerating machine (1834), printed bank notes in England, printed first penny postage stamps (1840), Perkins Medal (BITAS C&EN74 CDOAB DAB GEAPIT GI IIA MEIA MWBD NC10 WWWIA)

PERKINS, Kendall, 1908–1987, American aeronautical and electrical engineer; corporate vice president of engineering and research (1967–1973) and (1975–1978) of McDonnell Douglas Corp. (1941–1978), aerospace technology, engineering management in design of aircraft and space vehicles, NAE (1970) (Bridge MT5 NC10 WWIE)

PERLIS, Alan J., 1922–1990, American computer scientist and mathematician; scholar and researcher in design of computer programming languages, numerical analysis, productive in definition of Algorithmic Language (ALGOL), at Massachusetts Inst. of Technology (MIT) (1948–1949), Aberdeen Proving Grounds (1951), on faculty at Purdue Univ. (IN) (1952–1956), Carnegie Inst. of Technology (PA) and director of Computation Lab. (1956ff), NAE (1977) (AMOS Bridge MT6 PS WWWIA)

PERRINE, Frederic Auten Combs, 1862–1908, American electrical engineer; faculty and administration at Stanford Univ. (1893–1900), electric power and transmission, designed first 60 kilowatt transmission line (BITAS CDOAB DAB NC19 WWWIA)

PERRONET, Jean-Rodolphe, 1708–1794, French civil engineer; director of École des Ponts (1749–1794), inspector general (1750) and head of Corps des Ponts (1763), known for stone arch bridges as at Pont de Neully (1774) and Pont de la Concorde (1791) (DOSB EIH GEAPIT MWBD PS TEIH)

PERRY, John Howard, 1895–1953, American chemical engineer; chemist for American Agricultural Chemicals Co., US Bureau of Mines (USBM) (1923–1925), research at E. I. DuPont Co. (1925ff), assisted in development of McGraw-Hill Book Co. series of chemical engineering books including Perry's Chemical Engineering Handbook (1934), sulfuric acid research, role of platinum as a catalyst (ACCE PS)

PERRY, Stuart, 1814–1890, American engineer and inventor; numerous patents including noncompression gas engine (1844), improved bank lock (1846), apparatus for stereoscopic pictures (1859), patents relating to horsepower (1862–1863), milk cooler (1864), saw mill (1865), velocipede (1869), rotary hay feeder (1879) (ANB CDOAB DAB MEIA WWWIA WWWIS)

PERRY, William Alfred, 1853–1916, American engineer and manufacturer; worked for Worthington & Co. (1885), manufacture of steam pumps, directed construction of first propeller ferry boat in harbor of NYC, served as partner, vice president, and director of Union Ferry Co. (1868–1896) (MEIA WWWIA)

PESTEL, Eduard C., 1914–1988, German mechanical engineer; mechanics, biomechanics, computer modeling, eco-sciences, elasto-mechanics, head of engineering and director of planning for Leybold K. Kim Osaka (1942–1947), faculty at Technical Univ. of Hannover (1948–1977) as professor, director, dean and rector, university reform, government official (minister for science and art), futurology, fNAE (1981) (Bridge MT7)

PETER of COLECHURCH, fl. 1175, English civil engineer; built stone bridge across Thames River (England) on artificial islands (1176–1209) (GEAPIT PS TEIH)

PETERS, Edward Dyer, 1849–1917, American military, mining and metallurgical engineer; first employed as millman and assayer then became superintendent and metallurgist of Caribou Silver mine (CO) (1869–1872), territorial assayer for southern Colorado (1872), designed, built, and operated Mount Lincoln smelting works (1872–1874), went into medicine then returned to mining (1880ff) in New Jersey, Montana, and Ontario, faculty of metallurgy and mining at Columbia Univ. (1901) and Harvard Univ. (1904) (BITAS DAB WWIA WWWIA)

PETERS, Richard, 1810–1889, German American civil engineer and agriculturalist; various jobs with different railroads as surveyor and engineer, with Georgia Railroad (1835ff) as superintendent, then formed and became director of Atlantic Steam Mills (GA) (1850ff) for flour production, formed Harden Co. (1855ff) for nursery to improve southern plants (ANB CDOAB DAB)

PETERSON, Allen M., 1922–1994, American electrical engineer; faculty at Stanford Univ. and Stanford Research Inst. (SRI) (1954–1994), technology of monitoring land, sea, and atmosphere at a distance, radar systems, digital signal processing, microprocessors, logic design, NAE (1973) (Bridge MT8)

PETERSON, Dean F., 1913–1989, American civil and agricultural engineer; faculty at Colorado State Univ. (1948–1957), faculty and dean of engineering at Utah State Univ. (1957–1973), development and efficient use of water resources for food production, numerous overseas activities, NAE (1974) (Bridge MT5 WWIE)

PETIT, Karel Louis, 1909–1995, Belgian agricultural engineer; design and construction of rural structures, Univ. of Ghent (ante 1979) (44 yr.) serving as rector (8 yr.), internationally recognized, president of CIGR (CIGRArch Res2)

PETRIE, William, 1821–1908, British electrician and inventor; magnetism and electricity (1840), electric lighting (1846–1853), invented self-regulating arc lamp (1847), electochemical processes (DNB PS WWW)

PETROV, Nikolay Pavlovich, 1836–1920, Russian mechanical engineer; taught at Nikolaevskaya Engineering Academy and at St. Petersburg Technical Inst., derived equations for motion of wheels with and without braking, determined center of gravity of a train during braking (1878), hydrodynamic theory of lubricants, Petrov law of friction (CDOSB DOSB PS)

PETT, Phineas of Chatam, 1570–1647, English shipbuilder; father of Peter Pett [1610–1670] of Chatham, shipbuilder, work led to larger and faster ships (DNB GEAPIT)

PETTIT, Joseph Mayo, 1916–1986, American electrical engineer; faculty at Univ. of California (1940–1942), in WWII worked on electronic countermeasures in Army Air Corps (1942–1945), Airborne Instruments Lab. (1945–1947), faculty and dean of engineering at Stanford Univ. (1947–1972), president of Georgia Inst. Tech (1972–1986), consultant, NAE (1967) (MT3 WWIE WWWIA)

PHILBRICK, George Arthur, 1913–??, American electrical engineer and inventor; research engineer at Foxboro Co. (MA) 1936–1942, founder and president of Philbrick Research, Inc. (MA) (1946ff), during WWII with OSRD, regulatory mechanisms, computing devices (AMOS)

PHILIP, John Robert, 1927–1999, Australian civil and agricultural engineer, soil scientist, and applied mechanics; research and management in land and water in Commonwealth Scientific and Industrial Research Organization (CSIRO) (Australia) (1961–1992) including several contractual projects, Queensland Water Supply Commission (1962), became head of agricultural physics section of Plant Industry (1959) that became the Division of Environmental Mechanics (1970) that he headed for 20 yr., Philip filtration model, irrigation, multiphase flow in porous media, groundwater hydrology, environmental mechanics, geotechnical engineering, fNAE (1995) (AMAWOS Bridge MT10 WWWIS)

PHILLIPS, Édouard, 1821–1899, French mining engineer, physicist and mathematician; metal bridges, leaf springs on railroad cars (1850), spiral spring (1861), Phillips curves (1861), watch spring (1865), École Centrale and later at École Polytechnique (1864ff) (Paris) (EE WWWIS)

PHILLIPS, Horatio Frederick, 1845–1912, English aeronautics specialist and inventor; aviation, designer of wing sections with curved upper and lower sections that provided basis of airfoil design (1884, 1891) (ITCOF MWBD PS WWWIS)

PHILLIPS, Samuel Cochran, 1921–1990, American electrical engineer; US Air Force (until 1975), director of Intercontinental Ballistic Missiles program (ICBM) (1959–1963), vice president of TRW Defense Systems Group (1975ff), leadership in development of Minuteman weapon system, leadership in development of the Apollo program (1969–1972), National Security Agency (NSA), NAE (1971) (Bridge MT5 WWIE WWWIA)

PHILON (PHILO), fl. 250–230 BC, Greek military engineer and philosopher; known for work on catapults, comprehensive treatment of mechanics, pneumatic devices, town fortification (DOSB GEAPIT WWWIS)

PHYFE (FIFE), Duncan, 1768–1854, Scottish American cabinet maker; to US (c.1783), furniture maker who developed a unique style, had workshop in NYC, many designs used mahogany wood (MWBD NC19 1000Y WWWIA)

PICARD, Charles Émile, 1856–1942, French mathematician; professor at Sorbonne (1898ff), secretary of Academie des Sciences (1917–1941), Picard theorem (1879), harmonic vibrations, successive approximation for proving existence of integrals of differential equations, fNAS (1903) (DOSB MWBD WWWIS)

PICCARD, Jean Felix, 1884–1962, Swiss American chemist, aeronautical and electrical engineer; to US (1916), twin brother of Auguste Piccard [1884–1962], known for work on stratosphere using balloons for altitude studies and records, invented bathyscape (1948), faculty at Univ. of Minnesota (1936ff), with wife Jeannette Ridlon Piccard [1895–1981] made balloon ascent to altitude of 57,564 ft. (1934), Jacques Piccard [b. 1922] son of Auguste Piccard pioneered in under water depth studies setting a record in dive (1953) (ANB CDOAB DAB IAI MWBD NC47 PS S:TLAW WWWIA)

PICCARD, Jeannette Ridlon, 1895–1981, American woman aeronautics specialist and clergy; wife of Jean Felix Piccard [1884–1962], stratosphere balloon flight (1934), housing secretary (1942–1943), ordained priest in Episcopal Church (1974ff), numerous consulting tasks including Aerospace Research Department of General Mills, Waldorf Paper Products (MN), Office of Naval Research, NASA, nurses-aid for American Red Cross (1944–1946) (PS WWWIA)

PICK, Lewis Andrew, 1890–1956, American military and civil engineer; commissioned in army during WWI and WWII in Corps of Engineers (until 1953), in charge of construction of Ledo (India) or Stilwell road from India into Burma in WWII, was Missouri River engineer (1942–1943) (1945–1949) (ANB CDOAB DAB EB NC46 WWWIA)

PICTET, Raoul-Pierre, 1846–1929, Swiss physicist; studied low-temperature reactions, liquefaction of oxygen, nitrogen, and carbon dioxide at about the same time as Louis-Paul Cailletet [1832–1913] (1877), developed compression refrigeration system (1874), professor at Geneva (1879–1886) (MWBD WWWIS)

PIDGIN, Charles Felton, 1844–1923, American statistician and inventor; invented materials and instruments for quick tabulation of statistics, author (CDOAB DAB NC13)

PIERCE, George Washington, 1872–1956, American applied physicist; faculty at Harvard Univ. (1901–1940), discovered several phenomena involving electroacoustics, radiation, piezoelectricity, electromagnetic waves, wireless radio, motion impedance, antenna, submarine detection, chief of Harvard High Tension Electrical Lab. (1914ff), 53 patents, NAS (1920) (ANB BM33 TFE&I WWWIA WWWIS)

PIERRE DE CHELLES, fl. 1300, French architectural engineer; with Jehan de Chelles [fl. 1260] responsible for enlarging and rebuilding of Notre Dame Cathedral (GEAPIT)

PIERRE DE MARICOURT, fl. 1269, French military engineer and scientist; responsible for improvement of and perhaps discovery in Europe of magnetic compass, referred to by Roger Bacon [c.1220–1292] as greatest living scientist (GEAPIT)

PIGFORD, Robert L., 1917–1988, American chemical engineer; mass transfer, distillation, solvent extraction, research engineer at DuPont (1941–1947), faculty at Univ. of Delaware (1947–1975), Univ. of California (1975–1985), returned to Univ. of Delaware (1985–1988), NAE (1971), NAS (1972) (BM65 MT4 WWIE WWWIA)

PIKARSKY, Milton, 1924–1989, American civil engineer and inventor; faculty at Illinois Inst. of Technology (1979–1984), City College of New York (1984ff), and director Inst. of Transportation Systems, developed urban transportation systems, New York Central Railroad (1944–1956), developed urban works, Commission of Chicago Public Works (1964–1973), Chicago Transit Authority (1973–1975), worked with numerous consulting groups, NAE (1973) (Bridge MT4 WWIE WWWIA)

PIKE, William Abbot, 1851–1895, American mechanical engineer; faculty at Univ. of Maine (1871–1880), professor of engineering and then dean of engineering at Univ. of Minnesota (1880–1892), consulting engineer, designed coal docks for railroad, later city water work system (BISAT MEIA NC21)

PILCHER, Percy Sinclair, 1866–1899, English aeronautical engineer; pioneer in construction and flight of heavier than air aircraft (1895ff), especially gliders from which he died as a result of a glider crash (MWBD)

PILYUGIN, Nikolai Alekseevich, 1908–1982, Russian electrical engineer; graduate of Moscow Higher Technical School (1935), automatic controls, worked in several scientific and research organizations, USSR Academy of Sciences (1960) (PSOCE SMOS TTOS)

PIPER, Arthur M., 1898–1989, American civil engineer, meteorologist and geologist; first with Idaho Bureau of Mines and Geology, then with USGS (1926–1968), worked in groundwater studies (1926–1935) after which he was in charge of groundwater studies in northwestern states (1935–1941), worked on military water supply problems in US and Pacific, after WWII worked as hydraulic engineer and hydrologist, was one of first four people identified as hydrologists by USGS (also George Ferguson [b. 1906], Henry Beckman, and Keith Jackson), studied byproducts of nuclear explosions (1961ff), after which he retired (1968) then worked part time (8 yr.) (HOWRD WWWIA)

PIPPARD, Alfred John Sutton, 1891–1969, British civil engineer; during WWI worked on aircraft structures in air department of Admiralty, did consulting in partnership with Alan Grant Ogilvie [1887–1954] (1919–1922), faculty and administration at Univ. College (1922–1928), Bristol (1928–1938) then Imperial College (1938–1956) (DNB PS RS16 WWW)

PIRELLI, Giovanni Battista, 1848–1932, Italian industrialist; established first rubber factory in Italy (1872), manufacture of electric cable (1884), automobile tires (1899), business expanded by sons (MWBD)

PISTOLKORS, Aleksandr A., 1896–??, Soviet electrical engineer; theory of antennae, radio, worked in radio laboratories (1926–1942), taught at Leningrad Electrotechnical Inst. and Leningrad Inst. of Communication Engineers (1931–1945), Moscow Inst. of Communication Engineers (1945–1950), member of USSR Academy of Sciences (1946ff) (PS SMOS WWWIS)

PITMAN, Sir Isaac, 1813–1897, English inventor; invented a shorthand method of writing called the Pitman system (1837), first introduced in Great Britain then introduced in United States by his brother Benn Pitman [1822–1910] (1852) that was the major rival of the Gregg system (John Robert Gregg [1867–1948]) which predominated in US (DNB PS SAI MWBD WWWIS)

PITOT, Henri, 1695–1771, French civil and hydraulic engineer; fluid mechanics and hydraulics, invented Pitot tube (1735) to measure velocity of flow of water, public works director included draining of swamps (1740) and superintendent of Canal Midi (France), built aqueduct (DOSB EE EIH FNIE GEAPIT MWBD MWGD PS TEIH WWWIS)

PITTS, Hiram Avery, 1800–1860, American inventor; blacksmith, invented improvements in hand chain pump (1832), improved horsepower treadmill (1832), with brother John Pitts patented chain band for power transmission (ME) (1834), grain threshers and fanning mill (1837), manufacture of threshers (1847), invented machines for harvesting hemp, corn and cob mills, established Chicago-Pitts plant (IL) to manufacture threshers (1853), invented Buffalo Pitts thresher (1877) that had an endless apron feed (CDOAB DAB MEIA NC13 TGH WWWIA)

PIXII, Hippolyte, 1808–1835, French inventor; built first practical electricity generator (1832), instrument maker (EIH GI TBDOS(Po))

PLANCK, Max Karl Ernst Ludwig, 1858–1947, German physicist; professor at Munich (1880–1885), Kiel (1885–1889), Berlin (1889–1928), radiation from black bodies (1900) called Planck law, quantum theory (1900ff), thermodynamics, conservation of energy, entropy, widely referenced in engineering literature, Nobel prize (1918), fRS (1926), fNAS (1926) (BEST IAI L&M MWBD PA RSObit SAI SAI-MP S:TLAW TGS WWWIS)

PLASKETT, John Stanley, 1865–1941, Canadian engineer and astronomer; instrument design, telescope construction, observatory construction, director of Dominion Astrophysical Observatory (BC) (1917–1935) (TBDOS(Po) MWBD WWWIS)

PLATT, Wilbur Osborne, 1860–1934, American machinist and inventor; machinist who advanced to president (1917) of Joseph Reid [1843–1917] Gas Engine Co. (PA) (1878ff), president of other companies such as Reid Land and Development Co. and Frick-Reid Supply Corp., invented an internal combustion engine (1920), horizontal power transmission wheels, vaporizers and igniters, bearings, governors, gas-purifying apparatus (MEIA)

PLESSET, Milton S., 1908–1991, American engineering scientist; faculty at California Inst. of Technology (1948–1978), worked in applied physics, two-phase flows, thermal hydraulics and safety of nuclear reactors, Rayleigh-Plesset equation on bubble dynamics, at Douglas Aircraft Co. during WWII, Univ. of Rochester (5 yr.), Univ. of California-Los Angeles (UCLA) (1977–1988), NAE (1979) (Bridge MT6 WWWIA)

PLINY the Elder (aka Gaius Plinius Secundus), 23–79, Roman military officer; military service, officer in cavalry, ship building, ship maintenance (EITAW GEAPIT PS TBDOS(Po) TGS WWWIS)

PLUNKETT, Roy Joseph, Jr., 1910–1994, American chemist and inventor; worked on Manhattan project during WWII, DuPont (1936–1975), promoted to manager of various operations within company, helped develop freons for refrigerants, discovered Teflon (1938) (ANB FAL HOCS&T IAI NYPL NYTObit WWWIA)

POE, Orlando Metcalfe, 1832–1895, American soldier and engineer; during Civil War was field officer (1861–1863), chief engineer in US Army (1864–1865) reaching rank of brigadier general, active in railroad construction and harbor improvement in Great Lakes area (1866ff) (CDOAB DAB NC6 WWWIA)

POETTMANN, Frederick H., 1919–1995, American petroleum engineer; faculty at Colorado School of Mines (1983–1995), Phillips Petroleum Co. (1946–1955), Marathon Oil Co. (1955–1983), development of optimum oil reservoir production, leader in secondary and tertiary oil recovery, NAE (1978) (Bridge MT8 WWWIA)

POINCARÉ, Jules Henri, 1854–1912, French engineer, mathematician, physicist and inventor; topology, worked on electromagnetic theory of light, electric oscillations, mechanics, celestial mechanics, three-body problems, time-keeping, philosophy of science, recognized importance of work by Albert Einstein [1879–1955] on relativity, Poincaré theorem (1890), École Polytechnique (1873–1879), faculty of Caen Univ. (1879–1881), Univ. of Paris (1881ff), fNAS (1898) (BEST DOSB DOT EB HOFL RHW MWBD WWWIS)

POISEUILLE, Jean-Louis-Marie, 1799–1869, French physician and physiologist and inventor; blood-flow fluid mechanics, flow of viscous fluids, Hagen-Poiseuille equation (1840) (German engineer Gotthilf Heinrich Ludwig Hagen [1797–1884]), used mercury manometer for pressure measurement (CEOS&T CFR DOSB FNIE HOHT L&M MWBD OED PS WWWIS)

POISSON, Simeon-Denis, 1781–1840, French mathematician and physicist; worked on several problems related to engineering, developed Poisson ratio (1830) related to elasticity of materials, Poisson distribution (1837) used in statistics and probability (BDOS DOSB DOT FINE L&M MWBD RHW WWWIS)

POLHEM, Christopher, 1661–1751, Swedish inventor; invented machines for mining, built a water-powered factory to minimize human muscle (1700), recognized the importance of division of labor for productivity, built a minting machine for England (BEST GEAPIT PS WWWIS)

POLIKARPOV, Nikolai, 1892–1944, Russian aeronautical engineer; developed first modern monoplane fighter aircraft, became director of State Aeronautical Factory No. 1, designed new fighter planes (1927ff), his low wing monoplane with fully retractable landing gear (Sikorsky-16) (1934) that became pattern for others (ITCOF PS)

PONCELET, Jean-Victor, 1788–1867, French mathematician and engineer; military engineer (1815–1825), fortifications, improved undershot waterwheel (1824) called Poncelet wheel, professor of mechanics and applied mechanics at Metz (1825–1835) and at Paris (1838–1848), founded and published on projective geometry, developed principle of duality (DOSB EE EIH MWBD TBDOS(Po) WWWIS)

POND, Irving Kane, 1857–1939, American architect and civil engineer; structural engineer, after serving with William Le Baron Jenney [1832–1907] (1879–1880) and then for Solon S. Beman (1880–1886), with brother Allen B. Pond [1858–1929] set up consulting office of Pond & Pond (1886ff), numerous buildings for housing, public buildings, and education, 50 yr. in construction business (ANB CDOAB DAB MWBD NC18 NYTObit WWWIA)

PONTE, Antonio da, 1512–c.1595, Italian architect and engineer; designed and built two famous bridges in Venice, the Rialto Bridge (1587) and the Bridge of Sighs (1589) (MWBD)

PONTON, Mungo, 1802–1880, Scottish inventor; discovered that sunlight rendered potassium dichromate insoluble that formed the basis of photoengraving (1839) (MWBD)

POPE, Franklin Leonard, 1840–1895, American electrician and inventor; development of telegraphy (1865ff), partner of Thomas A. Edison [1847–1931] (1869–1870), practical application of railroad block signal that was invented by Thomas S. Hall [1827–1880] (BITAS CDOAB NC7 WWWIA)

POPEIL, Samuel J., 1915–1984, American inventor and manufacturer; with younger brother Raymond Popeil founded and was president of Popeil Brothers (1938ff) that moved to Chicago (IL) (1941ff) and that made numerous household gadgets such as peelers and slicers and devices named Veg-O-Matic and Pocket Fisherman (ANB)

POPKOV, Valerii Ivanovich, 1908–??, Soviet electrical engineer; high-voltage technology, electric discharge in gases with high voltage, electric filters, long-distance transmission of electricity, worked at All-Union Technical Inst. of Energetics (1932–1936), Inst. of Energetics at USSR Academy of Sciences (1953ff) (PS PSOCE SMOS WWWIS)

POPOV, Aleksandr Stepanovich, 1859–1905, Russian physicist and electrical engineer; constructed device for receiving electromagnetic signals (1895), device further developed for ship-to-shore communications (1898), considered by Russians as the inventor of the radio (BDOS DOSB GI MWBD WWWIS)

POPOV, Yevgenii Pavlovich, 1914–??, Soviet electromechanical engineer; theory and practice of automatic controls, remote controls, Air Force Engineering Academy at Leningrad (1943ff) where he headed the Department of Automation and Remote Control (1949), USSR Academy of Sciences (1960ff) (PS PSOCE SMOS WWWIS)

PORSCHE, Ferdinand, 1875–1951, Bohemian engineer; automobile design and manufacturing, synchronizing of gearboxes, torsion bar suspension (TBDOS(Po))

PORTA, Giambattista Della, 1535–1615, Italian natural philosopher; cryptography, optics, telescope development, wrote 20 books (1589) in sixteenth century that furthered education leading to experimentation and application (DOSB EIH GEAPIT MMAH WWWIS)

PORTAIL—see DuPortail

PORTER, Charles Talbot, 1826–1910, American lawyer, engineer and inventor; modification of Watt governor called Porter governor (1858), steam indicator (1863), developed jet condenser for engine (1864), engine manufacturing at Hewes & Phillips (NJ) (1875ff), consulting engineer (1883) (EE MEIA NC20)

PORTER, Edwin Stanton, 1870–1941, American inventor; director-producer-inventor of cinemas (1896ff), worked with Cary & Moen Co. (NY) (1890), with Thomas A. Edison [1847–1931] (until 1909), then with several others, invented simplex projector manufactured by Precision Machine Co. of which he was president (1916–1925), International Projector Co. (1925ff) (ANB NC30)

PORTER, Holbrook Fitz-John, 1858–1933, American mechanical and industrial engineer; worked for Delameter Iron Works (NY) (1878–1882), New Jersey Steel and Iron Co. (1882–1902), Columbia Univ. (1888–1890), Cary & Moen Co. (NY) (1890–1894), Bethlehem Steel Co. (1896–1902), vice president of Nernst Lamp Co. (1902–1905), proponent of employee representation in management, invented vertical firewall, consulting engineer (1905ff) (CDOAB MEIA NCB WWWIA)

PORTER, Rufus, 1792–1884, American artist and inventor; constructed camera-obscura (1820), painter (1824), invented cord-making machine (1825), founded Scientific American magazine (1845), electroplating (1845), inventions included clock, steam carriage, wind-driven corn sheller, rotary plow, engine lathe, rotary engine, fire alarm, washing machine, revolving rifle patent sold to Samuel Colt [1814–1862] (DAB MEIA NC7 WWWIA WWWIS)

POST, Charles Williams, 1854–1914, American manufacturer; established sanitarium (1891), developed, manufactured, and distributed various health and prepared food products including coffee substitute (Postum) (1885) and others such as Grape Nuts (1897), Post Toasties, Bran Flakes, et al. (AIE MWBD NC25)

POST, Wiley, 1899–1935, American aviator; flew around the world with navigator Australian Harold Charles Gatty [1903–1957] (1931), made first solo flight around the world (1933), designed a pioneer version of pressure suit, with Will Rogers [1879–1935] killed in airplane crash in Alaska (1935) (MWBD WWWIA)

POSTEL, Jonathan B., d. 1998, age 55, American computer scientist and inventor; faculty at Univ. of California-Los Angeles (UCLA), helped create and administer Internet including address system for Internet, helped develop Arpanet (1964) that was supported by DOD that became operational (1969), developed the Internet Assigned Numbers Authority (IANA) (FAL NYTObit TTOT)

POTTER, Andrey Abraham, 1882–1979, Russian American mechanical engineer; General Electric Co. (1903–1905), faculty and dean of engineering at Kansas State Univ. (1905–1920), dean of engineering at Purdue Univ. (IN) (1920–1953), dean of engineering at Michigan State Univ. (1953–1954), author of textbooks, specialties in heat, power, tractors, and design (AE61 NUC WWIE WWWIA WWWIS)

POTTER, William Bancroft, 1863–1934, American electrical engineer and inventor; electrical equipment particularly for traction and lighting, Thomas-Houston Co. as machinist (1887–1892), General Electric Co. (1892ff), more than 130 patents (CDOAS DAB NC13 WWWIA)

POULSEN, Valdemar, 1869–1942, Danish electrical engineer and inventor; invented magnetic wire recorder (1897), invented recording telephone (1898), with Reginald Fessenden [1866–1932] invented radio telephone (1900), varying magnetic field used to reproduce sound wave, invented arc generator used in wireless telegraphy and telephony (1903) (BEST I&T16 MWBD PS WA WWWIS)

POWELL, John Wesley, 1834–1902, American naturalist, military engineer and ethnologist; served in Civil War becoming a major in which he lost his right arm, taught in Illinois (1865–1867), explorer of western rivers, led expedition (1869–1875) through Colorado and Grand Canyon along with brother Walter Henry Powell [1842–1915], climbed and named Mount Powell (CO) (elevation 13,534 ft), US Geological Survey (USGS) (1875–1894) becoming director (1881), described basis for national park system, pioneer work in linguistics and ethnological study of Indians, Smithsonian Institution (1879–1902) where he was director of Bureau of Ethnology, instituted several publication services, helped establish Cosmos Club (DC) (1878), Lake Powell (AZ, UT) (1964) behind Glen Canyon dam named after him, NAS (1880) (BM8 HOWRD MWBD MWGD NC3 SAI-MP TNAS WWWIA WWWIS)

PRAGER, William, 1903–1980, German American mechanical engineer; hardness of metals, faculty at Karlsruhe Inst. of Technology (1933), Univ. of Istanbul (1933–1941), Brown Univ. (RI) (1941–1963) (1968ff), Univ. of California-

San Diego (1965–1968), consultant, NAE (1965), NAS (1968) (ANB L&M MT2 WWWIA WWWIS)

PRANDTL, Ludwig, 1875–1953, German physicist and aerodynamicist; professor at Göttingen (1904–1953) including director following Max Karl Ernst Planck [1858–1947] of Max Planck Inst. for Fluid Mechanics (1925ff), boundary layer (1904), studies on wing theory and design (1918–1919), supersonic flow, turbulence, wind tunnel design, Prandtl number (1910, 1922) used in heat transfer for problems involving heating and cooling by convection, theory of plasticity, meteorology (BDOS EB FNIE L&M MWBD PS TBDOS(Po) WWWIS)

PRATT, Francis Ashbury, 1827–1902, American inventor; with Amos Whitney [1832–1928] formed Pratt & Whitney Co. (1865) of which he was president (1865–1898), manufacture of machine tools, invented metal planing machine (1869), gear cutter (1884), milling machine (1885), consulting engineer (ANB CDOAB DAB MEIA MWBD NC27)

PRATT, John, 1831–1900, American inventor; in England patented writing machine called ptereotype (1866), in US patented typewriters (1868)(1882), sold second typewriter to Hammond Co. (NY) (DAB MEIA)

PRATT, Nat W., 1852–1896, American mechanical engineer and inventor; with Babcock and Wilcox Co. (1871ff) becoming president (1893), 26 patents relating to steam boilers, 3 patents relating to ordnance, consulting engineer for Dynamite Gun Co. (MEIA)

PRATT, Perry W., 1914–1981, American aeronautical and mechanical engineer; test engineer of fighter and bomber engines, gas turbine engine development, jet propulsion for Pratt and Whitney Aircraft (30 yr.) becoming vice president and chief scientist (1958–1971), NAE (1967) (MT2 WWWIA)

PRATT, Thomas Willis, 1812–1875, American civil engineer and inventor; invented bridge and roof truss called Pratt truss (1844) (CDOAB DAB MWBD NC22 WWWIA)

PREECE, Sir William Henry, 1834–1913, British electrical engineer; engineer in post office telegraphic system (1870–1899), contributed to railway signaling and wireless telegraphy, worked with Guglielmo Marconi [1874–1937], introduced first telephones in Great Britain (DNB MWBD WWW WWWIS)

PREGEL, Boris, 1893–1976, Russian American civil, electrical and mechanical engineer; involved in mining and mineral industry and related products in France and US, president of Canadian Radium and Uranium Corp. (NYC) (1940–1960), chairman of the board of Conrad Precision Industries (1960–1974), consulting engineer, numerous national and international activities (NC59 PS WWIA WWWIA)

PRESCOTT, George Bartlett, 1830–1894, American electrical engineer; chief electrician for Western Union (1862–1882), known for writings on telegraph, telephone and electrical subjects (CDOAB DAB NC5 WWWIA)

PRIEST, Edward Dwight, 1861–1931, American electrical engineer; specialist in railroad electrification, designed motors for use on elevated railroads (CDOAB DAB)

PRIESTLEY, Joseph, 1733–1804, English clergyman, chemist and inventor; discovered oxygen (1774), decomposition of ammonia by electricity (1781), studies on electricity, invented soda water, lived last 10 years of his life in the US (ASAI BEST BITAS C&EN78 DAB DNB IAI IIA MWBD NC6 SAI S:TLAW TEIH WBE WWWIA WWWIS)

PRIESTMAN, William Dent, 1847–1936, British inventor; developed heavy oil engine called Priestman engine (1886) (EE)

PRINCE, Frederick Henry, 1859–1953, American investment banker; sponsored plan to consolidate all major trunk railroad lines in US, concentrated on meat packing becoming chairman of executive committee of Armour and Co. (1934) (CDOAB NC45 WWWIA)

PRINDLE, Karl E., d. 1998, age 95, American chemist and inventor; DuPont Cellophane Co. (NY), produced marketable cellophane, developed a coating for cellophane to increase resistance to moisture, DuPont got license to manufacture (1923), employed by Dobeckmun Co. that was purchased by Dow Chemical Co. (1957), also invented Lurex—a synthetic thread (FAL NYTObit)

PRITCHARD, Donald William, 1922–1999, American meteorologist, oceanographer and coastal engineer; hydrodynamics of estuaries and coastal waters, storm induced waves, innovative applications of benefits to the environment and to society, weather officer with US Army Air Corps (1942–1946), Chesapeake Bay Inst. at Johns Hopkins Univ. (1949–1978) including chair of oceanography (20 yr.), administrator of Marine Science Research Inst. at State Univ. of New York at Stonybrook (1978ff), NAE (1993) (Bridge MT9)

PRITCHARD, Wilbur L., 1923–1999, American electrical engineer; contributed to microwave and satellite technology as applied to communication and direct broadcast, with Philco Corp. (1943–1946), Raytheon Co. (1946–1962), vice president COMSAT (1967–1973), president of Satellite System Engineering (1974–1987), president of W. L. Pritchard & Co., NAE (1995) (Bridge MT9 WWIE WWWIA)

PRITIKIN, Nathan, 1915–1985, American inventor and nutritionist; during WWII invented device for Norden bombsight, invented device for weather forecasting, did projects for General Electric Co. and Bendix Co., sold his electronics company and 36 patents, as a self-taught nutritionist he led in promoting diets low in fat, salt, and cholesterol that influenced the American diet (ANB NYTObit WWWIA)

PROCTER, William Cooper, 1862–1934, American manufacturer; son of William Alexander Procter [1817–1874] who with William Gamble founded Procter & Gamble firm (1837) later incorporated as Procter & Gamble Co. (1890), James N. Gamble [1836–1932] was his brother-in-law and helped build the company, company was founded to use slaughterhouse wastes to produce oil for lamps, soaps, and a variety of products such as Ivory soap (1879), Crisco, Oxydol, Dreft, and a variety of cotton seed oil products, William Cooper Procter joined P&G (1883) and succeeded his father William Alexander Procter as president (1907ff) then was chairman of the board (1930ff), company recognized for developing profit-sharing benefits with employees (ANB NC25,40 NYTObit TIA WWWIA)

PRONY, Gaspard-Clair-François-Marie-Riche de, 1755–1839, French mathematician, engineer and inventor; director of survey (1791), inspector of roads and bridges (1805–1839), absorption dynamometer (1821) called Prony brake, improvement of roads and drainage (CFR DOSB EE GEAPIT NYPL PS MWBD WWWIS)

PRUTTON, Carl Frederick, 1898–1970, American chemical engineer and inventor; faculty at Case Western Reserve Univ. (OH) (1920–1948), including head of chemical engineering (1936–1948), vice president of Mathieson Chemical Co. (1949–1953), Food Machinery Corp. (FMC) (1954–1960) including vice president (1956ff), numerous patents, consultant, NAE (1966) (MT1 NC55 PS WWIE WWWIA)

PTOLEMY of Alexandria, Claudius (aka Claudius Ptolemaeus), c.90–170 AD, Egyptian Greek astronomer and geographer; espoused the theory that the sun, planets and stars revolved around the earth (Ptolemaic system), wrote on geometry, optical phenomena, music, and maps (CDOS MWBD RHW SAI-MP TBDOS(Po) TEIH TGS WWWIS)

PULLMAN, George Mortimer, 1831–1897, American inventor; cabinet maker (1848–1855), contractor (1855–1859), storekeeper (1859–1963), with associates designed railroad car with folding upper berth (1864), and lower berth (1865), organized Pullman Palace Car Co. (IL) (1867) and devised dining cars (1868), chair cars (1875), and vestibule cars (1887) (AIE ASAI BDOTHOT CDOAB DAB MWBD NC11 NYPL WA WWWIA WWWIS)

PUPIN, Michael Idvorsky, 1858–1935, Austrian American physicist and inventor; professor of electromechanics at Columbia Univ. (NYC) (1901–1931), invented multiplex telegraphy (1894), the Pupin coil, developed a means of short-exposure X-ray photography, inductance line for long-distance telephony, during WWI worked on detection of submarines, 34 patents, NAS (1905) (AIE ANB BM19 MWBD NC26 TEE WWWIA WWWIS)

PURDY, Corydon Tyler, 1859–1944, American civil engineer; structural engineer, steel framing for high buildings, structural design of large and complex buildings, in partnership with Lightner Henderson [1866–1916] (1893–1916), then chairman of the board of company that followed the partnership (ANB CDOAB DAB NC12,C WWWIA)

PURVIS, William B., 19th century, Black American inventor; invented hand stamp (1883), invented a fountain pen with an internal container for ink (1890) (AAI)

PUTT, Donald L., 1905–1988, American aerospace engineer; Army Air Corps (1928–1958), Air Materiel Command, director or research and development (1948–1952), vice commander then commander of the Air Research and Development Command (1952–1954), deputy chief of development at headquarters (1954–1958) (IWAP)

PYE, David Randall, 1886–1960, British mechanical engineer; Oxford Univ. (1909ff), at Air Ministry (1925), provost of Univ. College (1943–1951), aeronautics, engines, heat and energy (DNB PS RS7 WWW WWWIS)

PYTHAGORAS of Samos, c.569–500 BC, Greek philosopher, astronomer, and mathematician; study of sound particularly from strings, relationship of sides of right triangle, Pythagorean theorem (5th century BC) (DOSB DOT EB FNIE HOFL MWBD PS SAI-MB TGS WWWIS)

PYTHEOS (aka PYTHIUS or PYTHIOS), 4th century BC, Greek architect; designed and organized schools, built temple and Mausoleum of Halicarnassus, which is one of wonders of the ancient world (EIH HOEIC MWBD)

Q

QUARLES, Donald Aubrey, 1894–1959, American electrical engineer; served in WWI discharged as a captain, research specialty in electricity transmission engineering with Western Electric Co. (1919–1952) that became a part of Bell Telephone Labs., president of Sandia Corp. (1952–1955), entered government service (1953ff) in US Department of Defense (DOD) (NC62 PS WWWIA)

QUINBY, Isaac Ferdinand, 1821–1891, American natural philosopher; army officer, mathematics professor, instructor US Military Academy (1845–1847), natural and experimental philosophy at Univ. of Rochester (NY) (1951–1861), in Civil War (1861–1863), US marshall (NY) (1869), surveyor of Rochester (1885–1889), official at Soldiers Home (NY) (1879–1886) (BITAS DAB WWWIA)

R

RABI, Isidor Isaac, 1898–1988, Polish Austrian American physicist and inventor; to US (1899), faculty at several universities, primarily at Columbia Univ. (NYC) (1929–1964), magnetic resonance imaging (MRI) (1931), microwave, radar, NAS (1940), Nobel prize (1944), first person to use MRI on living tissue was Raymond V. Damadian [b. 1916], further work on MRI was by Swiss Richard R. Ernst [b. 1933] for which he received the Nobel prize (1991) (BM62 CDOS MWBD NCH PA RHW TPA WWWIA WWWIS)

RABINOVICH, Isaak Moiseevich, 1866–??, Soviet structural engineer; kinetic method of structural mechanics, methods of statically indeterminate structures, taught at a number of universities and technical colleges in Moscow, was professor at Military Engineering Academy (1932ff) and at the Engineering Construction Inst. (1933ff), USSR Academy of Sciences (1946ff) (PA PS SMOS)

RABINOW (RABINOVICH), Jacob, 1910–1999, Russian American electrical engineer and inventor; to US (1921), invention and development of devices in computers, power transmission, post office automation, self-regulation of clocks, consultant to US National Bureau of Standards (NBS) (1938–1954), president of Rabinow Engineering Co. (1954–1964), vice president of Control Data Corp. (1964–1972), 226 patents, NAE (1976) (Bridge TEOUT WWIE)

RAFTER, George W., 1851–1907, American civil engineer; began engineering career (1873) as surveyor, engineer with Ft. Worth (TX) water works (1876–1883), Rochester (NY) (1883–1890) water works, New York state (1893ff) water supply and control, sewage disposal (BITAS CDOAB DAB NC12)

RAHIYA, Michael, Jr., 1914–1973, American chemical engineer; worked on purification of water with Joseph E. Seagram & Sons (1937–1944), Diehl Pump and Supply Co. and consultant, formed his own company that specialized in water treatment, tank fabrication and erection and boiler repair serving as president (until 1973), worked for breweries and distillers and developed Dri-Molass—a dry molasses used for animal feed (NC58 PS)

RAINS, Gabriel James, 1803–1881, American military engineer; graduate US Military Academy (1827), served in Seminole war and war with Mexico, commissioned in Confederate Army (1861) becoming brigadier general, expert in explosives and land mines (ANB NC4 WWWIA)

RAINS, George Washington, 1817–1898, American military engineer and inventor; brother of Gabriel J. Rains [1803–1881], graduate of US Military Academy (1842), in war with Mexico, in US Army (until 1856), was president of Washington Iron Works (NY) (1856) and Highland Iron Works (NY), commissioned as a major in Confederate Army in charge of gunpowder and explosives, after war served as professor and dean of Medical College of Georgia (until 1894) (ANB CDOAB DAB WWWIA)

RAJCHMAN, Jan A., 1911–1989, Swiss American electrical engineer; computers and computer memories, computer logic devices, digital random access, read only memories, optical devices, electronic imaging for medical purposes, vice president of Information Sciences, RCA Labs., 116 patents, NAE (1966) (Bridge MT5 WWIE WWWIA WWWIS)

RAMELLI, Agostino, 1530–1608, Italian military engineer; instruments and mechanisms, made a variety of engines of war, served in French military, invented rotary pump with slotted cylinder with four wings eccentric to cylinder with shell (1588) (EE GEAPIT TEIH PS)

RAMEY, Henry J., Jr., 1925–1993, American chemical and petroleum engineer; Mobil Oil Co. (1952–1963), Texas A & M Univ. (1963–1966), faculty at Stanford Univ. (CA) (1966ff), transient well testing, enhanced recovery of crude oil, transient flow in porous materials, natural gas production, application of reservoir principles to geothermal energy recovery, NAE (1981) (Bridge MT7 WWIE WWWIA)

RAMÓN Y CAJAL, Santiago, 1852–1934, Spanish physiologist; developed theories on action of cells and brain, discovered laws on structure and action of nerve cells, brain and spinal cord (1889), staining methods, structure of retina of eye, on faculties at Valencia (1884–1887), Barcelona (1887–1892), and Madrid (1892–1922), Cajal Inst. in Madrid founded in his honor (1921), Nobel prize (1906), fNAS (1920) (BEST CDOS MWBD RHW WWWIS)

RAMSAY, Erskine, 1864–1953, American mining engineer and inventor; with father invented numerous devices for coal mining, pioneer in use of coal byproducts, founder of Pratt Consolidated Coal Co. (1904) and Alabama By-Products Corp. (1924), over 40 patents (CDOAB DAB WWWIA)

RAMSDEN, Jesse, 1735–1800, English instrument maker and inventor; astronomical instruments, opened shop (1762), patents for improvements to the sextant, theodolite, barometer, micrometer, devised mural circle (DNB GEAPIT MWBD WWWIS)

RAND, James Henry, 1859–1944, American inventor and business man; devised system of business folders, dividers, and tabs (c. 1880), established firm of Rand Ledger Co. succeeded by son James Henry Rand, Jr. [1886–1968] (1925ff), later merged to form Remington Rand Co. (1927) (CDOAB DAB NC54 WWWIA)

RAND, William Bradford Witchell, 1902–1988, American geologist and equipment development; offshore geological exploration, Shell Oil Co., Union Oil Co. (1946ff), formed Submarex Co. to do offshore exploration and drilling, first to do rotary hole drilling in Pacific Ocean (1951) and Atlantic Ocean (1958), consulting, NAE (1973) (MT5)

RANDOLPH, Isham, 1848–1920, American civil engineer; consultant in railroads, canals, and land reclamation, supervised construction of Chicago drainage canal as chief engineer of Chicago Sanitary District (1893–1907) (BITAS CDOAB DAB NC19 WWWIA)

RANDOLPH, Lingan Strother, 1859–1922, American mechanical engineer; test engineer for the Lake Erie and Western Railroad, superintendent of Florida Railway and Navigation Co. (1885), Baltimore & Ohio Railroad (1890), Baltimore Refining Co. (1892), invented tank for electrolytic separation of metals, professor and later also dean of engineering at Virginia Polytechnical Inst. (1893ff), designed heat and light plants, consulting engineer (MEIA NC19 WWWIA)

RANKINE, William John Macquorn, 1820–1872, Scottish civil engineer and physicist; professor of civil engineering (Glasgow) (1855–1872), metal fatigue of railroad axles, steam engine theory, Rankine cycle (1859), pressure and stability of retaining walls known as Rankine law (c. 1860), Rankine-Hugoniot law (Pierre Henry Hugoniot [1851–1887]) (BEST DNB FNIE L&M MWBD PS TBDOS(Po) TEIH WWWIS)

RANNIE, William Duncan, 1914–1988, Canadian American mechanical engineer; to US (1938), Northrop Aircraft Co. (1941–1946), faculty at California Inst. of Technology (1946ff), jet propulsion, three dimensional flow, stall, distortion in turbomachinery, turbulent heat transfer, NAE (1979) (Bridge MT4 WWWIA)

RANSOME, Ernest L., 1844–1917, English American civil engineer; to US (1870), associated with the Pacific Stone Co., pioneered in reinforcement, mixing equipment and construction systems, formed several companies with his name such as Ransome Construction Co. (BDOACE)

RANSOME, Robert, 1753–1830, English inventor; self-sharpening cast iron plow (1785), built all-metal plow made of interchangeable parts (1808), built gas supply for town in England (1818) (BDOTHOT DNB WA)

RATEAU, Camille Edmond Auguste, 1863–1930, French civil engineer and inventor; teacher and designer and manufacturer of turbine engines, invented the pressure-stage impulse turbine, reported to be the first to apply turbochargers to aircraft engines (1916ff) (CDOSB DOSB MWBD PS WWWIS)

RAUTENSTRAUCH, Walter, 1880–1951, American industrial engineer; taught at Columbia Univ. (1906–1945) formed the first department of industrial engineering in the US, disciple of Frederick W. Taylor [1856–1915] on scientific management theories and practices (ANB CDOAB DAB WWWIA)

RAYLEIGH, Lord (aka John William Strutt), 1842–1919, English physicist; director of Cavendish Lab. at Cambridge Univ. (1879–1884), Royal Inst. (1887–1905), wave motion, wavelengths in black body radiation, sound waves, earthquake waves, help in determining absolute measurements in electricity and magnetism, FRS (1904), chancellor of Cambridge Univ. (1908–1919), derived equation for bubble growth (1917), several physical laws named after him, fNAS (1898), Nobel prize (1904) (BEST DOSB FNIE L&M MWBD PS RSObit WWWIS)

RAYMOND, Arthur E., 1899–1999, American aeronautical engineer; aircraft designs, worked on design and development of Douglas Aircraft DC2, DC3 and DC8, helped Douglas Aircraft Co. (1925–1960) move from the age of propellers to jet power, retired (1960) as vice president of Douglas Aircraft Co., Aerospace Corp. (1960–1971), consultant, NAS (1950), NAE (1964) (Bridge WWIE WWWIA)

RAYMOND, Charles Walker, 1842–1913, American military engineer; graduate of US Military Academy (1865) and served in military service and retired as brigadier general (1904), surveyed and made map of Yukon River (AK and NW Canada) (1860), supervised river and harbor improvements, chair of board of engineers for constructing the Pennsylvania Railroad (PRR) tunnels under the Hudson River (CDOAB DAB MWGD WWWIA)

RAYMOND, Rossiter Worthington, 1840–1918, American mining engineer; brother of Charles W. Raymond [1842–1913], US Commission of Mining Statistics (1868–1876), associated with Cooper, Hewitt & Co., editor of American Journal of Mining (BISAT CDOAB DAB NC8 WWWIA)

REA, Samuel, 1855–1929, American civil engineer; expert in railroad construction, president of Pennsylvania Railroad (PRR) (1913–1925) after serving as chainman then assistant engineer (1875) (1879ff), and a brief period with Pittsburgh & Lake Erie Railroad (ANB CDOAB DAB NC15 WWWIA)

READ, Nathan, 1759–1849, American inventor; developed light steam boiler, developed double-acting steam engine, constructed paddle wheel boats (1789), Salem Iron Factory (MA) (1796ff) where iron cables, anchors, ship materials manufactured, patented nail cutting and heading machine (1798), farmer (MA) (1795) (ME) (1807ff), served in government (CDOAB DAB GEAPIT MEIA NC4 WWWIA WWWIS)

RÉAUMUR, René-Antoine-Ferchault de, 1683–1757, French physicist and inventor; prepared an opaque white glass called Réaumur porcelain, worked on new methods of steel manufacturing, thermometer with a scale of 80 divisions between freezing and boiling (1730), explained digestion as a chemical process (BEST GEAPIT WA WWWIS)

REBER, Louis Ehrhart, 1858–1948, American mechanical engineer; established department of mechanical arts (1884) became professor (1887) then dean at

Pennsylvania State Univ., Pennsylvania Commission (1889ff), director of extension at Univ. of Wisconsin (1907ff), during WWI served in various educational and personnel areas (MEIA NC37 WWWIA)

REBOUCAS, André, 1838–1898, Black Brazilian engineer and inventor; served as an officer in engineer corps (1864–1870), invented submersible exploding device called torpedo for destroying enemy ships, taught at Polytechnical School in Rio de Janeiro (1870ff) (BI

REDFIELD, Casper Lavater, 1853–1943, American mechanical engineer and inventor; machinist, draftsman and designer for National Machinery Co. (OH) of nut- and bolt-making machines, automatic steam engines and power plants, sawmill machinery, prolific writer on many subjects, researcher, approximately 60 patents (NCC PS WWWIA)

REDINGTON, Rowland Wells, 1924–1995, American mechanical engineer; research at General Electric Co. (1951–1989), imaging technology, use for medical purposes such as use of CAT scanner, helped develop use of magnetic resonance, NAE (1986) (MT8 WWWIA)

REECH, Ferdinand, 1805–1884, French marine engineer; director of École du Génie (1831ff), noted for formulation of hydraulics-model law of gravitational similitude (CDOSB DOSB PS)

REED, Chester O., 1885–1940, American agricultural engineer; field machinery, recognized as an outstanding teacher, faculty at Univ. of Illinois, representative and advertising manager of Samson Tractor Co. (a division of GMC) then became head of executive council (1920), faculty at Ohio State Univ. (1922ff) (AE21)

REES, Eberhard F. M., 1908–1998, German American engineer; design and development of rocket engines, guided missiles, launch vehicles for manned spacecraft, worked in rocketry in US (1945ff), US Army Ballistic Missile Agency (1956–1960), director of NASA Marshall Space Flight Center (1970–1973), NAE (1973) (Bridge MT9 WWIE)

REES, James, 1821–1889, Welsh American inventor; founder of engine and boat works in Pittsburgh, popularized the stern-wheeler and made several improvements (CDOAB DAB NC23 WWWIA)

REESE, Abram, 1829–1908, Welsh American inventor; Cambria Iron Works (PA) (1854ff), positions with numerous iron works (PA, MO), numerous patents including streetcar rail-rolling machines (1859), rivet and bolt machine (1860), horseshoe-making machine (1867), pinch bar (1871), universal rolling mill (1892), at least 15 patents (ANB CDOAB DAB MEIA)

REESE, Isaac, 1821–1908, Welsh American inventor and manufacturer; brick manufacturer in Pittsburgh (PA), produced Reese Silica Brick (1892), which was resistant to effect of high temperatures (CDOAB DAB)

REESE, Jacob, 1825–1907, Welsh American metallurgist and inventor; brother of Abram Reese [1829–1908] and Isaac Reese [1821–1908] produced several inventions related to manufacture of iron and steel, perfected an open hearth process (1881) (BITAS CDOAB DAB)

REICHENBACH, Georg Friedrich von, 1772–1826, German instrument maker; devised machine for making precision instruments (1796), telescope combining transit and the mural circle, invented rifled cannon, invented a water pressure pump, engaged in building hydraulic machinery (1820ff) (GEAPIT MWBD PS WWWIS)

REID, David Boswell, 1805–1863, Scottish American engineer; ventilation engineer, public health and safety, became physician in Scotland, to US (1855), Univ. of Wisconsin (1859ff), inspector of military hospitals (1863) (CDOAB DAB WWWIA)

REID, Joseph, 1843–1917, Scottish American inventor and manufacturer; machinist, first in Scotland then to Canada (1863) then to US, worked for Baldwin Locomotive Works (PA), then Atlantic and Great Western Railroad (PA) (1876), and W. J. Innes & Co. (1877), established own shop that specialized in refining supplies, designed and patented oil burners, formed Reid Burner Co. (1885), built natural gas engine (1894), formed other companies with his name (MEIA)

REISSNER, Eric, 1913–1996, German American civil engineer; to US (1936), faculty at Massachusetts Inst. of Technology (1936–1969), with industry and government in applied structural mechanics, particular contributions to determining the strength and stability of aircraft structures, plates, shells, unsteady thermodynamics, Univ. of California-San Diego (1969ff), consultant, NAE (1976) (Bridge MT9 WWIE WWWIA)

REMINGTON, Eliphalet, 1793–1861, American manufacturer; firearms manufacturer, established factory (1828), first made rifles, made Remington pistol (1847ff), expanded to make agricultural implements (1856ff), with son Philo Remington [1816–1889], who took over business of E. Remington & Sons and expanded to make sewing machines (1870ff) and typewriters (1873), units later sold to pay debts (ASAI MWBD NC9 PS WWWIA WWWIS)

REMINGTON, Philo, 1816–1889, American mechanical engineer; son of Eliphalet Remington [1793–1861], breech-loading rifle, sewing machines (1870ff), and typewriters (1873), became president of Remington & Sons (1861–1889) (IIA MEIA MWBD NC9 TBDOS(Po) WWWIA WWWIS)

RENARD, Charles, 1847–1905, French military engineer; with Arthur Krebs [1847–1935], built first dirigible La France which flew an 8-kilometer circle (1884) (MWBD WWWIS)

RENAULT, Louis, 1877–1944, French industrialist; founded automobile manufacturing company Renault Frères (1899) (MWBD WWWIS)

RENNIE, John, 1761–1821, Scottish civil engineer; drained fens, constructed and improved harbors, designed several bridges, designed docks and breakwater, two sons were civil engineers—George Rennie [1791–1866] and Sir John Rennie [1794–1874]—who carried on many of their father's activities (DNB GEAPIT LOTE MWBD PS SFLOE TEIH WWWIS)

RENWICK, Edward Sabine, 1823–1912, American mechanical engineer and inventor; superintendent of Wyoming Iron Works (PA) (1845), erected blast furnace for manufacture of pig iron, with lawyer Peter H. Watson (DC) patented self-binding reaper (1851), patented valve for steam engine (1856), ship drive (1868), incubators, chicken brooders and raised chickens as a business (1877–1886), consulting engineer and patent expert (1855ff), approximately 20 patents (ANB BITAS CB CDOAB DAB MEIA NC11 WWWIA)

RENWICK, Henry Brevoort, 1817–1895, American civil engineer; with US government (1837–1840), US Boundary Commission (1840–1847), examiner in US Patent Office (1847–1852), inspector of steam vessels in NYC (1853–1870) (BDOACE CDOAB DAB NC11 WWWIA)

RENWICK, James, Sr., 1792–1863, English American engineer; professor of natural philosophy at Columbia Univ. (NYC) (1820–1853), consultant for Morris canal, commissioner for Northeast boundary (1840) (BITAS BDOAS(El) CDOAB DAB EIH NC11 PS WWWIA WWWIS)

REPSOLD, Adolf, 1806–1871, German instrument maker; with brother George Repsold had a family instrument business called A. and G. Repsold, made small transits for Fredrich Wilhelm Bessel [1784–1846], and an instrument for Edinburgh observatory (1831), made lamp system for lighthouses (1833), made meridian circle for several clients, particularly related to astronomical instruments, his sons followed him in the business (DOSB PS)

REPSOLD, Johann Georg, 1770–1830, German instrument maker; designer and builder of astronomical instruments, established shop in Hamburg (c. 1860), constructed portable transit instruments, made a meridian circle (1818), made mountings for telescopes (CDOSB DOSB GEAPIT PS WWWIS)

REQUA, Mark Lawrence, 1865–1937, American mining engineer; worked on Central Pacific Railroad (1880s), president of Eureka and Palisade Railroad (1898–1906), member of mining firm of Requa, Bradley and Mackenzie, worked with Herbert C. Hoover [1874–1964] (1905ff), active in politics and government (CDOAB DAB MWBD NC27 WWWIA

RESNIK, Judith A., 1949–1986, American electrical engineer and astronaut; Radio Corp. of America (RCA) (1970–1974), biomedical engineer at National Institutes of Health (NIH) (1974–1977), systems engineer with Xerox Corporation (1977–1978), astronaut with NASA (1978–1986), second US woman to fly in space (1984), died in explosion of Challenger space mission (1986) (AWIS)

RESSEL, Josef Ludwig Franz, 1793–1857, Austrian inventor; perhaps the first to develop a screw propeller for ships (1829) (MWBD)

REULEAUX, Franz, 1829–1905, German mechanical engineer; considered founder of modern kinetics, strength of materials, machine design, professor at Swiss Federal Polytechical Inst. (1856–1864), Gewerbe Inst. in Berlin (1864–1896) serving as professor mechanical engineering, rector of Technische Hochschule at Charlottenburg, prolific writer covering broad spectrum of topics (CDOSB DOSB PS)

REYNOLDS, Edwin, 1831–1909, American mechanical engineer and inventor; machinist and superintendent at several firms, Corliss Steam Engine Co. (RI) (1861), Edward P. Allis Co. (WI) (1877), developed Reynolds-Corliss Engine, W. J. Chalmers Co. (WI) (1891), patented several devices for use on steam engine, received over 40 patents, consulting engineer (1905ff) (BITAS CDOAB DAB MEIA NC2 WWWIA)

REYNOLDS, Osborne, 1842–1912, British engineer and physicist; research in condensation and heat transfer between solids and liquids, pumps and turbines, wave and tidal motions in rivers, mechanical equivalent of heat, multistage turbine pump (1875), nondimensional representation of resistance to flow in channels called Reynolds number (1879) (1883), efficiency of screw propellers, theory of lubrication (1886), turbulence (1889), Reynolds-Froude brake (1897) (William Froude [1810–1879]) (CFR DOSB EE FNIE L&M MWBD PS TBDOS(Po) WWWIS)

REYNOLDS, Samuel Godfrey, 1801–1881, American inventor; tanner, developed and patented improved machinery for making nails, pins, spikes (CDOAB DAB WWWIA WWWIS)

RIBAUCOUR, Albert, 1845–1893, French engineer and mathematician; naval engineer at Rochefort Naval base (1870–1878), Aix-en-Provence (1878–1886), after 1886 was chief engineer in Algeria (1886–1893) where he worked on railroads, harbors, construction of bridges and canals, his work in mathematics done in spare time and dealt with congruencies of circles and spheres (DOSB PS)

RIBEIRO SANTOS, Carlos, 1813–1882, Portuguese engineer and geologist; Army military engineer (1834ff) promoted to commissioned officer (1837) and later promoted to general, advised on geology and political matters, began research (1840) and became director of coal mines (1852), organized the national mining service, prepared plan for supplying Lisbon with water (1854), founded commission of geological works (1857), secretary of council of mines (1859), superintendent of copper mines (1874), active in political affairs (DOSB PS)

RICARDO, Harry Ralph, 1885–1974, English mechanical engineer; chairman and technical director of Ricardo and Co., Ltd., consulting engineers (1919–1964), Rendel, Palmer and Tritton (1906–1915), Air Ministry (1916–1918) doing research in high-speed internal combustion engines, developed two-cycle gasoline engine called Dolphin, established toluene number to rate detonation of fuels later replaced by octane number, worked with engineers Oliver Thornycroft [1885–1951] and David Randall Pye [1886–1966], engaged in dealing with locomotives, steam plant, hydraulics and pneumatic equipment, FRS (1929) (DNB IAI PS RS22 TBDOS(Po) WWW WWWIS)

RICCI, Ostilio, 1540–1603, Italian military engineer and mathematician; gave instruction to Galileo (c. 1583), studied flow of streams in vicinity of Bologna (c. 1590), taught at Florence Academy of Design (1593ff), directed construction of fortifications (1597) (DOSB PS)

RICE, Calvin Winsor, 1868–1934, American electrical and mechanical engineer; with Thomson-Houston Electric Co. continuing with its successor the General Electric Co. (1890–1895), after position with mining, telegraph and meter and testing, became secretary of ASME (1906–1934), leader in establishing the Engineering Societies Building in NYC (CDOAB DAB NC25,A NYPL WWWIA)

RICE, Charles De Los, 1859–1939, American mechanical engineer and inventor; shop worker, machine construction, tool making, worked for several firms such as Caligraph Typewriter Co. (PA) (1885), Yost Writing Machine Co. (CT) (1888), Pope Manufacturing Co. (NJ) (1890), Underwood Typewriter Co. (NJ) (1901), patented devices for typewriter, machine tools, bearings, belt gearing and pneumatic wheel, received over 75 patents (MEIA NC35)

RICE, Edwin Wilbur, Jr., 1862–1935, American electrical engineer; first with the American Electric Co. that was taken over by Thomson-Houston Co. that merged with General Electric Co. (1893), served as technical director of General Electric Co. and as its president (1913–1922), over 100 patents (ANB CDOAB DAB NC3 PS WWWIA)

RICE, George Samuel, 1866–1950, American mining engineer; private engineering practice, mine safety (1908), chief mining engineer of US Bureau of Mines (1910–1937), recognized as world leader in research and promotion of safe mining practice (CDOAB DAB NC38 WWWIA)

RICE, George Staples, 1849–1920, American civil engineer; worked on construction of water works and reservoirs (1870–1880), mining operations in Oregon and Colorado (1880–1887), returned to East Coast where he worked with water systems (1887–1891), private practice in engineering with George W. Evans [1876–1951]

formed Rice and Evans (1891–1900)(1910–1914), worked on transportation and subway construction (1914–1920) (BDOACE BITAS NC12 WWWIA)

RICE, Richard Henry, 1863–1922, American engineer and inventor; considerable experience as engine and machine designer, General Electric Co. (1903ff) where he directed work on development of steam turbine, designed first turbo-blower for blast furnace in US (CDOAB DAB NC44 WWWIA)

RICE, Stephen O., 1907–1986, American physicist and electrical engineer; Bell Telephone Labs. (1930–1972) where he worked on modulation theory, radio wave propagation, noise and channel interference, NAE (1977) (MT4 WWIE WWWIA)

RICH, Ben R., 1925–1995, American aeronautical and mechanical engineer; unique aeronautical and thermodynamic concepts in aircraft design, development of classified high-performance military aircraft systems including U-2 plane and Stealth fighter, headed research at Lockheed (1950ff) and at Skunk Works (1975–1991), NAE (1981) (Bridge MT8 WWWIA)

RICHARD of Wallingford, c.1292–1336, English mechanical engineer and mathematician and inventor; invented clock that showed canonical hours (six times per day), invented a mechanical planetarium (Albion), work involved power to a gear train (GEAPIT WWWIS)

RICHARDS, Charles Brinckerhoff, 1833–1919, American mechanical engineer and inventor; before and during Civil War worked for Colt's Armory (CT), patented exhaust valve and cut-off governors for steam engine, drum type indicator for high-speed steam engine (1860) with Charles T. Porter [1826–1910], professor at Yale Univ. in dynamic engineering (1884ff) (BITAS CDOAB DAB EE MEIA NC25 WWWIA)

RICHARDS, Ellen Henrietta Swallow, 1842–1911, American woman chemist; instructor at Women's Lab. (1876–1884), mining engineering, sanitation, spontaneous combustion, explosion, water analysis (AWIS AWIT NC7 NWITPS WOS WWWIA WWWIAH-S&T WWWIS)

RICHARDS, Francis Henry, 1850–1933, American mechanical engineer and inventor; Stanley Rule & Level Works (CT), later with Pratt and Whitney (CT), received several hundred patents (1875ff) for machinery relating to tool making, sewing machines, printing press, cotton gins, button manufacturing, vending machines, coke furnaces, 350 patents by 1896 plus others, consulting engineer and patent advisor (1898ff) (MEIA NC7)

RICHARDS, Joseph William, 1864–1921, English American metallurgist; received first Ph.D. granted by Lehigh Univ. (PA) (1893), faculty at Lehigh Univ. (1887ff), authority on metallurgy of aluminum, electrometallurgy, metallurgical-chemical calculations (ANB CDOSB NC13 WWWIA)

RICHARDS, Robert B., 1914–1988, American civil engineer; with DeLeuw, Cather Organization (1936–1979) serving as engineer, president, director, chairman and president, NAE (1970) (MT8)

RICHARDS, Robert Hallowell, 1844–1945, American mining engineer; after graduating from Massachusetts Inst. of Technology (MIT) (1868) served on faculty of MIT where he stressed laboratory work and organized mining and metallurgical laboratories (1871), was professor of mining engineering and head of department (1873–1914), invented or improved a number of machines for ore dressing (CDOAB DAB NC12 WWWIA)

RICHARDSON, Frederick Denys, 1913–1983, English metallurgist; Royal Navy in WWII working on magnetic mines, British Iron & Steel Assoc. after WWII, faculty at Imperial College (1950–1976), fNAE (1976) (DNB MT2)

RICHART, Frank E., Jr., 1918–1994, American civil engineer; faculty at Univ. of Illinois (1946–1948), Harvard Univ. (1948–1952), Univ. of Florida (1952–1962), and Univ. of Michigan (1962–1986), soil dynamics, foundations for vibrating machinery, engineering problems related to vibration, consultant, NAE (1969) (Bridge MT8 WWWIA)

RICHE de PRONY—see PRONY, Gaspard F. de

RICHTER, Charles Francis, 1900–1985, American seismologist; faculty at California Inst. of Technology (1927–1970), with German Beno Gutenberg [1889–1960] developed the Richter scale measure of the magnitude of an earthquake (1927, 1935) (ANB BEST CB75 CDOS IAI L&M MWBD PS SAI-MP S:TLAW WA WWWIA WWWIS)

RICKETTS, Palmer Chamberlaine, 1856–1934, American civil engineer; mechanics, professor at Rensselaer Polytechnic Inst. (NY) (1875–1886) (1892ff), bridge engineer with Troy & Boston Railroad (1886–1887), Rome & Watertown and Ogdensburg Railroad (1887–1891) (BDOACE CDOAB DAB MEIA NCB,26)

RICKOVER, Hyman George, 1900–1986, Polish American naval officer and engineer; served in US Navy beginning as an ensign (1922) and as vice admiral (1959), admiral (1973) in charge of development of nuclear powered navy with the first nuclear submarine (Nautilus) (1954), NAE (1967) (AG BEST IAI MT3 MWBD PS WWIE WWWIA WWWIS)

RIDDELL, John Leonard, 1807–1865, American physician and inventor; taught chemistry at what became Tulane Univ. (LA) (1836–1865), noted for invention of the binocular microscope (1851) (BDOAS(El) BITAS CDOAB DAB)

RIDDELL, John T., 1852–1917, Irish American mechanical engineer and inventor; machinist, mechanic, foreman who worked with Daft Electrical Co., Thomson-

Houston Electric Co. (MA) (1887), General Electric Co. (NY) (1895ff), with over 15 patents most of which involved making, devising, and improving electrical equipment, drilling and machinery devices (MEIA NC35)

RIDGWAY, Raymond, 1897–1947, American chemical engineer; after work with Aluminum Co. of America (Alcoa) and with his father's firm, joined Norton Co. (1922–1947) as researcher becoming associate director, established methods and equipment for making boron carbide, during WWII worked on atomic energy project, electrochemistry, granted 26 patents (ACCE DAB PS)

RIDGWAY, Robert, 1862–1938, American civil engineer; chainman and instrument man for Northern Pacific Railroad (1882–1884), New York Aqueduct Commission developing water supply for the city, senior assistant engineer with NYC Rapid Transit Railroad Commission (1900–1905), Board of Water Supply (until 1912), involved in several transits, water systems, tunnels, New York Port Authority, consulting engineer (CDOAB DAB NC39 PS)

RIEMANN, Georg Friedrich Bernhard, 1826–1866, German mathematician; at Univ. of Göttingen (1854ff), theory of complex functions, Riemann zeta function (1860), non-Euclidian geometry (CDOS DOT ME126 MWBD RHD WWWIS)

RIKER, Carroll Livingston, 1854–1931, American mechanical engineer and inventor; studied fluids at rest and in motion and currents in NYC harbor, built first refrigerated warehouses (1873–1875), designed machinery for refrigerating shop (1874), built pumping dredges (1887), torpedo (1898), built unsinkable freighters (WWI), erected first factory in US to make unfermented grape juice (1922), flood control, received about 20 patents (DAB MEIA)

RILEY, Howard Wait, 1879–1971, American agricultural engineer; chairman Manton Lord (1901–1904), Telpherage Co. (1904–1906), United Morse Chain Co. (1906–1907), faculty in agricultural engineering including department head at Cornell Univ. (1907ff) (WWIE(I))

RILLIEUX, Norbert, 1806–1894, Black American engineer and inventor; faculty at L'École Controle (FR) (1830), inventor of steam-operated cotton-baling process, vacuum-operating pan (1843), designed and installed multiple-effect evaporator (1846) first applied to sugar industry then applied to many industries, made it easier and less expensive to produce sugar, soap, glue, food products, developed practical sewage system for New Orleans (LA) (BDOBA BI BPOSAI ChemHerit20 IIA MEIA PS WA)

RIQUET DE BONREPOS, Pierre-Paul de, 1604–1680, French engineer; planned canal (Canal du Midi) to connect Atlantic Ocean and Mediterranean (1662), first to use explosive for blasting rock (GEAPIT MWBD)

RITCHEY, Harold, 1912–??, American chemical engineer; research chemist Union Oil Co. (1938–1941) (1946–1947), US Navy in WWII (1941–1946), fission reactor engineer General Electric Co. (1947–1949), developed rocket engines and solid propellants followed by management positions at Thiokol Corp. (1949–1974) becoming president (1964–1971) and chairman of the board and CEO (1971–1974) (S WWIA)

RITES, Francis M., 1858–1913, American mechanical engineer and inventor; worked for Lehigh & Hudson River Railroad (1881), Westinghouse Machine Co. (1883), patented 10 governors (1886ff) and at least three patents for compressors (1895–1903) (MEIA)

RITTENHOUSE, David, 1732–1796, American instrument maker and surveyor; designer of modern surveying equipment in which angle-measuring instruments began to replace chain surveying, workshop for clock making (1750–1760), constructed a variety of telescopes, explored the Western Reserve after the American revolution, director of US Mint (1792ff) (BDOAS(El) BITAS DOSB EIH GEAPIT PS)

RIVA-ROCCI, Sciopone, 1863–1937, Italian physician and inventor; sphygmomanometer apparatus for blood pressure measurement (1896) (DOSB FETCTW 1000Y)

RIX, Edward Austin, 1855–1937, American mechanical engineer and inventor; proprietor of Phoenix Iron Works (1883ff), established several shops and companies with his name, designed pneumatic machinery (1888), established Rix Compressed Air Co. (CA) (1890), use of compressed air as a power source as for air hammer, air compressor for dynamite gun (1896), variable-volume air compressor (1899), supercharger air compressor, received 35 patents (MEIA)

ROACH (ROCHE), John, 1813–1887, Irish American shipbuilder; to US (1830), foundryman and ironmaker, engine builder, led shift from wooden to iron vessels (1868ff), built marine engines and iron steamships and war vessels (CDOAB MWBD)

ROBB, Byron Burnett, 1882–1961, American agricultural engineer; drainage engineer for state of New York (1911–1915), faculty at Cornell Univ. (NY) (1911ff) becoming professor (1919) (WWIE(I))

ROBB, William Lispenard, 1861–1933, American electrical engineer; professor of physics at Trinity College (CT) (1885–1902), professor of electrical engineering and physics at Rensselaer Polytechnic Inst. (NY) (1902–1933) (CDOAB DAB)

ROBECK, Gordon G., 1923–1993, American environmental engineer; improvement of drinking water quality, director US EPA Drinking Water Division, commissioned corps of USPHS (1944–1957), water and sewage treatment investigating stream pollution and industrial water survey, NAE (1980) (Bridge MT7 WWWIA)

ROBERDAU, Isaac, 1763–1829, American military and civil engineer; assisted Pierre C. L'Enfant [1754–1825] in work on new city of Washington (DC) (1791–1792), engineering practice in Pennsylvania (until 1812), appointed major in topographical engineer corps of US Army (1813) becoming chief (1818) and brevet lieutenant colonel (1823) (ANB CDOAB DAB)

ROBERT de Courcy, fl. c.1300, French architect-engineer; one of builders of Rheims Cathedral (building took 270 yr. to finish), other builders were Jean D'Orbais [fl. 1212–1231], Jean de Loup [fl. 1231–1247], Gaucher de Rheims [fl. 1250–1259], and Bernard de Soissons [fl. 1263–1298] (GEAPIT)

ROBERT de Luzarches, fl. c.1220, French architect and engineer; credited with being the premier architect of Amiens Cathedral (with octagonal shape), of particular interest is the quality of the foundation and vaults (GEAPIT)

ROBERT, Henry Martyn, 1837–1923, American military engineer; constructed defense of Washington (DC), and Philadelphia (PA) during Civil War, authored respected Robert's Rules of Order (1876), retired as army brigadier general (1901) (ANB DAB ET MWBD NC27 1000Y WWWIA)

ROBERT, Nicholas-Louis, 1761–1828, French inventor; with brother Anne-Jean Robert [1758–1920] launched first hydrogen balloon (1783) and with Jacques-A.-C. Charles [1746–1823] ascended in the balloon (1783), invented first machine to produce continuous sheets of paper (1798) (GI MWBD PS)

ROBERTS, Benjamin Stone, 1810–1875, American military and civil engineer; graduate of US Military Academy (1835) and resigned (1839) later reentered military, chief engineer of railroad, then geologist, in militia and participated in numerous battles on the western front, in Civil War advanced to brigadier general (1865), taught at Yale Univ. and practiced law, invented breech-loading rifle (BDOACE CDOAB DAB)

ROBERTS, George Brooke, 1833–1897, American civil engineer; began as a rodman for the Pennsylvania Railroad (PRR) (1851), experience with several railroad systems then returned to Pennsylvania Railroad (1862ff) and became president (1880–1897) (ANB CDOAB PS)

ROBERTS, Sir Gilbert, 1899–1978, English civil engineer and inventor; developed all-welded ships in WWII, new designs and methods of building bridges with welded construction in Scotland, England, New Zealand, Ghana and Turkey, crane construction, audio telescope, with bridge department of Dorman Long and Co., Ltd. (1926–1935), with Sir William Arnold & Co. Ltd., Glasgow (1936–1949), partner with Freeman, Fox & Partners (1949–1969), consultant (MWBD TGE:B WWW)

ROBERTS, Nathan Smith, 1776–1852, American civil engineer; engineer in charge of building middle section of Erie canal (1816–1822) and western section of canal (1822–1825), other tasks for canals and railroads, for Baltimore & Ohio Railroad and federal government (BDOACE CDOAB DAB GEAPIT NC2 WWWIA)

ROBERTS, Richard, 1789–1864, Welsh engineer and inventor; toolmaker, invented screw-cutting lathe (1817), planing machine (1817), straight line mechanism for steam engine, founded the firm of Sharpe, Thomas and Co. (1828), built locomotives and associated products (1830), steam carriage (1834) and numerous machine tools (BDOTHOT EE GEAPIT PS TBDOS(Po) TEIH)

ROBERTS, Richard William, 1935–1978, American chemist and engineer and inventor; General Electric Research and Development (Canada) (1960–1973) (1977–1978), discovered iodine-based lubricants for difficult-to-lubricate metals, worked on high-vacuum systems and chemical kinetics, was director of NBS (1973–1975), with ERDA (1975–1977), three US patents, NAE (1977) (MT1)

ROBERTS, S., 19th century, English mathematician; three-bar mechanism for motion in free space (1875), various mechanical linkages and motion, Roberts-Chebyshev theorem (1878) (Pafnuti L. Chebyshev [1821–1894]) (EE L&M WWW WWWIS)

ROBERTS, Solomon White, 1811–1882, American civil engineer; canal and railroad expert, construction of first successful anthracite furnace for iron smelting in Lehigh Valley (PA) (1839, 1840) (CDOAB DAB)

ROBERTS, William Milnor, 1810–1881, American civil engineer; canal and railroad expert, chief engineer of Northern Pacific Railroad (1869–1879), chief engineer of all public works in Brazil (1879–1881) (CDOAB DAB)

ROBERTSON, Andrew, 1883–1977, English mechanical engineer; behavior of mild steel under stress, WWI in materials testing laboratory of Royal Aircraft Establishment, faculty and chair of mechanical and mining engineering at Univ. of Bristol (1919–1946) and dean (1924) until he retired (1946), design of machines, design of steel structures, FRS (1940) (DNB PS RS24 WWW)

ROBINS, Benjamin, 1707–1751, English mathematician and military engineer; invented ballistic pendulum (1742), whirling arm to measure resistance of a body to movement in a fluid, established basis for ordnance theory, improved rifling of gun barrels (DOSB EE GEAPIT MWBD PS)

ROBINSON, Albert Alonzo, 1844–1918, American civil engineer; with railroad companies, then with Atchinson, Topeka & Sante Fe Railroad (1871–1893) becoming general manager (1888), president of Mexico Central Railroad Co. (1893–1896) (BDOACE CDOAB DAB)

ROBINSON, Clark Shove, 1888–1947, American chemical engineer; began as a chemist with Sherwin-Williams Co. (1910), was assistant plant manager for Roessler & Hasslacher Chemical Co. (1911–1912), faculty at Massachusetts Inst. of Technology (1915–1947), with Walworth Manufacturing Co. (that made Stillsen wrenches), consultant, work in distillation, explosives, served in WWI and WWII advancing to colonel (1942) (HOCE NC35 PS SAI)

ROBINSON, Denis M., 1907–1994, American electrical engineer; underground transmission of electric power, in London with Callenders Cables Ltd. (1931–1939), Royal Air Forces on radar development (1939–1945), with John G. Trump [1907–1985] and Robert J. Van de Graaff [1901–1967] founded High Voltage Engineering Corp. (MA) as chief operating officer (1946–1970) and chairman of the board (1970ff), NAE (1970) (WWIE)

ROBINSON, Moncure, 1802–1891, American civil engineer; built original line of Philadelphia & Reading Railroad (1834–1836) (CDOAB DAB NC8 WWWIA)

ROBINSON, Stillman Williams, 1838–1910, American civil and mechanical engineer and inventor; faculty in mining at the Univ. of Michigan (1866–1870), Illinois Industrial Univ. (1870) including dean, railroad consultant (1887), at Ohio State Univ., bridge inspector (1880–1882), Wire Grip Fastener Co. and McKay Shoe Machinery Co. (1895ff), consulting, patents on shoemaking, wire grip fastener, granted 40 patents (BDOACE BITAS CDOAB DAB MEIA)

ROBINSON, William, 1840–1921, Irish American electrical engineer and inventor; invented railroad block signal (1872), invented radical car truck used on electric railways, improvements to steam turbines (CDOAB DAB)

ROBISON, John, 1739–1805, Scottish natural philosopher; mechanical arts, lecturer in chemistry at Glasgow Univ. (1766), professor of natural philosophy at Edinburgh Univ. (1773ff) (LOTE SFLOE)

ROBSON, J., 1833–1913, British inventor; two-cycle compression gasoline engine called Robson engine (1877) (EE)

ROCHELEAU, Robert F., 1920–1991, American chemical engineer; efficient use and protection of clean water resources, principal division consultant for E. I. du Pont de Nemours & Co. (1954ff), after serving in water and pollution control first in New Hampshire then in West Virginia, NAE (1981) (Bridge MT6)

ROCKWELL, Mabel MacFerran, 1902–1979, American woman electrical and aeronautical engineer; development of manufacturing processes for aircraft and underwater propulsion systems, designed a control system for missiles, electric power systems, worked for Southern California Edison Co., the Metropolitan Water District of Southern California on power systems including Boulder Dam, during WWII with Lockheed Aircraft Corp. and several government agencies, after WWII with Convair (AWIS NS NTCS TechRev)

RODD, Thomas, 1849–1929, English American civil engineer; US Navy during Civil War, rodman, surveyor and assistant for surveying Philadelphia (1866–1872), with Pennsylvania Railroad (PRR) (1872–1919), consulting engineer for several railroad companies, including reconstruction following Johnstown flood (PA) (1889) and structures for railroads, coal and ore handling (BDOACE NC36 WWWIA)

RODDIS, Louis H., Jr., 1918–1991, American nuclear engineer; graduate of USNA and officer in navy, participated in design of early nuclear projects such as USS Nautilus, Atomic Energy Commission (1955–1958), president of Penn Electric Co. (1958–1969), Consolidated Edison Co., nuclear power and propulsion, NAE (1967) (Bridge MT6 WWIE)

RODMAN, Thomas Jackson, 1815–1871, American inventor; US Military Academy graduate (1841), at West Point Foundry developed process for cooling cast iron guns from the inside, improved powder for artillery, commanded arsenal (MA) for making castings for heavy guns during the Civil War (ASAI CDOAB DAB NC4 WWWIA)

ROE, Sir Edwin Alliott Verdon, 1877–1958, English aircraft designer and manufacturer; first Englishman to build and fly an airplane (1908), with brother Humphrey Verdon Roe [1878–1949] established A. V. Roe and Co. (1910), founder and president of Sanders-Roe, Ltd. (1928ff), developed Avco series of aircraft, later developed and manufactured flying boats (DNB MWBD TBDOS(Po) WWW)

ROE, Kenneth A., 1916–1991, American mechanical engineer; design for application of new fields of technology in power production, participated in aerospace and defense programs, served in US Navy in WWII, chairman (1971ff) and CEO of Burns and Roe Enterprises, Inc., (31 yr.) becoming chairman of board (1971), NAE (1978) (Bridge MT6 WWIE)

ROEBLING, John Augustus, 1806–1869, German American civil engineer, industrialist and inventor; established factory in US to manufacture wire rope (1841), designed long-span suspension bridges—Niagara Falls (1855), Cincinnati (1867), designed Brooklyn bridge, his son Washington Augustus Roebling [1837–1926] associated with father (1857ff), succeeded father as chief engineer and completed construction of Brooklyn bridge (1883) after serving in the Civil War (ACE ANB BITAS CDOAB DAB DOSB EIH MWBD NC4 NYPL 1000Y PS TEIH TGS WWWIA WWWIS)

ROEBLING, Washington Augustus, 1837–1926, American civil engineer; served as assistant to his father John A. Roebling [1806–1869] (until 1861), officer in engineers of Union Army (1861–1865) leaving as brevet colonel, returned to his father's employee succeeding him as chief engineer of Brooklyn bridge project (1869–1883), president of John A. Roebling's Sons

Co. (NYC) (1876–c.1888) (CDOAB DAB DOSB EIH NC26 TEIH WWWIA)

ROEBUCK, John, 1718–1794, English physician and inventor; use of leaden chambers in manufacture of sulfuric acid (1746), set up vitriol works (1749), established ironworks (1759) and patented process for converting cast iron into malleable iron (1762) (DOSB GEAPIT GI MWBD PS TEIH TGS)

ROGERS, Fairman, 1833–1900, American civil engineer; survey work in helping to determine the Epping base line in Maine (1857), construction of roads and bridges, magnetism of iron vessels, served in Civil War, lecturer and professor at Univ. of Pennsylvania and at Franklin Inst. (PA), NAS (1863) (ANB BITAS BM6 NC11 PS WWWIA)

ROGERS, Henry J., 1811–1879, American electrical engineer and inventor; telegraph pioneer, inventor, associate of Samuel F. B. Morse [1791–1872] (CDOAB DAB NC4 WWWIA)

ROGERS, James Harris, 1856–1929, American inventor; several patents, the most valuable invention was his system of printing telegraphy (1887–1894) (CDOAB DAB NC21 WWWIS)

ROGERS, John Raphael, 1856–1934, American civil engineer and inventor; taught in public schools, civil engineer for railroads (IA, MO, WI) (1883–1885), formed Rogers Typograph Co. (1888ff), then to Mergenthaler Linotype Co (1895), received 400 to 500 patents many pertaining to linotype (ANB CDOAB DAB MEIA NCC WWWIA)

ROGERS, Thomas, 1792–1856, American inventor; blacksmithing, war of 1812, with John Clark formed Clark & Rogers Co. to make power looms (1819ff), organized a machine works to manufacture textile equipment (1828), locomotives and railroad equipment (1832), innovated locomotive-driving wheels (CDOAB DAB MEIA)

ROGERS, William Augustus, 1832–1898, American astronomer, mechanical engineer and inventor; professor at Alfred Univ. (1859ff), built and equipped observatory (1865), served in navy during Civil War, Harvard College (1869ff) where he studied and worked in astronomy, with George Meade Bond [1852–1935] constructed Rogers-Bond universal comparator (1885), micrometers, limits of precision, professor of astronomy at Colby Univ. (1886ff), NAS (1885) (BITAS BM46 MEIA NC9 WWWIA WWWIS)

ROGERS, Winfield Scott, 1853–1931, American mechanical engineer; draftsman, tool designer, worked for several companies making tools, sawing machines, railroad equipment, paper mill equipment, laundry machinery, ball bearings, roller bearings, vice president and general manager of Steamobile Co. (1901), president of Bantam Antifriction Co. (1903) (MEIA NC24)

ROHLICH, Gerald A., 1910–1992, American civil engineer; faculty at Carnegie Inst. of Technology (PA), Pennsylvania State Univ., Univ. of Wisconsin (1946–1972), Univ. of Texas (1972ff), water quality, industrial waste treatment, lake and stream pollution, NAE (1970) (MT3 WWIE TGS)

ROLLS, Charles Stewart, 1877–1910, English manufacturer and aviator; pioneer maker of automobiles, formed C. S. Rolls & Co. (1902), merged with Royce, Ltd. (Frederick Henry Royce [1863–1933]) forming Rolls-Royce, Ltd (1906), first person to fly across the English channel and back (1910), first English casualty of aviation (DNB MWBD TGS WWW)

ROMANS, Bernard, c.1720–c.1784, Dutch American civil engineer; to US (1784), survey work and botanical studies in Georgia and Florida, construction of fortification on Hudson River, commissioned captain in Pennsylvania artillery (1776–1786) (BDOAS(El) BITAS CDOAB DAB PS WWWIA WWWIS)

RONALDS, Francis, 1788–1873, British chemical engineer; electrical experimentation, transmitted and experimented with electrical impulses (1816), first to transmit signals from static electricity up to 8 miles as a forerunner of telegraph, devised system for continuously recording meteorological data, retired (1852) (DNB GEAPIT RS WWWIS)

RÖNTGEN (or ROENTGEN), Wilhelm Konrad, 1845–1923, German physicist, mechanical engineer and inventor; professor at several universities primarily in Munich (1900–1920), X-ray development from cathode-ray tube (1895), X-ray of human skeleton (1896), work led to discovery of radioactivity, studied elasticity, conduction of heat and piezoelectricity, Nobel prize (1901) (BEST GI IAI MWBD 1000Y SAI SAI-MP S:TLAW WWWIS)

ROOSEVELT, Hilborne Lewis, 1849–1886, American inventor; experimented with electrical devices, organ builder, cousin of Theodore Roosevelt [1858–1919] (CDOAB DAB NC26 WWWIA)

ROOSEVELT, Nicholas J., 1767–1854, American inventor and steam engineer; with others founded Soho Foundry (NJ) (1794) that built engines, rolling mills, guns for ships, engines for steamboats, patented vertical paddle wheels (1814), with Robert Fulton [1765–1815] built steamboat New Orleans (1811) (ANB CDOAB DAB EIH MEIA NC12 WWWIA)

ROOT, Elisha King, 1808–1865, American mechanic and inventor; machinist, lathe hand with Collins Co. (CT) (1832), invented improvements in axe-manufacturing machinery (1845), superintendent of Colt's Armory (1849–1865) becoming president (1862) using interchangeable parts, patented drop hammer (1853) (1858), several patents dealing with machine tools and die forging process (BITAS CDOAB DAB IIA MEIA NC18 WWWIA)

ROOT, L. Eugene, 1910–1992, American aeronautical engineer; design of aircraft and missiles, development of Polaris (world's first submarine-launched strategic missile), with Douglas Aircraft Co. (1934–1953), president of Lockheed Missiles & Space Co. and group vice president of Lockheed Corp., with Lockheed (1953–1969), NAE (1965) (Bridge MT6 WWIE)

ROSE, Albert, 1910–1990, American physicist and electrical engineer; photoconductivity, photosensitivity, television camera tubes, RCA Labs. (1935–1975), at numerous universities as visiting professor, NAE (1975) (Bridge MT6 WWIE)

ROSE, Philip S., 1872–1962, American agricultural engineer; faculty at North Dakota State Univ. where he taught thermodynamics and engineering mechanics (1899–1909) incorporating applications to agriculture, launched *Gas Review* for Clarke Publishing Co. (1909–1917) and editor and later managing editor of *Country Gentleman* (1917–1940) (AE43)

ROSENBAUM, Joe B., 1912–1987, American metallurgist; methods of processing minerals while abating pollution, US Corps of Engineers (1942–1946), chief metallurgist for US Bureau of Mines (USBM) (1941–1976), consultant (1977ff), NAE (1973) (Bridge MT4 WWIE)

ROSENBERG, Julius, 1915–1953, American electrical engineer; engineer with Army Signal Corps (1940–1945), with wife Ethel Rosenberg [1915–1953] executed for spying activities (1953) (CDOAB DAB)

ROSENBLUETH, Emilio, 1926–1994, Mexican civil and structural engineer; faculty and administration at City Univ. of Mexico (1959ff), design of structures especially earthquake engineering, consulting engineer (1956–1977), fNAE (1977) (Bridge MT8 WWIE)

ROSENGARTEN, George David, 1869–1936, American chemist and chemical engineer; raised and worked in family business Rosengarten and Sons which manufactured chemicals, became vice president (1901), company merged to form Powers-Weightman-Rosengarten, Inc. in which he served 22 yr. as vice president until it was obtained by Merck and Co. (1927) from which he retired as director (1929) (ACCE NC35 PS)

ROSEWATER, Andrew, 1848–1909, American civil engineer; railroad surveys, city engineer in Omaha (NE) (1868–1875)(1881–1887), consulting and design engineer of sewage systems for 25 cities (BDOACE)

ROSS, James Delmage McKenzie, 1872–1939, American electrical engineer; headed municipal power plant in Seattle (1911–1939), architect of public power system in US in 1930s and 1940s (CDOAB DAB WWWIA)

ROSSER, Thomas Lafayette, 1836–1910, American military engineer; during Civil War served Confederacy as engineer reaching rank of major general, in artillery and cavalry, later was chief engineer first of Northern Pacific Railroad and later of the Canadian Pacific Railroad, retired (1886) (CDOAB DAB NC49 WWWIA)

ROTCH, William, 1844–1925, American civil engineer; assistant to chief engineer at Massachusetts Water Works (1871–1880), consulting engineer with various railroad companies (1880–1891), president of wharf and storage companies (until 1922), consulting with mining and powder company (BDOACE NC15 WWWIA)

ROTHWELL, Richard Pennefather, 1836–1901, Canadian American mining engineer; worked in US (1864ff) primarily as consultant to operators of anthracite mines, co-founder of American Inst. of Mining Engineers (1871) and was co-editor (1874–1890) of Engineering and Mining Journal and editor (1890ff) (CDOAB DAB NC10 WWWIA)

ROUSE, Hunter, 1906–1996, American civil engineer; faculty and dean of engineering at Univ. of Iowa (1939–1972), after serving at Massachusetts Inst. of Technology, Columbia Univ. (NYC), and California Inst. of Technology, Soil Conservation Service (1936–1939), hydraulics and fluid mechanics, NAE (1966) (Bridge WWIE)

ROUSSEAU, Harry Harwood, 1870–1930, American naval officer and engineer; naval officer reaching rank of rear admiral, with George W. Goethals [1858–1928] (1907–1914) on Isthmian Canal Commission in charge of design and construction, in charge of shipyard division of Shipping Board during WWI (CDOAB DAB)

ROWLAND, Thomas Fitch, 1831–1907, American mechanical engineer and inventor; designed engines and machinery for Allaire Works (1852) and Morgan Iron Works (1853), in partnership with others founded and renamed company Continental Works (1860) building iron ships, pipes, gunboats, ferryboats (1870), steam engines, and boilers, then incorporated as Continental Iron Works (1887) (CDOAB DAB MEIA NC12)

ROWLEDGE, Arthur J., 1876–1957, English engineer; consulting engineer for Rolls Royce Co., FRS (1941) (PS RS4 WWW)

ROYCE, Sir Henry (Frederick Henry), 1863–1933, English engineer and manufacturer; founded engineering firm of Royce, Ltd. (1884), built three experimental cars (1903), joined with Charles Stewart Rolls [1877–1910] to form Rolls-Royce, Ltd. (1906), manufactured automotive vehicles, developed engine electrical distributor (DNB MWBD PS TBDOS(Po) TGE WWW WWWIS)

ROYLE, Vernon, 1846–1934, American mechanical engineer and inventor; engraver with Heber Wells (1868–1877), with father's machine business (1877ff) after which he developed numerous patents relating to engraving, textiles, circular loom, tubing machines, machines to apply insulation to wire (MEIA NC12)

RUDENBERG, Reinhold, 1883–1961, German American electrical engineer; to England (1936), to US (1939), test engineer at Siemens-Schuckwert Works (1908–1913), faculty at Berlin Technische Hochschule (1913–1927), at Harvard Univ. (1939–1952), developed 60-MVA turbine generator (CDOSB DOSB PS)

RUGAN, Henry Fister, 1857–1916, American mechanical engineer; worked for several railroads, at Rose Polytechnic Inst. (IN) (1882), Texas and Pacific Railroad (TX) (1885), chief engineer of sugar plantations (1890), faculty at Tulane Univ. (LA) (1895–1898), cast iron specialist (MEIA)

RÜHMKORFF, Heinrich Daniel, 1803–1877, German French instrument maker and inventor; founded shop in Paris (1855), Rühmkorff induction coil, thermoelectric battery (CDOSB DOSB PS TBDOS(Po))

RUMELY, William Nicholas, 1858–1936, American mechanical and agricultural engineer and inventor; son of blacksmith Meinrad Rumely [1823–1904] with brother John Rumely [1853–1931] formed M and J Rumely Co. (IN) and Rumely Engine Works that became Advance Rumely Co., manufactured threshing machines and steam-traction engines, worked in father's company (1879ff), president of company (1904ff), president of several companies including Illinois Thresher Co., at least 27 patents on threshers, separators, kerosene tractors (1910) developed by John Secor, Allis-Chalmers Co. acquired Advance Rumely Co. (1931), several company names associated with Rumely (AE18 MEIA Res10 WWWIA)

RUMFORD, Count (aka Benjamin Thompson), 1753–1814, American British physicist and inventor; relationship of mechanical motion and heat, determined the mechanical equivalent of heat, inventions included a double boiler for cooking, a drip coffee pot, and a kitchen range, invented photometer (ASAI BEST FNIE)

RUMSEY, James, 1743–1792, American inventor; served in Revolutionary War, superintendent of construction of canal (1785), building of steamboat (1787) patented (1791) in US and England, projects on sawmill, grist mill, pumps (ANB BITAS CDOAB DAB EIH MEIA MWBD NC5 PS WWWIA)

RUNGE, Carl David Tolmé, 1856–1927, German applied mathematician; developed series formula (1887), spectroscopy, with Wilhelm Martin Kutta [1867–1944] developed Runge-Kutta method for solving differential equations (1895, 1901), Runge theorem (1885), Runge rule, Runge vector (CDOSB CEOS&T EB EDOP EOM EOS&T L&M WWWIS)

RUSCH, Hubert, 1903–1979, Austrian and German civil engineer; pre-stressed concrete, reinforced concrete, properties of concrete, with construction firm (1926–1948), professor of civil engineering at Technical Univ. of Munich (1948ff), fNAE (1977) (MT2)

RUSKA, Ernst August Friedrich, 1906–1988, German electrical engineer and inventor; electron optics, vacuum tubes, cathode rays, faculty at Technical Univ. of Berlin (1931ff), with Max Knoll [1897–1969] invented electron microscope (1933) based on ideas of Frenchman Louis-Victor-Pierre-Raymond Broglie [1892–1987], worked with Fernsch Co. (1933–1937), Siemens Co. (1937–1955), Max Planck Inst. (1949–1974), work led to development of atomic force (ATM) microscope and worked with Gerd Benning and Swiss Heinrich Rohrer to develop the scanning tunneling microscope, Nobel prize (1986) (ABES BEST CDOS GI IAI MWBD NYT-A PS SATF WWWIS)

RUST, John Daniel, 1892–1954, American inventor and manufacturer; invented first cotton-picking machine (1928) (1933) joined by his brother Mack D. Rust (1933), started Southern Harvester Co. (LA) (1932), succeeded by Rust Cotton Picker Co. (TN) (1936), later manufactured and sold by Allis-Chalmers Manufacturing Co. (ANB CDOAB I&T19 NC42)

RUTHERFORD, Sir Ernest, 1871–1937, New Zealand British physicist and instrument developer; radiation, identified alpha and beta rays (1896), with Hans Geiger [1882–1945] developed apparatus to measure number of particles emitted by radiation, McGill Univ. (1898–1907), Univ. of Manchester (1907–1919), director of Cambridge Cavendish Lab. (1919ff), Nobel prize (1908), fNAS (1911) (BEST DNB MWBD SAI SAI-MP S:TLAW TGS WBE WWWIS)

RUTLEDGE, Philip C., 1906–1990, American civil engineer; design of foundations for buildings, bridges, industrial plants, and waterfront structures, professor and department chairman at Harvard Univ., Purdue Univ., and Northwestern Univ. (1933–1952), partner in consulting firm (1952ff), consulting engineer, NAE (1968) (Bridge MT5 WWIE)

RUUD, Edwin, 1854–1932, Norwegian American mechanical engineer; worked with locomotive works, foundry, Westinghouse Electric Co. (1887–1905), developed automatic gas water heater, meters, gas engine, Ruud Manufacturing Co. that was previously James Hay Co. (1905ff), purchased Humphrey Co. (1913) and continued invention, design and development of control and manufacture of water heaters, over 200 patents (MEIA NC24)

RYAN, Harris Joseph, 1866–1934, American electrical engineer; high-voltage phenomena, alternating current, electrical measurements, electrical machinery, transformers, faculty at Univ. of Nebraska (1887–1889), Cornell Univ. (1889–1905), Stanford Univ. (1905–1931), following retirement worked on hearing aids, NAS (1920) (ANB BM19 CDOAB DAB NC26 PS)

RYAN, Jack, 1926–1991, American electrical engineer and inventor; Raytheon Co. (1948–1955), worked on Sparrow and Hawk missiles, vice president for research and development at Mattel, Inc. (1955ff), defense contractor, invented Barbie doll and several electrical/mechanical toys (ANB)

RYAN, Walter D'Arcy, 1870–1934, American illuminating engineer; flood lighting, high-intensity street lighting (CDOAB DAB)

S

SABINE, Wallace Clement Ware, 1868–1919, American physicist; architectural acoustics, professor at Harvard Univ. (1905ff), Boston Symphony Hall designed using his principles (1908), Sabine law (1910)—the unit of sound absorbing ability, founder and dean of Harvard Graduate School of Applied Science (1906–1915) (BEST BISAT L&M MWBD NC27 SAI WWWIA WWWIS)

SAEGMULLER, George Nicholas, 1847–1934, Bavarian American mechanical engineer and inventor; to US (1870), became partner in Fauth & Co. (DC) making astronomical and engineering instruments (1870ff), merged with Bausch & Lomb Optical Co. as vice president (1905), invented astronomical implements and invented fire control equipment for US Navy (MEIA)

SAINT-VENANT, Adhemar Barre de, 1797–1886, French civil and agricultural engineer; at the forefront in building bridges, roads, and canals, contributed to theory of machines, role of shear in deformation of bodies, dynamics of fluids (1834), shearing stress and shearing strain (1837–1842), Paris paving service (1839–1843), chair of agricultural engineering at Institute in Versailles (1848ff) where he wrote on erosion control, ponds, drainage, and the shape of plow shares (GEAPIT L&M)

SAITO, Shinroku, 1919–1994, Japanese aeronautical engineer; ceramic science and technology advancement, faculty and administration at Tokyo Inst. of Technology (1945–1981), government, served as chair of Kanagawa Academy of Science and Technology (1981ff), fNAE (1994) (Bridge MT8)

SALK, Jonas Edward, 1914–1995, American physician and scientist; developed vaccine using killed virus for prevention of poliomyelitis (1955), at Univ. of Michigan (1944–1947), Univ. of Pittsburgh (PA) (1947–1963), director of Salk Inst. at Univ. of California-San Diego (1963–1975), Russian American Albert Bruce Sabin [1906–1993] developed a vaccine with weakened virus (1957) that could be taken orally (CDOS GI MWBD 1000Y RHW VNSE WWWIA WWWIS)

SALT, Sir Titus, 1803–1876, English manufacturer; wool stapler, wool spinning (1834), developed spinning machine for coarse wool, developed method of manufacturing alpaca (1836), built model manufacturing town called Saltaire (1853) (MWBD)

SALVADORI, Mario G., 1907–1997, Italian American civil engineer; design and analysis of shell and high-rise structures, numerical analysis in engineering, faculty at Columbia Univ. (NYC) (1940–1990), chairman Weidlinger Associates (1953ff), consulting engineer, NAE (1983) (Bridge WWIE)

SAMUELY, Felix James, 1902–1959, British civil engineer; leader in design and construction of all-welded steel structures, tubular construction, built on cork to reduce vibration, consulting engineer beginning with Berger and Samuely, Ltd. (1929ff) (TGE:B)

SANCHINI, Dominick J., 1926–1990, American mechanical and aerospace engineer; design, development, and operation of space shuttle main engine, vice president of Rockwell International Corp., NAE (1984) (Bridge MT6)

SANGALLO, Giuliano da, c.1445–1516, Florentine architect and engineer; military engineer, assisted in designing and building St. Peter's cathedral, engineer in Florentine army against Naples (1478), built facade projects for S. Lorenzo (1515), his brother Antonio Sangallo [1455–1535] was an architect and engineer who built several churches and palaces (MWBD)

SANGER, Eugen, 1905–1964, Austrian rocket engineer; directed a rocket program in Germany (1936–1945), with French armament ministry (1946–1954), head of jet propulsion work at Stuttgart (1954–1963), designed a stratosphere rocket plane (1933) (MWBD)

SANMICHELI, Michele, 1484–1559, Italian architect and engineer; military engineer, worked on fortifications for Italian cities, chief architect for cathedral of Orvieto (1509–1528), built churches and palaces (MWBD)

SANTI, Gina P., d. 1997, age 81, American engineer and inventor; US Air Force, developed and perfected pilot ejection system (1949), most of career at Wright Patterson Air Force Base (OH), NACA (1943–1947) (FAL NYTObit)

SANTOS-DUMONT, Alberto, 1873–1932, Brazilian aviation pioneer; in France he built and flew cylindrical balloon with gasoline engine (1898), erected aircraft station at Neuilly (1903), first flight in Europe of box-kite-like powered craft (1909), built monoplanes called grasshopper the forerunner of modern light planes (MWBD)

SARGENT, Frederick, 1859–1919, English American electrical engineer; to US (1880), designed marine engines (1880), E. P. Allis & Co. (WI) (1882), Western Edison Light Co. (IL) (1884), with various Edison companies, consulting engineer (1890ff), designed central generating plants throughout the world (CDOAB DAB MEIA)

SARGENT, James, 1824–1910, American inventor; lock expert, formed Sargent and Greenleaf Co. (1864), manufactured nonpickable lock, perfected time lock, manufactured railway signals, automatic fire alarms, glass-lined steel tanks (CDOAB DAB NC22 WWWIA)

SARNOFF, David, 1891–1971, Russian American communications executive; to US (1900), telegraph operator who picked up distress signal from Titanic (1912), research and administration of Radio Corp. of America (RCA) as commercial manager (1919) to president (1930–1947) to chairman of the board (1947–1970), a pioneer in radio and television broadcasting (MWBD 1000Y WWWIA)

SATTERLEE, Francis Le Roy, 1881–1935, American radiologist and inventor; took first X-ray in US (1896), invented first X-ray shield for X-ray operators, took X-rays of mouth and teeth (MWBD)

SAUGRAIN de VIGNI, Antoine F., 1763–1820, French instrument maker; scientific expeditions to Mexico (1786), to US (1787), perfected and sold ink, barometers, thermometers, and glass tubes, manufactured instruments and chemicals, produced friction matches (ACCE BDOAS(El) BITAS DAB PS WWWIS)

SAUNDERS, Owen Alfred, 1903–1993, English mathematician and mechanical engineer; heat-transfer research and design, lubrication, rocket research, improved output of piston engines (1940ff), faculty and administrator at Imperial College in London (1946ff), vice chancellor at the Univ. of London (1967ff), fNAE (1979) (MT8 WWWIA WWWIS)

SAUNDERS, William Lawrence, 1856–1931, American mechanical engineer and inventor; engineer with National Storage Co. (NJ) (1878ff), officer including president of Ingersoll-Rand Co. (NJ), numerous patents dealing with mining and quarrying, use for compressed air, involved in political and government activities (CDOAB DAB MEIA NC26 WWWIA)

SAUSSURE, Horace Bénédict de, 1740–1799, Swiss physicist and inventor; credited with developing an improved hygrometer using a human hair to measure atmospheric humidity (1783), when a hair is moved from 100 to 0 percent relative humidity it shrinks about 2.5% of its length, Leonardo da Vinci [1452–1519] used a ball of string that became wetter and heavier as the relative humidity increased (BEST EB WWWIS)

SAUVAGE, Pierre-Louis-Frédéric, 1785–1857, French mechanical and naval engineer; shipbuilder (1811), stone-cutting business (1821), patented screw for propelling ship (1832) plus several other items that failed to become economically successful (GEAPIT)

SAUVEUR, Albert, 1863–1939, Belgian American metallurgist; faculty at Harvard Univ. (1899–1939), credited with founding physical metallurgy, specialist in microstructure of metals, heat treatment of metals, NAS (1927) (ANB BM22 MWBD NC29 WWWIS)

SAVAGE, Arthur William, 1857–1938, Jamaican American manufacturer and inventor; to US (1886), founder of

Savage Arms Co. (NY) (1893), inventor of dirigible torpedo, made improvements in magazine rifles (MWBD)

SAVAGE, John Lucian, 1879–1967, American civil engineer; engineer with US Reclamation Service (1903–1908) (1918–1945) including construction of big dams in western US, in Idaho division working on canals, dams, reservoirs, consulting engineer (1908–1918), NAS (1949) (ANB BM49 CDOAB DAB EIH PS)

SAVAGE, Warren Fairbank, 1922–1988, American metallurgical engineer; faculty of metallurgy at Rensselaer Polytechnic Inst. (NY) and supervisor of welding laboratory (1942–1957), president of Duffers Association (1957ff), welding and joining, resistance and fusion welding, international consultant, NAE (1981) (AMOS MT5)

SAVART, Félix, 1791–1841, French physician and physicist; faculty at College of France (1828ff), studied hearing, invented toothed wheel (Savart wheel) for determining vibrations, research on resonance, polarized light (Savart plate), whirling motion of cyclones, built polariscope, studied force of jets (EB MWBD WWWIS)

SAVERY, Thomas, c.1650–1715, English military engineer and inventor; first commercially successful steam engine for pumping water (1698), Savery fire engine (1702), vacuum and steam condensate, invented ship odometer (ASAI BEST CBD EIH GEAPIT MISAT MWBD PS TBDOS(Po) TEIH)

SAVIT, Carl Hertz, 1922–1996, American mathematician and seismic expert; geophysical exploration, signal detection, data handling, Litton Industries (1948–1960), Western Geological Co., consultant, NAE (1995) (AMOS Bridge)

SAWYER, Sylvanus, 1822–1895, American inventor; machine shop work (1839ff), worked on rattan machinery with several patents, organized American Rattan Co. (MA) (until 1855), manufactured weapons (1864–1867), manufactured watchmakers' tools (1876ff) (CDOAB DAB MEIA NC4 WWWIA WWWIS)

SAWYER, Walter Howard, 1867–1923, American civil engineer; hydraulic and sanitary expert, supervised water power development in Lewiston (ME) and river storage plant (1902ff), invented a number of devices related to specialty (CDOAB DAB)

SAX, Antoine-Joseph (known as Adolphe), 1814–1894, Belgian maker of musical instruments; instructor at Paris Conservatory (1857), with his father Charles-Joseph Sax [1791–1865], also a maker of musical instruments, developed the saxhorn (1845), invented saxotromba (1845), invented the saxophone (1840, pat. 1846) (GI MWBD WWWIS)

SAXTON, Joseph, 1799–1873, American mechanician, instrument maker, and inventor; built electric generator and electric motor (c. 1830), standard weighing machinery for US Mint (PA) (1837–1843), built balances to check weights of coinage, superintendent of weights and measures of US Coast Survey (1843–1873), invented among other items a deep sea thermometer, self-regulating tide gauge, an immersed hydrometer, NAS (1863) (BDOAS(El) BM1 CDOAB DAB I&T15 MEIA MWBD NC9 PS WWWIA)

SAYBOLT, George M., ??–1924, American inventor; Saybolt universal viscometer used to obtain viscosity based on flow through orifice (1886), Saybolt Fural viscosimeter used for heavier oils, time of efflux termed Saybolt seconds universal (SSU) is a measure of viscosity used in US, the Redwood (Boverton Redwood [1846–1919]) viscosimeter (c. 1886) used in England, and the Engler (Carl Oswald Viktor Engler [1842–1925]) viscometer used in Germany (EE EOS&T OED WWW WWWIS)

SAYRE, Robert Heysham, 1824–1907, American civil engineer; engineer corps involved in surveys and construction of railroads, coal transportation particularly connecting to mines (1843–1852), and for inclined planes (1840ff), worked on enlargement of Morris Canal (1840), held several positions becoming president and chief engineer of South Pennsylvania Railroad (1852–1888) except (1882–1885), director of several companies (ANB BDOACE CDOAB DAB)

SCHADE, Henry Adrian, 1900–1992, American mechanical engineer; USNA (1923) serving in the US Navy (1923–1949) including director of Naval Research Laboratory (NRL) (1945–1949) retiring as commodore, professor of mechanical engineering and chairman of Naval Architecture at the Univ. of California (1949–1969), NAE (1973) (MT7 WWIE)

SCHADE, Otto Heinrich, Sr., 1903–1981, German American electrical engineer; radio engineering at Atwater Kent Radio Co. (1926–1931), Radio Corp. of American (RCA) (1931–1968), television, circuits, tubes, optical and electrical imaging devices, consultant, NAE (1977) (MT2 WWIE)

SCHAEFER, Vincent Joseph, 1906–1993, American physicist and inventor; General Electric Research Lab. (1933–1954), director of research for Munitalp Foundation, Inc. (1954–1959), consultant (1959–1964), faculty at the State Univ. of New York at Albany, director of Atmospheric Sciences Research Center, cloud seeding and atmospheric phenomenon (HOCS&T WWIA WWWIS)

SCHAWLOW, Arthur Leonard, 1921–1999, American physicist and inventor; Bell Telephone Labs. (BTL) (1951–1961), faculty at Stanford Univ. (1961–1991), inventor with Charles Hard Townes [b. 1915] and Theodore Harold Maiman [b. 1927] of maser (1953), optical maser (1958), the first laser was built by Theodore H. Maiman (1960), NAS (1970), Nobel prize (1981), National Medal of Science (1991) (BEST BM83 CDOS PA RHW TTOT WWWIA)

SCHEINER, Christoph, 1573–1650, German astronomer and inventor; joined Jesuit Order (1595), one of first to observe existence of sunspots, physiology of eye, invented a pantograph (1603, 1631) for copying plans on different scales, sundials and their construction (CDOSB DOSB MWBD NYPL)

SCHICHAU, Ferdinand, 1814–1896, German engineer and shipbuilder; constructed first screw-driven ships in Germany (1855), with Carl H. Ziess [1816–1888] built first gunboats with compound engines for German navy (1879) (MWBD WWWIS)

SCHICK, Jacob, 1877–1937, American inventor and manufacturer; in mine operations, explored in Alaska (1897), in Spanish-American war in Philippines (1898–1910) serving in several capacities in construction and quartermaster corps, WWI (1916–1919), invented machine to fill gas masks, invented pencil sharpener (1921), magazine razor (1923), dry razor, electric dry razor (1923), started Schick Dry Razor Co. (1930ff), also invented improvements to diesel engine, tooth paste tube and cap, card-shuffling machine, driving mechanism for motor vehicles (NC30 NYPL S&TF WA)

SCHICKARD, Wilhelm, 1592–1635, German mathematician; developed calculating machine (1624) in which six vertical cylinders converted numbers to logarithms in the process (DOSB GI). See Babbage.

SCHIEREN, Charles Adolph, 1842–1915, German American inventor; to US (1856), leather belting manufacturer, secured patents for high speed machinery (1887–1888), involved in politics in Brooklyn (NYC)—mayor (1893–1895) (CDOAB DAB)

SCHMITT, Otto Herbert, 1913–1998, American bioengineer, electrical engineer, and inventor; physiology and zoology, development of bioengineering and biophysics in interdisciplinary academic and research activities, faculty at Univ. of Minnesota (1939–1983), trigger circuits, nerve impulse mechanisms, stereovector cardiography, bioelectricity, and electronic circuitry, NAE (1979) (AMOS Bridge MT10)

SCHNEIDER, Charles Conrad, 1843–1916, Saxony American mechanical and civil engineer; to US (1867), with locomotive works then bridge and construction company (1867–1973), designer with Delaware Bridge Co. of NY, designed several bridges over rivers and structural work (1878–1886), followed by several consultancies with railroads, bridge works, government of Japan, studied collapse of Quebec bridge (BDOACE)

SCHNEIDER, Herman, 1872–1939, American civil engineer; faculty at Lehigh Univ. (1899–1903), dean of engineering at Univ. of Cincinnati (OH) (1906–1939) including acting president (4 yr.), inaugurated cooperative system of teaching and education (CDOAB DAB)

SCHNEIDER, Joseph-Eugène, 1805–1875, French industrialist; with his brother Adolphe Schneider [1802–1845] organized and developed ironworks, built first steam locomotive (1838), and first river steamboat (1840) in France (MWBD)

SCHNELLER, George Otto, 1843–1895, German American inventor; to US (c. 1860), patented and made machinery to manufacture and insert brass corset eyelets (1880–1888) (CDOAB DAB)

SCHNYDER, Walter Angelo, 1914–1977, American chemical engineer; at Hoffman-LaRoche Inc. (1942–1977) making vitamins for 12 yr., plant production of a new sulfa drug, then director of fine chemicals (1962), and later manager of fine chemical operations (1971), after serving as assistant to vice president for chemical production (ACCE NC59 PS)

SCHRÖDINGER, Erwin, 1887–1961, Austrian physicist; wave mechanics of subatomic particles, professor at Univ. of Zurich (1921–1927), Univ. of Berlin (1927–1933), Dublin Inst. for Advanced Studies (1940–1956), Schrödinger wave equation (1926), Nobel prize (1933) (BEST MWBD RHW SAI-MP TGS WWWIS)

SCHROEPFER, George John, 1906–1984, American civil and mechanical engineer; sanitary engineering, Minneapolis/St. Paul Sanitary District first as asst. chief engineer then chief engineer and superintendent (1933–1945), faculty at the Univ. of Minnesota (1945ff), NAE (1981) (MT3)

SCHUBAUER, Galen B., 1904–1992, American physicist and aeronautics; self-excited oscillations in laminar boundary layers, turbulent flow phenomena, Schubauer-Skramstad experiments in laminar instability, entire career at US National Bureau of Standards (1929–1968) where he became chief of the fluid mechanics branch, NAE (1980) (Bridge MT7)

SCHUHMANN, Reinhardt, Jr., 1914–1996, American metallurgist; extractive metallurgy, thermodynamics of high-temperature processes, applications to process analysis and design, faculty and department head then distinguished professor at Purdue Univ. (IN) (1959ff), NAE (1976) (Bridge WWIE)

SCHULTEN, Rudolf, 1923–1996, German mechanical engineer; reactor design and construction, nuclear engineer at Heisenberg Inst. (1953), Brown Boveri & Cie (BBC) in Mannheim (1956–1964), Technical Univ. of Aachen (1964–1986), fNAE (1978) (MT9)

SCHULTZ, Carl Emil, 1909–1966, American manufacturer and inventor; after serving as lawyer and being counsel to Reconstruction Finance Corp. (RFC) and Michigan State Banking Commission (1936–1943) he and his brothers-in-law developed a way to simplify and speed the analysis of steel to determine its carbon and sulfur content, served as president of the Lab. Equipment Co. (LECO) to make instruments (1936ff), formed Leco Plating Co., served US Navy in WWII (NC51 PS)

SCHUYLER, James Dix, 1848–1912, American civil and hydraulic engineer; Kansas Pacific Railroad (1869–1873), with other railroad companies, irrigation studies for California (1877–1882), design of sea wall, dams including Hemet dam (1891–1895), built water supply systems in western US, consulted on numerous international projects including Canada, Mexico, Japan, and Panama (BDOACE CDOAB DAB NC18 WWWIA)

SCHWAN, Judith Alecia, 1925–1996, American woman chemical engineer and inventor; developed photographic products including emulsions, from research to management at Eastman Kodak Co. (NY) (1950–1987) becoming assistant director of Kodak Research Labs., 21 patents, NAE (1982) (MT10 WWWIA)

SCHWARTZWALDER, Karl, 1907–1975, American ceramic engineer; with General Motors Corp. (GMC) (1931–1972) including chief ceramic engineer, director of research and development, NAE (1970) (MT1 TGS WWIE)

SCHWARZ, Berthold, 14th century, German monk and inventor; built cannon with tubular barrel, used gunpowder in weapon (1313), probably preceded by Chinese (GI MWBD)

SCISSON, Sidney E., 1917–1990, American engineer; US Corps of Engineers (1939–1943), underground storage of petroleum products and gas in US and worldwide, Pate Engineering (1945–1948), chairman of Fenix & Scisson, Inc. (1948ff), consultant, NAE (1977) (Bridge MT6)

SCOATES, Daniels, 1882–1939, American agricultural engineer; faculty at Montana State College (1910), Mississippi A & M College as professor and head of department (1911–1919), head of department at Texas A & M Univ. (1919–1939), author of texts and laboratory manuals, long-time editor for McGraw-Hill Book Co. (AE30)

SCOTT BLAIR, George William, 1902–1987, British rheologist; Rothamsted Experiment Station (10 yr.), fellow of Rockefeller Foundation at Cornell Univ., faculty at Univ. of Reading (1937–1967), head of chemistry and later physics at National Inst. for Research in Dairying, leader in industrial rheology, introduced biorheology (WWW)

SCOTT-PAINE, Hubert, 1891–1954, English aircraft and marine engineer; built first circular hull for flying boat (1913), first plane with cabins for pilot and passengers (1915), first international flying boat service (1920), designed gunboats, torpedo boats, and related items (MWBD)

SCOTT-RUSSELL, John, 1808–1882, English inventor; Scott-Russell straight line motion mechanism for direct-acting steam engine, although credited to him the invention was first patented by William Freemantle (1803) (EE)

SCOVEL, Sylvester Henry (aka Henry Sylvester), 1869–1905, American engineer and journalist; draftsman, worked on process for extracting fuel from bituminous slack, plumbing contractor and automobile dealer in Cuba, correspondent in Cuba for several US newspapers (1895–1899), involved in Cuban revolution (ANB CDOAB DAB NC2 WWWIA)

SEARLES, William Henry, 1837–1921, American civil engineer; worked in a series of positions as engineer for railroads, military defense (OH), consulting engineer for railroads and canals, American Pier and Columns Co., civil and consulting engineer (OH) (BDOACE WWWIA)

SEAVER, John Wright, 1855–1911, American mechanical and civil engineer; machinist, employed by iron works, formed Seaver and Kellogg Co. (1875), built first steel cars in US, assistant engineer for Kellogg Bridge Works (1876), chief engineer Iron City Bridge Works (PA) (1880), formed Wellman-Seaver Engineering Co. (1886ff), with J. E. Moore consulting engineers (1906ff) (MEIA)

SEBAN, Ralph A., 1917–1993, American mechanical engineer; heat transfer, boundary layer, separated flows, freezing and ice formation, faculty at Univ. of California-Berkeley (1946ff) including department chairman (4 yr.), NAE (1978) (Bridge MT7)

SECHLER, Ernest Edwin, 1905–1979, American mechanical and aeronautical engineer; construction of lightweight structures, elasticity, airplane design, development of missiles and large booster rockets, involved in construction at California Inst. of Technology (1930ff) where he was professor (1940ff) and worked with the aircraft industry, consulting, NAE (1979) (AMOS MT2)

SEE, Horace, 1835–1909, American mechanical engineer and inventor; employed in machine shops, shipbuilder shops, superintendent of George Snyder Machine Works (PA), served in Civil War (1862ff), to Cramp and Sons (PA) (1870) becoming superintendent of engineering (1879), consulting marine engineer (NYC) (1889), four patents (CDOAB DAB MEIA NC2 WWWIA WWWIS)

SEE, James Waring, 1850–1920, American mechanical engineer and inventor; participated in Civil War, established shop for iron works, later chief engineer, from foreman to chief engineer at Niles Works (OH), consulting mechanical engineer (1876), six patents (DOSB MEIA PS)

SEEBECK, Thomas Johann, 1770–1831, German physicist and inventor; worked on theory of color, discovered thermoelectricity the flow of electricity between conducting materials at different temperatures called Seebeck effect (1821), thermocouple (1822) (EDOP MWBD NYPL SAI WWWIS)

SEED, Harry Bolton, 1922–1989, English American civil engineer; faculty at Univ. of California-Berkeley (1950ff), foundation engineering, soil mechanics as related to earthquake engineering, previously at London Univ. and Harvard Univ., NAE (1970), NAS (1986) (BM66 Bridge MT5 WWIE)

SEGLER, Georg K. R., ??–1978, German agricultural engineer; director of Deere & Co. in West Germany, faculty at Stuttgart Hohenheim Univ. (ante 1975), director of Inst. fur Agrartechnik (AE59)

SEGUIN, Marc, 1785–1875, French engineer; with brother Camille Seguin [d. 1852] constructed first wire cable suspension bridge over Rhone River (1824), major role in first railroad in France (1824–1833), invented multiple-fire-tube steam engine boiler (1827) known as Seguin boiler, wrote treatises on several technology subjects (EE EIH GEAPIT MWBD PS RHW WWWIS)

SEIBERLING, Frank Augustus, 1859–1955, American inventor and manufacturer; founded Goodyear Tire and Rubber Co. (OH) (1899) then founded Seiberling Rubber Co. (OH) (1921ff), helped reorganize Goodyear (1927) (CDOAB NCC WWWIA)

SEIPPEL, Claude P., 1900–1986, Swiss American electrical engineer; Brown Boveri and Co. (BBC) (1923) (1928–1965) becoming senior executive, development, design, and testing multistage axial-flow compressors, gas turbine unit for electricity production, patents covered turbocharging, gas turbine governing, and pressure wave exchanger, work with Velox steam generator, consulting (1965ff), 38 patents, fNAE (1984) (MT4)

SEITZ, Charles Edward, 1891–1972, American agricultural engineer; organized first agricultural engineering curriculum east of Mississippi River at Virginia Polytechnic Univ. (1919), organized rural electrification movement in Virginia (1924), state committees on electrification and also for soil and water (AE49,53)

SELDEN, George Baldwin, 1846–1922, American lawyer and inventor; patent law, three-cylinder internal combustion engine using hydrocarbon fuel, used Brayton engine (George B. Brayton [1830–1892]), first gasoline engine automobile with front wheel drive (1895), invented gasoline motor propelled vehicle called road engine (1879, 1895), involved in legal battle with Ford Motor Co. (ASAI CDOAB DAB EIH ME125 MWBD NC20 PS WWWIA WWWIS)

SELLERS, Coleman, 1827–1907, American mechanical engineer and inventor; worked in Globe Rolling Mill (1846ff) becoming superintendent (1848), supervised locomotive works of Niles Company (OH) (1851), chief engineer of William Sellers & Co. (PA) (1856ff), professor of mechanics at Franklin Inst. (1881), consulting engineer (1886), chief engineer of Niagara Falls Power Co. (1890ff), and Canadian Niagara Power Co., patented mechanical devices (BITAS CDOAB DAB MEIA NC45 WWWIA)

SELLERS, Coleman, Jr., 1852–1922, American mechanical engineer; employed by William Sellers & Co.

(PA) (1886ff) becoming president (1919), State Commission of Navigation for Delaware River (MEIA NC45 WWWIA)

SELLERS, William, 1824–1905, American civil and mechanical engineer and inventor; organized Bancroft & Sellers (PA) (1848) later named William Sellers & Co. (1856) upon death of Edward Bancroft [d. 1855] for manufacture of machine tools, Sellers self-adjusting injector to supply water to boiler, standardized screw threads with 60 deg. angle first adopted by US government (1864) that 30 yr. later became international standard called Sellers screw thread also called Franklin Inst. screw thread, organized Edge Moore Iron Co. (1868), over 90 patents, NAS (1873) (ASAI BITAS CDOAB DAB EE MEIA NC7 TNAS WWWIA)

SENEFELDER, Aloys, 1771–1834, German inventor; invented lithography (1796), director of a royal printing office (1809), invented process of lithographing in color (1826), established school on lithography (GEAPIT MWBD NYPL PS SAI WWWIS)

SENMUT (also SENEMUT), fl. 1495–1475 BC, Egyptian architect, engineer and builder; erected obelisks that were then tallest (30 m.) and still remain, supervised quarrying, transportation, and erection (GEAPIT EIH TAE)

SERGEANT, Henry Clark, 1824–1907, American inventor and manufacturer; machinist, invented special machines for manufacturing wheel parts, formed firm of Sergeant & Cullingworth Co., organized Ingersoll Rock Drill Co., formed Sergeant Drill Co. (1884) then merged with Ingersoll (1886), approximately 20 patents with major contribution to rock drills (CDOAB DAB MEIA NC15)

SERPOLLET, L., 1859–1907, French inventor; invented Serpollet boiler—flash type, oil-fired boiler (1887) (EE)

SERRELL, Edward Wellman, 1826–1906, English American military and civil engineer; engaged in railroad and bridge design and construction (1845–1861), US Army (1848–1865) in topographical studies and construction work becoming colonel and brevet brigadier general, inventions included armor plate, gun carriages, coastal defenses, sand boxes for locomotives, long wire for telegraph (BDOACE CDOAB DAB WWWIA)

SESSIONS, Henry Howard, 1847–1915, American mechanical engineer and inventor; journeyman then master car builder for railroad companies (1865–1885), superintendent of Pullman Co. (IL) (1885), several patents dealing with railroad cars, vice president and director of Standard Coupler Co. (NY) (1896), at least 10 patents (CDOAB DAB MEIA)

SEVERSKY—see DE SEVERSKY

SEVERUD, Fred N., 1899–1990, Norwegian American civil engineer; building design and construction, consultant with Severud-Perrone-Szegezdy-Sturm (1928ff) particularly for brick and masonry, high-rise construction, large-span construction, NAE (1968) (Bridge MT5 WWIE)

SHANKLAND, Edward Clapp, 1854–1924, American civil engineer; with US Army engaged on the Missouri and Mississippi Rivers (1878–1883), with Wrought Iron Bridge Co. then Burnham & Root (1883–1888), opened office under his name, designed roofs and construction projects (BDOACE NC13 WWWIA)

SHARPLES, James, c.1751–1811, English American painter and inventor; to US (1793), made portraits of US notables, designed a steam carriage (MWBD WWWIA)

SHARPS, Christian, 1811–1874, American inventor; machinist, received 50 patents (1850–1874) including Sharps breech-loading rifle, mechanical superintendent of manufacturing works in Hartford (CT) where his rifle was made (1854ff) (MEIA NC5)

SHAW, Thomas, 1838–1901, American inventor; machinist, Cyclops Machine Works (PA) later superintendent, superintendent of William Butcher Steel Works (1867ff), worked in a wide variety of fields, received over 186 patents, devices to detect and record deadly gas were among his greatest contributions (1886–1890) (CDOAB DAB MEIA NC15 WWWIA WWWIS)

SHAY, Ephraim, 1839–1916, American inventor; sawmill proprietor, invented Shay geared steam engine used in logging and mining railroads (c.1877) (MWBD)

SHEA, Joseph F., 1926–1999, American aeronautical engineer; US Navy (1944–1947), various positions at Univ. of Michigan (1949–1950) (1953–1955), Bell Telephone Labs. (BTL) (1955–1959), and General Motors Corp. (GMC), developed ballistic missile technology, manned space flight with NASA (1962–1966) working on the Apollo mission, vice president for Polaroid Corp. (1967–1968), senior vice president of Raytheon Co. (MA) (1969ff), NAE (1971) (Bridge MT10 WWIE)

SHEA, Timothy E., 1899–1988, American electrical engineer; Bell Labs. (1921–1939), Columbia Univ. (1941–1945), founded US Navy underwater sound laboratory, also worked with Teletype Corp., Sandia Lab., Western Electric Co., NAE (1967) (WWIE)

SHEARER, John, 1843–1908, Australian engineer and inventor; introduced disc plow in Australia (1847), the Sovereign plow built by John Shearer & Sons (1877ff) became a great success, disc plow appeared in US (1893), brothers M. A. and J. M. Cravath invented disk plow in US (1842) but was not economically successful, a Mr. Hardy is also credited with inventing a disc plow in US (1896) (BOI NYT-A TTOT WWW)

SHEDD, Claude K., 1884–1966, American agricultural engineer; faculty at Univ. of Nebraska (1909–1911), Iowa State Univ. (1911–1925) with an appointment as

chief engineer of Nebraska Tractor Tests, extension engineer at Kansas State Univ. in charge of soil erosion, drainage, and irrigation (1925–1928), Univ. of Missouri (1928–1930), then engineer for USDA (24 yr.) (1930–1954) (AE47)

SHEDD, Joel Herbert, 1834–1915, American sanitary and hydraulic engineer; practiced civil engineering in Boston (MA) (1856–1869) and Providence (RI) (1869ff), worked on river and harbor improvements, as engineer for Providence designed and built city water works, sewage systems, large water power developments (CDOAB DAB WWWIA)

SHEN KUA, 1030–1093, Chinese engineer and astronomer; government service (1063), proposed calendar (1074), commissioner of provincial civil and military affairs (1077), wrote about magnetic compass, relief maps, origin of fossils (GEAPIT MWBD WWWIS)

SHERWOOD, Thomas Kilgore, 1903–1976, American chemical engineer; faculty at Massachusetts Inst. of Technology (1930–1969) to dean of engineering, areas of work in drying, heat and mass transfer, chemical reactions, dimensionless numbers, industrial processes, absorption and extraction, freeze drying, with OSRD (1940–1946), Univ. of California (1969ff), NAS (1958), NAE (1964) (BM63 MT1 PS SAI WWIE WWWIA WWWIS)

SHIH CH'I, fl. 300 BC, Chinese hydraulic engineer; finished building canal diverting water from Chang River to the Yellow River used for large-scale irrigation, one of the earliest recorded efforts (GEAPIT)

SHIH LU, fl. 220 BC, Chinese hydraulic engineer; built major canal for transporting troops and supplies called Magic Transport Canal, canal connected with south China (BDOTHOT GEAPIT)

SHILLING, Beatrice 'Tilly,' 1909–1990, American woman aeronautical engineer; designed a device that fit into the fuel line of an airplane to keep the flow constant, prize-winning motorcycle racer (Discover23)

SHOCKLEY, William Bradford, 1910–1989, English American physicist, electrical engineer, and inventor; Bell Labs. (1936ff), with Walter H. Brattain [1902–1987] and John Bardeen [1908–1991] developed transistor (1947) that led to the transistor radio (1953), director of research on weapons with US DOD (1955–1963), faculty at Stanford Univ. (1963ff), NAS (1951), Nobel prize (1956) (BEST BM68 GI IAI RHW TEIH TPY WWWIA WWWIS)

SHOENBERG, Sir Isaac, 1880–1963, Russian British electrical engineer; installed radio stations in Russia, to England (1914) where he worked for Electrical and Musical Industries (EMI) and headed research group (1931–1935) then became director of EMI (1955ff), principal inventor of high-definition television system (DNB MWBD WWW)

SHOLES, Christopher Latham, 1819–1890, American inventor; worked with number- and letter-writing machines (1864ff), emphasis on letter-writing machines (1867ff), with Carlos Glidden [1834–1877] and Samuel W. Soule received patent for typewriter (1868), sold rights to Remington Arms Co (1873) that developed Remington typewriter based on the patent, developed QWERTY keyboard (1872) based on work of Christopher L. Sholes [1819–1890] and James Densmore [1820–1889], at least five patents (AIE ANB ASAI CDOAB DAB GI GO-FYI IIA LAI MEIA MWBD NC3 (NYPL 1000Y PS SAI TTOT WA WBE WWWIA WWWIS)

SHORT, Sidney Howe, 1858–1902, American electrical engineer and inventor; faculty at Univ. of Denver (1880–1885), invented numerous devices for arc lighting and electric traction systems, with United States Electrical Co., Short Electric Railway Co. (1889) that was sold to General Electric Co. (1889) after which he went to England, was issued over 500 patents on improvements of electrical machinery and lighting (BITAS CDOAB DAB NC13 WWWIA WWWIS)

SHOTOKU Taishi, 573–621 AD, Japanese government official and builder; built system of highways, erected Buddhist temples, known for land reclamation (GEAPIT MWBD)

SHOUPP, William E., 1908–1981, American physicist and electrical engineer; nuclear engineering, ocean engineering, natural resources, Western Electric Co. (1938–1973) retiring as vice president, consultant (1973ff), NAE (1967) (MT2 WWIE WWWIS)

SHRAPNEL, Henry, 1761–1842, English artillery officer and inventor; Royal Artillery (1779), served in various colonies, inspector of artillery (1804), invented shrapnel shell (1784), appointed major general (1819) and commander of the Royal Artillery (1827), fuses, range tables, improved mortars and howitzers (BDOTHOT DNB MWBD 1000Y PS RHW TBDOS(Po) WA WWWIS)

SHREVE, Henry Miller, 1785–1851, American entrepreneur and steamboat captain; inaugurated fur trade from St. Louis to east (1807), skipper of second steamboat on Mississippi River, established feasibility of steamboat operation on the Mississippi and Ohio Rivers (ANB EIH MWBD)

SHREVE, Randolph Norris, 1885–1975, American chemical engineer; in charge of manufacturing chemicals at Mallinckrodt Chemical Works (1902–1904) (1907–1911), with Lamar Chemical Works (1911–1915), built and operated chemical plants (1915–1919), faculty at Purdue Univ. (IN) (1930–1961) and emeritus (1961ff), studied chemical industries of several countries (NC58 PS WWWIA)

SHULTZ, John Ernest, 1914–1973, American electrical engineer and inventor; worked for several electric and electronic companies, US Navy in WWII (1943–1946)

particularly on radar for submarines and radar control of projectiles, Eltron (1946–1949), Lear Co. (1949–1951), General Motors Corp. in AC Spark Plug division (1951–1967), worked on ballistic missiles, guidance systems, patents on in-tank fuel pump along with several other patents (NC56 PS)

SHUNK, William Findlay, 1830–1907, American civil engineer; midshipman US Navy (1846–1850), Pennsylvania Railroad (1851–1856), US Department of State, railroad surveys including intercontinental railroad, elevated railroad for Manhattan (NYC) (BDOACE NC2 WWWIA)

SIBERT, William Luther, 1860–1935, American military engineer; after graduation from US Military Academy assigned to engineer corps, improvement of river and harbor facilities and construction, Sault Ste. Marie Canal (MI) (1892–1894), several responsibilities in Philippines, Panama Canal, China flood prevention, WWI officer serving in France, promoted to major general, Alabama Dock Commission (1923) and Boulder Dam Commission (1928) (ANB BDOACE CDOAB DAB NC35 WWWIA)

SICKELS, Frederick Ellsworth, 1819–1895, American civil engineer and inventor; Sickles steering gear for ships (1860), built bridges in western US (1873), consulting engineer for National Waterworks Co. (1890), chief engineer for Kansas City (MO), received about 30 patents mostly for shipbuilding devices, consulting (ANB BDOACE CDOAB DAB EE I&T15 MEIA NC15 WWWIA)

SIDELL, William Henry, 1810–1893, American civil engineer; engineer and surveyor with NYC and harbor (1833–1837), US Hydrographic Survey (1837–1839), engineer with various railroads (1839–1846), chief engineer with various railroads (1839–1846), chief engineer with Isthmus of Panama (1846–1849), US Army (1861–1870) retired (1870) as brigadier general (BDOACE CDOAB DAB WWWIA)

SIEBE, Augustus, 1788–1872, German inventor; in England invented diving suit (1819) (MWBD)

SIEMENS, Ernst Werner von, 1816–1892, German electrical engineer and industrialist; one of four brothers who were electrical engineers, invented electroplating process (1842), gold and silver plating, invented dial telephone (1846), laid underground telegraph cable (1847), founded Siemens & Halske Co. for manufacturing telegraph equipment (1847), with brother Karl Siemens [1823–1883] laid cables under water and invented self-excited generator (1866) (EIH FNIE MWBD RHW TBDOS(Po) TGE)

SIEMENS, Karl Wilhelm (Sir Charles William), 1823–1883, German English electrical engineer and industrialist; brother of Ernst Werner Siemens [1816–1892], to England (1842), invented regenerative steam engine (1847), water meter (1851), improved process of making steel (1861), laid cable under water including from England to US (1875) (BEST MWBD PS SAI)

SIKORSKY, Igor Ivanovich, 1889–1972, Russsian American aeronautical engineer; to US (1919), built first helicopter (1909), tested biplane (1910), built cabin biplane (1912), flew first multimotor plane (1913), founded Sikorsky Aero Engineering Corp. (1923) that later became a part of United Aircraft Corp. (1929), which he headed (until 1957), developed first practical American helicopter (1939), developed XR-4 first successful military helicopter (1942) NAE (1968) (ANB CDOAB EIH LAI LIME MT1 MWBD NYPL 1000Y RHW TBDOS(Po) WWWIA WWWIS)

SILVER, Bernard, d. 1963, age 38, American inventor; with Norman J. Woodland began work on bar codes (1948) while at Drexel Inst. Technology (PA), vice president of Electro Nite, Inc., patent filed by both (1949), Norman Woodland went to IBM, patent sold to Philco (1962), bull's-eye bar code by RCA (1971), Universal Product Code (UPC) (1973) using parallel lines, first used commercially (1974) in Ohio (GI I&T8 ITTC NYTObit)

SILVER, Robert Simpson, 1913–1997, Scottish mechanical engineer; with several companies worked on improving warship pumps, boilers, desalinators with G. & J. Weir, Ltd. during WWII and (1956–1962), Gas Research Board, Imperial Chemical Industries (ICI), Federated Foundries, and the John Brown Land Boilers, Ltd., on the faculties at the Heriot-Watt Univ. in Edinburgh (1962–1967) and then the Glasgow Univ. (1967–1979), fNAE (1979) (MT10)

SILVER, Samuel, 1915–1976, American physicist and electrical engineer; faculty at Univ. of Oklahoma (until 1943), microwave antennas, microwave optics, radio-astronomy, space sciences, MIT radiation laboratory (1943–1946), Naval Research Lab., faculty at Univ. of California (1947ff), NAE (1968) (MT1 WWIE WWWIA WWWIS)

SILVER, Thomas, 1813–1888, American civil engineer and inventor; patented a marine steam engine governor (1855), numerous devices patented including hoisting apparatus, lamp, grain dryer, gas burner, lubricator, etc., received over 50 patents (CDOAB DAB MEIA NC6 WWWIA)

SIMJIAN, Luther, d. 1997, age 92, American inventor; electronics and optics, developed reflectone, invented flight simulator, ATM, computerized indoor golf practice range, exercise bike, more than 200 patents (FAL)

SIMMONS, Hezzleton Erastus, 1885–1954, American chemist and chemical engineer; faculty in rubber chemistry, director of evening division and later president of Univ. of Akron (OH) (1910–1951), during WWII was chief of rubber branch of War Production Board (1942–1946), chemical and physical properties of

Hevea rubber, rubber chemistry, rubber manufacturing, vulcanization (ACCE NC43 PS)

SIMPSON, James Hervey, 1813–1883, American civil engineer; after US Military Academy commissioned in artillery (1832) transferred to topographical engineers (1838) engaged in engineering projects in east and south (1838–1860), involved in road construction and coast survey, in Civil War, fortification and government wagon roads, retired as brigadier general (1880) (BDOACE CDOAB DAB WWWIA)

SIMPSON, Michael Hodge, 1809–1884, American inventor and manufacturer; developed machine for burring wool (patented by Frenchman Samuel Couillard) (1837), manufactured bunting (1839ff) and carpets, civic-minded industrialist (CDOAB DAB NC10 WWWIA)

SIMS, Gardner Chace, 1845–1910, American mechanical engineer; served as machinist, draftsman, and superintendent of New York Central and Hudson River Railroad, in partnership with Pardon Arrington built high speed steam engine, chief engineer in US Navy during Spanish American war becoming lieutenant commander, after war was engineer with US Army Transport Service, then Providence Police Commissioner (1902), later president of William A. Harris Steam Engine Co. (MEIA)

SIMS, Winfield Scott, 1844–1918, American inventor; first to apply electricity to torpedo propulsion, invented pneumatic field gun, a dynamite gun, several ordnance devices (CDOAB DAB WWWIA WWWIS)

SINCLAIR, Angus, 1841–1919, Scottish American mechanical engineer; marine engineer, Erie Railroad, locomotive engine with Burlington, Cedar Rapids and Northern Railroad, established methods of fuel economy and smoke prevention (MEIA WWWIA)

SINCLAIR, Donald B., 1910–1985, Canadian American electrical engineer; General Radio Co. (1936ff) from engineer to president, radar, countermeasures, guided missiles, aeronautics, NAE (1965) (MT3 WWIE)

SINGER, Isaac Merrit, 1811–1875, American inventor; machinist, innovative marketing, rock-drilling machine (1839), metal- and woodcarving machine (1849), an improved sewing machine (1851) and organized with Edward Clark the I. M. Singer Co. (1851) to manufacture it, over 20 patents (AIE ANB ASAI CDOAB DAB MEIA MWBD NC30 PS RHW TBDOS(Po) WWWIA WWWIS)

SINGLETON, Henry E., 1916–1999, American electrical engineer; lightweight inertial navigation systems, digital electronic control systems, co-founder and executive of Teledyne, Inc. (1960–1990), previous work at Hughes Aircraft, North American Aviation, and Litton Industries, several patents on gyro-stabilized and precision platforms and accelerometers, NAE (1979) (Bridge MT9)

SINK, Mary Virginia, 1913–??, American woman chemical engineer; engineer with Chrysler Corp. (1936ff), project engineer in smog abatement, supervised chemical laboratories and research, worked with radiographic and spectrographic instruments (CB64 PS WWOAW)

SKALAK, Richard, 1923–1997, American civil engineer; US Navy in radar (1944–1946), faculty in civil engineering and mechanics at Columbia Univ. (NYC) (1946ff), fluid dynamics, water hammer, surface waves, blood flow, NAE (1988) (MT9 WWIE WWWIA)

SKILLING, John Bower, 1921–1998, American civil engineer; pioneer in building engineering, structural design, chairman of Skilling Ward Magnusson Barkshire, Inc. (with firm for 23 yr.), NAE (1965) (Bridge WWIE WWWIA)

SKINNER, Halcyon, 1824–1900, American inventor; designed and constructed hand loom for Alexander Smith for carpet manufacture (1849ff), retained as mechanical expert and consultant (40 yr.), at least six patents for weaving carpets (CDOAB DAB MEIA NC5 WWWIA)

SKINNER, Le Grand, 1845–1922, American mechanical engineer and inventor; machinist, Remington Arms Manufacturing Co. (NY), built portable steam engines, formed partnership with Thomas Wood to build steam engines (1875–1883), incorporated business serving as president (1885ff), with inventions in oiling systems, governor, valves, held 33 patents (MEIA NC19)

ŠKODA, Emil von, 1839–1900, Czech engineer and industrialist; took over a machine works and organized Škoda Works (1866), known for manufacture of military equipment particularly large cannon and artillery (MWBD)

SLATER, Samuel, 1768–1835, English American industrialist; to US (1789), cotton machinery, established factory of Almy, Brown & Slater (RI) (1793), credited as founder of the American cotton manufacturing industry (AIE ASAI BITAS GEAPIT IIA MEIA MWBD NC24 PS RHW WWWIS)

SLAYTER, Russell Games, 1896–1964, American chemical engineer and inventor; consulting engineer for Owens-Illinois Glass Co. (1931–1938), vice president in charge of research and development serving as vice president of Owens-Corning Fiberglass Corp. (1938ff) (NC50 PS WWWIA)

SLICHTER, William P., 1922–1990, American materials scientist and engineer; US Army (1943–1946), development of specialized polymeric materials for telecommunications including photoresists, cross-linked dielectrics, and flame-resistant polymers, nuclear magnetic resonance spectroscope, executive director of research for AT&T Bell Labs. (1950ff), NAE (1976) (Bridge MT5 WWIE WWWIA)

SLIFER, Hiram Joseph, 1857–1919, American mining and civil engineer; Mexican National Construction Co.

(1879–1882), assistant engineer of Philadelphia Division of Pennsylvania Railroad (PRR) (1882–1891), served in engineering capacities with several eastern railways and Panama Railroad & Steamship lines as president (1912), served in France in WWI (BDOACE WWWIA)

SLOAN, Alfred Pritchard, Jr., 1875–1966, American industrialist; recognized for organization and management of large systems, administrator of General Motors Corp. (GMC) as president (1923–1937), as chairman (1937–1956) (MWBD NCE 1000Y WWWIA)

SLOAN, Robert Imlay, 1837–1901, American civil engineer; quartermaster department of Army of the Potomac (1861–1864), surveys for railroads, engineer for Trenton (NJ) for four years, Metropolitan Elevated Railroad (1876–1890), chief engineer for Chicago South Lake Elevated Railroad (1870ff), worked for engineering firm (BDOACE)

SLOCUM, Samuel, 1792–1861, American inventor and manufacturer; invented machine for sticking pins in paper so as to package them for sale (1841) (CDOAB DAB NC7 WWWIA)

SMEATON, John, 1724–1792, English civil engineer; instruments used in navigation, rediscovered hydraulic cement (1756), improvements in windmills and watermills (1759), lighthouse reconstruction (1759), Smeaton waterwheel, pumping engines, harbor construction (1774), structural engineering, bridges, considered as founder of civil engineering in Great Britain (EE EIH GEAPIT LOTE MWBD PS RHW TBDOS(Po) TEIA TGE:B WWWIS)

SMILES, Samuel, 1812–1904, English historian of science and technology; studied and practiced medicine, editor, writer, and lecturer (1838ff), with Leeds and Thirst Railway (1845–1854) then secretary of South Eastern Railway (1854–1866), president of National Provident Inst. (1866–1871), concept of model engineer, transportation and materials handling technology, wrote multiple volumes on lives of engineers (DNB LOTE MWBD SFLOE)

SMILLIE, Ralph, 1867–1960, American civil engineer; first worked for NYC subway system, during WWI was in US Navy, assistant engineer advancing to design engineer for Holland Tunnel (1919–1928), transit engineer for Newark (NJ) (1929–c. 1934) design of Lincoln Tunnel (1934–1945) with many innovations, engaged by Port Authority for tunnels (1944–1950), partner in consulting engineering firm of Smillie and Griffin (1950ff) (ANB CDOAB DAB)

SMITH, Apollo M. O., 1911–1997, American aeronautical engineer; aerodynamics and rocketry, pioneered in advances in computational aerodynamics, during WWII was chief engineer at Aeroject Engineering Corp., most of career at McDonnell Douglas Corp., NAE (1989) (Bridge WWIE WWWIA)

SMITH, Charles Shaler, 1836–1886, American bridge engineer; assignments with several different railroads for tracks, bridges and buildings, captain of engineers in Confederate Army, partner in an engineering firm that became Baltimore Bridge Co. becoming president and chief engineer (1869–1877) (BDOACE CDOAB DAB EIH NC3 WWWIA)

SMITH, Francis Hopkinson, 1838–1915, American mechanical engineer, artist and writer; initially worked in a foundry in NYC, established himself as an engineer and his firm was in construction work for 30 yr., several marine contracts for government including the foundation for the Statue of Liberty, lighthouse, breakwater works, did painting and made literary contributions (ANB CDOAB DAB NC5)

SMITH, Sir Francis Pettit, 1808–1874, English inventor; patent on a screw propeller (1836) for steamship, constructed screw-propelled steamer for British navy (1839), Smith propeller (1856), was advisor to British Admiralty (ante 1850) (DNB EE EIH MWBD WWWIS)

SMITH, Georg E. P., 1875–1975, American civil and agricultural engineer; designed early steel bridges in Arizona, faculty and department head at Univ. of Arizona (1906–1944), studied groundwater, irrigation, designed water and sewer systems (AE56 WWIE(1) WWWIA)

SMITH, Gustavus Woodson, 1822–1896, American military and civil engineer; US Military Academy (1842) where he taught (1844–1846) (1849–1854), employed by iron works and NYC street commission, Confederate Army where he was promoted to major general and commanded field troops (1861–1865), superintendent of Southeast Ironworks (1866–1869), Kentucky Insurance Commission (1870–1876) (BDOACE CDOAB DAB NC7 WWWIA)

SMITH, Hamilton, 1840–1900, American mining engineer; engineer and superintendent for building and connecting railroads in Indiana, Kentucky, California, and Nevada, authority on hydraulic mining, designed and built large dams, consulting engineer for several mining companies including El Calto Mine in Venezuela (1881–1885) (BDOACE BITAS CDOAB DAB NC12 WWWIA)

SMITH, Harold Babbitt, 1869–1932, American electrical engineer; faculty at Univ. of Arkansas (1892), head of department at Purdue Univ. (IN) (1893–1896), then to Worcester Polytechnic Inst. (MA) (1896ff), consulting engineer, Westinghouse Electric and Manufacturing Co. (1905ff) (CDOAB DAB WWWIA)

SMITH, Horace, 1808–1893, American inventor and manufacturer; worked with Daniel Baird Wesson [1825–1906] (1853ff), patented a revolver with Wesson (1854), metal cartridge (1854), manufactured repeating rifle and revolver (1857ff), executive head of Smith and Wesson Co. (1857–1873) (CDOAB DAB GI MEIA MWBD NC10 WWWIA)

SMITH, Jesse Merrick, 1848–1927, American mechanical engineer; designed and built blast furnaces (1873ff), surveyed and opened mines, consulting engineer (1880ff), designed power plants, railway lighting and electric plants, apparatus for heating (MEIA NC22 WWWIA)

SMITH, Jonas Waldo, 1861–1933, American civil engineer; water supply expert, first operated a small water works system, assistant engineer at Holyoke Water Power Co. (1887–1890), assistant engineer then chief engineer for East Jersey Water Co. (12 yr.), chief engineer of Aqueduct Commission of NYC then to Board of Water Supply, responsible for dams, reservoirs, and tunnels (1903–1922) (CDOAB DAB WWIE(I))

SMITH, Leslie John, 1881–1968, American agricultural engineer; initiated and headed farm mechanics at Michigan State College (1906–1909), Manitoba Univ. in Canada (1909–1920), head of department at Washington State Univ. (1921–1949) with laboratory building named after him, instrumental in forming and serving as secretary (23 yr.) of Washington Farm Electrification Committee (AE49 NUC WWIE(I))

SMITH, Levering, 1910–1993, American naval engineer; numerous capacities in US Navy (1932–1977) including naval ordnance, missile testing, and strategic systems for US Navy, development of solid propellants, director of strategic projects, retired as an admiral, consultant, NAE (1965) (Bridge MT7 WWIE)

SMITH, Oberlin, 1840–1926, American mechanical engineer; established shop for jobbing (1863), joined with John B. Webb [1841–1912] (1864) to organize Ferracute Machine Co. (1873), developed die presses, over 75 patents (MEIA NC12 WWWIA)

SMITH, Robert L., 1923–1995, American civil engineer; faculty and chairman of department at Univ. of Kansas (1962ff), Iowa Resources Council (1952–1955), Kansas Water Resource Board (1956–1962), water resources planning, urban hydrology, numerous consulting assignments, NAE (1975) (Bridge MT8 NC48 WWIE WWWIA)

SMITH, Wilbur S., 1911–1990, American electrical engineer; traffic engineering and ground transportation, consultant with Wilbur S. Smith Management (30 yr.), NAE (1968) (Bridge MT6 WWIE WWWIA)

SMITH, William, 1769–1839, British civil engineer and geologist; participated in surveying, inspection and construction of canals and mines (1794–1799), developed procedure for identifying coal strata, later he worked in drainage, irrigation, and coal-mine engineering, designed sea defenses in South Wales (Australia) (GEAPIT)

SMITH, William Farrar, 1824–1903, American military and civil engineer; graduate of US Military Academy, assigned to topographical engineers, numerous surveys including upper Great Lakes, Mexican border, Florida ship canal, numerous tasks with Union Army during Civil War becoming major general (until 1867), president of International Ocean Telegraph Co. (1865–1873), practice of civil engineering (CDOAB DAB WWWIA)

SMITH, William Sooy, 1830–1916, American civil engineer; graduate of US Military Academy (1853), served in artillery, resigned after work on harbor improvements on Lake Michigan, built bridges for railroads, served in Civil War retiring as brigadier general (1861–1864), practiced as civil engineer in Chicago and for US government, developed freezing method of putting foundations in quicksand and soft soils, invented first pneumatic caisson and designed fireproof buildings (DOAB DAB EIH WWWIA)

SMITHSON, James (aka James Louis Macie, ante 1801), 1765–1829, English scientist and philanthropist; chemistry, mineralogist, carbonate of zinc named smithsonite also known as calamine as a result of his contributions, estate left to establish Smithsonian Institution in US (1846) (BITAS EB MWBD RHW NC3 WWWIA WWWIS)

SNOW, Jonathan Parker, 1848–1934, American civil engineer; faculty at Thayer School (NH) (1877–1889), bridge engineer then chief engineer for B&M Railroad (1888–1911), consulting engineer (1911ff) (ANB CDOAB DAB WWWIA WWWIS)

SODERBERG, Carl Richard, 1895–1979, Swedish American mechanical engineer; to US (1919), citizen (1927), worked for Westinghouse Electric and Manufacturing Co. (1922–1938) becoming manager of the turbine division, faculty and administration at Massachusetts Inst. of Technology (1938ff) becoming dean of engineering (1954–1965) and later headed academic departments, consultant with marine industry, 18 patents, NAS (1947), NAE (1974) (ANB BM75 CB58 MT2 WWIE WWWIA WWWIS)

SOLEIL, Jean-Baptiste-François, 1798–1878, French instrument maker; worked with Augustin-Jean Fresnel [1788–1827] (1823–1827) in construction of lens, use of lens for lighthouses, optical research, invented apparatus to measure the interaxial angle in biaxial crystals and an improved saccharimeter, retired (1849) (DOSB PS WWWIS)

SOLVAY, Ernest, 1838–1922, Belgian industrial chemist and inventor; invented Solvay process for making soda ash (1861), established Solvay & Cie in Belgium (1863), process licensed to US Solvay Process Co. formed (1884) that eventually became a part of Allied Chemical & Dye Corp. (1920) (MWBD NC19 RHW TodChem11 WWWIS)

SOMERSET, Edward (aka Marquis of Worcester), 1601–1667, English inventor; mechanical experimenter (1628ff), attempted to show perpetual motion using a large wheel (DNB EIH GEAPIT MWBD NCF WWWIA)

SOMERVELL, Brehon Burke, 1892–1955, American military officer and builder; graduate of US Military Academy (1914), logistics, WWI, headed WPA in NYC (1936), in charge of construction of the Pentagon in record time (1941–1943), retired as general (CDOAB WWWIA)

SOMMEILLER, Germain, 1815–1871, French engineer; developed compressed-air drill (1861), used drill to build Mont Cenis tunnel between France and Switzerland (1870) (EIH MWBD NYPL)

SONG IYONG, fl. 1664, Korean horological engineer; engaged in clock making, built western-style clock for King Hyojong probably a verge-and-foliot escapement regulating the action of a falling weight or spring (GEAPIT)

SOOYSMITH, Charles, 1856–1916, American civil engineer; foundations, developed freezing process for excavating unstable soil (BITAS CDOAB DAB WWWIA WWWIS)

SOPWITH, Sir Thomas Octave Murdoch, 1888–1989, British aircraft designer and industrialist; established Sopwith Aviation Co. (1912) that produced the Camel, Pup, and Strutter military planes during WWI, established what became the Hawker Sideley Aircraft Co. (1920) that produced the Hawker Hurricane plane and later the Harrier vertical take-off jet fighter (MWBD WWWIS)

SOSTRATOS, 3rd century BC, Greek architect and builder; built lighthouse at Alexandria (c.280 BC) that was later identified as one of the Seven Wonders of the World (EIH GEAPIT MWBD TEIH)

SOUDERS, Mott, Jr., 1904–1974, American chemical engineer; thermodynamics, separation processes, and engineering economy, Shell Development Co. (1937–1940), Director of Development in San Francisco (1940–1950) and Emeryville (CA) (1950–1962), consultant (1931–1935) (1963ff), NAE (1970) (MT1 WWIE)

SOUFFLOT, Jacques-Germain, 1713–1780, French architect and engineer; worked in Paris for government restoring, building, and consulting on structures (1756ff), application of statistics to building construction, materials testing, innovative design of Saint Genevieve (Panthéon Francais) with open structure without massive piers to support the dome (1764), used iron clamps to hold stone and concrete (EOWB GEAPIT MWBD PS)

SOUTHER, John, 1816–1911, American inventor; designer and builder of locomotives, steam engines, and heavy machinery, partnership Lyman & Souther (MA) (1846ff) then Globe Works (MA), after Civil War firm renamed John Souther & Co. (1865) as builders of machine tools and machinery, New England Dredging Co. (1870) (MEIA)

SOUTHERN, J., 1758–1815, British engineer and inventor; forge engines (c. 1782), developed Southern steam indicator to show power that a steam engine developed (1796) (EE TGE:B)

SOWERS, George F., 1921–1996, American civil engineer; geotechnical engineering, dams, tunnels, earth construction, foundations, engineering consultant of Law Consulting Group (1947ff), hydraulic research, faculty at Georgia Inst. of Technology (1947ff), NAE (1994) (Bridge WWIE WWWIA)

SPAGHT, Monroe E., 1909–1993, English American chemist and engineer; manager of research, design, construction, and operation in the petroleum and petrochemical industry, managing director of Royal Dutch/Shell Group Co. (1933–1953), director of Stanford Research Inst. (1953–1970), NAE (1969) (Bridge MT7 WWIE)

SPANGLER, Henry Wilson, 1858–1912, American mechanical engineer; graduate of USNA (1878), faculty and administration at Univ. of Pennsylvania (1884ff), marine engineering, designer for World's Columbia Exposition (1893ff) (BITAS CDOAB DAB MEIA)

SPARKS, William Joseph, 1904–1976, American chemist and inventor; with Sherwin Williams Co. (1926), DuPont (1929), US Department of Agriculture (1940), Esso Research (1940ff), 125 US patents, NAE (1967) (ANB MT1 WWIE WWWIA)

SPENCER, Christopher Miner, 1833–1922, American inventor; journeyman, machinist, toolmaker (1849), Colt Armory (CT), Spencer Repeating Rifle Co. (1860), made repeating rifle (1860), formed Billing & Spencer Co. (1869–1874) manufacturing drop forges, formed other companies, seven patents (ANB CDOAB DAB MEIA NC22 WA)

SPENCER, Percy Le Baron, 1894–1970, American inventor and engineer; superintendent of American Radio and Research Corp. (1915–1918), manager of field engineering at Signal Co. (1920–1925), Raytheon Manufacturing Co. (MA) (1925–1970) serving in microwave division and several management positions, holder of 100 patents on electronics (GI WWWIA)

SPERRY, Elmer Ambrose, 1860–1930, American electrical and mechanical engineer, inventor, and industrialist; invented dynamo and new type of arc lamp (1879), organized Sperry Electric Co. (1880), organized Sperry Electric Railway Co. (1890) later sold to General Electric Co., gyroscopic compasses and stabilizers for ships and airplanes, control theory, organized Sperry Gyroscope Co. (1914) which he headed (until 1929), high-intensity searchlight (1918), with Clinton P. Townsend developed a chlorine detinning process to recover tin from old cans and scrap and the Townsend process (1900) for manufacture of pure caustic soda from salt, device to detect rail fissures electromagnetically, founded eight manufacturing companies, over 400 patents, NAS (1925) (AG AIE ANB BEST BM28

CDOAB DAB HFP I&T1999 MEIA MWBD NC23 PS TBDOS(Po) TTOS WWWIA WWWIS)

SPERTI, George Speri, 1900–1991, American electrical engineer; developed electric motor for large industries (1921), Univ. of Cincinnati (OH) (1923–1935), co-founder and director of research of Basic Science Research Lab. (1925–1935), Sperti Lamp Corp. (1930–1940), research professor to president of St. Thomas Inst. (OH) (1955ff), irradiation to produce vitamin and to kill microorganisms, sun lamp, preservation of orange juice, holder of 150 patents (ANB CB40 PS WWWIA WWWIS)

SPILHAUS, Athelstan Frederick, 1911–1998, South African American meteorologist, oceanographer, and engineer; to US (1931), devised Bathythermograph (1938) for underwater research (modified by Allyn Collins Vine [1914–1994] for antisubmarine warfare), faculty of New York Univ. (1937–1949), USAAF (1943–1946), faculty and dean of Inst. of Technology at Univ. of Minnesota (1949–1966), numerous industry and government assignments including Union of South Africa Defense Services, Junkers Airplane Works, Sperry Gyroscope Co., Massachusetts Inst. of Technology, and Woods Hole Inst., consultant (AMOS CB NYTObit WWWIA WWWIS)

SPILSBURY, Edmund Gybbon, 1845–1921, English American mining engineer; supervised mines in England and continent, explored for lead and zinc around Great Lakes (1870), introduced Harz system of ore dressing (1873), worked in various iron works in US, managing director of Trenton Iron Co. (NJ) (1888ff), consulting engineer (1897) (BITAS CDOAB DAB MEIA)

SPORN, Philip, 1896–1978, American electrical engineer; electric power systems, utility engineering, transmission, protection of electrical systems, distribution facilities, American Electric Power Co. becoming president (1920–1961), consultant (1961–1971), NAS (1962), NAE (1965) (ANB MT1 PS WWIE WWWIA WWWIS)

SPRAGUE, Frank Julian, 1857–1934, American electrical engineer and inventor; organized Sprague Electric Railway & Motor Co. (1884), manufactured electric motors, was pioneer in electric trolley system the first of which was in Richmond (VA) (1887), devised and installed high-speed automatic building elevators, invented multiple-unit system of train control (1895), developed high-tension direct-current electric railway system (ANB CDOAB DAB MWBD NC24 PS WWWIA WWWIS)

SPRAGUE, Robert Chapman, 1900–1991, American electrical engineer; US Navy (1920–1928) in construction, entrepreneur in inventing electronic components and developing a major electronics company, president and chairman of Sprague Electric Co. (58 yr.), NAE (1985) (Bridge MT6 NCI WWIA WWWIA)

SPRENGEL, Hermann Johann Philipp, 1834–1906, German English chemist, physicist, and inventor; to

England (1859), carried out research at Oxford Univ. and universities in London, researched high explosives, invented the Sprengel pump (1865) an air pump using liquids, developed pyknometer to determine the density and expansion of liquids, FRS (1878) (CBD DNB MWBD PS WWWIS)

SQUIER, George Owen, 1865–1934, American electrical engineer; graduate of US Military Academy (1887), appointed as officer in artillery with several assignments including signal corps, laying of cable telegraph system, promoted to major general, in charge of US Army air service (1916–1918), numerous studies on magnetism, wave system of telegraphy, invented monophone for broadcasting and Quickaid kit for army and Red Cross, NAS (1919) (ANB BM20 CDOAB DAB NC24 PS WWWIA WWWIS)

STALEY, Cady, 1840–1928, American civil engineer; worked on building of Central Pacific Railroad, sewage studies, professor and dean of faculty of Union College (NY) (1868–1886), president of Case School (OH) (1886–1902), consultant (CDOAB DAB NC11 WWWIA)

STANFORD, Leland, 1824–1893, American railroad builder and politician; sold mining supplies, promoted Central Pacific Railroad (1863–1869) of which he was president (1863–1893), founded forerunner of what became Stanford Univ. (1885ff), served as governor of California (1861–1863) and in the US Senate (1885–1893) (ANB EB MWBD NC2,15 WWWIA)

STANHOPE, Charles, 3rd Earl, 1753–1816, English inventor and politician; constructed calculating machines (c.1777), perfected a process of stereotyping (1805), invented a monochord for tuning musical instruments, a steam carriage, first hand operated iron printing press (1798), invented a Stanhope microscopic lens (DNB GEAPIT LOTE MWBD SFLOE WWWIS)

STANIER, William Arthur, 1876–1965, British mechanical engineer; his father W. H. Stanier spent lifetime in railroad engineering, with Great Western Railroad (1892–1932) and moved through several positions to become principal assistant to chief mechanical engineer (1922), became chief mechanical engineer at London Midland and Scottish Railroad (1932ff), noted for improvement of efficiency and reliability of railway operation (DNB PS RS12 WWW)

STANLEY, Francis Edgar, 1849–1918, and twin brother Freelan Oscar Stanley [1849–1940], American inventors and manufacturers; invented photographic dry plate process (1883), built a steam powered automobile (1897, pat. 1900), founded and managed Stanley Motor Carriage Co. (1898–1917), built the Stanley Steamer, which established speed records (1906) (AIE ANB ASAI CDOAB DAB MEIA MWBD NC18 PS WWWIA)

STANLEY, William, 1858–1916, American electrical engineer and inventor; invented transformer (1885), installed alternating current distribution systems (1886ff) and a watt meter, invented two-phase electric motors (BITAS CDOAB DAB EIH HOCS&T MWBD NC24 NYPL WA WWWIS)

STANTON, Thomas Ernest, 1865–1931, English civil and mechanical engineer; fluid flow in pipes, hydrodynamics, aerodynamics, resistance to flow, hardness testing, demonstrator at Owens College (1891–1896), senior lecturer at Union College, Liverpool (1896–1899), professor at Union College, Bristol (1899–1901), superintendent of National Physical Lab. (NPL) (1901–1930) (CDOSB DOSB PS WWW)

STARBIRD, Alfred D., 1912–1983, American civil and military engineer; from US Military Academy commissioned in US Army Corps of Engineers (1933–1971) retiring as a lieutenant general, involved in numerous defense activities in Europe, Asia, and US, Atomic Energy Commission, nuclear tests, and antiballistic missiles (ABM), sentinel systems manager, NAE (1973) (MT3 WWIE WWWIA)

STARLEY, James, 1831–1881, English inventor; invented and manufactured several kinds of sewing machines (1857ff), manufactured improved bicycle with spoked wheel (1874), differential gear with chain drive (1877), and tricycle (DNB MWBD NYPL 1000Y WWWIS)

STARR, Eugene C., 1901–1988, American electrical engineer; high-voltage research with General Electric Co. (1923–1927), faculty at Oregon State Univ. (1927–1954), chief engineer with Bonneville Power Administration (BPA) (OR) (1954–1961), consulting engineer, NAE (1977) (MT5 WWIE)

STARR, John Edwin, 1860–1931, American mechanical engineer and inventor; built refrigerating plants and cold storage warehouses in various cities, organized Starr Engineering Co. (1898ff), at least five patents on improving refrigerating systems (MEIA)

STARRETT, Laroy S., 1836–1922, American engineer, architect, and inventor; after 12 yr. farming devised meat chopper (1865), negotiated interests to manufacture then later purchased and was superintendent (10 yr.) of the Athor Manufacturing Co. (MA), invented washing machine (1865) and butter worker (1865), invented several hand tools such as calipers, rules, levels, gauges, steel tapes, plumb bobs, hacksaw, company incorporated as L. S. Starrett Co., of which he was president (1912–1922) (DAB NC18)

STARRETT, Paul, 1866–1957, American builder; was president of George A. Fuller Co. (1903–1922), cofounder with brother Ralph Starrett [1868–1930] and president of Starrett Brothers, Inc. (1922ff) that built several major NYC buildings (MWBD NCD WWWIA)

STARRETT, William Aiken, 1877–1932, American military and civil engineer, architect and builder; with

George A. Fuller Co. becoming superintendent (1899), vice president of Thompson-Starrett Co. (1901–1913), designed and built several commercial buildings, partner in architectural firm of Starrett & Van Vleck (1910–1918), formed construction firm of Starrett Brothers, Inc. (1922) with brothers Paul Starrett [1866–1957] and Ralph Starrett [1868–1930] built Empire State Building (NYC) (1931) (CDOAB DAB MWBD NC24 WWWIA)

STAUFFER, David McNeely, 1845–1913, American civil engineer; engraving of prints, autographs, wrote about engraving on copper and steel (1907) (CDOAB DAB NC9 WWWIA)

STEARNS, Frank Ballou, 1878–1955, American inventor and manufacturer; founded F. B. Stearns Co. (1898) makers of Stearns-Knight automobiles, introduced internal combustion engine (1912), left company (1918), worked on diesel engines which he sold to US Navy (1935), had at least 16 patents (CDOAB DAB NC45)

STEARNS, Frederic Pike, 1851–1919, American civil engineer; planned improvements to Charles River basin, engineer for Massachusetts Board of Health (1895–1907), saw completion of water supply for Boston and vicinity (MA), served as consulting engineer for several municipal water supply projects and for Panama Canal (CDOAB DAB NC14 WWWIA)

STEARNS, Irving Ariel, 1845–1920, American mining engineer; managed coal interests for the Pennsylvania Railroad (PRR) (1885–1897), made several improvements to mining and processing of anthracite coal, headed several coal and iron companies (ante 1915) (CDOAB DAB NC11)

STEIN, Earl Roscoe, 1886–1957, American chemical engineer; Midland division of Dow Chemical Co. (1911ff) responsible for chlorine cells (5 yr.), head of calcium-magnesium products section (30 yr.), mayor of Midland (MI) (1936–1940) (NC48 PS)

STEINMAN, David Barnard, 1886–1960, American civil engineer; renowned bridge designer and builder with worldwide reputation, built over 400 bridges worldwide including suspension bridge in Brazil then the largest in South America (1922–1926), Carquinez Strait Bridge (CA) (1923–1927), St. John's Bridge in Portland, (OR) (1929–1931), George Washington Bridge (NYC), Harbor Bridge in Sydney (1932), Triborough Bridge (NYC), Thousand Islands International Bridge (NY, Canada) (1938), Mackinac Bridge (MI) (1957), consulting engineer (1923ff) (ANB CB CDOAB DAB ET26 MWBD NCF PS WWWIA WWWIS)

STEINMETZ, Charles Proteus, 1865–1923, German American electrical engineer and inventor; to US (1889), consulting engineer with General Electric Co. (1893ff), professor at Union College (NY) (1902ff), worked on electric machinery, promoted alternating current phenomena, developed hysteresis laws, light-

ning arresters for high-power transmission lines, 20 cent postage stamp named for him (1983), over 200 patents (ANB ASAI BEST CDOAB DAB DOSB MWBD NC23 1000Y PS RHW TEIH TGS WWWIA WWWIS)

STEPHENS, Charles W., 1930–1990, American electrical engineer; special-purpose communications and information systems, spacecraft electronics, microelectronics, Bell Labs. (until 1954), vice president and general manager of TRW, Inc. (originally Ramo Wooldridge—a predecessor) (1957–1986), consultant (1986ff), NAE (1985) (Bridge MT5)

STEPHENSON, George, 1781–1848, English engineer and inventor; founder of railroads, devised miner's safety lamp (1815) (slightly preceding that of Humphry Davy [1778–1829]), built railway (1814), patented locomotive with steam blast (1815), flanged wheels, Stephenson locomotives pulled passenger cars on railroad (1825), construction engineer for railroads, his Rocket locomotive had tubular boiler (1829), used gauge of 4 ft. 8.5 in. width used in Great Britain for the Manchester to Liverpool Railroad line (1830) (BEST DNB EIH FNIE GEAPIT LOTE MWBD PS SAI SFLOE TBDOS(Po) TEIH TGE:B WWWIS)

STEPHENSON, George Robert, 1819–1905, English engineer; nephew of George Stephenson [1781–1848] for whom he worked, directed construction of railroads and bridges, proprietor of locomotive works (1859–1886) (DNB GI MWBD PS TEIH)

STEPHENSON, Robert, 1803–1859, English civil engineer; son of George Stephenson [1781–1848], mining operations in South America (1824–1827), manager of his father's locomotive works (1827ff), chief engineer for railroad (1833–1838), recognized for his railroad bridges in England and Canada (DNB EIH LOTE MWBD PS RHW SFLOE TBDOS(Po) TGE:B WWWIS)

STERN, Arthur C., 1909–1992, American environmental engineer; faculty at Univ. of North Carolina (1968–1978), assessing of air pollution, industrial hygiene, worked successively in air pollution for NYC (ante 1943), New York state (ante 1954), federal government as assistant chief for air pollution at US Department of Health, Education and Welfare Department (ante 1968), NAE (1976) (Bridge MT6 WWIE WWWIA)

STERNBERG, Eli, 1917–1988, Austrian American civil engineer; continuum mechanics and elasticity, faculty at Illinois Inst. of Technology (1945–1956), Brown Univ. (RI) (1957–1964), California Inst. of Technology (1964ff), several visiting professorships, NAE (1975), NAS (1979) (MT5 WWIE WWWIA)

STETEFELDT, Carl August, 1838–1896, German American metallurgist and inventor; to US (1863), invented Stetefeldt furnace, developed furnace for reducing sulphite ores containing gold and silver using chlorination process (BITAS CDOAB DAB WWWIA WWWIS)

STEVENS, Edwin Augustus, 1795–1868, American engineer and inventor; son of John F. Stevens [1749–1838], patented plow (1821), supervised Union Line Railroad (1825–1830), treasurer and manager of Camden & Amboy Railroad (1830), built armored vessels for navy, founded Stevens Inst. of Technology (NJ) (est. 1870) (CDOAB DAB MEIA MWBD WWWIA WWWIS)

STEVENS, John F., 1749–1838, American inventor and steam engineer; captain in Revolutionary War, secured federal legislation establishing first patent laws (1790), secured patent on first vertical steam boiler (1791), built screw-driven steamboat (1802), patented first multitubular boiler (1803), built first American-made steam locomotive (1825), two sons, Robert Livingston Stevens [1787–1856] and Edwin Augustus Stevens [1795–1868], were engineers and inventors who followed in their father's work (ANB ASAI BITAS CDOAB DAB EE EIH GEAPIT LAI MEIA MWBD PS WA WWWIA WWWIS)

STEVENS, Robert Livingston, 1787–1856, American engineer and inventor; designed and built steamboats and ferries, perfected first marine tubular boiler, president and chief engineer of Camden and Amboy Railroad and Transportation Co. (1830ff), built locomotives and armaments (AIE BITAS CDOAB DAB EIH GEAPIT LAI MEIA NC11 TEIH PS WWWIA WWWIS)

STEVENSON, Alan, 1807–1856, Scottish engineer; son of Robert Stevenson [1772–1850] who became an engineer, designed 10 lighthouses and introduced prismatic rings (DNB MWBD WWWIS)

STEVENSON, David, 1815–1886, Scottish engineer; son of Robert Stevenson [1772–1850] who became an engineer, worked on improvement of rivers in Scotland and England, built beacons and lighthouses, introduced use of paraffin (1870) (DNB EIH MWBD)

STEVENSON, Robert, 1772–1850, Scottish civil engineer and inventor; designed and built lighthouses (1772–1843), invented intermediate and flashing lights and the hydrophore, three sons were engineers: Alan Stevenson [1807–1865], David Stevenson [1815–1886], and Thomas Stevenson [1818–1887] (DNB MWBD RHW WWWIS)

STEVENSON, Thomas, 1818–1887, Scottish engineer; son of Robert Stevenson [1772–1850], invented azimuthal condensing system, devised Stevenson screen for thermometers (1864), contributed to meteorology (DNB MWBD)

STEVIN, Simon (aka Stevinus of Bruges), 1548–1620, Dutch mathematician, hydraulic engineer, and inventor; invented a drainage windmill (1588), supervised hydraulic works near Delft, built wind-driven land vehicle (1600), popularized the use of decimals (CDOSB EIH GEAPIT MWBD NYPL 1000Y TSOE WWWIS)

STEWART, Philo Penfield, 1798–1868, American clergyman and inventor; invented Oberlin stove (1834), deeded income to college that he helped establish, with John J. Shipherd [1802–1844] founded Oberlin College (OH) (1833) (CDOAB DAB EB MWBD NC2 WWWIA)

STEWART, Robert E., 1915–1983, American agricultural engineer; USAAF pilot and engineering officer in WWII (1941–1945), faculty at Univ. of Missouri (1953–1961), Ohio State Univ. (1961–1968) including department chairman, Texas A&M Univ. (1968ff), structural design, environmental control, thermoelectric calorimetry, NAE (1978) (AE64 MT3 WWIE WWWIA)

STIBITZ, George Robert, 1904–1995, American applied mathematician; recognized for his work in computer technology including binary arithmetic, particularly as applied to propeller vibration, machine tool design, automatic control, dynamics, memory indexing, floating-point arithmetic, operation from remote consoles, electronic music, and program-controlled computations, Bell Telephone Labs. (BTL) (1930–1941), National Defense Research Committee (1941–1946), visiting professor at Univ. of Vermont (1945ff), consultant (1945ff), NAE (1981) (AMOS Bridge WWWIA)

STILLMAN, Francis H., 1850–1912, American mechanical engineer; Cottrell Printing Press (RI) (1874), president of Watson & Stillman Co. (1883ff), president of Bridgeport Motor Co. (CT), hydraulic engineering and hydraulic machine tool construction (MEIA)

STILLMAN, Gregory E., 1936–1999, American electrical and computer engineer; MIT Lincoln Laboratory (1967–1975), faculty at Univ. of Illinois (1975–1999), developed methods of characterizing high-purity semiconductors, discovered the behavior of the discrete donor in gallium arsenide, NAE (1985) (Bridge MT9 WWIE)

STILLMAN, Thomas Bliss, 1852–1915, American chemist and chemical engineer; opened analytical laboratory (1879), faculty of Stevens Inst. of Technology (NJ) (1882–1909), interested in industrial practices, consulting (1909ff) and became member of firm Stillman and Van Sielen, city chemist (1911–1915) (ACCE BITAS DAB NC16 PS WWWIA)

STILLWELL, Lewis Buckley, 1863–1941, American electrical engineer; railroad electrification and electric power transmission, helped develop alternating current theory and 60-cycle frequency, Westinghouse Electric Co. (1886–1897), Niagara Power Co. (NY) (1897–1900), consulting engineer (1900ff), NAS (1921) (BM34 PS WWWIA)

STINE, Charles Milton Atland, 1882–1954, American chemist and chemical engineer; developed several products and processes for organic chemicals at E. I. DuPont (DE) (1907–1954) advancing from researcher to chemical director (1924) to vice president and member of executive committee (1930–1945) remaining as a director of synthetic dye processes, managed development of new families of fibers, films, plastics, paints, and elastomers, manager of research, DuPont named animal research laboratory after him (ACCE CB40 DAB DOSB NCE,47 PS WWWIA WWWIS)

STIRLING, Allan, 1844–1927, Scottish American mechanical engineer and inventor; several positions as draftsman, superintending engineer, mechanical specialist, established boiler business in Canada with Rathbun Co. (1885ff), established Stirling Boiler Co. (NYC) (1888–1918), Stirling four-drum boiler (1892), four boiler patents (MEIA NC21)

STIRLING, Robert, 1790–1878, Scottish clergyman and inventor; church pastor (1824ff), invented Stirling closed regenerative thermodynamic cycle hot air engine (1816) (EB EE FNIE MWBD RHW)

ST. JOHN, Isaac Munroe, 1827–1880, American civil engineer; construction superintendent with Blue Ridge Railroad in Georgia (1855–1861), engineering duty with Confederate Army (1861–1865) including mining bureau advancing to brigadier general, followed by service as chief engineer of railroad, city engineer of Louisville (KY), consulting engineer (ANB BDOACE CDOAB DAB WWWIA)

STODDARD, Joshua C., 1814–1902, American inventor; invented and patented (1855) a steam calliope that he manufactured, devised and patented hay-raking implements (1861, 1870, 1871) (CDOAB DAB NC8 WWWIA)

STODDARD, William Osborn, 1835–1925, American journalist and inventor; strong supporter of Abraham Lincoln [1809–1865], became his secretary (1861ff), received nine mechanical patents for telegraph, manufacturing, and railroad enterprises (ANB CDOAB DAB MWBD NC7 WWWIA WWWIS)

STODOLA, Aurel B., 1859–1942, Hungarian mechanical engineer; designed steam engines for a company in Prague (1886–1892), chair of thermal machinery at Technische Hochschule of Zurich (1892–1929), authority on steam turbines, automatic control, and flow of steam through Laval nozzles (CDOSB DOSB PS WWWIS)

STOEK, Harry Harkness, 1866–1923, American mining engineer; organized and headed mining engineering at Univ. of Illinois (1909–1923), worked at changing occupation from rule-of-thumb to an engineering science (CDOAB DAB WWWIA)

STOKES, Sir Frederick Wilfrid Scott, 1860–1927, English civil engineer and inventor; improvements in breakdown cranes, railway work, hydraulic and refrigerating machinery, and ordnance, invented Stokes trench mortar (1915) (MWBD WWWIS)

STOKES, Sir George Gabriel, 1819–1903, British mathematician and physicist; developed theory of viscous fluids (1845–1850), deduced equation known as Stokes law (1845) representing motion of a sphere falling through a viscous fluid, also worked on fluorescence (1852), sound, and light, several laws named after Stokes (BEST DNB DOT FNIE L&M MWBD RHW WWW WWWIS)

STONE, Archie Augustus, 1893–1977, American agricultural engineer; head of department at Long Island Agriculture and Technical Inst. (NY), OPA during WWII, head of National Central Univ. of Nanking, in executive post with International Harvester Co. (IHC), author and co-author of several books on machinery and power for agriculture (AE58 NUC)

STONE, Charles Augustus, 1867–1941, American electrical engineer; with electrical engineer Edwin Sibley Webster [1867–1950] formed firm of Stone & Webster (1889), after Stone worked for Thomas Edison Welding Co. and then C&C Motor Co., Inc., developed earliest high-voltage power transmission facilities in northeast, took over development and operation of public utility properties (ANB CDOAB DAB NC33 TEIA WWWIA)

STONE, Charles Pomeroy, 1824–1887, American military and civil engineer; ordnance officer, served in Union army as brigadier general during Civil War (1861–1864), after war served as mining, canal, and construction engineer, lieutenant general in Egyptian army (1870–1883) (ANB CDOAB NC11 WWWIA)

STONE, John Stone, 1869–1943, American radio and communication engineer and inventor; Bell Telephone Co. (1890–1899), contributed to improvement of long-distance telephone service, high-frequency circuits, electrical resonance, direction-finding systems, established Stone Wireless Telegraph Syndicate (1900–1913), consultant, obtained approximately 100 radio patents sold to Lee De Forest [1873–1961] (ANB CDOAB DAB MWBD NC34 TFE&I WWWIA WWWIS)

STONE, Roy, 1836–1905, American civil engineer; in Civil War (1861–1864) in several battles, injured and taken prisoner, became brevet brigadier general, after war promoted various transportation developments, active in good roads movement within US Department of Agriculture (USDA) where he was engineer (for 10 yr.), with time off for participation in Spanish American war, with USDA (ante 1901) (ANB BITAS)

STORROW, Charles Storer, 1809–1904, American civil engineer; construction engineer, manager of Boston & Lowell Railroad (1832–1845), engineer for Essex Co. (1846ff) that planned and built city of Lawrence (MA) (ANB CDOAB DAB NC21)

STOTT, Henry Gordon, 1866–1917, Scottish American electrical engineer; to US (1891), engineer for Buffalo Power and Light Co. (NY) (1891–1901) responsible for power plants and transmission lines for Manhattan Railroad Co. and Interboro Rapid Transit (NYC) (CDOAB DAB NC14 WWWIA)

STOTT, Louis Lathrop, 1906–1964, American inventor and manufacturer; Beryllium Corp. (1932–1942) advancing from salesman to vice president, with others founded Atlantic Plastics, Inc., where he was secretary-treasurer (1942–1946), organized Polymer Corp. (1946–1964) where he was president and director, was a pioneer in extruding nylon, introduced centrifugal castings and sintering techniques, held approximately 25 patents (NC51 PS)

STOUT, Oscar Van Pelt, 1865–1935, American civil and agricultural engineer; engineer in department of railroad system (1886–1890), city engineer (1890–1891), faculty and administration including dean of engineering (1912–1920) at Univ. of Nebraska (1890–1920), resident hydrographer for Nebraska, cooperative irrigation for USDA and California (1922ff), engineer corps in WWI, received first McCormick Gold Medal by ASAE (1932) (AE16 NC26 WWWIA)

STOUT, William Bushnell, 1880–1956, American engineer; founder and president of Stout Engineering Labs. (1919ff) engaged in research in automotive, airplane, and locomotive fields, built first all-metal plane (1922), started company to build airplanes (1922–1925), established passenger airline service (1926–1929), developed stainless steel and all-welded airplane engines (MWBD WWWIA WWWIS)

STRAHL, William Egbert, 1919–1974, American chemical engineer; US Army in WWII, with Strahl & Pitch (1948–1974) becoming successively treasurer, vice president, and executive vice president, also formed Babylon Chemicals Co. (1952–1974) becoming president (1971), refiners and blenders of waxes (NC58 PS)

STRATTON, James Hobson, 1898–1984, American military engineer; graduated from US Military Academy and assigned to Corps of Engineers from which he retired as brigadier general (1949) after 30 yr. service, joined engineering and architectural firm of Tibbetts-Abbett-McCarthy-Stratton (TAMS) as a partner (1949–1967), soil mechanics, construction of earth dams, military airfields, engineering for landing fields in Europe in WWII, NAE (1981) (MT3)

STRATTON, Julius Adams, 1901–1994, German American physicist and electrical engineer; faculty at Massachusetts Inst. of Technology (MIT) including Radiation Lab. (1940) and administration serving as president of MIT (1966–1971), chairman of board of Ford Foundation (1966–1971), served on presidential committees and commissions, NAS (1950), NAE (1964) (Bridge CB63 MT8 PS WWWIA WWWIS)

STRAUSS, Joseph Baermann, 1870–1938, American engineer; bridge designer and builder, started engineering firm (1904), made improvements in bridge design and construction, designed and built over 400 bridges including Columbia River Bridge (WA) (1930), George Washington Bridge (NY) (1931), Golden Gate Bridge (CA) (1937) (ANB CDOAB DAB MWBD NCB)

STREET, Robert, 18th century, British inventor; patent for an internal combustion engine using piston and a system of levers (1794), used engine to raise water (EB MWBD)

STRELETSKII, Nikolai S., 1855–??, Russian structural engineer; taught at Moscow Technical College (1915–1933), Moscow Inst. of Construction Engineers (1933ff), engaged in and directed studies on bridge structures (1918–1930), USSR Academy of Sciences (1931) (PS SMOS WWWIS)

STRICKLAND, William, 1787–1854, American architect and engineer; several buildings including Masonic Temple (1810), commercial and government buildings in Philadelphia (PA), worked with US mints, designed and built the Delaware breakwater, encouraged construction of Pennsylvania Railroad (PRR), wrote on engineering and architecture (ANB CDOAB DAB MWBD WWWIA)

STRINGFELLOW, John, 1799–1883, British aeronautical engineer and inventor; built steam-powered model airplanes, with William Samuel Henson [1805–1888] formed Aerial Transit Co. (1842), amid failures he continued to experiment and improve devices (BDOTHOT DOSB(Wi) GEAPIT WWWIS)

STRINGFELLOW, Victor T., 1902–1989, American geologist and hydrologist; taught at Washington Univ. (MO), Oklahoma A&M, and New Mexico Univ., with USGS (1930–1972), limestone terranes and groundwater hydrology, studied artesian aquifer (FL), studied saltwater encroachment in aquifers (HOWRD)

STROBEL, Charles Louis, 1852–1936, American civil engineer; specialist in steel-framed structures, designed steel structures for several skyscrapers, originator of Z-bar and designed standard I-beam, consultant for Burnham & Root and for Adler & Sullivan (ANB CDOAB DAB WWWIA)

STRONG, Charles Lyman, 1826–1883, American mining engineer; began as a bookkeeper (8 yr.), first superintendent of Gould & Curry mines (NV) (1860–1864), engaged in farming, reentered mining business (1874–1883) (CDOAB DAB NC17 WWWIA)

STRONG, Harriet Williams Russell, 1844–1926, American woman farmer, engineer, and inventor; pioneered California walnut industry, early advocate of flood control for irrigation, developed device for opening windows, patented an irrigation system (c.1883), flood control and water supply, invented device for impounding debris and water in hydraulic mining (1887, 1893) (AIE ANB AWIT CDOAB DAB NC17)

STROUD, William, 1860–1938, English inventor and industrialist; professor of physics at Leeds (1885–1909), with Scottish engineer Archibald Barr [1855–1931] invented range finders and instruments, chairman of Barr & Stroud, Ltd. (1931ff) (MWBD WWW)

STROWGER, Almon Brown, 1839–1902, American military engineer and inventor; developed first commercially successful automatic telephone-switching system (1887–1889), formed Strowger Automatic Exchange Co. (1891), introduced dial system (1896), company became a part of General Telephone and Electronics (GTE) Co. (BDOTHOT WA)

STRUMINSKII, Vladimir V., 1914–??, Soviet mechanical engineer; Central Aero-Hydrodynamics Inst. (1941ff), USSR Academy of Sciences (1958), aerodynamics, formulated theories on boundary layers (PS SMOS WWWIS)

STRUTT, John William—see RAYLEIGH, Lord

STUART, Charles Beebe, 1814–1881, American civil and military engineer; railroad construction in New York State (1833–1849), proposed suspension bridge over Niagara Falls (NY) (1855), supervised construction of Navy dry docks (1849–1853), soldier in Union Army during Civil War (CDOAB DAB WWWIA)

STUART, Francis Lee, 1866–1935, American civil engineer; specialist in railroad and transportation problems, worked with the Baltimore & Ohio Railroad for four different periods (1884–1887) (1892–1897) (1900–1905) (1910–1915), Nicaragua Canal Commission (1897–1900), was chief engineer of two railroads, private practice (1915ff) (ANB CDOAB DAB NC27 WWWIA)

STUART, Herbert Akroyd, 1864–1927, English inventor and engineer; Akroyd Stuart engine (1886–1889) that embodied principles of compression ignition and injection of fuel, use of crude oil for engine (1890) (DNB EB EE E64 MMAH WWWIS)

STUDEBAKER, Clement, 1831–1901, American wagon manufacturer; with brother Henry Studebaker [1826–1895] founded H. & C. Studebaker Co. (1826–1895) for making wagons, carriages, and buggies, the largest such company in the world, with other brothers (John, Peter and Jacob) formed Studebaker Brothers Manufacturing Co. with Clement Studebaker as president (1868), experimented with automobiles (1897ff) with their manufacture beginning after his death (MWBD NC23 WWWIA)

STUMM, Werner, 1924–1999, Swiss American environmental engineer; water quality control, corrosion, water chemistry, education of environmental engineers, faculty at Swiss Inst. for Water Resources and Water Pollution Control (EAWAG) a research institute of ETH Zurich (1952–1956) (1970–1988), Harvard Univ. (1956–1970), US citizen (1968), NAE (1991) (AMOS Bridge MT10)

STURGEON, William, 1783–1850, English physicist, electrical engineer, and inventor; electromagnet (1823), devised strong electromagnet (1835), process of amalgamating zinc plate of battery with mercury film (1830), electromagnetic rotary engine (1832), invented the commentator, invented first moving-coil galvanometer (1836), improved voltaic battery, thermoelectricity (DNB GEAPIT MWBD NYPL TBDOS(Po) WWWIS)

STURTEVANT, Benjamin Franklin, 1833–1890, American inventor; built machines to peg boots and shoes (1856), patents dealing with cutting veneers (1859), making wooden pegs (1862), Conway Co. (NH), rotary exhaust fan (1867), manufactured blowers used around the world made at Jamaica Plain (MS) (1878) (CDOAB DAB MEIA)

STUTZ, Harry Clayton, 1876–1930, American automobile manufacturer; founder and president of Stutz Motor Co. (1913) that built racing and sports cars one of which was the well known Stutz Bearcat (1913–1919), helped organize the H. C. S. Motor Car Co. (IN) (1919) (MWBD WWWIA)

STYRIKOVICH, Mikhail Adolfovich, 1902–??, Soviet heat engineer; builder of steam power stations, Leningrad Province Scientific Research Power Inst. (1927ff), Moscow Inst. of Energetics (1939ff) of Academy of Sciences, member of USSR Academy (1941) and academician (1964), specialized in diesel generator units, steam boiler pressures, movement of steam mixtures through pipes (PS PSOWE SMOS WWWIS)

SUITS, Chauncey G., 1905–1991, American electrical engineer; studied electric arc behavior, process for making synthetic diamonds, US Forest Products Lab. (WI) (1929–1930), vice president and director of research for General Electric Co. (1930–1965), during WWII with National Defense Research Committee (NDRC) (1942–1946), consultant, NAS (1946), NAE (1964) (Bridge MT6 NCJ WWIE WWWIA WWWIS)

SULMAN, Henry Livingstone, 1861–1940, English metallurgist; joined with H. F. K. Picard and formed consulting practice (1898ff), invented several processes for extraction of gold, with Picard developed froth flotation process for concentrating ores (DNB EB MWBD)

SUMMERFIELD, Martin, 1916–1996, American physicist and aeronautical engineer; president and chief scientist of Princeton Combustion Research Lab., Inc. (1951ff), faculty at Princeton Univ. (NJ) (1949–1952), rocketry, combustion research, infrared research, rocket propellents, combustion, heat transfer, international literature on aeronautics and astronautics, Aerojet Corp. (1943–1945), NAE (1979) (AMOS Bridge WWWIA)

SUMNER, Eric E., 1924–1993, Austrian American electrical and mechanical engineer; pioneered in the concepts of digital transmission and led in implementation of application, vice president of operations and planning at AT&T Bell Labs. (1948ff), NAE (1985) (AMOS Bridge MT7 WWWIA)

SUNDBACK, (Otto Frederick) Gideon, 1880–1954, Canadian Swedish American engineer and inventor; to

US (1905), the zip fastener (1893), Westinghouse Electric Corp. (PA), president and CEO of Lightning Fastner Co., Automatic Hook and Eye Co. (1908ff), separable fastener (1913, 1917), Hookless Fastener Co. using name of Talon later adopted as company name, B. F. Goodrich Co. used name zipper, about 10 US patents plus several foreign patents (GI ITTC NC28,40,E TEOUT WOI WWWIA WWWIA-S&T)

SUOMI, Verner E., 1915–1995, American space scientist, engineer, and inventor; faculty at Univ. of Wisconsin (1948–1986) except for assignments with National Science Foundation (NSF) and US Weather Bureau, directed institute of space science and engineering, sonic anemometry, turbulence, energy budgets, invention of radiation-sensing devices, invented spin scan cameras for space exploration, contributed to space technology, development and use of meteorological instruments, NAE (1966) (Bridge MT8 WWIE WWWIA)

SUPPES, Max M., 1856–1916, American mechanical engineer and inventor; helped develop first automatic mill for rolling T-rails (1879), numerous patents for making railroad rails, patented refrigeration system (1900), apparatus for crushing and handling coal, manufacture of coke (1901), and apparatus for blast furnace and rolling mills (1903), after work with several manufacturing plants became general manager of Lorain Steel Mill (OH), about 15 patents (MEIA)

SU SUNG, 1020–1101, Chinese astronomer and horological engineer; official for clock making, had broad interests, clock not only gave time but gave position and movement of heavenly bodies, his clocks presented to emperor (1094), clock became involved in political battles (GEAPIT)

SUTRO, Adolph Heinrich Joseph, 1830–1898, German American mining engineer; planned Sutro Tunnel (NV) through Mount Davidson (over 20,000 ft long) and supervised its construction (1869–1878), sold his interest (1879) and went into real estate and politics (ANB MWBD NC21 WWWIA)

SUYYA, fl. 860 AD, Indian hydraulic engineer; developed schemes for controlling Jhelum River (India and Parkistan), developed plan for comprehensive water control and land reclamation, land productivity of rice increased as a result of his projects (GEAPIT MWGD)

SVEDBERG, Theodor H. E., 1884–1971, Swedish chemist and inventor; professor at Univ. of Uppsala (1912–1949), director of Gustaf Werners Inst. (1949–1967), developed high-speed centrifuges, ultracentrifuges (1924), assisted Arne Wilhelm Kaurin Tiselius [1902–1971] in development of electrophoresis methods to analyze proteins (late 1920s), Nobel prize (1926), fNAS (1945) (LOTCS MWBD PhamCent RHW TCOS WWWIA WWWIS)

SWAIN, George Fillmore, 1857–1931, American civil engineer; topographical engineering, faculty at Massa-

chusetts Inst. of Technology (1881–1887), Harvard Univ. (1909–1929), worked for several railroad and transit commissions including the Boston Transit Commission (1894–1918), NAS (1923) (BDOACE BM17 CDOAB DAB NC12 PS WWWIA WWWIS)

SWAN, Sir Joseph Wilson, 1828–1914, English chemist, electrician, and inventor; primitive electric light (1860), dry photographic plate (1871), bromide photographic paper (1879), carbon filament incandescent electric bulb (1880), formed Edison and Swan United Electric Co. (1883), method of forming fibers of nitrocellulose (1883), electroplating, cellular surfaced lead plate storage battery (EIH GI MWBD SAI TBDOS(Po) TGE:B WWW WWWIS)

SWASEY, Ambrose, 1846–1937, American mechanical engineer and inventor; Pratt & Whitney Co. (CT) (1869–1880), with Worcester R. Warner [1846–1929] established machine tool business (IL, OH) (1880ff), built instruments for astronomy, established Engineering Foundation (1924), NAS (1922) (ANB BM22 CDOAB DAB MEIA NCB,27 PS WWWIA WWWIS)

SWEARINGEN, Judson S., 1907–1999, American chemical engineer; Rotoflow Corp. (1946–1990) becoming president, several companies associated with his name, low-temperature gas separation, petroleum and gas, refrigeration, seals, pumps and centrifugal compressors, previously with San Antonio Refining Co. (TX) (1933–1938) and faculty at the Univ. of Texas (1939–1942), NAE (1977) (WWIE)

SWEET, John Edson, 1832–1916, American mechanical engineer; invented nail-making machine (1862), with Patent Bolt & Nut Co. (England), returned to US (1864), built bridges (1871), machine shop at Cornell Univ. (NY), tool-making devices, invented straight-line engine and organized Straight Line Engine Co. (1880) (BITAS CDOAB DAB MEIA NC13 WWWIA WWWIS)

SWIFT, Jonathan, 1667–1745, Irish clergyman and visionary writer; buildings, instruments, trade, generator of ideas, author involved in government, political, and religious activities (DNB MWBD TEE)

SWIFT, Joseph Gardner, 1783–1865, American military engineer; graduate of US Military Academy (1802), chief engineer of US Army (1812–1818), worked on railroad and harbor improvements, recognized as an engineer of distinction who greatly influenced others (ANB CDOAB DAB NC10 WWWIA)

SWIFT, William Henry, 1800–1879, American military and civil engineer; brother of Joseph G. Swift [1783–1865], graduate of US Military Academy (1818) and commissioned in artillery (1819), employed in railroad and canal work and then captain in the topographical engineers (1828), was principal assistant in topographical bureau (1843–1849), served as president of railroads in eastern US (ANB CDOAB DAB WWWIA)

SWIGER, William F., 1916–1996, American civil engineer; structural design in geotechnical, earthquake, and offshore engineering, hydroelectric power plants, Stone and Webster Engineering Corp. (1942–1981), consultant (1981ff), NAE (1980) (Bridge WWIE)

SWINBURNE, James, 1858–1958, Scottish electrical engineer and inventor; locomotive work, dynamos, used words 'rotor' and 'stator,' developed lights, plastics, electric-wire insulation, a durable and tough lacquer to protect metals (Damard), watt-hour meter, with Leo H. Baekeland [1863–1944] developed durable materials, granted over 100 patents (BDOS DNB PA PS RHW RS5 TBDOS(Po) TPA)

SWINTON, Ernest Dunlop, 1868–1951, English engineer and inventor; British army (1888ff), WWI (1914–1917), suggested that tractor tracks (as on Holt caterpillar tractors) be added to armored vehicles making military tank (1914) first used in battle (1916), he coined word 'tank,' became general in British army commandant of Royal Tank Corps (1934–1938), lectured at Oxford Univ. on military history (1925–1939), wrote extensively on military affairs (DNB EB S&TF WWW)

SWITZER, Robert, d. 1997, age 83, American chemist and inventor; invented fluorescent dyes and paints, with his brother Joseph Switzer invented Day-Glo paint, after WWII with his brother formed Day-Glo Color Corp. (ante 1985) (FAL NYTObit)

SWOPE, Gerard, 1872–1957, American electrical engineer; beginning level at General Electric Co. as a helper (1893) advancing to president (1922–1939, 1942–1944), member of many government boards and advisory councils (CDOAB ANB MWBD NC45 WWWIA)

SYMINGTON, William, 1763–1831, Scottish engineer; with brother built working model of steam road carriage (1786), steam engine (1787), used engine to propel boat (1788), Symington engine—direct-acting horizontal steam engine (1801), built steam-driven paddle wheel to propel boat (1802) (EE EIH GEAPIT MWBD RHW WA WWWIS)

SYMONS, Thomas William, 1849–1920, American military engineer; US Military Academy graduate (1874), served on survey expeditions under George M. Wheeler [1842–1905] (1876–1879), and with Mississippi River commission and other government agencies, in charge of Buffalo (NY) breakwater (CDOAB DAB NC19 WWWIA)

SZEBEHELY, Victor G., 1921–1997, Hungarian American aerospace and mechanical engineer; served on the faculties of several universities including McGill Univ., Pennsylvania State Univ., Virginia Polytechnic Inst., then with Univ. of Maryland and George Washington Univ. (DC) while doing research at the David Taylor Model Basin (DC) (1951–1957), US citizen (1954), General Electric Co. space sciences (1957–1963), Yale Univ. (CT) (1963), Szbehely number, orbital mechanics, trajectory calculations, long-term satellite stability, Univ. of Texas (1968ff) including department chairman, NAE (1982) (Bridge MT10 WWIE WWWIA)

SZEKELY, Julian, 1934–1995, Hungarian American chemical and materials engineer and inventor; advanced materials processing technology, transport phenomena in materials processing, use of mathematical modeling and optimization of complex dynamic systems, environment and energy problems, developed mini-mills for steel making, faculty at Imperial College, to US (1966), State Univ. of New York-Buffalo (1966–1975), Massachusetts Inst. of Technology (MIT) (1975–1995), eight patents, NAE (1982) (Bridge MT8 WWIE WWWIA)

T

TACCOLA, Jacopo Mariano (aka Marianus Jacobus), 1381–1453, Italian military engineer; started as a sculptor then studied mechanisms (1427), drew but not known whether he invented chain transmission system, compound crankshaft, wrote about suction pump, caisson, aqueducts, watermills, windmills, and mining under defenses, superintendent of streets in Sienna (1442) (AHOEAT GEAPIT PS TEIH)

TACTICUS—see AENEAS THE TACTICIAN

TAGLIABUE, Giuseppe, 1812–1878, Italian American inventor; instrument maker in New York (1831ff), supplied instruments to US Geodetic Survey (USGS), patented improvements on hydrometer, mercury barometer and other scientific devices (ANB BITAS CDOAB DAB WWWIA)

TAINTER, Charles Sumner, 1854–1940, American inventor; invented several sound-recording devices, photophone (1880) with Alexander Graham Bell [1847–1922], with Chichester A. Bell developed and manufactured the graphophone (1881), dictaphone (EB MWBD WWWIS)

TAKEDA, Shingen, 1521–1573, Japanese warrior and water control expert; led military forces and captured land in central Japan, associated with system of water control to improve flood control and provide water to improve productivity of land (GEAPIT MWBD)

TALBOT, Arthur Newell, 1857–1942, American civil engineer; assistant engineer for construction of Santa Fe Railroad and Denver and Rio Grand Railroad, faculty and established engineering experiment station (1904) at the Univ. of Illinois (1885–1926), developed formula for computing size of culverts based on rainfall rates, worked on municipal problems, stresses on railroad tracks and ties (ANB BDOACE BDOTHOT DAB MWBD WWWIA WWWIS)

TALBOT, William Henry Fox, 1800–1877, English physicist and inventor; after a political career, obtained a picture in a camera and made it permanent, process of photography using 'negative' and 'positive' (1835), experimented with photography and invented Talbotype (1841), developed methods by which length of exposure greatly reduced (1851), flash photography (1851), interested in archaeology and was one of first to decipher ancient tablets found in Assyria (ASAI BEST BDOS(Wi) DNB EB EDOP L&M MWBD NYPL PS WA WWWIS)

TALCOTT, Andrew, 1797–1883, American soldier and civil engineer; graduated from US Military Academy (1818) and served in engineer corps (until 1836), chief engineer for a number of railroads, located and built railroad from Veracruz to Mexico City (1857–1867), developed Talcott method for survey (1834) and for determining latitude (CDOAB DAB EB GEAPIT WWWIA WWWIS)

TANDY, Charles David, 1918–1978, American merchandiser and electronics industry; became lieutenant commander in US Navy, began in leathercraft industry (1950), then expanded to Tandy Leather Co., acquired American handicrafts, established Tandy Corp. (1961), entered consumer electronics by acquiring Electronic Crafts (1962), incorporated Radio Shack (1963), acquired Leonard Department store complex (1967), introduced TR-80 small computer (1979) (ANB WWWIA)

TANI, Itiro, 1907–1990, Japanese aeronautical engineer; faculty at Univ. of Tokyo (1932ff), aerodynamics, fluid dynamics, pioneering contributions to theory and application of boundary layer, technical administration, international cooperation, fNAE (1979) (Bridge MT6)

TAO HSUN, fl. 1250, Chinese civil engineer; hydraulic engineering, built over 200 bridges, designed and built harbors and sea walls (GEAPIT)

TAQI AD-DIN, 1526–1585, Syrian Islamic horologist; specialist in mechanical clocks, clock-driving mechanisms, first scholar in Islamic and perhaps Ottoman world to give attention to mechanical clocks (GEAPIT)

TARLTON, Robert J., 20th century, American inventor; invented a community (PA) mountain-top antenna to serve people's houses (CATV) (community antenna television system) (1950), led to many developments in education and entertainment (NTBOT PS)

TARTAGLIA, Niccolò (aka Nicola Fontana), 1499–1557, Italian mathematician and inventor; studied flight of cannon balls, invented gunner's quadrant, formulated Tartaglia rule (AHOEAT MWBD TEIH WWWIS)

TARTANTO—see ARCHYTAS OF TARAS

TATHAM, William, 1752–1819, English American civil engineer; clerk in trading house of Carter & Trent (VA), with Watauga Assoc. (TN), in Revolutionary War (1776–1781), lieutenant colonel of militia (NC), lawyer,

geographer, map maker, irrigation, navigation, superintendent of construction of Wapping docks (England) (1801–1805), survey in US, draftsman for US Department of State (5 yr.) (BDOACE CDOAB DAB DNB WWWIA)

TATLIN, Vladimir, 1885–1953, Soviet builder; painter, sculptor, architect, constructor, built monument for 3rd International Fair in Paris (1920)—a machine that would rotate within an exposed steel frame, in Paris (1913–1914), designer of theaters (1934–1952) (EB MWBD)

TATLOW, Richard Henry III, 1906–1993, American civil engineer; design of structures, planning, partner in consulting firm (1929–1940), officer with Corps of Engineers (1941–1946), chairman and president of Abbott Merkt & Co. (1946–1984), NAE (1967) (Bridge MT7 WWIE)

TAYLOR, Brook, 1685–1731, British physicist, mathematician, and meteorologist; credited for developing calculus of finite differences (1715), Taylor theorem (1715), Taylor expansion series, secretary of Royal Society (1714–1718) (DNB DOT EOM IAOSTAI GLOSB MWBD WWWIS)

TAYLOR, David Watson, 1864–1940, American naval architect and engineer; USNA (1881), US Navy (1885–1923), advanced study in construction at William Cramp & Sons Iron Works (PA), Bureau of Construction and Repair (DC), chief of construction at navy yards (CA), director of navy's Experimental Model Basin (DC) (1899–1914) in which considerable work resulted in data used to improve and systemize hull designs, worked on propellers, promoted to rear admiral (1914–1922), chief constructor for navy, the large experimental ship basin was named after him (1937), NAS (1918) (ANB BM22 MWBD NC30 PS WWWIA WWWIS)

TAYLOR, Frederick Winslow, 1856–1915, American industrial and mechanical engineer and inventor; Midvale Steel Co. (PA) (1878–1890) chief engineer (1884), introduced time and motion study (1881), patented largest steam hammer in US (1890), with J. Maunsel White [1856–1912] developed heat treatment of tool steel, general manager of Manufacturer's Investment Co. (PA) (1890), consultant in shop management and manufacturing costs (1893), efficiency expert with Bethlehem Steel Co. (PA) (1898), known for scientific management, at least 100 patents mostly related to machine tools and related devices (ASAI BEST BITAS CDOAB DAB EIH MEIA MWBD NC23 1000Y PS TEIH TGS TTOS WWWIA WWWIS)

TAYLOR, Geoffrey Ingram, 1886–1975, British mathematician and inventor; hydrodynamics, turbulence, effect of rotation on fluid flow with development of the Taylor-Couette flow representation (1923?) that led to the Couette viscometer, meteorology, aerodynamics, dislocation principle in crystals (1934), fNAS (1945) (CEOS&T DNB EDOP EOAP WWW WWWIS)

TAYLOR, Harry, 1862–1930, American military engineer; graduate of US Military Academy (1884) and commissioned in engineers, worked on river and harbor development on East Coast (10 yr.), Philippines (1903–1905), chief engineer of several eastern projects, in WWI responsible for engineering in France, became chief engineer and major general (1924ff), then vice president of W. & A. Fletcher Co. (ANB NC23 WWWIA)

TAYLOR, Stevenson, 1848–1926, American mechanical and naval engineer; designed machinery for steamers (1869), head and vice president of W. & A. Fletcher Co. (1883ff), and officer of other shipbuilding companies, president of American Bureau of Shipping (1916), lieutenant commander US Naval Reserve (1917) (CDOAB DAB MEIA NC23)

TEDESKO, Anton, 1903–1994, Austrian American civil engineer; design of structures, arenas, air terminals, bridges, construction, with Roberts & Schaefer Co. (1955–1970) as vice president (until 1967), established consulting engineering office (1967ff), NAE (1967) (Bridge MT8 WWIE WWWIA)

TEEPLE, John Edgar, 1874–1931, American chemical engineer; instructor at Fremont Normal College (NE) (1894–1898), faculty at Cornell Univ. (1898–1903) for graduate work and advanced degrees, with J. T. Baker Chemical Co. and director of Industrial Labs., consulting engineer and chemist (1908–1931) in a wide range of projects and companies such as borax, boric acid, potassium chlorate, production of turpentine (NC27,B TEE WWWIA)

TELFORD, Thomas, 1757–1834, Scottish civil engineer; built canals, roads, and bridges in Scotland, surveying, improved Scottish harbors, Gota canal between Baltic Sea and North Sea (1808–1810), Menai Strait (Wales) suspension bridge (1819–1826) (DNB EIH GEAPIT LOTE PS SFLOE TBDOS(Po) MWBD MWGD TEIH TGE:B TGS WWWIS)

TERMAN, Frederick Emmons, 1900–1982, American electrical engineer; electronics, radio, electronic measurement, faculty at Southern Methodist Univ. (1925–1937), dean and provost at Stanford Univ. (1937–1965), consultant in engineering education and communication systems (1966ff), NAE (1964), NAS (1946) (ANB BM74 MMOS MT2 NCG PS TEIH WWWIA WWWIS)

TERRY, Eli, 1772–1852, American inventor; established business for making and repairing clocks (1793), use of water to drive tools in clock making (1800), established company of Terry, Thomas & Hoadley (CT) (1807–1810) in which he built business based on one-day shelf clocks (CDOAB DAB IIA MEIA NC6 WWWIA)

TERZAGHI, Karl Anton von, 1883–1963, Austrian American civil and mechanical engineer; to US (1912), US citizen (1943), reinforced concrete construction (1912ff), engineer for Adolph Baron Pittel (until 1908), surveys in Croatia (1908–1910), dam studies in US under Frank H. Nowell [1864–1958] (1910–1913), set up soil testing laboratory at Roberts College (Turkey) (1918–1925), faculty at Massachusetts Inst. of Technology (MIT) (1925–1929), Technische Hochschule Vienna (1929–1938), Harvard Univ. (1946–1956), Terzaghi principle (1948) that was used worldwide, founder of field of soil mechanics, consulting positions on dams, airports, foundations, mechanics of soil settlement (BDOACE DOT L&M LC MWBD WWWIA WWWIS)

TESLA, Nikola, 1856–1943, Croatian American electrical engineer and inventor; to U. S. (1884) citizen (1894), his work helped make alternating current practical, polyphase motor and method of power distribution, Tesla coil (1891), made transmission of high-voltage electricity possible through development of transformers, alternating current motor (1888), high-frequency coil, worked for Edison Works, George Westinghouse [1846–1914] supported his early work, Tesla Co., Inc. (NY), Tesla Electric Light Co. (NJ) (1887), magnetic flux density called tesla named after him, 111 patents (AIE ANB BEST BDOS CBD CDOAB DAB DOSB EIH FNIE GI IAI MWBD NC6 NYPL PA PS SAI TBDOS(Po) TCPONT TFE&I TGS WWWIA WWWIS)

THACHER, Edwin, 1839–1920, American civil engineer; worked for railroads before and during Civil War moving from general engineering work to bridge construction, assistant engineer of Louisville Bridge Co. (1868ff) then Louisville Bridge & Iron Co. (9 yr.), computing engineer followed by chief engineer of Keystone Bridge Co. (1879–1887), invented Thacher cylindrical slide rule (1881), devised Thacher bar used in reinforced concrete construction, consulting (CDOAB DAB NC7 WWWIA)

THAN-STON-RGYAL-PO, 1385–1464, Tibetan engineer; design of bridges, suspension bridge across Brahmaputra River (1420) (GEAPIT)

THAYER, Sylvanus, 1785–1872, American mechanical and military engineer; US Army Corps of Engineers (1808ff), superintendent of US Military Academy (1817–1833), worked on harbor fortifications and improvements (1833–1863), founded and endowed Thayer School of Engineering (1867) at Dartmouth College (NH) (ACE CDOAB DAB MWBD NC7 TEIH WWWIA)

THEIS, Charles V., 1900–1987, American civil engineer and geologist; USGS (1930–1985), was a staff scientist, involved in groundwater movement, radioactivity of water, nuclear waste disposal, worked with Atomic Energy Commission, recognized for Theis equation (1935) (HOWRD)

THEODOROS of Samos, fl. 575 BC, Greek architect and inventor; believed to have invented methods of polishing gems, brought to Greece techniques for making bronze statues in parts, invented tools such as stonecutter, level, square, and perhaps rule, designed foundation for temple in swampy soil (GEAPIT)

THEOPHILUS, fl. 1100, German metallurgist; metal and metalworking craftsman particularly on how to make, paint, and decorate walls and panels, including glass as well (GEAPIT)

THEREMIN, Leon, 1896–1993, Russian inventor and musician; electronic musical synthesizer first called therema (1922), had commercial success in US (1927–1938) then returned to USSR (1938), during WWII worked in military laboratory, among his inventions were a radio-controlled aircraft, tracking systems for ships and submarines, television system, listening device, professor of acoustics at Moscow Conservatory of Music (1945ff) (HOCS&T RHW)

THIELE, Ernest W., 1895–1993, American chemical engineer; McCabe-Thiele diagram (c.1927) (Warren Lee McCabe [1899–1982]), Thiele modulus, Thiele catalyst effectiveness factor that helped develop chemical engineering science, associate director of research at Standard Oil Co. (1925–1960), faculty at Univ. of Notre Dame (1960–1970), 29 patents, NAE (1980) (Bridge MT8 WWWIS)

THIMONNIER, Barthelemy, 1793–1859, French tailor and inventor; a tailor in Paris, patented sewing machine with chain stitch (1830) in Paris, made improvements in England (1845) and US (1848), first made of wood then metal (ASAI EB GEAPIT GI MWBD 1000Y)

THIOUT, Antoine, 1692–1767, French instrument maker; clockmaker (1724ff), developed early fusee machine, developed screw-cutting machine, machine tools (BDOS(Wi) PS)

THOMAS, Charles Allen, 1900–1982, American chemical engineer; formed Thomas & Hochwatt Lab., worked on Manhattan project (1943–1946), head of Clinton Lab. (1946–1948), Monsanto Co. (1947–1970) serving in executive positions including president, CEO and chairman of the board (1951–1965), NAS (1948), NAE (1964) (BM65 MT2 NCI WWWIA WWWIS)

THOMAS, Seth E., 1785–1859, American manufacturer; mass production, clock manufacture in firm of Terry, Thomas & Hoadley (1807), established his clock and manufacturing factory (1812), organized Seth Thomas Co. (1853), which was continued by his son Seth Thomas, Jr. [1816–1888] (IIA MWBD NC26,53 WWWIA)

THOMAS, Sidney Gilchrist, 1850–1885, British metallurgist and inventor; discovered method of eliminating phosphorus from iron in Bessemer process (1875), known as Thomas-Gilchrist process (Percy Carlyle Gilchrist [1851–1935]) (DNB EIH TBDOS(Po) MWBD WWWIS)

THOMPSON, Albert R., 1879–1947, American engineer and inventor; field engineer of Anderson-Barygrover

Manuf. Co. (CA) (1902–1928), chief engineer and other management positions for Food Manufacturing Corp. (1928–1945), developed automatic food processing equipment for cooking and sterilizing food and other devices for food processing, awarded over 200 patents, the pressure cooker was invented (1679) by Frenchman Denis Papin [1647–c.1714] (ASMENews24 NC36)

THOMPSON, Almon H., 1839–1906, American topographical engineer; in Civil War as officer in infantry, school superintendent (1865–1870) in Illinois, topographical engineer for John W. Powell [1834–1902] (1870–1878), geographer for US Geological Survey for both field and office work (BDOAS(El) BITAS PS WWWIA)

THOMPSON, Benjamin (aka Count Rumford), 1753–1814, American English scientist and inventor; English loyalist during Revolutionary War, served in Bavaria (1784–1795), in England (1795ff) introduced improvements in heating and cooking equipment such as double boiler, kitchen range, and drip coffee pot, studies on heat and friction (1798ff) (ASAI BEST BDOAS BITAS CDOAB IIA MWBD SAI WWWIA WWWIS)

THOMPSON, Charles Oliver, 1836–1885, American engineer; introduced shop practice in engineering instruction while at Worcester Polytechnic Inst. (MA) where he was principal (1868–1882), president of Rose Polytechnic Inst. (IN) (1883ff) (CDOAB DAB WWWIA)

THOMPSON, Erwin William, 1859–1935, American mechanical engineer; designed and operated cottonseed mills in southern states (GA, NC, SC), auditor and construction engineer for Southern Cotton Oil Co. (NYC), chief engineer of D. A. Tompkins Co. (NY), then representative of US Department of Commerce for cotton products (1913–1930) (MEIA WWWIA)

THOMPSON, Joseph W. 1833–1909, American inventor; designed and invented machinery for engines and machine work, served in Civil War, designed blowers for blast furnaces (1866) for firm that became Buckeye Engine Co. (OH) (1871), designed governor and valves for steam engines, Thompson indicator, about 10 patents (MEIA WWWIA)

THOMPSON, Sanford Eleazer, 1867–1949, American civil engineer; supervised construction of paper mill (ME) (1889ff), worked with Frederick W. Taylor [1856–1915] in industrial engineering and scientific management, consultant, management of underground bituminous coal mines, served on presidential commissions (for US Presidents Warren G. Harding [1865–1923] and Herbert C. Hoover [1874–1964]) (BDOACE WWWIA)

THOMPSON, Silvanus Phillips, 1851–1916, English physicist and electrical engineer; professor at Bristol (1878–1885), principal at Finsbury Technical College (1885–1916), worked on electricity and magnetism (DNB MWBD TFE&I WWW WWWIS)

THOMPSON, William Boyce, 1869–1930, American mining engineer and financier; mine operator, mining entrepreneur, Thompson Investment Co. (1897), Wall Street activities with Harry F. Sinclair [1876–1956], endowed and president of Boyce Thompson Inst. for Plant Research (NY) (1919) (dedicated 1924), involved in politics, director of Federal Reserve Bank of New York (1914–1919), head of American Red Cross Mission to Russia (1917), economically successful in scavenging copper from mines, also involved in mining of gold, silver, lead and zinc, and coal, led efforts to fund Belgian Relief Fund headed by Herbert C. Hoover [1874–1964], established Boyce Thompson Southwestern Arboretum (AZ) (ANB CDOAB DAB MWBD(in WBD) NC22 NUC WWWIA)

THOMSON, Elihu, 1853–1937, English American electrical engineer, industrialist, and inventor; to US (1858), partner in Thomson-Houston Electric Co. (1883) which was merged with Edison's company to form General Electric Co. (1892ff), invented electric welding (1886), three-phase generators, high-frequency transformer, high frequency generator (1890), centrifugal cream separator, watt meter, street arc lamp, consultant to General Electric Co. (1892ff), at least 700 patents (ANB BEST CDOAB MWBD NCB,27 SAI TBOF TFE&I WWWIA)

THOMSON, James, 1822–1892, British physicist, civil engineer, and inventor; his younger brother was Lord Kelvin (William Thomson [1824–1907]), James Thomson was known for the Thomson vortex water wheel with pivoted guide vanes (1850) that was named after him, blowing fans, river currents, atmospheric currents, jet pumps for land drainage, invented centrifugal pump, faculty at Queens College (1857–1873) and Glasgow (Scotland) (1873–1889), FRS (1877) (BEST DNB EE TBDOS(Po) WWWIS)

THOMSON, John, 1853–1926, Scottish American mechanical and civil engineer and inventor; to US (1854) with parents, patented watch escapement (1877), differential screw (1881), bench vice and wrench (1882), improved meters (1883), improved printing press (1887ff), disk water meter (1887), produced Thomson printing press, became patent attorney, organized John Thomson Press Co. (1890) and Thomson Meter Co. (1890), over 200 patents (DNB MEIA NC13 WWW)

THOMSON, John Edgar, 1808–1874, American civil engineer; worked on surveying crew and later in railroad construction for Pennsylvania on canals and railroads, chief engineer and later president of Georgia Railroad (1847ff) extending railroad and assisting others in development of railroads (ANB DAB WWWIA)

THOMSON, Sir Joseph John, 1856–1940, English physicist and inventor; professor at Cambridge Univ. (1884–1918), discovered electron (1897) and showed cathode ray deflection in an electric field (1897), Royal Inst. London (1905–1918), during WWI board of invention and research (1914–1918), master of Trinity College Cambridge (1918–1940), flow of electrons through gases, studied radioactivity, Nobel prize (1906), FRS (1908), fNAS (1903) (BEST CDOS CDOSB DNB DOSB EIH LOTCS MWBD RHS RSObit SAI TGS WWW WWWIS)

THOMSON, Reginald Heber, 1856–1949, American civil engineer; surveying, assistant city engineer/surveyor later city engineer of Seattle (WA) (1882–1911), surveys for hydroelectric power, sewers, rail service, water supply, handling of refuse, municipal lighting, first engineer for Port of Seattle (WA) (1912), consultant on bridges, bridge foundations, irrigation, roads (BDOACE WWWIA)

THOMSON, Robert William, 1822–1873, Scottish engineer; worked with George Stephenson [1781–1848] on railroads, invented rubber tire (1845) used on vehicles (BEST DNB GI MWBD PS WWWIS)

THOMSON, Sir William Marcus (aka 1st Baron Kelvin), 1824–1907, Irish mathematician, physicist, and inventor; educated as an engineer, developed and used mechanical models and analogues, professor at Glasgow Univ. (1846–1899), heat and energy, expressed second law of thermodynamics as entropy (1850), proposed Kelvin temperature scale, invented mirror galvanometer (1858), invented tide predictor, harmonic analyzer and numerous other devices, consultant in laying the Atlantic cable (ASAI BEST CDOS CDOSB DNB EIH IAI L&M MWBD SAI WWW?)

THON, J. George, 1912–1988, Austrian American civil engineer; hydraulics, with William Halcrow and Partners in England (1941–1948), Pioneer Services & Engineering Co. (1949–1951), Bechtel Corp. (1951ff) including chief engineer then manager of Division of Engineering (1967ff), NAE (1975) (MT4 WWIE)

THORNTON, William, 1759–1828, British West Indies American architect; to US (1787), won competition for design of public buildings (1789), Philadelphia Library and National Capitol (1792), steamboat experiments with John Fitch [1743–1798] (1788–1790), Commissioner of Washington (DC) (1794–1802), superintendent of patent office (1802–1828), eight patents (CDOAB DAB MWBD WWWIA)

THORP, Frank Hall, 1864–1932, American chemist and chemical engineer; leader in development of chemical engineering then called industrial chemistry at Massachusetts Inst. of Technology (MIT) (1893–1916), was chairman of chemical engineering following death of Lewis Mills Norton [1855–1893] (HOCE PS WWWIA)

THORP, John, 1784–1848, American inventor; many patents and developments related to cotton loom, spinning, and twisting, with Silas Shepard patented power loom (1816), important in transition from hand to machine cotton manufacture, including retting machine (1828), first power gang loom (1829), machine builder (1830ff) (AIE ANB CDOAB DAB MEIA WWWIA WWWIS)

THRELFALL, Richard, 1861–1932, English physicist and chemical engineer; professor at Sydney Univ. (1886–1899), developed laboratories and instrumentation, with Messrs. Albright & Wilson Chemical Manufacturers, Ltd. (1899–1932), production of phosphorous, electrolytic manufacture of sodium chlorate, electric generators, flow of gases in tubes and ducts (DNB FRSObit PS WWW)

THURBER, Charles, 1803–1886, American inventor; with his brother-in-law Ethan Allen manufactured firearms (1836–1856), established firm Allen & Thurber (MA) (1836–1856), patented hand printing machines (1843), writing machine for blind (1845) (CDOAB DAB MEIA MWBD WWWIA)

THURSTON, Robert Henry, 1839–1903, American civil and mechanical engineer; engineer with US Navy in Civil War, professor and head of department of mechanical engineering at US Naval Academy (1865), Stevens Inst. of Technology (NJ) (1866–1869), secretary of US Board of Test for Iron, Steel, and other Metals (1875ff), director of college at Cornell Univ. (NY) (1885), numerous technical publications (ACE ANB BITAS CDOAB DAB DOSB MEIA NC4 WWWIA)

TILGHMAN, Benjamin Chew, 1821–1901, American inventor and manufacturer; after studying law joined with brother Richard A. Tilghman [1824–1899] in developing a process that decomposed animal fat to make glycerin and other products (1844–1861), served as officer in Civil War (1861–1866) as brigadier general of volunteers, worked at converting wood pulp to paper (1867), developed sandblast business using steam (1871) and later used compressed air, formed Patent Sand Blast Co., Ltd. in US and England (AIE ANB LAI MWBD NC15 WWWIA)

TILYOU, George Cornelius, 1862–1914, American entrepreneur and inventor; developed entertainment and recreational facilities such as Coney Island and Steeplechase Park, publicized and used Ferris wheel (George Washington Ferris [1859–1896]), designed and patented devices and rides for his facilities (ANB DAB MWBD NC16 WWWIA)

TIMBY, Theodore Ruggles, 1822–1909, American inventor; invented a revolving gun turret (1843) later used on Monitor (1861), invented a form of dry dock, patents received for barometer (1862), turbine water wheel (1869), gun carriage (1871) (CDOAB DAB NC9 WWWIA WWWIS)

TIMKEN, Henry, 1831–1909, German American inventor and manufacturer; to US (c.1838), established carriage manufacturing company (MO) (1855), invented special carriage spring (1877), taper roller bearing (1898), organized Timken Roller Bearing Axle Co. (1898) (CDOAB DAB MWBD NC30 WWWIA)

TIMOFEEV, Pyotr Vasilevich, 1902–??, Soviet electrical engineer; electronics, All-Union Electrotechnical Inst. (1928ff), taught at Moscow Univ. and other colleges, institutes, and universities, USSR Academy of Sciences (1953ff) (PS PSOCE SMOS WWWIS)

TIMOSHENKO, Stepan (Stephen) Prokfyevich, 1878–1972, Russian American engineer and scientist; widely known author of engineering books covering elasticity, strength of materials, elastic stability, plates and shells, advanced dynamics, engineering education, also a student in Germany, England, and Soviet institutions (ante 1922), to US (1922) and became citizen, joined Westinghouse (1922), on faculty at Univ. of Michigan (1927–1936), at Stanford Univ. (1936–1944), NAS (1940) (ANB BM53 FNIE NC57 NYObit PS WWWIA WWWIS)

TING HUAN, fl. 180 AD, Chinese mechanical engineer and instrument maker; developed household devices such as incense burner, chimes, responsible for zoetrope lamp and entertainment devices (GEAPIT SACIC)

TINNERMAN, Albert H., 1879–1961, American inventor and manufacturer; started Tinnerman Range Works exclusively for making stoves (1913–1939) then Tinnerman Products, Inc. (1939ff), designed and patented speed nut (1924) to assemble stoves especially for porcelain parts, then applied to electronics, was president of firm (1939–1953) then chairman of the board (until 1961) (NC52 PS)

TISELIUS, Arne Wilhelm Kaurin, 1902–1971, Swedish chemist and inventor; Inst. of Advanced Studies at Princeton Univ. (NJ) (1934–1935), developed absorption chromatography and electrophoresis (1940), faculty at Uppsala Univ. (1938–1968), improved technology for protein analysis, invented electrophoresis, Nobel prize (1948), fNAS (1949) (CDOS CDOSB LOTCS RHW TBDOS(Po) MWBD WWWIS)

TOBIAS, Charles W., 1920–1996, Hungarian American chemical engineer; current distribution in electrolytic cells, mass transfer in electrode process, batteries, fuel cells, faculty at Univ. of California at Berkeley (1947ff), development of electrochemical engineering using quantitative methods, NAE (1983) (AMOS Bridge WWWIA WWWIS)

TODD, Frederick Henry, 1903–1992, English American marine engineer; naval architecture and marine engineering, National Physical Laboratory (England) (1928–1940), various activities in hydrodynamics in England and US, scientific advisor to David Taylor Model Basin (DC) of the US Navy, NAE (1965), Gibbs & Cox (1969–1972), consulting engineer (1972ff) (Bridge MT7 WWIE)

TODD, Meryl L., 1909–1994, American agricultural engineer; Trane Co. (1931–1936), began private practice

(1937), during WWII involved in construction of Naval Air Station (1942–1943), continued own firm and later Todd-Hedeen International (1945ff), designed and supervised agricultural commodities processing plants, recognized for early contribution to development of grain storage aeration (Res2)

TOKUGAWA, Yoshimune, 1684–1751, Japanese government and instrument maker; revamped government and became fief of Kii (1705), astronomical instrument maker for palace observatory, use of telescope for surveying, helped develop new land and planting new crops (GEAPIT MWBD)

TOMPION, Thomas, c.1639–1713, English instrument maker and inventor; clockmaker in England for royal observatory (1676), with Robert Hooke [1635–1703] made watch with balance spring (1675), patented cylinder escapement (1695), made watchmaking improvements, made barometer and sundials (BDOS(Wi) DNB MWBD PS)

TOMPKINS, Daniel Augustus, 1851–1914, American engineer; after working in iron and steel industry (1882) promoted, designed, and built cottonseed oil mills throughout southern US, favored community financing and founding of cotton mills, established textile schools (CDOAB DAB WWWIA)

TORRES QUEVEDO, Leonardo, 1852–1936, Spanish civil engineer; invented Telekino (1902), Center for Study of Aeronautics created for him (1904), he developed several calculating devices such as chess playing machine (1912), electrical-mechanical calculator (1920) (CDOSB DOSB PS WWWIS)

TORRICELLI, Evangelista, 1608–1647, Italian physicist and inventor; relationship of atmospheric pressure (weight) to lifting water with a pump, Torricellian vacuum was name given to vacuum in an upended tube of mercury, with Vincenzo Viviani [1622–1703] invented first barometer (1643), professor at Florentine Academy (1642ff) (BEST CDOS EIH FNIE GI MWBD NYPL RHW SAI TBDOS(Po) TEIH TGS WWWIS)

TOSELLO, Andre, ??–1982, Brazilian agricultural engineer; founder of School of Agricultural and Food Engineering at State Univ. of Campinas, Univ. of Sao Paulo (1935ff), considered father of agricultural engineering in Brazil, 22 patents in Brazil (AE64)

TOTTEN, George Muirson, 1809–1884, American civil engineer; assistant engineer for various eastern canals (1827–1831), assistant engineer for Delaware & Raritan (NJ) (1831–1835), road and railroad construction (PA, VA, NC) (1835–1843), worked in Colombia (SA), Panama as chief of US Army Corps of Engineers (1878), Venezuela, consulting engineer (BDOACE CDOAB DAB NC18 WWWIA)

TOTTEN, Joseph Gilbert, 1788–1864, American military engineer; graduated from US Military Academy

and commissioned in Corps of Engineers (1805), after resigning rejoined the Corps (1808), engaged in construction of New York harbor, chief engineer of the army in different units during war with England (1812–1814), sea coast defenses, Board of Engineers (1819ff) as a colonel, chief of engineers (1838ff) and became brigadier general (1863) when the engineering and topographical services were combined, NAS (1863) (ANB BITAS BM1 CDOAB DAB NC4 PS WWWIA)

TOWER, Zealous Bates, 1819–1900, American military engineer; involved in Mexican War (1847), in Civil War (1861–1864) as brigadier general in Union army, was superintendent at US Military Academy (1864), engineer in charge of defenses (CDOAB DAB NC4)

TOWLE, William Mason, 1851–1930, American mechanical engineer; partner of Mauley & Towle (1877), instructor at Rose Polytechnical Inst. (IN) (1885), Buckeye Engine Co. (OH) (1888), with Sibley College at Cornell Univ. (NY) (1889–1892), Pennsylvania State Univ. (1892–1902), Smith College of Syracuse Univ. (NY) (1902–1907), Clarkson College of Technology (NY) (1907–1917), all in mechanical arts and industrial engineering (MEIA NC24)

TOWN, Ithiel, 1784–1844, American architect and inventor; designed churches and US customhouses (NY), patented design for truss bridge (1820), built several bridges (EIH GEAPIT MWBD WWWIA)

TOWNE, Henry Robinson, 1844–1924, American mechanical engineer; during Civil War worked with Port Richmond Iron Works (PA) supervising erection and installation of machinery on ships, William Sellers & Co. (1867), Yale Lock Manufacturing Co. first as manager then president when firm became Yale and Towne Manufacturing Co. (1868–1916) where he retired as chairman of the board (ACE CDOAB DAB MEIA NC21 WWWIA)

TOWNE, John Henry, 1818–1875, American marine engineer; partner in Merrick & Towne (PA) designing marine engines and centrifugal sugar machines (1838–1849), consulting engineer building gas works (1849–1861), partner with I. P. Morris, Towne & Co. (1861) operating iron works, endowed Towne Scientific School at Univ. of Pennsylvania (CDOAB DAB MEIA NC24 WWWIA)

TRAPEZNIKOV, Vadim A., 1905–??, Soviet electrical engineer; automation and telemechanics, Moscow Technical Inst. (1928ff), USSR Academy of Sciences (1941ff) (PS PSOCE SMOS WWWIS)

TRAUTWINE, John Cresson, 1810–1883, American civil engineer; worked mainly with railroads (until 1844), surveys for Panama Railroad (1849–1853), explored Isthmus of Panama for ship canal, authored popular *Engineers' Pocket Book* (1871) (BDOAS(El) BITAS CDOAB DAB NC5 PS WWWIA)

TRAUTWINE, John Cresson, Jr., 1850–1924, American civil engineer; son of John Cresson Trautwine [1810–1883] who was also civil engineer, assisted father in revising books on civil engineering, water hammer, flow of water over weirs, chief engineer for Bureau of Water in Philadelphia (PA) (1895–1899), consultant on projects on water distillation, location of electric car lines, public water supplies, city water prospects (BDOACE CDOAB WWWIA)

TREADWELL, Daniel, 1791–1872, American mechanical engineer and inventor; built screw-making machine (1812), nail-making machine, improved printing press, patented power press (1826), patented machine for hemp and flax spinning called Eypsey (1831), patented machines for spinning, roping flax (1834), rope cordage, hatcheting flax and hemp (1834), also patent on steam-engine condensing system (1841), and for cannon construction (ANB BITAS CDOAB DAB MEIA NC10 PS WWWIA)

TREDGOLD, Thomas, 1788–1829, British mechanical and civil engineer; testing of materials particularly strength of timber and cast iron, originated units for describing performance of boilers and engines, the British thermal unit (Btu) for heat and the foot-pound delivered per bushel of coal for efficiency studies, railroad engineering—published one of first books on subject in Europe (1825) (ASAI GEAPIT)

TRÉSAGUET, Pierre-Marie-Jérôme, 1716–1796, French civil engineer; inspector general (1775), road building methods used by other European countries, worked in government (EIH GEAPIT MWBD PS RHW TBDOS(Po) TEIH)

TREVITHICK, Richard, 1771–1833, English engineer and inventor; designed steam engines (1796), steam engine called Cornish engine to pull passenger trains on railroad (1801), followed by its successor the Penydarren engine (1804), high-pressure steam engine threshing machine (1812), worked also in Peru introducing steam engines for pumps in the silver mines (ASAI BEST EIH FNIE GEAPIT GI MWBD NYPL PS RHW SS1660 TBDOS(Po) TEIH TGE:B TGS WWWIS)

TRIMBLE, Isaac Ridgeway, 1802–1888, American military and civil engineer; military service (1822–1832) in artillery and ordnance, worked on railroad as chief engineer (1832–1861), served in Confederate Army as brigadier general (1861–1865) (BDOACE CDOAB DAB NC4 WWWIA)

TROLAND, Leonard Thompson, 1889–1932, American physiologist, physicist, engineer, and inventor; taught physiology at Harvard Univ. (1916–1929), chief engineer of Technicolor Motion Picture Corp., invented and developed multicolor process for motion pictures, physiology of vision (CDOAB DAB MWBD WWWIA)

TROUGHTON, Edward, 1753–1835, English instrument maker; known for accuracy and uniformity of instruments, emphasis on design and manufacturing of measuring instruments and mountings, in partnership with brother invented beam compass and hydrostatic balance, later entered in partnership with William Simms [1793–1860] (1826ff), produced a variety of navigational and astronautical instruments, FRS (1810) (DNB GEAPIT WWWIS)

TROWBRIDGE, William Pettit, 1828–1892, American mechanical engineer and inventor; US Military Academy graduate assigned to Topographical Engineers (1848–1862) making numerous surveys (coast of ME, VA, Pacific Ocean) and installation of instruments, vice president and general manager of Novelty Iron Works (1865), professor at Sheffield Scientific School of Yale Univ. (CT) (1877ff), NAS (1872) (BDOAS(El) BITAS BM3 CDOAB DAB MEIA NC4,14 PS WWWIA WWWIS)

TRULLINGER, Robert William, 1889–1955, American civil and agricultural engineer; practice in civil engineering (1910–1912), specialist in agricultural engineering research and administration of USDA (1910–1946) becoming chief of the Office of Experiment Stations, irrigation and drainage, WWI research in artillery, John Deere gold medal awarded by ASAE (1941) (AE35 WWWIA)

TRUMP, John George, 1907–1985, American electrical engineer; faculty at Massachusetts Inst. of Technology (MIT) (1933–1985) and director of High Voltage Lab. while at MIT (1947–1973), director of British branch of Radiation Lab. (1944–1945), medical application of X-rays and electrons, dielectric strength of compressed gases and vacuum, installation of electric cables, NAE (1977) (MT3 WWWIA)

TSAI LUN (CAI LUN), c.57–c.121 AD, Chinese inventor; usually credited with the invention of paper (c.105 AD) when he was director of Imperial workshops, work probably based on previous work in China, results spread to Europe (BDOTHOT BEST GI NYPL SACIC WA)

TSCHEBOTARIOFF, Gregory, 1899–??, Russian American civil engineer; officer in Russian artillery in WWI (1916–1920), Technische Hochschule Berlin-Charlottenburg (1921–1925), use of reinforced concrete for consulting engineers in Paris (1924), with Paul Kossel & Co. (1925–1927), construction work for government of Egypt (1929–1933), Egyptian Univ. (1933–1937), faculty at Princeton Univ. (1937ff), specialized in soil mechanics and foundations (BDOACE)

TSWETT, Mikhail Semenovich, 1872–1920, Russian Italian botanist and inventor; invented adsorption chromatography (1906), lecturer in Russia (1896), docent at Warsaw Univ. (1901–??), Warsaw Technical Univ. (1908–1917), to Russia and director of Botanical Gardens at Yuryev Univ. (1918–1920) (CDOS LOTCS TBDOS(Po) WWWIS)

TUCKER, Stephen Davis, 1818–1902, American inventor; apprentice, foreman, experimenter, partner with R. Hoe & Co. (NYC) (1834–1893) dealing with printing presses, inking, cutting, folding paper, over 100 patents on printing presses, some with Richard M. Hoe [1812–1886] (CDOAB DAB MEIA)

TULL, Jethro, 1674–1741, British lawyer, agriculturalist and inventor; invented first practical machine seed drill (1701), use of manure and practice of pulverizing soil, taught farmers how to maximize crop yields (DNB GEAPIT GI MWBD NYPL 1000Y PS RHS TBDOS(Po) TGS)

TU MAO, fl. 25 AD, Chinese military engineer; during restoration of central government strengthened northern frontier including beacon stations, transport depots, and roads (GEAPIT)

TUPOLEV, Andrei Nikolayevich, 1888–1972, Soviet aeronautical engineer; leading aircraft designer, medium and heavy bombers including TU-104 (1955) and TU-114 (1956), with his son Aleksei Tupolev [1926–2001] developed world's first supersonic passenger jet (TU-144), director of Central Aerodynamic Inst. (1918–1935), lieutenant general of General Engineering Technical Service, USSR Academy of Sciences (1933ff) (Discover22 MWBD PSOCE SMOS WWWIS)

TUPPER, Earl Silas, 1907–1983, American inventor and merchandiser; developed mail order business, worked at DuPont in 1930s as chemical engineer, formed Tupper Plastics Co. (1938), devised an injection molding machine and followed with successful sales of plastic bottles and later (1945) with bowls, cups, dishes and food storage containers with leakproof seals, developed home sales, formed Tupperware Corp. (1946), sold company to Rexall Drug Co. (1958) (AIE ANB C&EN80 IAI I&T17 NYPL)

TURING, Alan Mathison, 1912–1954, British computer scientist, mathematician and engineer; member of team that cracked German enigma code in WWII, National Physical Lab. (NPL) (1945–1948), Univ. of Manchester (1948ff), Turing machine (calculator), helped develop field of artificial intelligence (CDOS DNB IAI MWBD 1000Y RHW TGE:B TGS WWWIS)

TURNBULL, William, 1801–1857, American military engineer; US Military Academy (1819), commissioned in artillery but detached to topographical engineers, advanced to major (1838), devoted career to harbor improvements, topographical studies, water control, government buildings, worked primarily in Great Lakes and vicinity, construction of aqueduct for Chesapeake and Ohio Canal (DC), served as engineer to Gen. Winfield Scott [1786–1866] in Mexican War (1847) (ANB CDOAB DAB MWBD NC12 WWWIA)

TURNER, Arthur William, 1894–1951, American agricultural engineer; started career as a tractor demonstrator with Sandusky Tractor Co. (OH) (1917), served in WWI, extension engineer in soil erosion control at Iowa State Univ. and professor (until 1927), International Harvester Co. (1927), provided leadership in strengthening engineering in USDA after WWII and served as director of engineering research (1944–1951) (AE32 WWIE)

TURNER, Claude Allen Porter, 1869–1955, American civil engineer and inventor; worked in engineering departments of several bridge companies in New England, Ohio, Connecticut, Pennsylvania, and Minnesota, with Gillette Herzog Co. until the formation of the American Bridge Co., founder and president of Turner Construction Co. (1902), patented flat-slab construction (1905), granted at least 30 patents (BDOACE EIH WWWIA)

TURNER, Walter Victor, 1866–1955, English American civil engineer and inventor; to US (1888), first as textile expert then with Santa Fe Railroad (1897–1903), became expert in air brakes, directed engineering activities of Westinghouse Air Brake Co. (1903ff), held more than 400 patents (CDOAB DAB NC18 WWWIA)

TURRIANO, Juanelo, 1500–1585, Italian mechanical and hydraulic engineer; clockmaker, watchmaker, provided water to Toledo (Spain), developed walking troughs to carry water, wrote classic manuscript on inventions and machines (GEAPIT)

TUVE, Merle Antony, 1901–1982, American physicist and inventor; magnetism, short-pulse propagation, Carnegie Inst. of Washington (DC) (1926–1966), during WWII helped develop radar along with others particularly Robert A. Watson-Watt [1892–1973] and British colleagues first used to measure the height of the ionosphere, developed proximity fuse used in WWII, seismic wave propagation in Earth's crust, during WWII with National Defense Research Committee (NDRC) and Applied Physics Lab. at Johns Hopkins Univ. (MD) (1940–1946), NAS (1946) (ANB BM70 IWAP MWBD TBOF TNAOS WWWIA WWWIS)

TU YU, fl. 250 AD, Chinese mechanical civil engineer; designed and built bridge over Yellow River, developed new tools and designs for water mills, introduced trip hammer (c.260) for powdering materials (GEAPIT)

TWEEDDALE, William, 1823–1900, Scottish American civil engineer; in office of city engineer of Troy (NY) and on railroad construction (until 1854), bridge engineer and contractor in Chicago (1855–1860), moved to Iowa for construction of bridges and buildings (1860ff), during Civil War was promoted to brevet colonel (1860–1865), bridge and building construction (until 1880) (BDOACE NC5)

TWINING, Alexander Catlin, 1801–1884, American civil engineer and inventor; railroad engineer for Hartford & New Haven Railroad (1834–1839), professor at Middlebury College (VT) (1839–1849), invented method of manufacturing ice on commercial scale (1853)

(ANB BDOAS(El) BITAS CDOAB DAB NC19 PS WWWIA)

TWYMAN, Frank, 1876–1959, English electrical engineer and inventor; Adam Hilger, Ltd. (1898–1946) becoming general manager (1904), director of Hilger and Watts (1948–1952), optical instruments, wave-length spectrometer (1902), simplified spectro-chemical analysis, quartz spectrograph (1910), Twyman-Green interferometer (1918–1923) (DNB PS RS5 WWW)

TYNDALL, John, 1820–1893, Irish physicist and inventor; educator, professor then superintendent of Royal Inst. (1853–1857), investigated transmission, radiation and absorption of heat, Tyndall effect—diffusion of light through molecules and dust (BEST CDOS DNB MWBD RHW TEE WWWIS)

TYTUS, John Butler, Jr., 1875–1944, American inventor; after working in a paper mill and for a bridge maker he began with American Rolling Mill Co. (Armco) (1904ff) and developed continuous unending sheet similar to continuous roll of paper with which he had worked, first plant using his process (1924), he worked with designing and building several new such plants and became superintendent of new plant in Zanesville (OH) (ANB CDOAB)

U

ULRICH, Max Julius, 1856–1914, German American mechanical engineer; in Germany with W. Uhland (1873), built mining machinery with Sachsenberg Brothers (1874), designed pumps with Haddick & Rethe (1879), to US (1881), superintendent of Ulrich Engine Co. (MA) (1882–1894), invented cut-off motion for duplex steam pumps, designer with Deane Steam Pump Co. (MA) (1892ff), taught courses in hydraulics, pumping, and machine design for International Correspondence School, designed oil engine for De La Vergne Machine Co. (1912) (MEIA)

UNWIN, William Cathorne, 1838–1933, British mechanical and civil engineer and inventor; strength of boiler flues, properties of saturated steam, continuous mechanical brakes for railroad cars (1859), fatigue of iron girders (1860–1862), mirror extensometer for measuring Young modulus, construction of turbines and waterwheels, works manager of Williamson & Brothers (1862–1866), Fairbairn Engineering Co. (1866–1872), professor at Royal Indian Engineering College (1872–1884), professor and dean of school that became London Univ., Imperial College (BDOTHOT DNB EE WWW WWWIS)

URE, Andrew, 1778–1857, British (Scottish) chemist and inventor; medical doctor (1801), professor of natural philosophy at Univ. of Strathclyde (Glasgow) (1804–1830), lectured in chemistry and mechanics, consulting chemist in London (1830ff), Ure bimetallic thermostat (1830), author in geology (1829), wrote on manufacturing and industrial regulations (1835), on gravity, latent heat of vapors (CDOSB DNB DOSB EE NYPL WWWIS)

UREY, Harold Clayton, 1893–1981, American chemist and developer; heavy water (1931), methods of separating isotopes, radioisotopes, during WWII worked on development of atomic bomb, on faculties at Columbia Univ. (NYC) (1929–1945), Univ. of Chicago (IL) (1945–1958), Univ. of California-La Jolla (1945ff), and other universities, NAS (1935), Nobel prize (1934) (BEST BM68 CDOS MWBD NCE RHW TGS WWWIA WWWIS)

V

VALTURIO, Roberto, 1405–1475, Italian military engineer (technologist); served as secretary to Pope Eugene IV [1431–1447], became private secretary to rulers, wrote (1455–1460) on subject of military engineering and published (1472) (CDOSB DOSB GEAPIT NYT-A PS)

VAN DE GRAAFF, Robert Jemison, 1901–1967, American physicist, mechanical engineer, and inventor; research associate (1931–1934) then research and teaching at Massachusetts Inst. of Technology (MIT) (1934–1960), built high-voltage electrostatic generator (1931), with others founded High Voltage Engineering Corp. (1946ff) (BEST CDOAB CDOS MWBD RHW TBDOS(Po) WWWIA WWWIS)

VAN DEPOELE, Charles Joseph, 1846–1892, Belgian American electrical engineer and inventor; practical electric traction (1874), patented electric railway system (1883), sold patents and became employee of Thomson-Houston Electric Co. (1888), patented electric generator (1880), carbon commutator brush (1888), and several other items including a gearless electric locomotive (1894) (BITAS CDOAB DAB PS MWBD NC13 WWWIA WWWIS)

VANDERBILT, Cornelius, 1794–1877, American entrepreneur and financier; established freight and passenger service between Staten Island (NY) and NYC (1810–1818), captain on ferry line between Brunswick and NYC (1818–1829), financed shipping and consolidated railroads on East Coast and Midwest, built Grand Central Terminal (NYC) (1873), left considerable money to fund what became Vanderbilt Univ. (TN) founded (1873), his son William Henry Vanderbilt [1821–1885] was president of New York Central Railroad, who had four sons three of whom were involved in railroad industry—Cornelius Vanderbilt [1843–1899], William K. Vanderbilt [1849–1920], Frederick W. Vanderbilt [1856–1938] (AIE IIA MWBD NC6,34 1000Y WWWIA)

VAN DER WAALS, Johannes Diderik, 1837–1923, Dutch physicist and inventor; professor at Univ. of Amsterdam (1877–1907), studied the nature of gaseous and liquid phases and their interaction, van der Waals equation (1873), developed gas property equations used in engineering, Nobel prize (1910), fNAS (1913) (BEST CDOS L&M MWBD SAI WWWIS)

VAN DER ZIEL, Aldert, 1910–1991, Dutch American electrical engineer; N.V. Philips Co. (1934–1937), faculty at Univ. of British Columbia (1947–1950), then Univ. of Minnesota (1950ff), noise in electronic devices, solid-state physics, graduate education in engineering, NAE (1978) (AMOS Bridge MT5 WWWIA)

VAN DYKE, John Wesley, 1849–1939, American executive and inventor; with various Standard Oil Co. units (1873–1911), president of Atlantic Refining Co. (1911–1927), patented inventions relating to sulfur elimination, patented inventions relating to railroad tank cars, with W. W. Irish invented tower still for refining (CDOAB DAB WWWIA)

VANONI, Vito August, 1904–1999, American civil engineer; soil conservation work with US Department of Agriculture (USDA), faculty at California Inst. of Technology (1940–1974), hydraulics particularly of open channel flow, sedimentation mechanics, soil conservation, application to construction and maintenance of engineering structures particularly of harbors and coasts, consultant, NAE (1977) (AMOS Bridge WWIE)

VAN UITERT, LeGrand Gerard, 1922–1999, American chemist and inventor; US Navy (1940–1946), synthesis and crystal growth of new magnetic and optical materials, use of these materials in devices, luminescence, supervisor of Solid State Material Synthesis Group at AT&T Bell Labs. (1952ff), NAE (1981) (AMOS Bridge WWWIA)

VAN VALKENBURG, Mac Elwyn, 1921–1997, American electrical engineer; academic experience at Massachusetts Inst. of Technology (MIT) (1943–1946), Univ. of Utah (1946–1955), and Stanford Univ. (1949–1951), faculty and engineering college dean (1984–1988) at Univ. of Illinois (1955ff), circuit theory, beacon electronics, antennas, servomechanisms, computer science, network analysis and synthesis NAE (1973) (AMOS Bridge WWIE WWWIA)

VARIAN, Sigurd Fergus, 1901–1961, American electrical engineer and inventor; initially involved in flying and barn storming, began working in electronics (1935ff), with brothers Eric Varian and Russell H. Varian [1898–1959] who had over 100 patents [1898–1959] and William W. Hansen [1909–1949] developed klystron, resonator (1939), they formed Varian Associates (1948) of which he was president then chairman (1948–1959) to make microwave tubes, later developed satellite borne magnetometer for measuring earth's magnetic field, faculty at Stanford Univ. (1937–1940) (1946–1959), engi-

neer at Sperry Gyroscope Co. (1940–1946) (BDOTHOT LOTCS MWBD NC49 NYTObit TP WWWIA)

VASCO COSTA, Fernando, ??–1996, Portuguese engineer; distinguished contributions to the theory and practice of ship berthing, mooring, maritime structural design, consulting engineer with Harbour Works, Lisbon, Portugal, fNAE (1989) (NAEAnnouncement)

VAUBAN, Sébastien Le Prestre de, 1633–1707, French military and civil engineer; joined engineering military activities (1655), Corps de Génie founded (1675) during reign of Louis XIV, Jean-Baptiste Colbert [1619–1683] was French minister who founded the Corps, chief engineer in charge of several sieges gaining recognition for devising tactics for capturing cities and regions for Louis XIV, designed fortifications, was marshal of France (1703), invented socket bayonet (BDOS(Wi) BDOTHOT GEAPIT MWBD PS WWWIS)

VAUCANSON, Jacques de, 1709–1782, French inventor; built many automata, many improvements in silk-weaving machinery (1741ff) such as the loom of Joseph-Marie Jacquard [1752–1834] (BDOS(Wi) GEAPIT MWBD PS WWWIS)

VAUCLAIN, Samuel Matthews, 1856–1940, American inventor; after apprenticeship with Pennsylvania Railroad (PRR) shop, became assistant foreman, foreman and general superintendent of Baldwin Locomotive Works (PA) (1877ff), partner of Burnham, Williams & Co. (1896ff), munitions activities in WWI, numerous patents on engines and machines to make engines and locomotives (CDOAB DAB MEIA NCE,33 WWWIA)

VEKSLER, Vladimir Iosifovich, 1907–1966, Soviet physicist and inventor; contributed to the development of the sychrocyclotron (1940s) also work done by Edwin M. McMillan [1907–1991] and Ernst O. Lawrence [1901–1958] (BEST IAI MWBD WWWIS)

VENN, John, 1834–1923, English mathematician and inventor; at Cambridge Univ. (1857ff) where he developed explanation and representation on chance and logic, representation of concepts for Venn diagrams named after him (DNB DOSB DOT HOFL MWBD WWW)

VENTURI, Giovanni Battista 1746–1822, Italian physicist, engineer, and inventor; did engineering for a duke, studied flow of water in pipes and open channels, fluid mechanics, observed phenomenon known as Venturi effect, Venturi meter or tube to measure the flow of fluids (gases and liquids) was named after him (1888 or 1894) by hydraulics engineer Clemens Herschel [1842–1930] (EE FNIE MWBD WWWIS)

VERANTIUS, Faustus (or Fauston Veranzio), 1551–1617, Italian engineer; interested in mechanisms, published book on new machines (1575) including windmills, dredger, truss bridges, suspension bridge—those of his creation not known (CDOSB DOSB EIH GEAPIT PS)

VERBECK, William, 1861–1930, American inventor; on faculty of military academies (CA, NY), then head of Manlius School, an officer in New York National Guard, pioneer in Boy Scout movement, received six patents relating to photography (CDOAB DAB NC27 WWWIA)

VERMUYDEN, Cornelius, 1595–c.1683, Dutch British hydraulic engineer; drainage engineer, to England (c.1621), became British subject (1633), repair of Thames River embankments (1621), planned and coordinated drainage systems for large areas, reclamation of Great Fens (1630–1637) and (1649–1652), used gravitation system with canals (BDOS(Wi) DNB GEAPIT LOTE MWBD PS SFLOE SS1660 TEIH)

VERNE, Jules, 1828–1905, French technological writer; success in writing technological adventure and science fiction such as *A Journey to the Center of the Earth* (1864) and *Twenty Thousand Leagues Under the Sea* (1870) (EB MWBD 1000Y)

VERNIER, Pierre, c.1580–1637, French mathematician and inventor; instrument maker, Vernier measuring scale (1631), invented Vernier caliper (1631) (CDOS GLOSB MWBD NYPL PS RHW TBDOS(Po) WWWIS)

VER SNYDER, Francis L., 1927–1989, American metallurgical engineer and inventor; US Army (1943–1945), research at General Electric Co. (1950–1961), invention, development and application of directionally solidified and single-crystal high-temperature superalloys, airfoils, director of research for materials at United Technologies Corp. (1961ff), NAE (1981) (AMOS Bridge MT4)

VICAT, Louis Joseph, 1786–1861, French civil engineer; with Joseph Aspdin [1799–1855] developed water-resistant cements, hydraulic cement, pioneered in use of concrete to replace stone for bridge piers, his first was the bridge at Souillac then the Argentac bridge piers (EIH GEAPIT)

VIDIE, Lucien (Nantes), 1805–1866, French scientist and inventor; invented aneroid barometer (1844) (EE MWBD)

VIELE, Egbert Ludovicus, 1825–1902, American civil engineer; US Military Academy (1847), in military service (until 1883), employed by New Jersey and became chief engineer of Central Park (NYC), made topographical studies related to park, drainage, streets, etc., in Civil War where he became brigadier general (1861–1863), consulting practice (CDOAB DAB NC2 WWWIA)

VIEILLE, Paul Marie Eugène, 1854–1934, French engineer and inventor; smokeless powder (1884), investigated shock waves (MWBD PS WA)

VIGNOLES, Charles Blacker, 1793–1875, British civil engineer; commissioned in British Army (40 yr. service), built railroads in England, Ireland, Russia, Switzerland, invented flat-bottom railroad line (1830), with John Ericcson [1803–1889] developed a method of going up steep inclines, was state surveyor for South Carolina, professor of civil engineering at Univ. of London (1841ff) (DNB GEAPIT MWBD WWWIS)

VINCI, Leonardo de—see LEONARDO DE VINCI

VINE, Allyn Collins, 1914–1994, American physicist and oceanographer; oceanographic engineering, design of deep submersibles for research, applied solar energy, researcher at Woods Hole Oceanographic Inst. (1940ff), NAE (1982) (AMOS Bridge MT8 WWWIA)

VINTON, Francis Laurens, 1835–1879, American mining engineer; graduate of US Military Academy (1856), to France to study at School of Mines, taught mechanical engineering at Cooper Union (NY), in Civil War was a captain in infantry (1861–1863) who became brigadier general, chair of civil and mining engineering at Columbia Univ. (NYC) (1863–1876), consulting mining engineer in Colorado (1876ff) (CDOAB DAB NC7 WWWIA)

VITRUVIUS POLLIO, Marcus (aka Marcus Vitruvius Pollio), fl. 1st century BC, Roman architect and military engineer; building materials, city planning, invented Roman odometer (called hodometer), theater construction, classical architecture (AIn BEST EIH GEAPIT ME125 MWBD TEIH WWWIS)

VOGEL, Herbert Davis, 1900–1984, American civil and military engineer; hydraulics, after 30 yr. with Corps of Engineers retired as brigadier general (1954), development and director of US Waterways Experiment Station (1929–1934), numerous engineering assignments as district and division engineer, then chairman of board of Tennessee Valley Authority (TVA) (1954–1963), president of Tennessee River Assoc. (1963–1967), consulting engineer (1967–1984), NAE (1977) (AMOS MT3 NCI WWWIA)

VOLLUM, (Charles) Howard, 1913–1986, American physicist, electrical engineer, and industrialist; with Murdock Radio & Appliance Co. (1936–1941), served in US Army Signal Corps (1941–1945), with Melvin J. Murdock [b. 1917] founded Tektronix, Inc. (OR) and was president and chief engineer (1946ff), involved in development of oscilloscope and cathode-ray tube with many improvements, NAE (1977) (MT3 WWWIA)

VOLTA, Count Alessandro Giuseppe Antonio Anastasio, 1745–1827, Italian physicist and inventor; invented electrophorus (1775) a negative-positive electric charge device, composition of methane gas and isolated methane (1779), taught in high school then appointed professor at Univ. of Pavia (1779ff), constructed devices that would produce flow of electricity that were the first batteries called voltaic pile (1799) based first on salt solutions and later on different metals, driving force of electricity called volt named for him (AmSci91 BEST CDOS FNIE GEAPIT GI IAI MWBD NYPL 1000Y RHW SAI TGS WBE WWWIS)

VONNEGUT, Bernard, 1914–1997, American chemist and inventor; glass research with Hartford Empire Co. (CT) (1939–1941), Massachusetts Inst. of Technology (1941–1945), General Electric Co. Research Labs. (NY) (1945–1952), Arthur D. Little, Inc. (MA) (1952–1967), faculty at State Univ. of New York at Albany (1967–1985), icing conditions, cloud seeding with Vincent J. Schaefer [1906–1993], iron lung (1928) (GLOSB HOCS&T TIY(2) WWWIA WWWIS)

VON NEUMANN, John—see NEUMANN, John von

VOSE, George Leonard, 1831–1910, American civil engineer; engaged in railroad location and construction across US (1850–1859), editor of *American Railway Times* (1859–1863), professor at Bowdoin College (ME) (1872–1881), then at Massachusetts Inst. of Technology (MIT) (1881–1886), author of several engineering books (BDOACE CDOAB DAB)

VREELAND, Frederick King, 1874–1964, American mechanical and electrical engineer, inventor, and manufacturer; Crocker-Wheeler Electric Co. (1896–1900), became foremost innovator of radio devices, organized and president of Vreeland Apparatus Co. (NY) (1905–1964), at least 10 patents used by other companies in US and abroad (NC51 PS)

VUTZ, William, 1903–1976, American agricultural engineer; New Holland Division (PA) of Sperry Rand and successor companies (1953ff), various machines for agriculture, engineering consultant (1968–1976), held 33 US and foreign patents (AE58)

VYSHNEGRADSKY, Ivan Alekseievich, 1831–1895, Russian civil engineer and machines; taught at Artillery Academy at St. Petersburg and St. Petersburg Technological Inst., minister of finances (1888–1892), made significant contributions to theory of automatic control, Vyshnegradsky criteria (1877) defined stability conditions for regulating a system (CDOSB DOSB PS WWWIS)

W

WAALS, Johannes Diderik Van der—see VAN DER WAALS, Johannes Diderik

WADDELL, John Alexander Low, 1854–1938, Canadian American civil engineer and inventor; after experience in Canada, on faculty at Rensselaer Polytechnic Inst. (RPI) (NY) (1878–1880), chief engineer of Raymond & Campbell bridge engineers (1881–1882), professor at Imperial Univ. of Tokyo (1882–1886), to US and opened office on bridge design (1866–1869), partnership with Ira G. Hedrick [1868–1937] (1899–1907), partnership with John Lyle Harrington [1868–1942] (1907–1915) and his son (1915–1918), practiced alone (1920–1927), partnership with Shortridge Hardesty [1884–1956]

(1927–1938) as consulting engineer (ANB BDOACE DAB NC27,D WWWIA)

WAGGONER, Eugene B., 1913–1991, American engineering geologist; use of geologic experience in construction activities, Woodward-Clyde, Inc. (1960–1973) becoming chief executive (1967), consultant (1973ff), NAE (1987) (Bridge MT6)

WAGNER, Aubrey J., 1912–1990, American civil engineer; resource development of major river basin, and in navigation, smaller tributary aspects, Tennessee Valley Authority (TVA) (1934–1961) becoming chairman, consultant, NAE (1973) (Bridge MT6 WWIE WWWIA)

WAIT, James Richard, 1924–1998, Canadian American electrical engineer; electromagnetic theory, radio engineering, electronics, radar, Canadian Army (1942–1945), NBS of US Department of Commerce (1955ff), faculty at Univ. of Arizona (1980ff), consultant, NAE (1977) (AMOS MT9 WWIE WWWIA)

WAIT, William Bell, 1839–1916, American educator and inventor; superintendent of school for blind (NYC), invented New York Point System (1868) to assist blind students (a variation of the Braille system), and learning devices (CBOAB DAB NC2 WWWIA)

WAITE, Christopher Champlin, 1843–1896, American civil engineer; assistant and chief engineer and superintendent of water works and several railroads (1864–1881), assistant to president of railroad, vice president of Cincinnati Chamber of Commerce (BDOACE)

WAITE, Henry Matson, 1869–1944, American civil engineer; railroad maintenance and supervision (1890–1909), chief engineer of a coal company (1909–1912), chief engineer of Cincinnati (OH) (1912–1914), city manager of Dayton (OH) (1914–1918), WWI service in Europe, private practice, with federal Public Works Administration (PWA) (1934–1937), private practice (ANB CDOAB DAB NC34,D WWWIA)

WALKER, Eric Arthur, 1910–1995, English American electrical engineer; faculty including head of Underwater Sound Lab. at Harvard Univ. (1942–1945), professor and dean of engineering and president of Pennsylvania State Univ. (1945–1970), vice president of Aluminum Co. of America (Alcoa) (1970–1975), served on numerous government boards and commissions, NAE (1964) (AMOS Bridge MT8 NCI WWIE WWWIA WWWIS)

WALKER, Harry Bruce, 1884–1957, American agricultural engineer; topographer for Chicago, Burlington & Quincy Railroad (IL), organized department at Kansas State Univ. (1911–1928), head of department at Univ. of California (1928–1951), recognized as a leader who maintained high standards of teaching and research particularly in drainage and irrigation (AE38 NC47 WWWIA)

WALKER, John Charles, 1893–1960, American chemical engineer and inventor; Empire Gas and Fuel

Co. of Cities Service (1917–1958) serving first in research then general superintendent of chemical division (1937–1947) and as director of research and development, concerned with problems of production and transportation of petroleum products, emulsions, breaking machines, at least 40 US patents, NAS (1945) (BM77 NC48 PS NYTObit)

WALKER, Reuben Lindsay, 1827–1890, American civil engineer; engineer with Chesapeake and Ohio Railroad (1845–1861), in Civil War (1861–1865) becoming brigadier general and chief of artillery, engaged in farming after war, employed by railroads, superintendent of construction and building of state projects (BDOACE DAB WWWIA)

WALKER, William Hultz, 1869–1934, American chemical engineer and inventor; faculty at Pennsylvania State Univ., formed partnership Little & Walker (1900–1905), Massachusetts Inst. of Technology (1905–1921) where he established school of chemical engineering, director of division of industrial cooperative research (1920), during WWI had rank of colonel, built chemical processing plant, produced artificial silk, worked on utilization of waste sulfite, recovery of turpentine from wood (HOCE NCA PS WWWIA WWWIS)

WALLACE, Charles F., 1885–1964, American manufacturer and inventor; after service with Gerard Ozone Co. and Moore Filter Co. formed partnership of Wallace and Tiernan (1913) developed equipment and processes for water treatment, Wallace served as vice president and secretary (until 1954) and thereafter was officer on the board of directors, developed first chlorinator using chlorine gas to treat water, held several patents for water treatment (NC51 PS WWWIA)

WALLACE, John Findley, 1852–1921, American civil engineer; first chief engineer of the Panama Canal (1904–1905), executive of midwestern and western railroads, president of Westinghouse, Church, Kerr & Co., consulting engineers (CDOAB DAB NC10 WWWIA)

WALLACE, William, 1825–1904, English American manufacturer and inventor; established firm of Wallace & Sons in US (1848), constructed dynamos based on patent by Moses G. Farmer [1820–1893] (1872), built dynamos patented by Wallace (1872), demonstrated operation of arc lamp in series, built electroplating plant (CDOAB DAB MWBD WWWIS)

WALLER, Frederic, 1886–1954, American inventor and manufacturer; built first automatic printer and timer for optical printer, invented cinerama from which first picture was made (1952) (AIE ANB CDOAB NCG WWWIS)

WALLIS, Barnes Neville (aka Neville Barnes), 1887–1979, British marine and aeronautical engineer and inventor; trained as a marine engineer, changed from shipbuilding to airships and airplanes, with the Vickers Co. (1911–1945) where he designed airships and bouncing bombs for WWII, design work led to the Wellesley bomber used in WWII, assisted in development of the Concorde supersonic airliner and developed the swing wing aircraft, British Aircraft Corp. (1945–1971) where he was chief of aeronautical research (CDOS DNB RHW TBDOS(Po) TGE:B WWW)

WALSCHAERTS, Egide, 1820–1901, Belgian engineer; railroad engineer in workshop of state railways in Malines (1842ff) becoming shop foreman in Brussels (1844ff), exploited a patent taken out in France for a new system of steam distribution applicable to steam engines (Belgian law prohibited his making a profit), system became standard equipment in British locomotives (1844) (BDOS(Wi) PS)

WANG, An, 1920–1990, Chinese American computer engineer; engineer at Central Radio Works in China (1941–1945), to US (1945), developed magnetic core memory, owner, director, president of Wang Lab. (1951ff), development, manufacturer and application of computers, NAE (1982) (AIE ANB MT6 MWBD S: TLAW WWWIA)

WANG CHENG, fl. 1627, Chinese horological engineer; developed Chinese-style waterwheel linkwork escapement with Johannes Schreck that he called a wheel clepsydra (GEAPIT SACIC)

WANG CHING, fl. 70 AD, Chinese hydraulic engineer; reconstructed Pien Canal (70 AD), rebuilt Peony Dam (78ff AD), used 'flash locks' in canals, served as inspector general of rivers and dikes (GEAPIT)

WANKEL, Felix Heinrich, 1902–1988, German engineer and inventor; devoted career to developing a pistonless cylinderless engine (1926ff), founded research establishment (1924), during WWII with German Air Ministry, with Technische Entwicklungstelle in Lindau (1945ff) and became director (1960), worked for Bavarian Motor Works (BMW) developing engine with rotary valves (1929), prototype tested (1956), rotary automobile engine (1957) made in Japan and England, classified as a trochoid rotary internal combustion engine (BDOTHOT CDOS EE EOS MWBD PS RHW SAI TBDOS(Po) TGS WOI)

WARD, Aaron Montgomery, 1843–1913, American merchandiser; sold by mail order issuing his first catalog (1871) under name of Montgomery Ward & Co., made merchandise available to wide range of people particularly rural residents, catalog was one of most widely read publications in US (AIE MWBD NC13 1000Y WWWIA)

WARD, Charles, 1841–1915, English American mechanical engineer; installed gas works for which he became superintendent and general manager (1871), installed water tube boiler on coastal defense ship (1880ff), pioneered use of a screw propellers on river steam boats (MEIA)

WARD, George Gray, 1844–1922, English American cable engineer; telegrapher, to US (c.1876), superintendent of new cable company, general manager of Commercial Cable Co. (1884ff), handled aspects of laying cable across Pacific Ocean (1902–1906) (CDOAB DAB NC14)

WARING, Fred, 1900–1984, American band leader and inventor; began studying engineering but left school (1922) and began music career (Pennsylvanians) with numerous appearances at colleges, Broadway, and overseas, many recordings, led in development of Waring blendor (sic) (1937) (ANB EOET GI WWWIA)

WARING, George Edwin, Jr., 1833–1898, American agriculturalist, civil and sanitary engineer; leader in sanitation movement, sewage projects, street cleaning, clean water, consulting engineer (ANB CDOAB DAB NC6 PharmCent WWWIA)

WARNER, Edward Pearson, 1894–1958, American aeronautical engineer; taught at Massachusetts Inst. of Technology (MIT) (1920–1926), assistant secretary of Navy (1926–1929), edited magazine *Aviation* (1929–1934), served on national and international civil aviation organizations (until 1957) (ANB CDOAB DAB WWWIA)

WARNER, Roger S., Jr., 1907–1976, American chemical engineer; development engineer with Plastics Corp. (1931), Taylor Engineering (1932), consulting engineer (1932–1933), Kaysam Corp. (1935–1938), B. F. Goodrich Co. (1938–1942), US government activities (1942–1950), Arthur D. Little, Inc. (1950–1951), vice president of Cambridge Corp. (1951–1958), office of secretary of defense (DOD) (1958–1960), consulting engineer (1960ff), process development of latex, mechanical design, cryogenic equipment, satellite systems (AMOS PS WWWIA)

WARNER, Worcester Reed, 1846–1929, American mechanical engineer; machinist and foreman of gear cutting at Pratt & Whitney Co. (CT) (1866–1880), established machine manufacturing business at Ambrose Swasey [1846–1937] (1880–1900) building telescopes, incorporated as Warner and Swasey Co. (IL, OH) (1900ff) served as president and chairman of the board (CDOAB MEIA NC21,27 WWWIA)

WARREN, Cyrus Moors, 1824–1891, American chemist, manufacturer, and inventor; authority on hydrocarbon constituents of tar, manufactured tar roofing materials, developed process of fractional condensation (c. 1865), invented asphalt-purification process (CDOAB DAB NC10 WWWIA WWWIS)

WARREN, Gouverneur Kemble, 1830–1882, American military and civil engineer; appointed officer in Corps of Topographical Engineers after graduation from US Military Academy (1850), assistant engineer on survey of delta of Mississippi River (1850–1854), head of surveys, maps, and reconnaissance of western territories (until 1859), action in Civil War (1861–1865) becoming major general (ante 1882) in US Army, involved in study of telegraph lines, survey of battlefields, piers, and harbor works, NAS (1876) (ANB BDOACE BDOAS(El) BITAS BM2 CDOAB DAB MWBD NC4 PS WWWIA)

WARREN, Henry Ellis, 1872–1957, American electrical engineer and inventor; invented synchronous electrical clock (1914), devised master clock for General Electric Co. to control power frequencies, formed and was president of Warren Clock Co. (1919ff) that became a part of General Electric Co. (1948) and was sold to Timex (1978), received 135 US patents (ANB CDOAB NC46 NYTObit WWWIA)

WARREN, Russell, 1783–1860, American architectural engineer; began as a carpenter and builder, practiced architecture, recognized for designing federal style mansions in Bristol (RI) (1805–1825), then in Providence (RI), credited with designing the Warren truss used in steel bridge construction, private practice (1836ff) (CDOAB DAB MWBD WWWIA)

WASHBURN, Frank Sherman, 1860–1922, American civil engineer and entrepreneur; with Chicago and North Western Railroad (1883–1887) as a bridge engineer and as division engineer, spent 2 yr. inventing railroad devices, was superintendent of Union Stock Yard & Transit Co. (1889), with others formed consulting and construction firm (c.1890), built dams and reservoirs, water supply systems, organized Crescent Coal Co. (TN) and was general manager (1899–1914), developed hydropower and was president of Alabama Power Co. (1913–1915) (ANB NC32 WWWIA)

WASHBURN, Nathan, 1818–1903, American iron founder and inventor; patented an improved cast iron railroad car wheel (1849), founded firm to build the wheel and to build textile machinery, during Civil War developed a puddling process to make superior gun barrel (CDOAB DAB NCIO)

WASHINGTON, George, 1732–1799, American surveyor, military officer, and politician; surveyor (1748ff), military service (1752ff) receiving commission (1753), in Ohio region with military outpost in Pittsburgh, was a Virginia farmer and planter, served in Virginia House of Burgesses (1759–1774), member of Continental Congress (1774–1775), commanded Continental Army during revolution (1775–1783) with surrender of British (1781), president of Continental Convention (1789), first president of US (1789–1797), considered father of country (US) (CDOAB DAB EB EOWB MWBD NC1ff 1000Y WWWIA)

WATERMAN, Lewis Edson, 1837–1901, American inventor; perfected, improved, patented, and manufactured Waterman fountain pen with ink carried in barrel of pen (1884), successful replacement for dip pen (CDOAB DAB ITNC MWBD NC1 NYPL WWWIA)

WATKINS, George Benson, 1895–1966, American chemical engineer; faculty at Kansas State Univ. (1921–1924), Univ. of Michigan (1924–1927), research engineer and director of research at Libbey-Owens-Ford (1927–1955), fuel technology, ceramics, safety glass, flat gas products (AMOS NC53 PS)

WATKINS, John Elfreth, 1852–1903, American civil engineer; mining engineer for Delaware & Hudson Co. (1871–1872), assistant engineer of construction with Pennsylvania Railroad (PRR) (1872–1884), US National Museum (DC) (1884–1892) (1895–1903) (BDOACE BITAS CDOAB DAB WWWIA)

WATSON, Thomas Augustus, 1854–1934, American inventor; telephone technician, assisted Alexander Graham Bell [1847–1922] (1874–1877), became director of research for Bell Telephone Co. (until 1881), opened machine shop and shipyard, formed Fore River Ship & Engine Co. (MA) (1901) (AIE ANB IAI IIA MWBD NC27 PS WWWIA)

WATSON, Thomas John, 1874–1956, American electrical engineer and inventor; general manager of National Cash Register Co. (1899–1913), provided innovative leadership of Computing-Tabulating-Recording (CTR) Co. (1913–1956) that became International Business Machine Co. (IBM) (1924) of which he became president and later chairman (1949–1956) (ANB IAI MWBD NC47,E WWWIA)

WATSON, Thomas John, Jr., 1914–1993, American computer leader and manufacturer; served in WWII, under his leadership at International Business Machines (IBM) (1946ff) of which he became president (1952), began focus on personal computers (PC) available (1981), whereas previously the emphasis was on large main frame computers (ANB IAI NCI WWWIA)

WATSON, William, 1834–1915, American civil and mechanical engineer; lecturer at Harvard Univ. (MA) (1863), professor at Massachusetts Inst. of Technology (MIT) (1865–1873), elasticity and resistance of materials, plaster modeling of masonry problems, mathematical applications (BITAS CDOAB DAB MEIA NC12 WWWIA)

WATSON-WATT, Robert Alexander, 1892–1973, British (Scottish) engineer, physicist, and inventor; patented device for radiolocation using short-wave radio waves (1919) called radiolocator, early invention and development of radar (radio detection and ranging) (1935ff), National Physical Lab. (Great Britain), applications to weather, speed of vehicles, guided missiles, career in government service (1915–1952), FRS (1941) (BEST CDOS DNB DOSB GI IAI NYPL 1000Y RHW SAI S:TLAW TBDOS(Po) TGE:B TP WA WWW WWWIS)

WATT, James, 1736–1819, Scottish mechanical engineer and inventor; instrument maker, faculty at Univ. of Glasgow (1757ff), made improvements to Newcomen steam engine (1764ff), developed use of steam condenser for steam engine (1769), introduced concept of using steam on both sides of the piston, started business of manufacturing steam engines (1774), centrifugal governor for controlling speed of engines (1784, 1788), invented steam heat for space heating, invented a centrifugal governor, defined the horsepower as 550 foot-pounds per second (ASAI BEST BDOS CDOS DNB DOSB EE EIH FNIE GEAPIT LOTE MWBD PS RHW SAI SAI-MP SFLOE TBDOS(Po) TEIH TGE:B TGS WWWIS)

WAVERING, Elmer H., 1907–1998, American inventor; Motorola, Inc. (ante 1972) becoming president then CEO (1964–1972), pioneer in automotive electronics, built first commercial automobile radio, he and associates built first automotive alternator (1950s) (FAL NYTObit WWIA)

WAYNICK, Arthur Henry, 1905–1982, American physicist and electrical engineer; radio engineer with Meno Radio Co. (1922–1935), with Wayne State Univ. (1935–1940), Harvard Underwater Sound Lab. (1943–1945), faculty at Pennsylvania State Univ. (1945–1971), NAE (1969) (MT2 WWWIA WWIE)

WEAVER, Ira Adalbert, 1871–1965, American inventor and manufacturer; interested in better agricultural implements, first with Deere & Co. (IL) (1897–1900), then with Sattley Manufacturing Co. (1900–1916), with his brother Gailard E. Weaver [1883–1942] formed Weaver Manufacturing Co. (1910–1960) of which he was president, firm merged with Dura Corp. (1959), more than 100 US patents (NC52 PS)

WEAVER, Warren, 1894–1978, American physicist, mathematician, and inventor; professor at Univ. of Wisconsin (1920–1932), Rockefeller Foundation (1932–1959), involved in development of instruments such as ultracentrifuge, X-ray diffraction, electron diffraction, spectroscopy, radio isotopes, particle accelerators, suggested term 'molecular biology,' NAS (1969) (BM57 CB CDOAB PharmCent WWWIA WWWIS)

WEAVER, William Dixon, 1857–1919, American electrical engineer; US Navy (ante 1892), designed electrical equipment, editor of electrical technical journals (1893ff), organized professional societies (BITAS CDOAB DAB WWWIA)

WEBB, George Herbert, 1860–1921, American civil engineer; rodman, levelman, transitman for railroads (1880–1888), construction of railroads in Chile and Peru (1888–1893), private practice (1893–1897), with various railroads (1897–1917), chief engineer of Detroit River Tunnel Co. (1911), US Army Engineering Corps in WWI with service in France as lieutenant colonel (1917–1919) (BDOACE WWWIA)

WEBB, Harry Howard, 1853–1939, American mining engineer; involved in development of gold and diamond mines in South Africa (1895–1916), later with the development of potash and borax deposits in California (CDOAB DAB)

WEBB, John Burkitt, 1841–1912, American civil and mechanical engineer and inventor; formed Smith & Webb Manufacturing Co. (NJ) (1863), faculty at Univ. of Illinois (1873–1879), at Cornell Univ. (NY) (1881–1886), Stevens Inst. of Technology (NJ) (1886–1908), consultant (1908ff) during which time he patented steam engine indicator, viscous dynamometer (1888) and dynomophone (1900) (BITAS CDOAB DAB MEIA)

WEBER, Ernst, 1901–1996, Austrian American electrical engineer; to US (1930), faculty and administration at Polytechnic Inst. of Brooklyn (1930–1969) becoming president (1957–1969), recognized for his microwave research that was critical to the war effort, served on several government boards and councils, NAE (1964), NAS (1965) (Bridge MSAE WWWIA WWWIS)

WEBER, Eugene William, 1910–1989, American civil engineer; water resources planning and development, civilian engineer with Corps of Engineers in US Army (1931–1965), International Joint Commission between US and Canada (1948ff), consulting engineer, NAE (1979) (AMOS MT5 WWIE WWWIA)

WEBER, George Adam, 1848–1923, American civil and mechanical engineer; pioneer in developing base-supporting rail joint (1888), established Weber Rail Joint Co. (1889) which merged to form Rail Joint Co. (1905), invented insulated joint (1894) (MEIA)

WEBER, Harold Christian, 1895–??, American chemical engineer; consultant with Walker, Lewis, McAdams & Knowland (1919–1920), faculty at Massachusetts Inst. of Technology (MIT) (1920ff), US Army chemical corps (1940–1947), chief scientific advisor (1958ff), electrochemistry, petroleum, electronics, pulp and paper, thermodynamics (AMOS HOCE PS WWWIA)

WEBER, Wilhelm Eduard, 1804–1891, German physicist and inventor; professor of physics at Göttingen Univ. (1831ff), with Johann K. F. Gauss [1777–1855] constructed a practical system of telegraph (1834), introduced logical system of units for electricity (1846), with his brother Ernst H. Weber [1795–1878] studied the flow of fluids through tubes (BEST FNIE GI MWBD RHW WWWIS)

WEBSTER, Edwin Sibley, 1867–1950, American electrical engineer; with Charles Augustus Stone [1867–1941] founded Stone and Webster Co. (1889), specialists and builders of hydroelectric plants and high-voltage systems (CDOAB DAB NC38 TEIH WWWIA)

WEBSTER, George Smedley, 1855–1931, American civil engineer; assistant engineer with US Coast and Geodetic Survey (1876–1877), assistant engineer then chief engineer of Philadelphia (PA) (1877–1916) (1920–1921), designed and constructed several large bridges,

built first sewage plant in Philadelphia (PA) (1908), chief of Bureau of Filtration (1916–1920) (BDOACE NC25,A WWWIA)

WEBSTER, Joseph Dana, 1811–1876, American military engineer; officer in topographical engineers (1838–1854), was chief of staff to General Ulysses S. Grant [1822–1885] (1862–1863) as a brigadier general, and the chief of staff to General Philip H. Sheridan [1831–1888] (1863–1865) (ANB CDOAB DAB WWWIA)

WEDGWOOD, Josiah, 1730–1795, English potter and inventor; established own firm (1759), experimented with and developed glazes for pottery, developed cream colored earthenware (1765), named new factory Etruria (1769), invented jasper ware (1773–1780), invented pyrometer (c.1792), developed pyrometer for measuring heat in kilns (DNB EIH GEAPIT LOTE MWBD OED RHW SFLOE TEIH TGS WWWIS)

WEDGWOOD, Thomas, 1771–1805, English physicist and inventor; after working in family potteries (father was Josiah Wedgwood [1730–1795], first to suggest (1791) that all bodies become red hot at the same temperature, used light-sensitive properties of silver nitrate to produce image on glass or paper (1799) but could not preserve pattern, method of permanently fixing image was done by J.-N. Niepce [1765–1833] and Louis J. M. Daguerre [1789–1851] (BOI DOSB S&TF TWABOI WWWIS)

WEGMANN, Edward, 1850–1935, Brazilian American civil engineer; with NYC water supply system (30 yr.) (1884–1914), worked on design of Croton dam and Croton water system (ANB CDOAB DAB NC25)

WEIBULL, Ernst Hjalmar Waloddi, 1887–1979, Swedish materials scientist; authority on failure rate of materials, developed a statistical reliability that represents a cumulative distribution function to failure known as Weibull distribution, added importance of number of cycles of loading to failure, Weibull probability graph paper can be used to illustrate and estimate time to failure, Weibull statistics, at Bockamollan Brosarps Station Sweden (EOAP EOM L&M NUC VNSE WWWIS)

WEICK, Fred, 1899–1993, American aeronautical and mechanical engineer; designed simple and widely accepted small airplanes such as the Ercoupe (1936) and the Piper Cherokee (1950s), designed the tricycle landing gear which became a standard configuration, worked for the Bureau of Aeronautics (DC), propeller research for NACA (now NASA), and Engineering and Research Corp. (Erco) (1936), Atlas Aircraft Products Co. (NYC) (1946ff) (RHW)

WEIDLINGER, Paul, 1914–1999, Hungarian American civil engineer; structural engineering, design of steel and reinforced concrete structures, to Bolivia as chief engineer and professor of structural engineering at Saint Andrew Univ. (until 1943), Atlas Aircraft Production (NYC) (1943–1946), director of National Housing Agency (US) (1946–1949), principal of Weidlinger Assoc. (1949ff), NAE (1982) (WWIE)

WEI MENG-PIEN, fl. 340 AD, Chinese mechanical engineer; invented a wagon-mill which was a wagon equipped with rotating millstones and tilt-hammers used to hull and grind grains, designed and constructed devices such as south-pointing vehicles (GEAPIT SACIS)

WEITZEL, Godfrey, 1835–1884, American military engineer; US Military Academy graduate (1856) commissioned (1857), fortification of New Orleans (LA) (1855–1859), professor at US Military Academy (NY) (1859–1861), Civil War service (1861–1866) reaching rank of major general, after war engaged in river and harbor improvements, ship canals, lighthouses for Corps of Engineers (until 1882) (ANB BDOACE CDOAB DAB NC11 WWWIA)

WEI-YO, Ciao, fl. 10th century, Chinese inventor; invented canal locks (980) (NYPL)

WELCH, Ashbel, 1809–1882, American civil engineer; with Delaware and Raritan Canal becoming chief engineer (1835), design of Chesapeake and Delaware Canal, then with New Jersey Railroad (1854ff) (CDOAB DAB NC9 WWWIA)

WELLINGTON, Arthur Mellen, 1847–1895, American mechanical and civil engineer; surveyor, assistant engineer, locating engineer for several railroad companies (1866–1878), engineer and assistant general manager of New York, Pennsylvania & Ohio Railroad (1878–1881), editor of railroad magazines, consulting engineer (1893ff) (BITAS DAB MEIA NC11 WWWIA)

WELLIVER, Albertus D., 1934–1994, American mechanical engineer; radial flow compressors, research with Curtiss-Wright Corp. (6 yr.) engine-airframe integration, senior vice president of engineering and technology of the Boeing Co. (1962ff), NAE (1987) (Bridge MT8 WWWIA)

WELLMAN, Samuel Thomas, 1847–1919, American mechanical engineer and inventor; served in Civil War in artillery, employed by Nashua Iron Co. (NH), Siemens, and Bay City Iron Works (1873–1890), organized Wellman Steel Co (1890) and was president of Wellman-Seaver Engineering Co. (OH) (1896–1900), received 100 patents for machinery to manufacture iron and steel (BITAS CDOAB DAB MEIA WWWIA)

WELLS, Edward Curtis, 1910–1986, American mechanical engineer; aerospace application, research administration, with Boeing Co. (1930–1972) from draftsman to senior vice president, consultant (1972ff), NAE (1967) (AMOS ANB MT3 WWWIA)

WELLS, Joseph V. B. 1906–1987, American civil engineer; with US Geological Survey (USGS) (1929–1961),

at various locations involved in surveying, stream gauging, surface water programs, served as assistant director (1960–1961) (HOWRD)

WELSBACH, Carl Auer von (indexed under Auer by some), 1858–1929, Austrian chemist, engineer, and inventor; studied rare earth metals, invented Welsback gas mantle (1885) that burned with white incandescent flame in a gas, and a lighter flint which produced a hot metal spark when scraped, introduced first metallic filament for incandescent lamps (1898), invented mischmetal and Auer metal (1904) (BDOS(Wi) BEST MWBD PS RHW)

WELTY, Albert B., 1877–1977, American agricultural engineer; in charge of research engineering for International Harvester Co. (IHC), with J. I. Case and Belted Machinery Division of Emerson Brantingham Co., organized machine design course at Univ. of Illinois (AE58)

WENHAM, Francis Herbert, 1824–1908, English engineer and inventor; educated as a marine engineer, constructed first wind tunnel (1871) for aeronautical research with John Browning, specialized in screw propellers and high-pressure engines, built hot-air engine, interested in mechanical flight and lighter than air flight (BDOTHOT DNB)

WENNER-GREN, Axel Leonard, 1881–1961, Swedish industrialist and inventor; salesman for Swedish lamp company, worked for Laval Separator Co., designed a domestic vacuum cleaner (1912), formed Electrolux Co. (1919) to manufacture vacuum sweepers (1920) and later refrigerants, for US and Sweden established Wenner-Gren Foundation (1937) for research, large-scale development projects, philanthropist, invented monorail called Alweg (1952–1954) (BDOS(Wi) MWBD PS TBDOS(Po) TCBE)

WENTORF, Robert H., Jr., 1926–1997, American physical chemist, chemical engineer, and inventor; faculty at Rensselaer Polytechnic Inst. (NY), employee of General Electric Research Lab. (1951ff) for 37 yr., solid state physics, photochemical reactions, inventions in superpressure technology leading to new materials such as manmade diamond and Borazon, products and processes, NAE (1979) (AMOS Bridge MT9 WPostObit WWWIA WWWIS)

WERNWAG, Lewis, 1769–1843, German American bridge engineer; to US (c.1786), bridge designer and builder, use of cantilever (1811), early largest singlespan bridge (340 ft.) in Philadelphia (PA) (1812), developed plans for water works and dams in Philadelphia (1817) (CDOAB DAB EIH WWWIA)

WESSENAUER, Gabriel O., 1906–1990, American civil engineer; nuclear and environmental aspects of power generation, transmission and use of electric power, manager of power for Tennessee Valley Authority

(TVA) (1944–1970), consultant, NAE (1968) (Bridge MT6 WWIE)

WESSON, Daniel Baird, 1825–1906, American inventor and manufacturer; manufacturer of firearms (1846) assuming business upon death of brother (1850), helped develop Leonard pistol (MA), with Horace Smith [1808–1893] patented cartridge pistol (CT) (1854), organized Smith & Wesson Co. (CT) (1853), superintendent of Volcanic Arms Co. (1855), numerous patents relating to firearms while working with several companies (CDOAB DAB MEIA MWBD NC10 WWWIA)

WESTERGAARD, Harald Malcolm, 1888–1950, Danish American civil engineer; to US (1914), US citizen (1920), faculty at Univ. of Illinois (1916–1936), faculty at Harvard Univ. (MA) (1936ff) including dean of Graduate School of Engineering (1937–1946), specialty in plates and slabs, elasticity, and stress analysis (ANB CDOAB DAB NC42)

WESTINGHOUSE, George, 1846–1914, American mechanical engineer and inventor; served in US Army in Civil War, numerous patents relating to railroad equipment of which the most famous is the automatic compressed air brake (1869), organized over 30 corporations such as Westinghouse Air Brake Co. (PA) (1869), Union Switch and Signal Co. (PA) (1882), Westinghouse Machine Co., Westinghouse Electric Co. (1886), over 105 Westinghouse associated global companies, natural gas transmission, over 400 patents (AIE ANB ASAI BDOS(Wi) BEST CDOAB BITAS DAB EE EIH GI IAI MEIA MWBD NC15 NYPL PS RHW SAI-MP TEIH TGS WWWIA WWWIS)

WESTINGHOUSE, Henry Herman, 1853–1933, American mechanical engineer; worked for brother George Westinghouse [1846–1914] in some of his companies beginning with Westinghouse Air Brake Co. (PA) (1873ff), officer in several companies of his brother including Canadian, English, Australian, and French operations (MEIA WWWIA)

WESTON, Charles Valentine, 1857–1933, American civil engineer; assistant engineer with several railroad companies (1879–1888), construction of water tunnel under a portion of Lake Michigan, water tunnel under Chicago River (1888–1894), chief engineer of elevated rail line (1894–1901), started consulting firm of Weston Brothers with brother George Weston (1901–1911) becoming president and general manager, individual private practice (1911ff), operating manager of elevated railroad in Philadelphia (1918–1920), consulting engineer and director of Chicago Railroad Co. (1920–1933) (BDOACE NC25 WWWIA)

WESTON, Edward Brownell, 1850–1936, English American electrical engineer; to US (1870), US citizen (1923), with Mardock & Co. (1870), with George J. Harris built system for electroplating (1872), manufacture of dynamo machinery (1875), manufacture of incandes-cent lighting equipment, formed Weston Dynamo Electric Machine Co. (1877) that was purchased by United States Electric Lighting Co. (1881), founded Weston Electrical Instrument Co. (1888) (ANB BITAS CDOAB DAB EIH MEIA MWBD WWWIA WWWIS)

WESTON, William, c.1752–1833, English civil engineer; to US (1793) as engineer for Schuylkill and Susquehanna Canal, advised on water supply for NYC (1799), built Schuylkill canal in US, Middlesex project, served as advisor to precursor of the Erie Canal (NY), advised on coffer dam for Permanent Bridge (PA) (ASAI CDOAB DAB EIH TEIH WWWIA)

WETHERILL, Samuel, 1821–1890, American chemical engineer and inventor; founded Gilbert, Wetherill, Baxter & Co. (1853), then Lehigh Zinc Co. later absorbed by New Jersey Zinc Co. (1897), involved in mineral recovery and chemical manufacture, patented special retorts (1854ff), participated in Civil War (1861–1864) (ACCE BITAS CDOAB DAB NC7 PS WWWIA)

WEYERHAEUSER, Frederick, 1834–1914, German American industrialist; to US (1852), first located in Pennsylvania, then Illinois (1856), to Minnesota (1891), purchased several lumber companies from Minnesota to the Pacific Northwest and became known as the lumber king, president of several companies, president of Weyerhaeuser Timber Co. succeeded by his son Frederick Edward Weyerhaeuser [1872–1942] (1914) (AIE ANB MWBD NC14 WWWIA)

WEYMOUTH, Frank Elwin, 1874–1941, American civil and hydraulic engineer; Isthmian Canal Commission (1899–1903), US Bureau of Reclamation Service (1903–1924), private practice, chief engineer and general manager of Colorado River Aqueduct (1929–1941) (ANB CDOAB DAB WWWIA)

WHEATON, Elmer P., 1909–1997, American aeronautical engineer; marine engineering, development and administration of aerospace programs, Douglas Aircraft (1934–1962), developed the commercial diving helmet called the Nautilus, after several assignments became vice president of research and development at Lockheed Missiles and Space Co. (1962–1974), consultant including Marine Development Assoc., Inc. (1974ff), NAE (1967) (Bridge MT10 WWIE)

WHEATSTONE, Charles, 1802–1875, English physicist and inventor; professor at King's College in London (1834ff), research in electricity, light, and sound, crude microphone (1827), invented concertina (1829), constructed electrical distribution system to bind several distant clocks to a central control clock (1830s), speed of electricity in a conductor (1834), with William Fothergill Cooke [1806–1879] devised an electric telegraph (1837), stereoscope (1838), chronoscope (1840), invented Wheatstone bridge (1843) (BEST CDOS DNB EIH GI MWBD NYPL RHW WA WWWIS)

WHEELER, Frederick Meriam, 1848–1910, American mechanical engineer and inventor; hydraulic and marine engineer, with George F. Blake Manufacturing Co., then International Steam Pump Co., patented pumping systems, valves for condensing, feed water apparatus (1887), organized Wheeler Condenser & Engineering Co. (NJ), Ludlow Valve Manufacturing Co. (NY), seven patents (MEIA)

WHEELER, George Montague, 1842–1905, American military engineer and surveyor; US Military Academy (1866), topographical studies, made Wheeler survey west of 100th meridian (1871–1888), supervised and published the results of his survey, retired as major (1888) (ANB BDOAS(El) BITAS CDOAB DAB PS WWWIA)

WHEELER, Harold Alden, 1903–1996, American electrical engineer; development of frequency modulated (FM) radio, television, microwave, antenna technologies, guided missile systems, subsurface radio communications, Hazeltine Corp. (1924–1946) (1959ff), president of Wheeler Labs. Inc. (1957–1968), consultant, 180 US patents, NAE (1986) (AMOS Bridge MT9 PS WWIE WWWIA WWWIS)

WHEELER, Nathaniel, 1820–1893, American manufacturer and inventor; manufactured rotary-hook sewing machine patented by Allen B. Wilson [1824–1888] (1851), invented wood-filling compound (1876, 1878), invented a home-ventilating system (1883) (CDOAB DAB NC9 WWWIA)

WHEELER, Schuyler Skaats, 1860–1923, American engineer, manufacturer and inventor; with Francis Bacon Crocker [1861–1921] founded Crocker-Wheeler Co. (1888), served as president (1889ff), invented electric fire engine system (1885), electric elevator (1885), with others developed standard electric motor (1886) and electric fan (1882) (CDOAB DAB MEIA NC41 www.factmoniter.com WWWIA WWWIS)

WHEELER, William, 1851–1932, American civil engineer; professor of mathematics and civil engineering at Imperial Agricultural College at Sapporo (Japan) (1876–1877) then president (1877–1880), worldwide consultant on water works in Boston (MA) (1880ff) (CDOAB DAB)

WHINFIELD, John Rex, 1901–1966, English chemist and inventor; Calico Printers' Assoc., with J. T. Dickson [b. c.1920] invented Teryline (1941), member of team that developed rayon known in US as Dacron by DuPont (1946ff) (BDOS(Wi) BDOTHOT DNB NYPL PS)

WHIPPLE, Amiel Weeks, 1816–1863, American military and topographical engineer; US Military Academy (1849), survey of US–Mexico boundary (1849–1853), Pacific Railroad surveys (1853–1856), in Civil War as Union brigadier general (1862) and was mortally wounded (1863) (BDOAS(El) BITAS CDOAB DAB NC10 PS WWWIA)

WHIPPLE, Squire, 1804–1888, American civil engineer and inventor; bridge builder, Whipple truss of trapezoidal form used in bridge construction, author of technical book on bridge construction and treatment of stresses (ANB BDOAS(El) BITAS CDOAB DAB EIH MWBD NC9 PS WWWIA)

WHISTLER, George Washington, 1800–1849, American military engineer; US Military Academy (1819) commissioned in artillery, assisted in location of Baltimore & Ohio Railroad (1828–1833), chief engineer of Western Railroad of Massachusetts (1833–1842), consulted with Russian government on railroads, fortifications, docks, and bridges (BITAS CDOAB DAB NC9 TEIH WWWIA)

WHITAKER, Milton C., 1870–1963, American chemical engineer and inventor; worked as mechanic and millwright at Univ. of Colorado, Columbia Univ. (NYC) (1899ff) becoming professor and head of chemical engineering, Welsbach Co. (1902–1910), consulting engineer (1910ff), American Cyanide Co. as vice president (1930–1947), held 60 US patents (NC50 PS WWWIA)

WHITBY, Kenneth Thomas, 1925–1983, American mechanical engineer; small-particle technology, particle size analysis, air filtration, combustion, faculty at Univ. of Minnesota (1946ff), research with US Navy (1946–1959), consultant, NAE (1978) (AMOS MT2 WWIE WWWIA)

WHITE, Alfred Holmes, 1873–1953, American chemical engineer and inventor; at Univ. of Illinois (1893–1896), faculty in chemical technology in engineering at Univ. of Michigan (for 46 yr.) where he was head of chemical engineering (1914–1942) retiring (1943), responsible for developing the chemical engineering curriculum adopted (1898), held 14 patents on wide range of chemical technologies (ACCE NC49 PS WWWIA)

WHITE, Canvass, 1790–1834, American civil engineer; surveyor, canal construction, Erie canal construction (1816–1825), designed canal locks, made concrete from local limestone, hydraulic concrete, worked on canals in eastern US (1825–1834) (ANB ASAI CDOAB DAB EIH GEAPIT MWBD NC12 TEIH WWWIA)

WHITE, Edward Higgins II, 1930–1967, American aeronautical engineer and astronaut; test pilot, Air Force officer, fighter pilot, entered astronaut corps (1962), became first person to walk in space with maneuvering unit that was preceded by Soviet Aleksey Leonov [b. 1934] with Pavel I. Belyayev [1925–1970] (1965), Gemini 4 flight, killed in training accident, 5-cent postage stamp issued in his honor (1967), also killed in the training accident were astronauts Virgil I. Grissom [1926–1967] and Roger B. Chafee [1935–1967] (EB GSA MWBD NYT-A TBOF WWWIA)

WHITE, Samuel Stockton, 1822–1879, American dentist and inventor; manufacturing of artificial teeth by firm

Jones and White (PA) (1845), instruments and appliances for dentists, flexible shaft for driving drills later adapted by many industries particularly automotive for speedometer cables (GI WWWIA)

WHITEHEAD, John Boswell, 1872–1954, American electrical engineer; with Westinghouse Electric and Manufacturing Co. (1893–1897) worked with manufacture and installation of generators and transformers for Niagara Falls Power Plant, faculty at Johns Hopkins Univ. (MD) (1904–1942) helped develop electrical engineering and was first dean of engineering (1919), did primary work in dielectric materials, NAS (1932) (ANB BM37 DNB NC46 NYTObit PS WWWIA WWWIS)

WHITEHEAD, Robert, 1823–1905, English engineer and inventor; invented self-propelled underwater torpedo (1866), established company to manufacture torpedoes (1872), developed servomotor to guide torpedo through the water (1876) (BDOS(Wi) GI MWBD PS RHW)

WHITMAN, Walter Gordon, 1885–1974, American chemical engineer; faculty and administration at Massachusetts Inst. of Technology (1917–1926) (1934–1961), research at Standard Oil Co. (1926–1934), gas absorption, corrosion, heat transfer, petroleum engineering, consulting engineer (1961ff), NAE (1974) (AMOS HOCE MT1 PS WWWIA)

WHITNEY, Amos, 1832–1928, American mechanical engineer; employed by Colt Firearms Co. (CT) (1849), organized Pratt & Whitney (CT) (1860) with Francis A. Pratt [1827–1902], manufactured guns, sewing machines, type-setters, employed by William A. Rogers [1832–1898] and George M. Bond [1852–1935] (1879), developed Rogers-Bond Universal comparator (MEIA MWBD NC32)

WHITNEY, Asa, 1791–1874, American inventor; worked for Mohawk and Hudson Railroad (1830–1839), canal commissioner for New York (1839–1842), manufactured locomotives in partnership with Matthias W. Baldwin [1795–1866] (1842–1846), formed Asa Whitney & Sons Co. to manufacture cast iron railroad car wheels (1849) (CDOAB DAB GEAPIT MWBD NC15 WWWIA WWWIS)

WHITNEY, Baxter D., 1817–1915, American mechanical engineer; operated factory for making tubs and pails (1836), built looms (1837), established factory and machine shop (1845), Whitney wood planing machine (1846), built Whitney gauge lathe (1857), gunstock machinery built during Civil War (MEIA)

WHITNEY, Charles Smith, 1892–1959, American civil engineer; worked successively for Stone & Webster, Gustav Lindenthal [1850–1935] and Othmar H. Ammann [1879–1965], with John Parkinson (1914–1917), with Corps of Engineers in France in WWI (1917–1919), numerous building and construction projects with architectural firms (1920ff), pioneered ultimate strength design, arch, shell,

folded plate and cable-suspended long-span roof structure of reinforced concrete (BDOACE WWWIA)

WHITNEY, Eli, 1765–1825, American mechanical engineer and inventor; cotton gin (1793), embroidery frame, perfected idea of interchangeable parts in gun manufacture (1798) with rudimentary assembly line for mass production (1798), developed machine tools such as drill press and milling machine, opened factory in Connecticut to produce firearms (AIE ANB ASAI BEST BITAS CDOAB DAB GEAPIT GI IAI IIA LAI MEIA MWBD NC10 NYPL 1000Y PS RHW SAI SAI-MP TGS WWWIA WWWIS)

WHITNEY, Eugene C., 1913–1998, American electrical engineer; electrical machines, design and building of many of the largest hydrogenerators in the world including those at Grand Coulee (WA), first with Westinghouse Electric Co. (1936), manager of Waterwheel Generator and Synchronous Condenser Engineering Co. (23 yr.), variable frequency generators, synchronous condensers, diesel and steam turbine generators, pumped storage projects, consultant (1975ff), NAE (1986) (Bridge MT10 WWIE WWWIA)

WHITTEMORE, Amos, 1759–1828, American inventor; set up own shop with others for manufacturing equipment for carding cotton (1795), equipment for bending wire and putting holes in leather, machine for cutting nails, with brothers formed William Whittemore & Co. (1800) (CDOAB DAB MEIA NC7 WWWIA WWWIS)

WHITTLE, Sir Frank, 1907–1996, British engineer; with Royal Air Force, designed and patented first modern turbojet engine (1930), formed Power Jets Ltd. (1936–1946) devoted to development of jet propulsion engines with a prototype (1937), with support of Air Ministry and Gloster Aircraft Co. developed experimental airplane (1939ff) which became the foundation of other companies after WWII, consultant to several companies, to US (1976), faculty of US Naval Academy (1976ff) as research professor, Draper Prize (co-recipient with Hans von Ohain [1911–1998] (1991), fNAE (1979) (Bridge CDOS EE EIH GI MT10 NYPL PS RHW SAI TGE:B TGS WWWIS)

WHITWORTH, Sir Joseph (First Baronet), 1803–1887, British mechanical engineer and inventor; mechanic and machinist, helped establish a rational system of screw threads (1844) known as Whitworth threads with a thread angle of 55 deg., made tools for making and measuring threads, set up business in Manchester (1833) that developed into Whitworth Works, produced machines for cutting, shaping, planing, gear cutting, formed Armstrong Whitworth firm (1893), method of making ductile steel for big guns, contributed to engineering education, FRS (1859) (BDOS(Wi) DNB EE FNIE MWBD PS RHW SAI TBDOS(Po) TEIH WWWIS)

WHORF, Benjamin Lee, 1897–1941, American linguist and chemical engineer; fire prevention authority for Hartford Insurance Co. (CT) (1918–1941), comparative linguistics (CDOAB DAB MWBD NC30)

WIEN, Wilhelm, 1864–1928, German physicist; professor at Univ. of Giessen (1899), at Univ. of Wurzburg (1900–1920), successor of Wilhelm C. Röntgen [1845–1923] at Univ. of Munich (1920ff), black body radiation (1892ff), Wien formula (1896) for describing energy distribution in radiation spectrum, Nobel prize (1911) (BEST CDOS MWBD RHW WWWIS)

WIENER, Norbert, 1894–1964, American mathematician; formulated theory of Brownian movement (1920), harmonic analysis, flow of information (1925), communication and control in mechanical devices and biological systems, feedback control applied to radar in WWII, faculties at Harvard Univ. (MA) and Univ. of Maine and Massachusetts Inst. of Technology (MIT) (1919–1960), NAS (1934) (BEST BM61 CB CDOS MSAE MWBD RHW TGS WWWIA WWWIS)

WIESNER, Jerome Bert, 1915–1994, American electrical engineer; communications theory and application, lasers, communication, radar, special assistant to President John F. Kennedy [1917–1963] for science and technology, faculty and administration as provost (1967–1971) then president (1971–1980) at Massachusetts Inst. of Technology (1946–1980), NAS (1960), NAE (1966) (AMOS ANB BM78 Bridge CBY HOCS&T MT8 PS WWIE WWWIA WWWIS)

WIGHT, Sedgwick North, 1879–1968, American engineer and inventor; various positions including signal inspector with Lake Shore & Michigan Railroad (1903–1910), General Railway Signal Co. (NY) (1910–1949) for which he was engineer of train operations at his retirement (1949), numerous inventions on control and safety of trains, over 90 US patents (NC54 PS)

WILCOX, Stephen, 1830–1893, American mechanical engineer and inventor; invented safety water tube boiler with inclined tubes (1856) working with George Herman Babcock [1832–1893] (1860ff) patented pumps, steam engines, water tube boiler, incorporated Babcock & Wilcox Co. (1867) the year that the Babcock-Wilcox boiler was patented for manufacture of boilers and steam engines, obtained approximately 50 patents (CDOAB DAB MEIA MWBD RHW WWWIA)

WILEY, Andrew Jackson, 1862–1931, American irrigation engineer; private irrigation projects throughout western US (1883ff), consultant to US Bureau of Reclamation (1902ff), worked on Roosevelt, Boulder, Coolidge, and Madden dams plus others, Canal Zone power plant and Columbia basin project (WA) (CDOAB DAB WWWIA)

WILEY, William Halsted, 1842–1925, American mechanical engineer; served in Union army (1862–1866), entered father's publishing firm of John Wiley & Sons (NYC) (1876), correspondent for periodical *Engineering* published in London (1885) (MEIA NC12 WWWIA)

WILHELM, Richard Herman, 1909–1968, American chemical engineer; faculty and administration at Princeton Univ. (NJ) (1934ff), catalytic reactor design, reaction rates in flowing system, fluidization, heat transfer, NAE (1968) (AMOS HOCE MHSE MT1 NC54 PS WWWIA)

WILKINSON, Alfred, 1845–1910, English American mechanical engineer; during Civil War (1862) served as engineer for US Navy, with Carr, Crawley & Devlin after war, consulting engineer (PA) (1876ff), invented Wilkinson automatic mechanical stoker and organized Wilkinson Manufacturing Co. (PA) (1891) (MEIA)

WILKINSON, David, 1771–1852, American inventor; lathe for cutting threads, slide rests for lathes (1798), established David Wilkinson & Co. (RI) (c.1800) that manufactured textile machinery and cannons, patent for canals, lock, gate (1835), considered by some as the father of the machine tool industry in US (ASAI CDOAB DAB IIA MEIA NC8 PS WWWIA)

WILKINSON, Jeremiah, 1741–1831, American farmer and inventor; forge master, perfected several devices to assist in manufacture of wool-cards, drew wire using horse power (ante 1776), nail manufacture from cold iron (c.1776) (CDOAB DAB NC8 PS WWWIA)

WILKINSON, John, 1728–1808, British inventor; invented Wilkinson boring engine for making canons and cylinders for steam engines (1775), cast first iron bridge in England (1779), first iron barge (1787), process for making lead pipe (1790) (BDOS(Wi) DNB EE EIH GEAPIT LOTE MWBD PS SFLOE TEIH TGS WWWIS)

WILLANS, Peter William, 1851–1892, British inventor; invented Willans steam engine—a high-speed, enclosed-valve steam engine (EE)

WILLARD, Simon, 1753–1848, American clockmaker; patented Willard timepiece (1802), two brothers Benjamin Willard [1743–1803] and Aaron Willard [1757–1844] also made clocks (IIA MWBD WWWIA)

WILLARD, Solomon, 1783–1861, American sculptor, architect, and inventor; made wooden model of completed US Capitol for Charles Bulfinch [1763–1844], designed and built buildings in Boston and built Bunker Hill monument (MA) (1825–1842) (ANB CDOAB MWBD NC4 WWWIA)

WILLCOX, Charles Henry, 1839–1909, American mechanical engineer and inventor; several patents on sewing machines while working with Willcox & Gibbs Co. of which his father James Willcox was partner, about 10 patents (MEIA NC19)

WILLENBROCK, Frederick Karl, 1920–1995, American electrical engineer; microwave and solid state electronics, US Navy (1943–1946), faculty and administration at Harvard Univ. (MA) (1950–1967), State Univ. of New York at Buffalo (1967–1970), National Bureau of

Standards (MD) (1970–1976) as director of the Institute of Applied Technology, faculty and dean of engineering at Southern Methodist Univ. (TX) (1976–1986), executive secretary of ASEE (1986–1989), assistant director of International programs at National Science Foundation (1989–1991), known for formulating and directing research programs applicable to national needs, independent consultant, NAE (1975) (AMOS Bridge MT10 WWIE WWWIA WWWIS)

WILLIAM OF SENS, d. 1180, French architect and cathedral builder; started out as a mason, responsible for selection of stone, foundation, overall plans for Cathedral of St. Elienne in Sens (1140ff), rebuilt Canterbury Cathedral in England after a fire (1174), contrasting stone on interior coupled with flying-stone buttresses in the Notre Dame (Paris), built cranes and hoisting equipment for construction (GEAPIT 1000Y TEIH)

WILLIAMS, Carbine, 1901?–1975, American inventor; blacksmith, Atlantic Coast Line Railroad, jailed for illegal operation of stills, while in prison he sketched, designed, manufactured a new style of automatic weapon, built six new guns, pardoned (1929), Ordnance Department recognized his inventions and he became famous for the invention of M-1 carbine rifle (1940) used extensively in WWII, he worked for several arms manufacturing companies (ANB)

WILLIAMS, Edwin F., c.1854–1914, American mechanical engineer and inventor; in infantry in Civil War, use of boilers and steam power in machine shop, sawmill, distilling, engineer for Cheyenne and Arapahoe Indians (1870), mills and mines with numerous companies as engineer or chief engineer for construction and installation of steam power equipment notably Hill, Clark & Co. (MA) (1879), Hartford Engineering Works (CT), William Tod & Co. (OH) (1887) and Fairbanks, Morse & Co. (WI), approximately 10 patents (MEIA WWWIA)

WILLIAMS, Frank Martin, 1873–1930, American civil engineer; engineer and surveyor for state of New York (1909–1910) (1915–1922), supervised barge canal, construction of Holland Vehicular Tunnel (1927) (Clifford M. Holland [1883–1924]) and Sacandaga reservoir (NY) (1930) (now Great Sacandaga Lake) (EB CDOAB DAB PS MWBD WWWIA)

WILLIAMS, Frederic Calland, 1911–1977, English electrical engineer and inventor; professor at Manchester (1946ff), invented radar aircraft identification system, developed Williams cathode tube storage for computer memory (1946) (DNB MSAE MWBD PS RHW RS24 WWW)

WILLIAMS, Gardner Fred, 1842–1922, American mining engineer; consultant for western mining regions (1875–1883), manager of De Beers Mining Co. in South Africa (1887–1905), improved working methods of diamond mines (CDOAB DAB NC9 WWWIA)

WILLIAMS, Gardner Stewart, 1866–1931, American civil engineer; assistant engineer of water works construction (ND) (1887), engineer at Russel Wheel & Foundry Co. (MI) (1893–1898), faculty and in charge of hydraulic laboratory (1898–1904) at Univ. of Michigan (1904–1911), consulting engineer (1911–1931) (BDOACE NC54 WWWIA)

WILLIAMS, Jesse Lynch, 1807–1886, American civil engineer; Miami & Erie Canal (1828), following positions in charge of canal routes in Ohio and Indiana, became chief engineer of Wabash & Erie Canal (1847–1876) (BDOACE DAB NC21 WWWIA)

WILLIAMS, John Shoebridge, 1790–1878, American civil engineer and surveyor; taught Quaker school (OH), worked in iron foundry (until 1819), helped develop Ohio Valley Transportation system, surveyed route for Chesapeake and Ohio Canal (1824–1825), engineer for National Roads (1826ff), turnpike construction and management, promoted railroad development (ANB)

WILLS, Childe Harold, 1878–1940, American engineer and metallurgist; as an early associate of Henry Ford [1863–1947] he designed and manufactured automobiles, became specialist in metallurgy, chief engineer and factory manager of Ford Motor Co. (1903–1919), manufactured Wills-St. Claire automobile (1919–1926) (ANB CDOAB NYTObit)

WILLSON, Frederick Newton, 1855–1939, American mechanical engineer; faculty and department chairman at what later became Princeton Univ. (NJ) (1880–1923), taught graphics, engineering drawing, and descriptive geometry (MEIA WWWIA)

WILSON, Allen Benjamin, 1824–1888, American inventor; patented several sewing machine developments, joined Wheeler, Wilson & Co. to manufacture developments, invented cotton picking machinery, photographic devices, gas-illuminating devices (CDOAB DAB MEIA NC9 WWWIA)

WILSON, Basil Wrigley, 1909–1996, South African American civil engineer; oceanographic engineering, coastal engineering, railroad and tracks, ship motion and mooring, design for tsunami protection, faculty at Texas A&M Univ. (1953–1961, National Engineering Science Co. (1961–1965), consulting engineer (1968ff), NAE (1984) (AMOS Bridge MT9 WWIE WWWIA WWWIS)

WILSON, Charles Thomson Rees, 1869–1959, Scottish physicist and inventor; designed chamber for studying clouds (1895–1899), studied electrical conduction in dust-free air using a gold leaf electroscope he designed (1900–1910), invented the Wilson cloud chamber (1911) for studying subatomic particles, professor at Cambridge Univ. (1925–1934), Nobel prize (1927) (CDOS DNB MWBD RHW WWW WWWIS)

WILSON, George Francis, 1818–1883, American chemist and inventor; established George F. Wilson & Co., with Eben Norton Horsford [1818–1893] which later became Rumford Chemical Co. of which he was president (until 1883) (CDOAB DAB NC WWWIA)

WILSON, James Harrison, 1837–1925, American military officer and engineer; chief topographical engineer in US Army in Civil War (1860–1865), was superintendent of navigation (1865–1870), advanced to major general, also served in Spanish-American War and in China (1900), retired (1915) (CDOAB DAB MWBD NC2 WWWIA)

WILSON, John Allston, 1837–1896, American civil engineer; topographer on surveys in Honduras, assistant engineer (1858–1864) and chief engineer (1868–1875) of Pennsylvania Railroad (PRR), involved in numerous extensions of PRR including buildings, roads, bridges, and mining operations (BDOACE WWWIA)

WILSON, Joseph Miller, 1838–1902, American civil engineer; engineering work on Pennsylvania Railroad (PRR) (1859–1867), with Philadelphia, Wilmington & Baltimore Railroad, organized firm of Wilson Brothers & Co. for engineering and architecture (1876) that included work on bridges, buildings, railroad facilities, tunnels, elevated railways, water supply in Philadelphia (1890) (BDOACE BITAS DAB NC7 WWWIA)

WILSON, Robert Erastus, 1893–1964, American chemical engineer, inventor, and industrialist; in WWI with US Bureau of Mines who as a captain directed research in gas masks, discharged as a major, after 3 yr. at Massachusetts Inst. of Technology (MIT) to Standard Oil Co. of Indiana (1922–1958) advancing through several positions with the parent company and its subsidiary including CEO and chairman of the board (1945–1958), commissioner of Atomic Energy Commission (AEC) (1960–1964), 90 patents, NAS (1947) (ACCE BM54 NC52,I PS WWWIA)

WILSON, Stanley DeWolf, 1912–1985, American civil engineer; foundation engineering, soil mechanics, highway engineering, California State Division of Highways (1933–1942), Army Corps of Engineers (1943–1946), faculty at Harvard Univ. (MA) (1948–1953), Shannon & Wilson consulting (1954ff), NAE (1967) (MT3)

WILSON, Thornton Arnold, 1921–1999, American aeronautical engineer; engineering and management of complex aerospace systems for commercial and military use, Boeing Co. (1943ff) becoming chairman and CEO, NAE (1974) (Bridge WWIE WWWIA)

WILSON, William Hasell, 1811–1902, American civil engineer; worked on railroad, engaged in general practice (1837–1857), resident engineer of Philadelphia & Columbia Railroad (1857) then chief engineer for Pennsylvania Railroad (PRR) from Philadelphia to Pittsburgh, constructed and was president of Altoona Gas Co. (PA) (1857–1871), headed real estate department for Pennsylvania Railroad (1874–1884), was president and director of several Pennsylvania railroads (CDOAB DAB WWWIA)

WINANS, Ross, 1796–1877, American inventor; assisted in construction of Baltimore & Ohio Railroad (1828), built locomotives (1829ff), Gillingham & Winans Co. (1834), invented wheel design for railroads, became independent locomotive builder (1840), retired (1860) (CDOAB MEIA NC11)

WINANS, Thomas DeKay, 1820–1878, American engineer and inventor; son of Ross Winans [1796–1877], worked for Harrison, Winans and Eastwick that supplied locomotives and rolling stock for US and Russia, cast iron for first permanent bridge over Neva River (Russia), director of Baltimore & Ohio Railroad, with his father designed hull for high-speed steamers (CDOAB DAB MWGD NC1 WWWIS)

WINCHE, Warren E., 1917–1983, American chemical engineer; at National Defense Research Council (NDRC) (1943–1945), development of nuclear energy, fuel element design, fuel reprocessing, DuPont (1951–1962) as researcher and manager of chemical separations work, Brookhaven National Lab. (1962ff) where he became deputy director (1979), 10 patents, NAE (1982) (MT2)

WINCHELL, Horace Vaughn, 1865–1923, American mining engineer; geologist for Anaconda Copper Co. (1898–1906), Northern Pacific Railroad (1906–1908), consulting geologist (1908–1921) (CDOAB DAB MWBD NC20 WWWIA)

WINSOR, Frederick Albert, 1763–1830, German inventor; pioneer of gas lighting in Great Britain and France, based on patent of French engineer Philippe Lebon [1767–1804] (1799) for thermolamp, started a gasworks (1807) in London, was granted several patents for gas furnaces and purifiers (GEAPIT MWBD RHW WWWIS)

WINTER, Frederick Charles, 1913–1966, American industrial engineer and inventor; with various industries as an industrial engineer, US Army Ordnance (1939–1947) becoming colonel, faculty at Columbia Univ. (NYC) (1946–1960), consultant for Agricultural Research Service of US Department of Agriculture (ARS of USDA) for grain handling, storage, and related facilities (NC53)

WINTER, George, 1907–1982, Austrian American civil engineer; structural engineering, elasticity, buckling, light-gauge building construction, steel construction, reinforced concrete, faculty at Cornell Univ. (NY) (1938–1970) including chair of structural engineering (1948–1970), consultant (1970ff), NAE (1970) (AMOS MT3 WWIE WWWIA)

WIRT, Frederick A., 1891–1956, American agricultural engineer; faculty at Univ. of Arkansas, professor and head at Univ. of Maryland, at Kansas State Univ. as extension specialist, sales manager of John Deere Plow Co. and with other implement firms, J. I. Case Co. (1922–1956) was editor of periodical *Case Eagle* (AE37 LC NUC)

WISELY, William Homer, 1906–1982, American civil engineer; worked in water and public health, in the Division of Sanitation Engineering of the Illinois Department of Public Health (1926–1940), engineer-manager of Urbana-Champaign Sanitation District (1940) and was part time and later full time with Water Pollution Control Federation (1943–1955), executive secretary of ASCE (1955–1972) (ANB WWWIA)

WITHROW, James R., 1878–1953, American chemical engineer; taught chemistry then became chairman of chemical engineering (1924–1948) at Ohio State Univ. (1907–1948), advocate of unit operations for teaching, covered spectrum of industrial concerns—sugar, petroleum, refining, electrochemistry, wood distillation, insecticides, and paints (ACCE AMOS NC42 PS WWWIA)

WÖHLER, August, 1819–1914, German inventor; Borsig engineering works in Berlin (1840–1843), construction of railroads, operation of locomotives, with Lower Silesia-Brandenburg Railroad and then imperial railway director (1874–1889), Wöhler laws (1870) for testing materials, machine for rotating fatigue test specimens (DOSB EE L&M)

WÖHLER, Friedrich, 1800–1882, German engineer and chemist; taught applied chemistry in Berlin (1825–1831), professor in Kassel (1831–1836), Göttingen (1836–1882), collection of minerals, preparation of artificial urea (1823), with Frenchman Henri-Étienne-Sainte-Claire Deville [1818–1881] developed boron nitride as an excellent lubricant for high-temperature applications, fNAS (1865) (DDIEI DOSB GMI MWBD RHW SAI TGS WWWIS)

WOLF, Frederick W., 1837–1912, German American mechanical engineer and inventor; designed engines for Guppy Machine Shop (Italy), to US (1866), master machinist with Pere Marquette Railroad (1867), consulting engineer working with breweries, started machine shop (1875) to install Linde ammonia ice machines, incorporated and president of Fred W. Wolf Co. (1887), manufactured refrigeration machinery, installed sugar beet factories (1897) (BDOAS CDOS MEIA RHW WWWIS)

WOLFF, Alfred R., 1859–1909, American mechanical engineer; consulting steam engineer (1880), made specialty of heating and ventilating equipment (1888ff), introduced use of cheesecloth filters (1894), helped develop humidistat (1902) (MEIA WWWIA)

WOLMAN, Abel, 1892–1989, American sanitary engineer; stream pollution, water supply, water and sewage treatment, faculty at Johns Hopkins Univ. (MD) (1937–1962), helped to develop standardized water chlorination, consulting engineer, NAS (1963), NAE (1965) (AMOS ANB BDOACE BM83 Bridge CB57 MT5 PS WWIE WWWIA WWWIS)

WOOD, Carlos C., 1913–1997, American aeronautical engineer; development of aircraft and missiles, California Inst. of Technology (1933–1937), Douglas Aircraft Co. (1937–1960), vice president of engineering, Sikorsky Aircraft Division of United Technologies Corp. (1960–1970) (Bridge MT9 WWIE)

WOOD, DeVolson, 1832–1897, American civil and mechanical engineer and inventor; faculty at Univ. of Michigan (1857ff), also taught mathematics and mechanics, invented steam rock drill (1866), air compressor, published several texts and references (BDOACE BITAS MEIA NC13)

WOOD, Henry Alexander Wise, 1866–1939, American inventor and manufacturer; made printing press improvements including Autoplate (1903), high-speed press feeding devices (CDOAB NC41 WWWIA)

WOOD, Ivan D., 1888–1979, American agricultural engineer; extension engineer for Nebraska (1914–1940) except for service in WWI, worked in water conservation and irrigation, district and area engineer for western states for Farm Security Administration (1940–1945), irrigation engineer for Soil Conservation Service (SCS) (1945–1954), federal irrigation specialist (1954–1957) (AE39,60 WWIE(I))

WOOD, James J., 1856–1928, Irish American engineer and inventor; patented arc-light dynamo (1880), floodlighting (1895), internal engines installed in submarine, machine for constructing cables for Brooklyn bridge, held 240 patents, consulting engineer (c.1885ff) (CDOAB DAB WWWIS)

WOOD, Jethro, 1774–1834, American inventor; plow improvements, patented cast iron plow (1814), invented iron plow that worked well in sand but not in clay, John Deere [1804–1886] made plow (1833) that scoured in clay using a discarded steel saw blade, introduced plow with changeable parts (1819), Charles Newbold (NJ) got first patent on a cast iron plow (1793) (ASAI DAB NC11 TTOT WWWIA)

WOOD, Walter Abbott, 1815–1892, American manufacturer and inventor; inventor and manufacturer of agricultural implements, produced wagons, wooden mowers, reapers and binders (1852ff), improved mowers and reapers originally patented by John H. Manny [19th century] (AEOT CDOAB DAB HFP NC6 WWWIA)

WOODBURY, Charles Jeptha Hill, 1851–1916, American industrial engineer; engineer and later vice president of Boston Manufacturers Mutual Fire Insurance Co. (MA) (1878–1894), assistant engineer of AT&T Co. (1894–1907), secretary of National Association of Cotton Manufacturers (1894–1916), authority on fire protection and mill construction (BITAS CDOAB DAB NC12 WWWIA)

WOODBURY, Daniel Phineas, 1812–1864, American military engineer; US Military Academy (1836), helped build forts in western regions and Key West, served in Civil War as brigadier general (1862ff) constructing roads, bridges, and siege works (BITAS CDOAB DAB NC1 WWWIA)

WOODCROFT, Bennett, 1803–1879, British inventor; invented Woodcroft propeller—single-threaded-increasing-pitch screw, 2 or 2 1/2 turns (DNB EE)

WOODRUFF, Theodore Tuttle, 1811–1892, American inventor; railroad car builder of seat and coach car (1856), sleeping car (1858), patented process for indigo manufacture (1872), coffee hulling machine, steam plow (ASAI CDOAB DAB NC14 WWWIA)

WOODRUFF, William N., fl. 19th century, American inventor; segment of a disc fitted into shaft known as Woodruff key (c.1890) (EE)

WOODS, Granville T., 1856–1910, Black American electrical engineer and inventor; contributions in field of electricity, electric relays, electromagnetic brakes, galvanic batteries, automatic cut-off circuits, egg incubator, steam boiler furnace (1884), telephone transmitter (1884), railway telegraph (1887), railroad safety devices, third rail for subways, with brother established Woods Electrical Co. (OH), referred to as Black Edison, more than 100 patents (AIE BDOBA BI BIS BITAS BPOSAI CBB PS SAI-MP S:TLAW TFM)

WOODWARD, Guy O., 1915–??, American irrigation engineer; irrigation engineer for sugar beet industry (8 yr.), extension specialist at Univ. of Wyoming (7 yr.), educational director of National Sprinkler Irrigation Assoc. (3 yr.), federal extension service (2 yr.), consulting with George D. Clyde [1898–1972] and Wayne O. Criddle [b. 1914] (NC53 WWIE)

WOODWARD, Richard J., 1907–??, American civil engineer; bookkeeper and food merchandizer (1935–1941), engineer in US Navy (1941–1945), faculty of Univ. of California (1945–1950), founding partner of Clyde & Woodward & Assoc. (UT) (1950ff) (George D. Clyde [1898–1972]) geotechnical engineering and consulting engineers and chairman of the board and president (1955–1973), president of Design Professionals Insurance Co. (1970–1972) (WWIE)

WOODWARD, Richard Lewis, 1913–1981, American sanitary and civil engineer; water and atmospheric pollution, radioactivity, water supply, primarily with United States Public Health Service (USPH) (1937–1963), Ohio River pollution survey (1939–1942), nuclear control in war areas (1942–1946), US Sanitary Engineering Center (1946–1963), consulting engineer (1963ff), NAE (1977) (MT2 WWIE WWWIA WWWIS)

WOODWARD, Robert Simpson, 1849–1924, American engineer and mathematical physicist; chief geographer of US Geological Survey (USGS) (1884ff), geophysical nature of deformation of earth's surface, cooling of the earth, methods of topographical mapping, professor of mechanics and mathematical physics at Columbia Univ. (NYC) (1893ff) becoming dean, followed by president of Carnegie Inst. of Washington (DC) (1904–1920) (CDOAB DAB NC13 WWWIA)

WOOLEY, John Cochran, 1884–1982, American agricultural engineer; faculty and department head at Univ. of Idaho, and at Univ. of Missouri, farm waste disposal, design, construction and maintenance of farm structures, fence post treatment, consultant (AE63 NUC)

WOOLF, Arthur, 1766–1837, British mechanical engineer; first worked as a millwright, patented a compound pumping action engine called Woolf engine (1803), engineer for brewery (10 yr.), established a steam engine factory, built steam engines for mines, worked at perfecting Cornish engine (DNB EE EIH GEAPIT TEIH WWWIS)

WORCESTER, 2nd Marquis of (aka Edward Somerset), 1601–1667, English inventor; began mechanical experiments (1628), worked with a large wheel attempting to produce perpetual motion (c.1638) (DNB EIH GEAPIT MWBD WWWIS)

WORCESTER, Joseph Ruggles, 1860–1913, American civil engineer; engineering department and chief engineer of Boston Bridge Works (1882–1894), private practice (1894–1906), member of J. R. Worcester & Co. (1906ff), consultant (BDOACE WWWIA)

WORK, Lincoln T., 1898–1968, American chemical engineer and inventor; faculty at Columbia Univ. (NYC) (1921–1940), director of research and development for Metal & Thermit Corp. (1940–1949), consulting engineer (1921–1940) (1949ff), cement, particle size, pigments, grinding, mixing, drying, chemical and metallurgical engineering (ACCE AMOS PS WWWIS)

WORTHEN, William Ezra, 1819–1897, American civil engineer; authority on pumping machinery and drainage, worked with Loammi Baldwin [1780–1838] and James B. Francis [1815–1892], Albany & West Stockbridge Railroad (1849ff) (CDOAB DAB IIA MWBD NC12 TBDOS(Po))

WORTHINGTON, Henry Rossiter, 1817–1880, American civil and mechanical engineer and inventor; became hydraulic engineer, patented numerous devices connected with hydraulics such as boiler feed pump (1840), propulsion of canal boats (1844), direct-acting steam pump (1848), duplex steam pump (1859), duplex water works engine (1861) (ANB CDOAB DAB EE MEIA NC28)

WORTHINGTON, Wayne, 1891–1980, American agricultural engineer; J. I. Case Co. (1910–1915) worked on farm machinery and tractors in Russia, Argentina and

Uruguay, Aultman and Taylor Machinery Co. (1915–1917) designed threshers, then chief engineer of Electric Wheel Co. (1917–1919), with Aultman and Taylor and Rumely Co. (1920–1929), Deere & Co. (1930–1959) becoming director of research (1948–1964), specialist in gears, instrumentation, hydraulics, with son formed Wayne Engineering Co. (1964) (AE35,62)

WREN, Christopher, 1632–1723, English architect and builder; credited with rebuilding London after great fire (1666), surveyor of King's Works (1669ff), solver of building problems, designed St. Paul's cathedral (London), faculty at Gresham College and Oxford Univ. (DNB EB GEAPIT 1000Y SAI-MP WWWIS)

WRIGHT, Benjamin, 1770–1842, American civil engineer; canal commission of NY (1811), chief engineer for construction of Erie Canal (1817–1825), of Chesapeake and Ohio Canal (1828–1831), of St. Lawrence Canal (1833), consulting engineer (ANB ASAI DAB EIH GEAPIT MWBD NC1 TEIH WWWIA)

WRIGHT, Frank Lloyd, 1867–1959, American architect; practiced in Chicago (1893ff), known for original and innovative designs of numerous dwellings and public buildings including Imperial Hotel in Tokyo (1915–1922) Guggenheim Museum (NYC) (1943–1959), probably received more awards and greater recognition than any architect of his time (ANB BDOTHOT EB MWBD 1000Y WWWIA)

WRIGHT, Horatio Gouverneur, 1820–1899, American military engineer and officer; entered army (1841), US Army brigadier general (1861) and later major general (1864), chief of engineers (1879) (ANB CDOAB DAB MWBD NC4)

WRIGHT, Orville, 1871–1948, American inventor; co-invented airplane with brother Wilbur Wright [1867–1912], formed Wright Cycle Co. (1892), built bicycles, experimented with kites and gliders, piloted first successful engine propelled heavier than air airplane (1903) called the Kitty Hawk Flyer, first airplane casualty was Lt. Thomas E. Selfridge [1882–1908] (1908), with brother organized American Wright Brothers Co. (1909), sold interest in airplane manufacturing company (1915), director of Wright Aeronautical Lab. (1915ff), NAS (1936), 6-cent US postage stamp issued in honor of the brothers (1949) (AIE ASAI BDOS BEST BITAS BM25 CB CDOAB CDOS DOSB EIHGI IAI I&T19 MWBD NC14,15 NYPL PS RHW SAI S:TLAW WWWIA WWWIS)

WRIGHT, Theodore Paul, 1895–1970, American aeronautical engineer; naval aircraft inspector in WWI (1917–1921), Curtiss Aeroplane and Motor Co. (Glenn H. Curtiss [1878–1930]) (1921ff) becoming president and chief engineer (1937ff) of the Curtiss Wright Corp. (1928ff), in WWII, administrator of Civilian Aviation Authority (CAA) (1944–1948), vice president of Cornell Univ. (NY) and president of Cornell Aeronautical Lab. (1948–1960) (AIE CDOS CDOAB DAB MWBD NCH RHW WWWIA)

WRIGHT, Wilbur, 1867–1912, American inventor; co-invented airplane, with brother Orville Wright [1871–1948], formed Wright Cycle Co. (1892), manufactured bicycles, experimented with kites and gliders, built plane that was first successfully piloted by brothers (1903) first by Orville followed by Wilbur on the same day (Dec. 17), built plane for US Army (1908), first casualty was army Lt. Thomas E. Selfridge [1882–1908] at Ft. Myer (VA) (1908), organized American Wright Brothers Co. (1909) (ANB ASAI BDOS BEST BM25 DOSB IAI IIA MWBD NC14,15 NYPL PS SAI SAI-MP S:TLAW TGS WWW? WWWIA WWWIS)

WU TE-JEN, fl. 1107, Chinese mechanical engineer; worked on south-pointing carriages based on differential gearing, could be considered as predecessor of present differential devices to power wheels (GEAPIT SACIC)

WYETH, Newell Convers, 1911–1990, American mechanical engineer and inventor; son of N. C. Wyeth [1882–1945] the illustrator, built hydroplanes in his youth, engineer with Delco Co. (OH) of General Motors Corp. (MI), then with DuPont (DE) (1936ff), moved from routine engineer to development laboratory and developed automatic methods to form dynamite cartridges, made improvements in machine efficiency, invented Typar for carpet backing, plastic (PET) bottles for carbonated beverages with Ronald Roseveare (1973), held 23 patents (ANB CB NC50 NYTObit WWWIA)

WYMAN, Horace, 1827–1915, American mechanical engineer and inventor; machinist Amoskeag Manufacturing Co. (NH) (1846), Hinckley Locomotive Works and Holyhoke Water Power Co. (1854–1860), superintendent then manager of Crompton Loom Works (MA) (1860–1901), patented several different looms with George Crompton [1829–1886], at least 10 patents (CDOAB DAB MEIA MWBD NC13)

WYNER, Aaron D., 1939–1997, American electrical engineer; information theory and with applications to communication and cryptography, served as full- and part-time faculty at Columbia Univ. (NYC) and Polytechnic Inst. of New York (PINY), with Bell Labs. (1963ff) and Lucent Technologies, NAE (1994) (Bridge MT9 WWIE WWWIA)

X

XERXES, c.519–465 BC, Persian king; king of Persia (486–465 BC), continued vast construction projects of canals, bridges and fleet of his father Darius I [550–486 BC] who also served as king (522–486 BC), involved in many battles including defeat of Greeks, after using Phoenician and Greek engineers hired principal engineer Harapolis (EB EIH TEIH MWBD)

Y

YABLOCHKOV, Pavel Nikolayevich, (aka Paul Jablochkov), 1847–1894, Russian electrical engineer and inventor; director of telegraph lines between Moscow and Kursk (1871–1875), invented Yablochkov candle (arc lamp) (1876) (MWBD)

YAGI, Sakae, 1904–1991, Japanese chemical engineer; faculty and administration at Univ. of Tokyo, leader in chemical engineering education, flame combustion, industrial furnaces, chemical reactor engineering, and later executive vice president of Chiyoda Engineering and Construction Co. (1965ff), fNAE (1981) (Bridge MT6)

YAH Jen-Yang, 13th century, Chinese inventor; credited with making artificial pearls and cultured pearls (SACIC)

YAITA KIMBEI, fl. 1550, Japanese swordsmith and gunsmith; his workshop reproduced European musket with matchlock, skill developed to produce the trigger-and-spring firing mechanism, basis of firearms development and production that spread throughout Japan and region (GEAPIT)

YALE, Linus, Jr., 1821–1868, American inventor; invented Yale lock (1851), patented pin tumbler cylinder lock (1861), manufactured locks, with John H. Towne [1818–1875] and Henry R. Towne [1844–1924] established Yale Lock Manufacturing Co. (CT) (1868), Englishman Robert Barron [18th century] invented tumbler lock, Charles Chubb [1779–1845] and John Chubb [1816–1872] formed firm of Chubb & Son (1820) that invented improvements on locks, Englishman Joseph Bramah [1748–1814] invented the tubular key and lock mechanism (1784) (AIE ANB BDOTHOT CDOAB DAB DNB GI MEIA MWBD NC9 NYPL WWWIA)

YANG CH'ENG YEN, fl. 200 BC, Chinese civil engineer; developed extensive defense system, known for siege warfare, Ch'Ang-An fortifications near today's city of Sian (EB GEAPIT MWGD)

YANG SU, fl. 587 AD, Chinese military engineer; prominent naval commander, designed and built (584) and commanded large fleet of war ships, built five battleships each 30 meters high each carrying 800 men and developed strategy for their use (GEAPIT SACIC)

YANG YEN, fl. 26 BC, Chinese hydraulic engineer; massive flood control devices, was director of engineering works and served as vice minister of the imperial court, improved navigation of upper part of Yellow River (GEAPIT)

YARNELL, David Le Roy, 1886–1937, American agricultural engineer; worked in hydraulic engineering, with US Department of Agriculture (USDA) (1909–1916) worked on drainage and flood control projects, worldwide recognition for his work on experimental hydraulics, at Univ. of Iowa hydraulics laboratory worked with Bureau of Agricultural Engineering (USDA) (1922ff), used models for research, flow of water through culverts, rainfall frequency-intensity data and charts (1935) (AE18 WWIE(I))

YARRANTON, Andrew, 1616–1684, English civil engineer; canal building, increasing navigation of rivers, construction of flood gates, weirs, bridges, wharves, locks (AHOT DNB GEAPIT)

YATES, Alden P., 1928–1989, American civil engineer; worked for Bechtel Group, Inc. (CA) serving as vice chairman of the group and chairman of its executive committee (1953–1989), effective in improving the understanding between engineers and public groups affected by public works, NAE (1986) (Bridge MT4 WWWIA)

YEH, Jen-Yang, 13th century, Chinese inventor; developed cultured pearls (SACIC)

YERKES, Arnold P., 1895–1956, American agricultural engineer; inspector of US Army ordnance (3 yr.), office of farm management of US Department of Agriculture (1913ff), during WWI assistant to Farm Equipment Administration in allocation of materials to industry, became captain in ordnance, International Harvester Co. (IHC) (1918–1950) where he edited publication on power farming and development of the farm tractor (AE37)

YEVELE, Henry, c. 1320–1400, English architect and military engineer; prominent master mason of his time, involved in rebuilding the most important structures in England such as Westminster Abbey, Canterbury Cathedral, etc., designed defensive works in Canterbury (DNB GEAPIT)

YI CH'ON, 1376–1451, Korean instrument maker; government official, minister of works (1420s), then Bureau of Astronomics where he helped design instruments—sundials, clocks, rain gauge, firearms, and rocket launchers (GEAPIT)

YIH, Chia-Shun, 1918–1997, Chinese American engineering mechanics; fluid mechanics and hydraulics, first worked in the National Hydraulics Lab. (China), Chinese Bridge Co., faculty at several universities (1948–1952) including Univ. of Michigan (1952–1988), numerous assignments with universities and research organizations worldwide, mechanics of fluids particularly flows with density stratification, developed general theory of the dynamics of nonhomogeneous fluid flow, NAE (1980) (Bridge MT10 WWWIA WWWIS)

YING SHUN-CH'EN, fl. 1071, Chinese hydraulic engineer; developed a self-regulating system of water control along canal to reduce silting and flooding particularly of the Grand Canal (GEAPIT SACIC)

YOUNG, James, 1811–1883, Scottish industrial chemist and chemical engineer; oil for industrial uses for which he set up experimental works (1848–1866), gas manufacture by distillation of low-temperature shales, oil technologist, president of Anderson College (1866–1877) (BDOS(Wi) CDOS DNB PS WWWIS)

YOUNG, James Frederick, 1917–1980, American mechanical engineer; systems analysis, system engineering and planning, new technologies for electric utility industry, General Electric Co. (1937ff) through several positions to manager of electric utility systems engineering, NAE (1967) (AMOS MT2)

YOUNG, Thomas, 1773–1829, English physician and engineering physicist; first to measure astigmatism (1800), study of vision, eye, and light (1801), discovered interference of light (1801), wave nature of light (1803), calculated the wavelength of visible light, proposed the idea that three colors of red, green, and blue could in various combinations give various colors, the Young-Helmholtz (Herman L. F. von Helmholtz [1821–1894]) three-color theory (1801, 1805), understanding of surface tension of liquids, presented Young modulus (1802)—a constant in equations defining behavior of elastic materials (BEST BDOS CDOS DNB DOT EB EIH FNIE L&M MWBD PS RHW WWWIS)

YU, 5th century BC, Chinese civil engineer; known as the Great Engineer, semi-legendary Emperor patron saint of hydraulic engineers, water control, irrigation, conservation (SACIC)

YU HSU, fl. 110 AD, Chinese civil engineer; built roads for wheeled vehicles on the northern frontier, cleared river of obstructions for navigation, used hot-cold technique to break rocks (GEAPIT SACIC)

YUWEN K'AI, fl. 610 AD, Chinese engineer and architect; improved accuracy of water clocks using a compensating tank for water, directed reconstruction of major cities, was director of Architectural and Engineering Department of the Imperial Palace (570s) and minister of works (30 yr.) including canals and carriages (GEAPIT SACIC)

Z

ZACHOS, John Celivergos, 1820–1898, Turkish American inventor; to US (1830), with several Ohio educational institutions in a variety of fields (1845ff), Cooper Union (NYC) (1871ff), patented machine for printing at high speeds (1876), author (CDOAB DAB WWWIA)

ZAVALISHIN, Dmitrii Aleksandrovich, 1900–??, Soviet electrical engineer; electrical machinery, electronics, semiconductors, professor of electrotechnology at Armed Forces School of Engineering (1941–1946), chair of electrical machines in Leningrad (1946–1959),

chief of laboratory at Inst. of Electromechanics of USSR Academy of Sciences (1959ff) (PS PSOCE SMOS WWWIS)

ZEEMAN, Pieter, 1865–1943, Dutch experimental physicist; faculty at Univ. of Amsterdam (1900–1935) and director of its Physical Inst. (1908ff) and director of the Zeeman Lab. (1923ff), Zeeman effect (1896) that is the splitting of the spectra lines based on suggestion of Hendrik A. Lorentz [1853–1928], Nobel prize (1902), fRS (1921) (BEST CDOS DOT EB MWBD RHW RSObit WWWIS)

ZEISS, Carl, 1816–1888, German manufacturer; workshop for building optical materials, founded Zeiss optical factory (1846), later taken over by his partner and director of research Ernst Abbe [1840–1905] (1888ff) (EB MWBD SAI)

ZENTMAYER, Joseph, 1826–1888, German American inventor and manufacturer; to US (1848), scientific instrument maker in Philadelphia (PA) (1853ff), improved microscope, invented improved photographic lens (1865) (BITAS CDOAB DAB NC13 WWWIA)

ZEPPELIN, Ferdinand Adolf August Heinrich von Graf, 1838–1917, German aeronaut, airship designer, and inventor; service with Union army in US Civil War (1863), retired as brigadier general in the German army (1891), constructed first airship of rigid type named after him (1900), established factory (1900), manufactured airships (1906ff) (BEST EB MWBD NYPL PS RHW SAI TGS WA WWWIS)

ZERONI, Moshe, ??–1996, Israeli agricultural engineer; professor at Ben-Gurion Univ. at Negav Israel, worked on environmental system engineering at the Jacob Blaustein Inst. for Desert Research (Res3)

ZEUNER, Gustav Anton, 1828–1907, German physicist and engineer; thermodynamics, mechanics, and hydraulics, director of Polytechnic Inst. Dresden (1873–1897) (MWBD PS WWWIS)

ZHAVORONKOV, Nikolai M., 1907–??, Soviet chemical engineer; Moscow Chemical-Technological Inst. (1930–1962) becoming director (1948ff), processes for separating liquid and gas mixtures by absorption, rectification, molecular distillation, and chemical ion exchange, USSR Academy of Sciences at Inst. of General and Inorganic Chemistry (PS PSOWE SMOS WWWIS)

ZHUKOVSKY, Nikolai Ergorovich, 1847–1921, Russian aeronautical engineer; study of flow of fluids, principles of flight, hydrodynamics (1880s), heavier than air flight (1890s), used wind tunnel for wing studies, applied Daniel Bernoulli [1700–1782] flow equations (1905), lift of airplane wings represented by Zhukovsky-Chaplygin postulate (1910), taught at Univ. of Moscow and the Moscow Technical School, head of Academy of Military and Aeronautical Engineering (named after him in 1922), recognized as the father of Russian aviation (CDOSB FETCTW)

ZIEGLER, Karl, 1898–1973, German chemist and inventor; with Giulio Nata [1903–1979] an Italian chemist and chemical engineer developed production of synthetic rubber, developed process to produce rigid polyethylene, was director of Kaiser Wilhelm Inst. (1943) (later Max Planck Inst.), Nobel prize for both (1963), professor at Heidelberg Univ. (1928), Halle (1936), Aachen (1947) (IAI LOTCS MWBD RHW WWWIS)

ZIMMERMAN, Oliver Brunner, 1873–1941, American mechanical and agricultural engineer; from high school teacher to professor at Univ. of Wisconsin (until 1905), worked on development and application of farm machinery, major at General Engineering Depot during WWI (1917–1919), engineering and experimental work with International Harvester Co. (IHC), mechanical design of tractors and implements, worked with National Bureau of Standards (NBS) (AE22 NCA WWIE(I))

ZINK, Carlton L., 1902–1994, American agricultural engineer; after teaching in Jacksonville (FL) and Univ. of Arizona, faculty of Univ. of Nebraska (1929–1943) during which he was engineer in charge of Nebraska Tractor Tests (1930–1945), development engineer with Firestone Tire and Rubber Co. (1943–1950), product planning and safety with Deere & Co. (1950–1968), consultant in farm safety (1968ff) (AE49 Res2 WWIE)

ZOLL, Paul Maurice, 1911–1999, American physician and inventor; with Harvard Medical School, Beth Israel Hospital (ante 1993), founded Zoll Medical Corp., during WWII served in armed forces (1939–1945), carried out first successful defibrillation using electric shock which led to pioneer pacemaker (1952), external defibrillator (1956) (BDOTHOT IAI NYTObit TIY WA WWWIA)

ZONCA, Vittorio, 1568–1603, Italian architect and mechanical engineer; catalogued and developed wide variety of mechanisms such as used for watermill, mills for finishing cloth and grinding dyestuffs, barbecue pit (AHOT GEAPIT)

ZONS, Frederick William, 1885–1960, American chemical engineer; faculty at Columbia Univ. (NYC) (1907–1909), with Thomas A. Edison [1847–1931] (1909–1911), consulting engineer on storage batteries, American Neon Light Co. (2 1/2 yr.), Westinghouse

Lamp Co. (PA) (1911–1917), New Process Gas Mantle Co. (1917–1919), consultant (1919ff), editor of journals (NC48 PS)

ZOTOS, John, 1932–1998, American mechanical engineer; research in computer derived mathematical models that describe chemical, mechanical, and physical properties of metal alloys, Raytheon Manufacturing Co. (MA) (1954), Watertown Arsenal Rodman Lab. (1955–1960), faculty of Northeastern Univ. (1960–1992) (WWWIA)

ZRAKET, Charles A., 1924–1997, American electrical engineer; information systems particularly as related to national defense, computer design, MIT Lincoln Labs. (1952–1958), advancing to president and CEO (1986–1990) of Mitre Corp. (1958–1990), NAE (1991) (Bridge MT10 WWIE)

ZUSE, Konrad, 1910–1995, German civil engineer; built a mechanical computer called Z1 (1934), binary calculator (1939), WWII service, worked at Henschel factory on aeronautical projects, formed Zuse KG Co. (1950) building and selling computers, the company was sold to Siemans AG (1969) and he returned to research, fNAE (1981) (BDOTHOT GI MT9 NYPL)

ZVONKOV, Vasilli V., 1891–??, Soviet transportation engineer; worked on transportation, communications, water economy, professor at Military Transport Academy (1935–1955), USSR Academy of Sciences (1955ff) (PS SMOS WWWIS)

ZWORYKIN, Vladimir Kosma, 1889–1982, Russian American electrical engineer, physicist and inventor; to US (1919), invented television camera (1923, 1924) while at Westinghouse Electric Corp. (PA) (1920–1929), iconoscope or TV camera (1923), developed electron microscope (with James Hillier [1915–2007]) (c.1946), pictures of biological materials (1957), to Radio Corp. of America (RCA) (1929–1954) where he was research administrator and vice president (1947), consultant (1954ff), over 120 patents, NAS (1943), NAE (1965) (AIE AMOS ANB BEST CB CDOS EB EIH IAI MT2 MWBD NYPL 1000Y RHW S:TLAW TGS TTOS WA WWWIA WWWIS)

ZYBACH, Frank L., 1894–1980, American farmer and inventor; mechanization of irrigation, invented self-propelled center-pivot irrigation system (1952) manufactured by Valley Manufacturing Co. (1954) the company name was changed to Valmont Industries, Inc. (NE) (1966), at least five patents (AE61 SciAm234)

APPENDICES

APPENDIX A

PEOPLE WITH 50 OR MORE PATENTS*
(People deceased before 2000)

Adler, Charles, Jr., 1899–1980	60 patents
Alexanderson, Ernst, 1878–1975	322 patents
Allen, Herbert, 1907–1990	over 150 patents
Babbitt, Benjamin Talbot, 1809–1899	at least 108 patents
Baekeland, Leo Hendrik, 1863–1944	over 100 patents
Ball, Albert, 1835–1927	approx. 135 patents
Bancroft, John Sellers, 1843–1919	over 100 patents
Barkan, Philip, 1925–1996	over 50 patents
Benerito, Ruth Rogan, 1916–?	over 50 patents
Berliner, Emile, 1851–1929	over 100 patents
Birdseye, Clarence, 1886–1956	approx. 300 patents
Black, Harold Stephen, 1898–1983	over 333 patents
Bogardus, James, 1800–1874	approx. 50 patents
Breyer, Frank Gottlub, 1886–1966	over 50 patents
Broughton, Donald B., 1917–1984	50 patents
Brown, Alexander, 1852–1911	over 100 patents
Brown, George H., 1908–1987	80 patents
Brown, Theo, d. 1971 at age 92	approx. 158 patents
Brown, Walter J., 1900–1972	over 150 U.S. & foreign
Browning, John Moses	over 128 patents
Burson, William Worth, 1832–1913	over 50 patents
Bush, Vannevar, 1890–1974	approx. 50 patents
Busignies, Henri G., 1905–1981	over 140 patents
Campbell, Andrew, 1821–1890	over 50 patents
Camras, Marvin, 1916–1995	over 500 patents
Capstaff, John George, 1879–1960	over 100 patents
Carrier, Willis Haviland, 1876–1950	over 80 patents
Carver, George Washington, 1864–1943	over 100 U.S. & foreign
Cave, Henry, 1874–1963	over 100 patents
Chubb, Lewis Warrington, 1882–1952	over 200 patents
Clark, Thomas Edward, 1869–1962	over 70 patents
Colley, Russell, age 97, d. 1996	65 patents
Conrader, Rudolph, 1859–1929	over 200 patents
Cottrell, Calvert Bryan, 1821–1893	over 100 patents
Coulter, Wallace Henry, 1913–1998	74 patents
Coxe, Eckley B., 1839–1895	over 100 patents

Crompton, George, 1829–1886 . . . .   over 200 patents
Crossthwait, David N., Jr., c. 1892–1976. . . . .   over 114 patents
Crowell, Luther Childs, 1840–1903. . . . .   issued 280 patents

Daniels, Fred Harris, 1853–1913. . . . .   over 151 patents
DeForest, Lee, 1873–1961. . . . .   over 300 patents
Dixon, Robert Munn, 1860–1918 . . . .   over 50 patents
Doane, William Howard, 1832–1915. . . . .   over 70 patents
Dodge, James Mapes, 1852–1915 . . . .   over 200 patents
Dow, Herbert H., 1866–1930. . . . .   over 100 patents
Draper, William Franklin, 1842–1910 . . . .   over 200 patents
DuPont, Francis Irenee, 1873–1942. . . . .   over 100 patents

Eckert, John Prosper, 1919–1995 . . . .   over 87 patents
Edison, Thomas Alva, 1847–1931. . . . .   1093 U.S. patents
Eichemeyer, Rudolf, 1831–1895 . . . .   over 150 patents
Ellis, Carleton, 1876–1941 . . . .   over 753 patents

Farnsworth, Philo Taylor, 1906–1971 . . . .   over 300 patents
Ferranti, Sebastian, 1864–1930 . . . .   176 patents
Ferris, Robert E., 1907–1970. . . . .   89 patents
Fessenden, Raymond A., 1866–1932. . . . .   over 500 patents
Field, Stephen D., 1846–1913 . . . .   over 100 patents
Fiske, Bradley Allen, 1854–1942. . . . .   over 60 patents
Folmer, William Frederic, 1861–1936.   over 300 patents
Frazier, John Earl, 1902–1985. . . . .   50 patents
Freeman, Benjamin William, 1890–1963 . . . .   over 60 patents
Frolich, Per Keyser, 1899–1977. . . . .   approx. 75 patents
Fuller, Levi Knight, 1841–1896. . . . .   approx. 50 patents
Fuller, Richard Buckminster, 1895–1983 . . . .   2000 U.S. & foreign patents

Gabor, Dennis, 1900–1979 . . . .   over 100 patents
Gally, Merritt, 1838–1916 . . . .   over 50 patents
Gerber, Heinz Joseph, 1924–1996. . . . .   issued 677 patents
Gernsback, Hugo, 1884–1967 . . . .   over 80 patents
Goddard, Robert Hutchings, 1882–1945 . . . .   214 patents
Goddu, Louis, 1837–1919 . . . .   about 300 patents
Golary, Marcel J. E., 1902–1989 . . . .   over 58 patents
Good, John, 1841–1908 . . . .   over 100 patents
Goodyear, Charles, 1800–1880 . . . .   approx. 60 patents
Gordon, George Phineas, 1810–1878 . . . .   over 50 patents
Gourdine, Meredith Charles, 1929–1998. . . . .   70 patents
Greist, John Milton, 1850–1906 . . . .   approx. 50 patents

Hall, Lloyd A., 1894–1971. . . . .   over 105 patents
Hanford, William Edward, 1908–1996.   120 patents
Harvey, Hayward, 1824–1893 . . . .   over 125 patents
Hawkins, Walter Lincoln, 1911–1992 . . . .   147 U.S. & foreign patents
Henry, Beulah Louise, 1887–1939. . . . .   52 patents
Hodgkinson, Francis, 1867–1949 . . . .   issued 101 patents
Holly, Birdsill, 1822–1894. . . . .   over 150 patents
Houdry, Eugene J., 1892–1962 . . . .   over 100 patents

House, Henry A., 1840–1930.	over 300 patents
Howard, Henry, 1868–1951	89 patents
Hull, August, 1887–1952.	91 patents
Hund, August, 1887–1952.	approx. 50 patents
Hyatt, John Wesley, 1837–1920	over 200 patents
Ipatiev, Vladimir N., 1867–1952	200 patents
Jacuzzi, Candido, 1903–1986 (31 US+21)	52 patents
Jenkins, Charles Francis, 1867–1934.	about 400 patents
Johnson, Eldridge Reeves, 1867–1945	70 patents
Johnson, Reynold Benjamin, 1906–1998	over 90 patents
Jones, Frederick McKinley, 1892–1961.	61 patents
Julian, Percy Lavon, 1899–1975	over 130 patents
Kelvin, Lord (William Thomson), 1824–1907	70 inventions (patents?)
Ketchledge, Raymond, W., 1919–1987	58 patents
Kettering, Charles F., 1876–1958.	over 185 U.S. patents
Klopsteg, Paul Ernest, 1899–1991.	approx. 50 patents
Lake, Simon, 1866–1945.	over 200 patents
Lamm, A. Uno, 1904–1989	about 150 patents
Lamme, Benjamin G., 1864–1924.	issued 162 patents
Leavitt, Frank McDowell, 1856–1928.	over 300 patents
Lemelson, Jerome, 1923–1997	issued 554 patents
LeTourneau, Robert G., 1888–1969	187 U.S. patents
Lewis, Wilfred, 1854–1929	over 50 patents
List, Hans, 1896–1996. . . . . :. . .	364 patents
Lyon, G. Albert, 1881–1961	approx. 1200 patents
Mauch, Hans Adolph, 1906–1984.	over 80 patents
Maxim, Sir Hiram Stevens, 1840–1916.	over 100 patents
McCoy, Elijah J., 1843–1929.	58 patents
McLean, William Burdette, 1914–1976.	approx. 50 patents
Mergenthaler, Ottmar, 1854–1899.	over 60 patents
Miller, Stewart Edward, 1918–1990	80 patents
Moore, Daniel McFarlan, 1869–1936	over 100 patents
Murray, Jerome, d. 1998, age 85	75 patents
Murray, Thomas Edward, 1891–1961	over 200 patents
Nobel, Alfred Bernhard, 1833–1896	over 100 patents
Nordberg, Bruno Victor, 1857–1924	70 patents
Norton, Charles Hotchkiss, 1851–1942.	over 100 patents
Noyes, LaVerne, 1849–1919	over 100 patents
Offner, Franklin F., 1911–1999	60 patents
Othmer, Donald Frederick, 1904–1995	over 150 patents
Painter, William, 1828–1906	85 patents
Parsons, Charles Algernon, 1854–1931.	over 300 patents
Pierce, George Washington, 1872–1956	53 patents
Potter, William Bancroft, 1863–1934	more than 130 patents

Rabinow, Jacob, 1910–1999. . . . .   226 patents
Rajchman, Jan (John) A., 1911–1989 . . . .   116 patents
Redfield, Casper Lavater, 1853–1948 . . . .   approx. 60 patents
Rice, Charles De Los, 1859–1939. . . . .   over 75 patents
Rice, Edwin Wilbur, Jr., 1862–1935 . . . .   over 100 patents
Richards, Francis Henry, 1850–1933. . . . .   over 350 patents
Rogers. John Raphael, 1856–1934. . . . .   400–500 patents
Ruud, Edwin, 1854–1932. . . . .   over 200 patents

Sharps, Christian, 1811–1874 . . . .   50 patents
Sellers, William, 1824–1905 . . . .   over 90 patents
Shaw, Thomas, 1838–1901 . . . .   over 186 patents
Short, Sidney Howe, 1858–1902 . . . .   over 500 patents
Silver, Thomas, 1813–1888 . . . .   over 50 patents
Simjian, Luther, d. 1997, age 92 . . . .   over 200 patents
Smith, Oberlin, 1840–1926 . . . .   over 75 patents
Sparks, William Joseph, 1904–1976 . . . .   125 patents
Spencer, Percy LeBaron, 1894–1970. . . . .   100 patents
Sperry, Elmer Ambrose, 1860–1930 . . . .   over 400 patents
Sperti, George Speri, 1900–1991. . . . .   over 150 patents
Steinmetz, Charles Proteus, 1865–1925 . . . .   over 200 patents
Stone, John Stone, 1869–1943. . . . .   approx. 100 patents
Swinburne, James, 1858–1958 . . . .   over 100 patents

Taylor, Frederick W., 1856–1915. . . . .   approx. 100 patents
Tesla, Nikola, 1856–1943 . . . .   111 patents
Thompson, Albert R., 1879–1947 . . . .   over 200 patents
Thomson, Elihu, 1853–1937 . . . .   over 700 patents
Thomson, John, 1856?–1926. . . . .   over 200 patents
Tucker, Stephen Davis, 1818–1902 . . . .   over 100 patents
Turner, Walter Victor, 1866–1917 . . . .   over 400 patents

Varian, Russell H. (w/bro.), 1898–1959 . . . .   over 100 patents

Warren, Henry Ellis, 1872–1957. . . . .   135 U.S. patents
Weaver, Ira A., 1871–1965. . . . .   100 patents
Wellman, Samuel Thomas, 1847–1919 . . . .   100 patents
Westinghouse, George, 1846–1914 . . . .   over 400 patents
Wheeler, Harold Alden, 1903–1996 . . . .   180 patents
Whitaker, Milton C., 1870–1963. . . . .   60 U.S. patents
Wilcox, Stephen, 1830–1893. . . . .   approx. 50 patents
Wilson, Robert Erastus, 1893–1964 . . . .   90 patents
Wood, James J., 1856–1928 . . . .   240 patents
Woods, Granville T., 1856–1910 . . . .   over 100 patents

Zworykin, Vladimir K., 1889–1982. . . . .   over 120 patents

* Based on biographical references cited in the text of the book. These references usually did not identify the country granting the patent. Inventions do not necessarily become patents. For example, Beulah Louise Henry is credited with over 100 inventions and obtained 52 patents. Also, Joseph Henry is credited with numerous inventions but he believed that scientific findings and developments belonged in the "public" domain and did not file patents.

APPENDIX B

NOBEL PRIZE RECIPIENTS INCLUDED IN BOOK

Name		*Year*
Alvarez, Luis Walter, 1911–1988		1968
Anderson, Carl David, 1905–1991	(J)	1936
Arrhenius, Svante August, 1859–1927		1903
Bardeen, John, 1908–1991	(J)	1956, 1972
Becquerel, Antoine Henri, 1852–1908	(J)	1903
Bergius, Frederich Karl R., 1884–1949	(J)	1931
Bohr, Niels H. D., 1885–1962		1922
Bosch, Karl, 1874–1940	(J)	1931
Bragg, William Henry, 1862–1942	(J)	1915
Braun, Karl Ferdinand, 1850–1918	(J)	1909
Bridgman, Percy Williams, 1882–1961		1946
Broglie, Louis Victor Pierre Raymond de, 1892–1987		1929
Calvin, Melvin, 1911–1991		1961
Cockcroft, John Douglas, 1897–1967	(J)	1951
Compton, Arthur Holly, 1892–1962	(J)	1927
Curie, Marie, 1867–1934	(J)	1903, 1911
Curie, Pierre, 1859–1906	(J)	1903
Dalen, Nils Gustaf, 1869–1937		1912
Einstein, Albert, 1879–1955		1921
Einthoven, Willem, 1860–1927		1924
Elion, Gertrude Belle, 1918–1999	(J)	1988
Ernst, Richard R., b. 1933		1991
Fermi, Enrico, 1901–1954		1938
Feyman, Richard P., 1918–1988	(J)	1965
Fleming, Alexander, 1881–1955	(J)	1945
Gabor, Dennis, 1900–1979		1971
Guillaume, Charles Edouard, 1861–1938		1920
Haber, Fritz, 1868–1934		1918
Hahn, Otto, 1878–1968		1944
Heisenberg, Werner Karl, 1901–1976		1932
Herzberg, Gerhard, 1904–1999		1971
Heyrovsky, Jaroslav, 1890–1967		1959
Hodgkin, Dorothy Mary Crowfoot, 1910–1984		1964
Kapitza, Petr Leonidovich, 1894–1984	(J)	1978
Koch, Robert, 1843–1910		1905
Langmuir, Irving, 1881–1957		1932
Lawrence, Ernest Orlando, 1901–1958		1939
Libby, Willard Frank, 1908–1980		1960

Marconi, Guglielmo, 1874–1937	(J)	1909
Michelson, Albert Abraham, 1852–1931		1907
Millikan, Robert Andrews, 1868–1953		1923
Natta, Giulio, 1903–1979	(J)	1963
Nernst, Walther Herman, 1864–1941		1920
Onsager, Lars, 1903–1976		1968
Ostwald, Friedrich Wilhelm, 1853–1932		1909
Pauling, Linus Carl, 1901–1994		1954, 1962
Planck, Max K. E. L., 1858–1947		1918
Rabi, Isidor I., 1898–1988		1944
Ramón y Cajal, Santiago, 1852–1934	(J)	1906
Rayleigh, Lord (John William Strutt), 1842–1919		1904
Röntgen, Wilhelm Konrad, 1845–1923		1901
Ruska, Ernst August Friedrich, 1906–1988	(J)	1986
Rutherford, Ernest, 1871–1937		1908
Schawlow, Arthur Leonard, 1921–1999	(J)	1981
Schrödinger, Erwin, 1887–1961	(J)	1933
Shockley, William Bradford, 1910–1989	(J)	1956
Stark, Johannes, 1874–1957		1919
Svedberg, Theodor H. E., 1884–1971		1926
Thomson, Joseph John, 1856–1940		1906
Tiselius, Arne W. K., 1902–1971		1948
Urey, Harold Clayton, 1893–1981		1934
Van Der Waals, Johannes Diderik, 1837–1923		1910
Wien, Wilhelm, 1864–1928		1911
Wilson, Charles Thomson Rees, 1869–1959	(J)	1927
Zeeman, Pieter, 1865–1943	(J)	1902
Ziegler, Karl, 1989–1973	(J)	1963

(J) joint award
The Nobel prizes were first awarded in 1901.

REFERENCES

AAFIS&T   African American Firsts in Science & Technology. 1999. Webster, Raymond B. Gale Group, Detroit, MI 462 p.
ISBN 0-7876-3876-5   QI41.W43

AAI   African-American Inventors. 1994. McKissack, Patricia and McKissack, Frederick. Milbrook Press, Brookfield, CT 96 p.
ISBN 1-56294-468-1   T39.M45

AAISMAI   African Americans in Science, Mathematics, and Invention. 2000. Spangenburg, Ray and Maser, Kit. Facts On File, Inc., New York, NY 254 p.
ISBN 0-8160-4806-1   Q141.S6285

ACCE   American Chemists and Chemical Engineers. 1976. 1994. Miles, Wyndham D. American Chemical Society, Washington, DC 544 p. vol. 2 (1994) Miles, W. D. and Gould, Robert F. Gould Book, Guilford, CT 345 p.
ISBN 0-8412-0278-8   QD21.A43

ACE   The American Civil Engineer 1852–2002. 2002. Wisely, William Homer, updated by Virginia Fairweather. American Society of Civil Engineers, Reston, VA 235 p.
LCCN 2002026092   TA1.W83

ACOAB   Appleton's Cyclopedia of American Biography. 1888–1918, 1918–1931, and other dates of printing. Reprint 1968. Wilson, James Grant, et al. 1918, 1931, 1968. D. Appleton, New York, NY 7 vol. + 6 vol. supp. Reprint by Gale. 1968.
LCCN 6701406   E176.A65 E48.04

AE   Agricultural Engineering. 1920–1994. American Society of Agricultural Engineers (ASAE), St. Joseph, MI, serial, includes monthly newspaper Within ASAE as a separate issue (1986–1993).
ISSN 0002-1458   S671.A3

AEOT   American Engineers of the Nineteenth Century—A Biographical Index. 1978. Roysdon, Christine and Khatri, Linda. Garland Publishing, Inc., New York, NY 247 p. (no index)
ISBN 0-8240-9827-7   TA139.R7

AEOTHOT   An Encyclopedia of the History of Technology. 1990. McNeil, Ian (ed.). Routledge, London and New York, NY 1062 p.
ISBN 0-415-01306-2   T15.E53

AG   American Genesis: A Century of Invention and Technological Enthusiasm.   1989. Hughes, Thomas Parke. Viking Penguin Books, New York, NY 529 p.
ISBN 0670814784   T21.H82

AHOI   A History of Invention. 1987. 2000. Williams, Trevor I. Facts On File, New York, NY 367 p. Updated by Schaaf, William E., Jr. 2000. Little, Brown & Co., London, UK 367 p.
ISBN 0-8160-4072-9   T15.W718

AHOI&T   American Heritage of Invention & Technology. 1985ff. Also listed under I&T. American Heritage, Inc. affiliated with Forbes, Inc., New York, NY.
ISSN 008-323   T1.A455

AHOT   A History of Technology. 1954–1958. Singer, Charles, et al. Oxford at Clarendon Press, Oxford, England In 8 vol.
LCCN 55008645   T15.S53

AI   American Inventors. 1934. 1963. 471 Hylander, Clarence John. Macmillan, New York, NY 216 p.
LCCN 34-11090   T39.H9

AIE   American Inventors, Entrepreneurs & Business Visionaries. 2002. Casey, Charles W., Jr. Facts On File, New York, NY 410 p.
ISBN 0-8160-4559-3   CT214.C29

AIN   Ancient Inventions. 1944. James, Peter and Thorpe, Nick. Ballantine Books, New York, NY 672 p.
ISBN 0-345364767   T16.J36

AITBF   An Index to Biographical Fragments in Unpublished Scientific Journals. 1973. Barr, Ernest Scott. The University of Alabama Press, University, AL 294 p.
ISBN 0-8173-9603-9   Q141.B29

AM   Ancient Memories. 1999. James, Peter and Thorpe, Nick. Ballantine Books, New York, NY 651 p.
ISBN 0-345-40195-6   CB311.J385

AMOS/AMAWOS American Men of Science. 1960. American Men and Women of Science (20th ed.). 1999.   Cattell, Jaques. Jaques Cattell Press, Inc., Tempe, AZ and R. R. Bowker, New Providence, NJ In 8 vol.
ISBN 0-8352-3463-0   Q141.A4

AmScient   American Scientist. 1913ff. Sigma Xi, The Scientific Research Society, Research Triangle, NC bimonthly serial
ISSN 0003-0996   LJ85.S502

ANB   American National Biography. 1999. (people deceased by 1996) Garraty, John A. and Carnes, Mark C. Oxford University Press, New York, NY In 24 vol.
ISBN 0-19-520634-5   CT213.A68E176.A62

AOAAH   Almanac of African American Heritage. 2001. Miles, Johnny H., et al. Prentice Hall, New York, NY 430 p.
ISBN 0-7352-0226-5   E185.A448

AOFP   Almanac of Famous People. 1998. Castronova, Frank V. Gale Group, Detroit, MI In 2 vol.
ISBN 0-7876-00-0041-X   CT104.B56

ASAI   American Science and Invention. 1954. Wilson, Mitchell A. Simon and Schuster, New York, NY 437 p.
LCCC No. 54-9812   Q125.W7914

AS&TI   Applied Science & Technology Index. 1958ff. Preceded by Industrial Arts Index (1913–1957). Howard, Joyce M. (ed.). H. W. Wilson Co., New York, NY
ISSN 0003-6986   Z7913.I7

AWIS   American Women in Science—1950 to present. 1998. Bailey, Martha J. ABC-CLIO, Inc., Santa Barbara, CA 455p.
ISBN 0-87436-921-5   Q141.B254

AWIT   American Women in Technology. 2000. Zierdt-Warshan, Linda, Winkler, Alan, and Bernstein, Leonard. ABC-CLIO, Santa Barbara, CA 384 p.
ISBN 1-57607-072   T36.Z54

AWOA   American Women of Achievement. 1990. James, Carrie. Chelsea House, New York, NY 111 p.
ISBN 1555466699   NA737.M68J36

BAC   Building a Century: Bechtel 1898–1998. Andrews McMeele Pub., Kansas City, MO 243 p.
ISBN 0-8362-5322-1   TA217.B4B43

B-ACGA   Breakthroughs—A Chronology of Great Achievements in Science and Mathematics, 1200–1930. 1985. Parkinson, Claire L. G. K. Hall & Co., Boston, MA 576 p. ISBN 0-8161-8706-1   CS125.P327

BCST   Black Contributors to Science and Energy Technology. 1979. U. S. Department of Energy, Washington, DC.

BDOACE   A Biographical Dictionary of American Civil Engineers. 1972. American Society of Civil Engineers, New York, NY Vol. 1 edited by a committee of ASCE 163 p.; Vol. 2, edited by Francis E. Griggs, Jr. 141 p. ISBN 0-87262-822-1   TA139.A53

BDOAS   Biographical Dictionary of American Science (17th–19th Centuries). 1979. Elliott, Clark A. Greenwood Press, Westport, CT 360 p. ISBN 0-313-20419-5   Q141.E37

BDOBA   The Biographical Dictionary of Black Americans. 1992. 1999. Kranz, Rachael C., Facts On File, Inc., New York, NY 190 p. Kranz, Rachael C. and Koslow, Philip I. (1999) 310 p. ISBN 0-816039038   E185.96.K73

BDOS BDOS(Wi) A Biographical Dictionary of Scientists. 1969. 1974. 1982. 1994. Williams, Trevor I. and Withers, Sonja (eds.) A&C Black, London 641 p. (1974) and Wiley Interscience, New York, NY 592 p. (1969) ISBN 0470946814   Q141.W62

BDOTHOT   Biographical Dictionary of the History of Technology. 1996. Day, Lance and McNeill, Ian. Routledge, London 844 p. ISBN 0-415-06042-7   T39.B49

BDSE   The Biographical Dictionary of Scientists: Engineers and Inventors. 1985. Also referred to as E&I below as a part of TBDOS(E&I) Abbott, David. Blond Educational of Frederick Muller, Ltd., London and Peter Bedrick Books, New York, NY 189 p. ISBN 0-87226-009-7 TA139.B56

Bent   The Bent. 1910ff. Published by the Tau Beta Pi Association (established 1885), Knoxville, TN now quarterly magazine. ISSN 0005-884X   LJ115.T383

BEOS   A Biographical Encyclopedia of Scientists. 1994 (2nd ed.). Daintith, John, et al. Market House Books, Bristol, England and Facts On File, Inc., New York, NY 935 p. In 2 vol. ISBN 0-7503-0287-9 (2 vol.)   Q141.B53

BEOS(Ol)   Biographical Encyclopedia of Scientists. 1998. Olson, Richard and Smith, Roger (eds.). Marshall Cavendish Inc., New York, NY 1460 p. In 5 vol. ISBN 0761470646 (set)   Q141.B532

BEST   Biographical Encyclopedia of Science & Technology. 1964. 1972. Asimov, Isaac. Avon Books, New York, NY 805 p. ISBN 0-380-00619-7   Q141.A74

BI   Black Inventors. Aaseng, Nathan. Facts On File, Inc., New York, NY 128 p. ISBN 0-8160-3407-9 T39.A36

BIOA   Black Inventors of America. 1989. Burt, McKinley, Jr. National Book Co., Portland, OR 149 p. ISBN 0-89420095X   T39.B87

BIS   Blacks in Science. 1977. Carwell, Hattie, Exposition Press, Inc., Hicksville, NY 96 p. ISBN 0-682-48911-5   Q141.C23

BM   Biographical Memoirs. 1877ff. National Academy of Sciences (NAS), Washington DC, published periodically (presently) by National Academies Press, Washington, DC. ISBN 0-309-7035X   ISSN 0077-2933   Q141.N2

BOPS   Black Pioneers of Science and Invention. 1970. Habra, Louis. Harcourt, Brace & World, Inc., New York, NY 181 p.
LC 77-109090   Q141.H2

Bridge   The Bridge. 1970ff. A quarterly publication of the National Academy of Engineering, Washington, DC
USPS 551-240

BIT   British Technology Index. Ante 1980. Renamed Current Technologies Index (CTI) 1981-1997, renamed Abstracts in New Technologies and Engineering (ANTE). 1997ff. Bowker Saur, Windsor Court, United Kingdom
ISSN 1367-9899 (new)   Z7913.B7

C&EN   Chemical & Engineering News. 1923ff. American Chemical Society, Washington, DC Serial published every other week
ISSN 0009-2347   TP1.C35

CB, CBY   Current Biography. 1940-1985; Current Biography Yearbook. 1986ff. Thompson, Clifford is latest editor (1999), published by H. W. Wilson, New York, NY
ISSN 0084-9499   CT100.C8

CBB   Contemporary Black Biography. 1999. Phelps, Shirelle (ed.). Gale Group, Farmington Hills, MI In 22 vol.
ISBN 0-7876-2419-5   E185.96.C66

CBD   Chambers Biographical Dictionary. 1984. 1997 (5th ed.) Thorne, J. O., Collocott, T. C., and Parry, Melanie (eds.) (5th ed.). W & R Chambers, Edinburgh 1493 p., Larousse Publishers, New York, NY 2008 p.
ISBN 0-550-16060-4   CT103.C4

CBE The Cambridge Biographical Encyclopedia. 1994. 1998. Crystal, David (ed.). Cambridge University Press, London and New York 1179 p.
ISBN 0521630991   CT103.C26

CDOAB   Concise Dictionary of American Biography. 1997 (5th ed.). Charles Scribner's Sons, New York, NY 1817 p. In 2 vol.
ISBN 0-684-80549-9   E176.C73

CDOS or TCDOS(Mi)   The Cambridge Dictionary of Scientists. 1996. Millar, David, et al. Cambridge University Press, Cambridge England and New York, NY 387 p.
ISBN 0-521561855X(hb)   Q125.G39

CDOSB   Concise Dictionary of Scientific Biography. 2000. American Council of Learned Societies, Charles Scribner's Sons, New York, NY 1097 p.
ISBN 0-684-80631-2   Q141.C55

CEOS&T   Concise Encyclopedia of Science and Technology. 1989. 1994. 1998. Parker, S. P. (ed.). McGraw-Hill Inc., New York, NY 2222 p.
ISBN 0-07-045512-0   Q121.M29

CFR   Channel Flow Resistance: Centennial of Manning's Formula. 1992. Yen, Ben Chie (ed.). Water Resources, Littleton, CO 453 p.
ISBN 0918334721   TC175.C36

ChemHerit Chemical Heritage. 1982ff. Chemical Heritage Foundation, Philadelphia, PA quarterly
ISSN 0736-4555   QD11.C477

COTHOS   A Chronology of the History of Science, 1450–1900. 1987. Cascoigne, Robert M. Garland Publishing, New York and London 585 p.
ISBN 0-8240-9106   Q125.G39

DAAS   Distinguished African American Scientists of the 20th Century. 1996. Kessler, James H., Kidd, J. S., Kidd, R. A., and Morin, K. A. Oryx Press, Phoenix, AZ 382 p.
ISBN 0-89774-955-3   Q141.K455

DAB   Dictionary of American Biography. 1937. 1964. 1996. Johnson, Allen (ed.). Charles Scribner's Sons, New York, NY In 20 vol. between 1928 and 1936; In 10 supp. vol. between 1944 and 1995. These volumes followed by American National Biography (ANB) in 1999
LCCN 44041895   E176.D563

DDIEI   Dictionnaire des Inventeurs et Inventions. 1966. Galiana, Thomas de and Rival, Michel. Larousse, Paris 822 p. 474
ISBN 2-03-750015-7   T39.D54

DNB   Dictionary of National Biography (British). 1885ff. Oxford University Press, London. Also DNB MP (Missing Persons). Compilation followed by Oxford Dictionary of National Biography. 2004 In 60 vol.
LCCN 95047330   DA28.D4   DA28.D525

DOANB   Dictionary of American Negro Biography. 1982. Logan, Rayford W. and Winston, Michael R. W. W. Norton & Co., New York, NY 680 p. (No index)
ISBN 0-393-015130   E185.96D53

DOI&D   Dictionary of Inventions and Discoveries. 1967. 1976. Carter, Ernest F. Crane, Russak & Co., New York, NY 208 p.
ISBN 0-8448-0867-9   T9.C335

DORS&E   The Directory of Russian Scientists and Engineers. 1994. Kholkin, Yuri I. and Mettee, Howard D. K&M Consultants, Ltd., Youngstown, OH 289 p.
ISBN 0-9645120-3-3   Q145.D56

DOSB   Dictionary of Scientific Biography. 1970. 1980. Scribners, New York, NY In 16 vol. + 2 supp. vol.
ISBN 0-684-16962-2 (set)   Q141.D5

DOT   Dictionary of Theories. 1993. Bothamley, Jennifer (ed.). Gale Research International, Detroit, MI 637 p.
ISBN 1-57859-045-0   AG5.D53

DOTHOS   Dictionary of the History of Science. 1981. Byrum, W. F., Browne, E. J. and Porter, Roy. Princeton University Press, Princeton, NJ 494 p.
ISBN 0-691-08271   Q125.B98

E-64   Engineering. 1964. Rapport, Samuel B. and Wright, Helen. Washington Square Press, Inc., New York, NY 378 p. (New York University Press in 1963).
LCCC No. 62-18019   TA15.R35

EAI   Engineers and Inventors. 1985. A book in a series of six of The Biographical Dictionary of Scientists, sometimes identified as BDOS(EAI). Abbott, David. 1986. Blond Educational of Frederick Muller Ltd., London and Peter Bedrick Books, New York, NY 189 p.
ISBN 0-584-70004-0   TA139.B56

EB   Encyclopaedia Britannica. 1961. 1998 (15th ed.) The New Encyclopaedia Britannica. Chicago, IL 29 vol. + 2 index vol.
ISBN 0-85229-633-0   AE5.E363

EDOP   Encyclopaedic Dictionary of Physics. 1961–1964. Thewlis, James (ed.). Macmillan, New York, NY 9 vol. + supp.
LCD 60-7069   QC5.E54

EE   Engineering Eponyms. 1965. 1975. Auger, Charles Peter. The Library Association, London, England 130 p. 122 p.
ISBN 0-85365-37-9   TJ9.A84

EI   Everyday Inventions. 1972. Hooper, Meredith. August & Robertson Publisher, London, England 138 p.
ISBN 0-207-12259-8   T15.H73

EITAW   Engineering in the Ancient World. 1978. Landels, J. G. University of California Press, Berkeley, CA 224 p.
ISBN 0-520-03429-5   T16.L36

EOAP   Encyclopedia of Applied Physics. 1991. Trigg, George L. (ed.). Wiley-VCH Verlag, New York, NY 23 vol. + index
ISBN 3-527-26841-3 (set)   QC5.E543

EOCT   Kirk-Othmer Encyclopedia of Chemical Technology. 1978. 1984. 1991. Grayson, Marin (ed.). John Wiley & Sons, New York, NY In 25 vol. + supp.
ISBN 0-471-52694-0   TP9.E685 475

EOET   Extraordinary Origins of Everyday Things. 1987. Panati, Charles. Harper & Row Publishers, Inc., New York, NY 463 p.
ISBN 0-06-055098-8   AG6.P37

EOM   Encyclopaedia of Mathematics. 1988-1994. Hazewinkel, M. (ed.). Kluwer Academic Publishers, Dordrecht/Berlin/London, The Netherlands In 10 vol. + 3 supp.
ISBN 1-55608-009-3 (set)   QA5.M3713

EOMEI   Encyclopedia of Modern Everyday Inventions. 2003. Cole, David J., Browning, Eve B., and Schroeder, Fred E. H. Greenwood Press, Westport, CT 285 p.
ISBN 0-313-31345-8   T20.C56

EOPS&T   Encyclopedia of Physical Science and Technology. 2002. Meyers, Robert A. Academic Press, San Diego, CA and London In 18 vol.
ISBN 0-12-227410-5   Q123.E497

EOS   Encyclopedia of Science. 1999. Facts On File, Inc., New York, NY In 2 vol.
ISBN 0-8160-4008-7   Q123.F334

EOS&T   Encyclopedia of Science and Technology. 1960. 1997 (8th ed.). Parker, Sybil P. (ed.). McGraw Hill Book Co., New York, NY In 20 vol.
ISBN 0-07-911504-7 (set)   Q121.M3

EOSTAS   Encyclopedia of Science, Technology, and Society. 1999. Volti, Rudi (ed.). Facts On File, Inc., New York, NY In 3 vol.
ISBN 0-8180-3123-1 (set)   Q121.V65

EOWB   Encyclopedia of World Biography. 1973. McGraw-Hill Book Co., New York, NY In 12 vol.
ISBN 07-079633-5   CT103.M27

EOWS   Encyclopedia of World Scientists. 2001. Oaks, Elizabeth H. Facts On File, Inc., New York, NY 450 p.
ISBN 0-8160-4130-X   Q141.O25

ET   Engineering Times. 1979ff. Monthly publication (except August and September), by the National Society of Professional Engineers, Alexandria, VA
ISSN 0195-6876   TA1.E862

FAL   Fame at Last. 2000. Ball, John C. and Jones, Jill. Andrews McMeel Publ., Kansas City, MO 407 p.
ISBN 0-7407-0940-2   CT220.B25

FEL   Famous Engineering Landmarks of the World. 1998. Berlow, Lawrence H. Oryx Pub., Phoenix, AZ 250 p.
ISBN 0-89774-966-9   TA15.B42

FETCTW   Five Equations that Changed the World. 1995. Guillen, Michael. Hyperion, New York, NY 227 p.
ISBN 0-7868-6103-7   QC24.5.G85

FFF   Famous First Facts. 1935. 1981. Kane, Joseph N. H. W. Wilson Co., New York, NY 1350 p.
ISBN 0-8242-0661-4   AG5.K315

FIOTEC   French Inventions of the 18th Century. 1952. McCloy, Shelby Thomas. University of Kentucky Press, Lexington, KY 212 p.
LCCN 52-005903   T26.F8M2

FNIE   Famous Names in Engineering. 1981. Carvill, James. Butterworth & Co., Ltd., London & Boston 93 p.
ISBN 0-408-00540-8   TA139.C37

GDIMS   Great Discoverers in Modern Science. 1955. Pringle, Patrick, et al. Harrap, London and Roy Publishers, New York, NY 206 p.
LCCN 55009317   TP39.P7

GEAPIT   Great Engineers and Pioneers in Technology. 1981. Turner, Roland and Goulden, Steven L. St. Martin's Press, New York, NY 488 p. 476
ISBN 0-312345747   TA139.G7

GI   Great Inventions. 2004. Dyson, James and Uhlig, Robert (ed.). Barnes & Noble Books, New York, NY 496 p. First published in UK as James Dyson's History of Great Inventions. 2001.
ISBN 0-7607-6141-8   Several books under this title in T13 to T48 series

GLOSB   Grolier Library of Science Biographies. 1997. Daintith, John and Gjertsen, Derek (ed.). Grolier Educational, Danbury, CT 10 vol.
ISBN 0-7172-7626-0   Q141.G84

GMI   Great Modern Inventions. 1991. Messadie, Gerald. Chambers, Edinburgh and New York, NY 236 p.
ISBN 0550-17001-4   T18.M4713

GO-FYI   Good Old-Fashioned Yankee Ingenuity. 1990. Harris, Henry. Scarborough House Publ., Inc., Chelsea, MI 304 p.
ISBN 0-8128-3142-X   E156.H375

GP   Great Projects. 2001. Tobin, James, Free Press, NY 322 p.
ISBN 07443210646   TA23.T63

GSA   Great Scientific Achievements—The Twentieth Century, 1994-1997. Smith, Roger (ed.). Salem Press, Pasadena, CA In 14 vol.
ISBN 0-89356-414-1 (set)   Q121.G72

GSEI&D   Groundbreaking Scientific Experiments, Inventions & Discoveries. 2002. Windelspecht, Michael. Greenwood Press, Westport, CT 270 p.
ISBN 0-313-31501-9   Q125.W7917

GTHP   A Guide to Hart-Parr, Oliver and White Tractors, 1901-1996. 1997. Gay, Larry. American Society of American Engineers, St..Joseph, MI 95 p.
ISBN 0929355873   TL233.6.H37G38

HBW   Human-Built World. 2004. Hughes, Thomas P. University of Chicago Press, Chicago, IL 223 p.
ISBN 0226359336   T14.5.H84

HFP   Historical First Patents. 1994. Brown, Travis. The Scarecrow Press, Inc., Metuchen, NJ 216 p.
ISBN 0-8108-2898-7   T223.P2B76

HIW   How It Works—The New Illustrated Science and Invention Encyclopedia. 1987. H. S. Stuttman, Inc., Pub., Westport, CT 3628 p. In 26 vol. + supp.
ISBN 0-87475-450-X   Q123.H73

HIW-E   How It Works—An Illustrated Encyclopedia of Science and Technology. 1978. Hancock, Ralph (ed.). Marshall Cavendish Ltd., New York and London In 22 vol.
ISBN 0-85685-522-7 (set)   T9.H74

HOCE   History of Chemical Engineering. 1980. Furter, William F. (ed.). American Chemical Society, Washington, DC 435 p.
LCCN 80-017432   QD1.A355 no. 190

HOCS&T   Handbook of Current Science & Technology. 1996. Bunch, Bryan. Gale Publishing Co., New York, NY 834 p.
ISBN 0-81031-9552-5   Q158.F.B84

HOEAT   A History of Engineering and Technology: Artful Methods. 1999. Garrison, Ervan G. CRC Press, Boca Raton, FL 347 p.
ISBN 0-8493-9810-X   TA15.G37

HOEIC   A History of Engineering in Classical and Medieval Times. 1984. Hill, Donald. Routledge Open Court Publishing Co., La Salle, IL 263 p.
ISBN 0-87548-4220   TA16.H55

HOFL   History of Formal Logic. 1961. 1970. Bochenski, I. M. Translated and edited from German to English by Ivo Thomas. English edition published by Univ. of Notre Dame Press, Notre Dame, IN and 477 Chelsea Publ., New York, NY 567 p.
LCCN 58-14183 (lst ed.)   BC15.B643

HOHT   History of Heat Transfer. 1988. Layton, Edwin T. and Lienhard, John H. (ed.). American Society of Mechanical Engineers, New York, NY 260 p.
LCCN 88179515   TJ260.H57

HOWRD   History of Water Resource Division of USGS—The Years of Change. 1996. Hudson, Hugh H., et al. United States Geological Survey, Washington, DC 559 p.
LCCN 90003638   GB701.H48

IAD   Inventors and Discoverers. 1998. Newhouse, Elizabeth L., (ed.). National Geographic Society, Washington, DC 320 p.
ISBN 0-870447513   T18.I57

IAI   Inventions and Inventors. 2002. Smith, Roger. Salem Press, Pasadena, CA 936 p. In 2 vol.
ISBN 1-58765-018-5   T20.I99

IAOSTI   Illustrated Almanac of Science, Technology and Invention. 1997. Francis, Raymond L. Plenum Press, New York, NY unpaged
ISBN 0-306-45633-8   Q158.5.F73

I&T   Invention & Technology—see AHOI&T (American Heritage of Invention & Technology)

IB   Invention Book. 1985. Caney, Steven. Workman Publishing Co., New York, NY 207 p.
ISBN 0-89480-076-0 (pb)   T48.C3

IIA   Invention in America. 1996. Bourne, Russell. Fulcrum Publishing Co., Golden, CO 138 p.
ISBN 1-55591-23-1   T21.B693

I&T   Invention and Technology, American Heritage of. 1985ff. An affiliate of Forbes, Inc., New York, NY quarterly
ISSN 8756-7296   T1.A455

IrrAge   Irrigation Age. 1891ff. Irrigation Age Co. and other publishers, Chicago, IL serial—no longer published
TC801.I7

ITCOF   In the Cause of Flight: Technologists of aeronautics and astronautics. 1981. Wolko, Howard S. Smithsonian Institution Press, Washington, DC 121 p.
LCCN 80-19749   TL539.W67

ITNC   Inventing the Nineteenth Century. 2001. Van Dulken, Stephen. New York University Press, New York, NY 218 p.
ISBN 0-8147-8810-6   T19.V36

ITTC   Inventing the 20th Century—100 Inventions that shaped the world. 2000. Van Dulken, Stephen. New York University Press, New York, NY 246 p.
ISBN 0-8147-8808-4   T20.V36

IWAP   In War and Peace. 2002. Stever, H. Guyford. John Henry Press, Washington, DC 382 p.
ISBN 0-309-08411-3   Q143.S743A3

IWWIE   International Who's Who in Engineering. 1984. Kay, Ernest (ed.). International Biographical Centre, Cambridge, England 589 p.
ISBN 0-900332-71-9   TA12.I575

L&M   Laws and Models. 1999. Hall, Carl W. CRC Press LLC, Boca Raton, FL 524 p.
ISBN 0-8493-2018-6   Q40.H35

LAI   Leading American Inventors. 1912. 1968. Iles, George. Henry Holt & Co., New York, NY and Books for Libraries Press, Freeport, NY 447 p.
LCCC 68008472   T39.I5 478

LATI   Los Angeles Times Index. 1972ff. University Microfilms, Ann Arbor, MI annual
ISSN 0742-4817   A121.L65N49

LDOS   Larousse Dictionary of Scientists. 1994. Muir, Hazel. Larousse Kingfisher Chambers, New York and Edinburgh 595 p.
ISBN 0-7523-0002-4   Q141.L36

LIME   Landmarks in Mechanical Engineering. 1997. History and Heritage Committee of ASME, New York, NY 364 p.
ISBN 1-55753-093-9   TJ23.L35

LOTCS   Luminaries of the Chemical Sciences. 2002. American Chemical Society, Washington, DC 146 p.

LOTE   Lives of the Engineers. 1861/1862. 1874. 1904. 1968. Smiles, Samuel. John Murray Publ., London and A. M. Kelley Publisher, New York, NY In 3 vol. (1968)
LCCN 68026161   TA139.S65

MADOI   Mothers and Daughters of Invention. 1993. Stanley, Autumn. Scarecrow Press, Metuchen, NJ and London 1116 p.
ISBN 0-8108-2586-4   T36.S73

ME   Mechanical Engineering (supplemented by ASME News). 1906ff. American Society of Mechanical Engineers, New York, NY monthly (recently)
ISSN 0025-6501   TJ1.A72

MEI   More Everyday Inventions. 1976. Hooper, Meredith. Angus and Robertson, London, England 127 p.
ISBN 0207956065   T15.H74

MEIA   Mechanical Engineers in America—born prior to 1861. A Biographical Dictionary. 1980. American Society of Mechanical Engineers, New York, NY 330 p.
LCCC No. 79-57364   TJ139.A47

MEOP, EOP   Macmillan Encyclopedia of Physics. 1996. Rigden, John S. (ed.). Simon & Schuster Macmillan, New York, NY In 4 vol.
ISBN 0-02-897359-3 (set)   QC5.M15

MISAT.   Milestones in Science and Technology. 1987. 1994. Mount, Ellis and List, Barbara A. Oryx Press, Phoenix, AZ 206 p.
ISBN 0897746716   Q199.M68

MMAH   Men, Machines, and History. 1948. 1966 (2nd ed.). Lilley, Samuel. International Publishers, New York, NY 352 p.
LCD 66-21951   T15.L5

MMOS   Modern Men of Science. 1966. 1968. Greene, Jay E. McGraw-Hill Book Co., New York, NY In 2 vol. This book followed by MSAE. 1980.
LCCN 66014808   Q141.M15

MNCS   Mid-Nineteenth Century Scientists. 1969. North, John David (ed.). Pergamon Press, New York, NY 190 p.
ISBN 0080132375   Q141.M43

MPSI   Macmillan Profiles, Scientists and Inventors. 1993. 1998. Culligan, Judy (ed.). Simon & Schuster, Inc., New York, NY 389 p.
ISBN 0-02-864983-4   Q141.S298

MSAE   Modern Scientists and Engineers. 1980. Parker, S. P. (ed.). McGraw-Hill Book Co., New York, NY In 3 vol.
ISBN 0-07-045266-0   Q141.M15

MT   Memorial Tributes, National Academy of Engineering (NAE). 1979ff. National Academy Press, Washington, DC Periodic publication
ISBN 0-309-02889-2   ISSN 1075-8844   TA139.N34a 479

MWBD   Merriam-Webster's Biographical Dictionary (including Webster's Biographical Dictionary). 1959. 1988. 1995. Merriam-Webster, Inc., Springfield, MA 1170 p.
ISBN 0-87779-743-9   CT103.M47

MWGD   Webster's Geographical Dictionary. 1969. G. and C. Merriam Co., Springfield, MA 1293 p.
LCD 70001646   G103.W45

MWPI   Men Who Pioneered Inventions. 1969. Poole, Lynn and Gray. Dodd, Mead and Co., New York, NY 207 p.
LCCN 69-017604   TC39.P64

NAW   Notable American Women. 1980. Sicherman, Barbara and Green, Carol Hurd. Belknap Press of Harvard Univ. Press, Cambridge, MA 773 p.
ISBN 0-674-62732-6   CT3260.N573

NBAS   Notable Black American Scientists. 1999. Krapp, Kristine. Gale Research, Farmington Hills, MI 349 p.
ISBN 0-7876-2788-5   Q141.N726

NC   The National Cyclopedia of American Biography. 1899 to 1984. James T. White & Co., Clifton, NJ In 63 vol. + A through M vol.
LCCC No. 21-21756   E176.E588

NOAIMA   Negroes of Achievement in Modern America. 1970. Flynn, James J. Dodd, Mead & Co., New York, NY 273 p.
LC 70-111911   E185.96.F55

NS   Notable Scientists: From 1900 to the Present. 2001. Narins, Brigham (ed.). The Gale Group, Farmington, MI 2633 In 5 vol.
ISBN 0-7876-1751-2 (set)   Q141.N728

NTBOT   New Trail Blazers of Technology. 1976. Manchester, Harland Frank. Scribner, New York, NY 214 p.
ISBN 0684147181   T39.M23

NT-CS   Notable Twentieth-Century Scientists. 1995. McMurray, Emily J. (ed.). Gale Research, Inc., New York, NY In 4 vol.
ISBN 0-8103-9181-3 (set)   Q141.N73

NUC   National Union Catalog (pre-1956). 1968–1981. Mansell, London and American Library Association, Chicago, IL In 754 vol.
LCCN 67-30001   Z881.A.U518

NWITPS   Notable Women in the Physical Sciences. 1997. Shearer, Benjamin F. and Barbara S.
Greenwood Press, Westport, CT 479 p.
ISBN 0-313-29303-1   Q141.N734

NWS   Notable Women Scientists. 1999. Proffitt, Pamela (ed.). Gale Group, Farmington, MI 668 p.
ISBN 0-7876-3900-1   Q141.N736

NYPL   New York Public Library, Desk Reference. 1998. Simon & Schuster-Macmillan Co., New
York, NY 1040 p.
ISBN 0-02-862169-7   AG6.N49

NYPLSDF   The New York Public Library Science Desk Reference. 1995. Barnes-Svarney, Patricia
(ed.). Stonesong Press at Macmillan of Simon & Schuster, New York, NY 668 p.
ISBN 0-02-860403   Q173.B25

NYT-A   New York Times Almanac—2000. 1999. Wright, John W., et al. Penguin, New York, NY 1004 p.
ISBN 0-14-05-1457-0   AY67.N5N48

NYTObit   The New York Times Obituaries Index. I. 1858–1968. II. 1969–1978. The New York Times,
New York, NY. 1970. 1980. 1136 p., 131 p. 480
ISBN 0-667-00598-6   CT213.N47

NYTBS   The New York Times Biographical Service. 1990ff. Published monthly by University Microfilms (UMI), Ann Arbor, MI (ceased publication in 2001)
ISSN 0161-2433   CT120.N45

OED   The Oxford English Dictionary. 1923. 1989. Murray, James A. H. , et al. (1923), Simpson, J. A.
and Weiner, E. S. C. (1989) Clarendon Press, Oxford, England In 12 vol. (1923), 20 vol. (1989)
ISBN 0-19-861186 (set)   PE1625.O87

OFTT   Obituaries from the Times. 1951–1975. The Times (of London). Newspaper Archives Development Ltd., Reading, England
ISBN 0-903713969   CT120.Q16

1000Y   1000 Years, 1000 People. 1998. Gottlieb, Agnes Hooper, et al., Kodansha America, New York,
NY 331 p.
ISBN 1-56836-253-6   CT103.A16

PA   Peoples Almanac, the 20th Century. 1999. Wallechinsky, David. The Overlook Press, New York,
NY 921 p.
ISBN 0-87951-944-4   AY64.W27

Pharm Cent   Pharmaceutical Century. 2000. American Chemical Society, Washington, DC 253 p.
(Supplement to ACS Pub.)

PHOAI   A Popular History of American Invention. 1924. 1975. Kaempffert, Waldemar. AMS Press,
New York, NY 2 vol.
ISBN 0404119212   T21.K5

PopMech   Popular Mechanics Magazine. 1902ff. Popular Mechanics Co., Chicago, IL monthly serial
ISBN 0032-4558   T1.P77

PopSci   Popular Science. 1872ff. Popular Science Publishing Co., New York, NY monthly serial
(two volumes/yr.)
ISSN 0161-7370   AP2.P8

Prism   ASEE Prism. 1991ff. American Society for Engineering Education, Washington, DC (serial)
ISSN 1056-8077   P61.A78

PS   Prominent Scientists. 1980. 1994 (3rd ed.). Pelletier, Paul. Neal-Schuman Publ., New York, NY 353 p.
ISBN 1-53570-114-0   Q141.P398

PSOCE   Prominent Scientists of Continental Europe. 1968. Turkevich, John and Ludmille, B. American Elsevier Publishing. Co., New York, NY 204 p.
ISBN 0444000461   Q141.T82

Res   Resource. 1994ff. American Society of Agricultural Engineers, St. Joseph, MI monthly
ISSN 1076-3333   S671.R4

RG   Readers' Guide to Periodical Literature. 1900ff. H. W. Wilson Co. Publ., New York, NY now annual publication
ISSN 0034-0464   A13.R48

RHW   Random House Webster's Dictionary of Scientists. 1997. Jenkins-Jones, Sara (ed.). Random House, New York, NY 552 p.
ISBN 0-375-70057-9   Q141.H88

ROT   Rules of Thumb for Engineers and Scientists. 1991. Fisher, David J. Trans Tech Publ., Ltd., Gulf Publishing Co., Houston, TX 242 p.
ISBN 0-872017869   Q199.F57

RS RSObit   Biographical Memoirs of the Fellows of the Royal Society (GB) (1955–1988), London, England vol. 1-33. Q41.L8476. Obituary notices of fellows of the 481 Royal Society (GB). 1932–1954. Harrison & Sons, London   Q41.L8474

SACIC   Science and Civilization in Ancient China. 1954ff. Needham, Joseph. Watts, London, England. Series of publications. University Press, Cambridge, England LCD 54-4723   DS721.N39 and the shorter version by Ronan, Colin and Needham, Joseph. In 3 vol.
ISBN 0-521-25272-5   DS721.N392

SAI   Scientists and Inventors. 1978. 1986. Feldman, Anthony and Ford, Peter. Facts On File, New York, NY 336 p.
ISBN 0871964104   T39.F37

SAIAD   Scientific American Inventions and Discoveries. 2004. Carlisle, Rodney. John Wiley & Sons, New York, NY 502 p.
ISBN 0-471-24410-4   T15.C378

SAI-MP   Scientists and Inventors—Macmillan Profiles. 1998. Culligan, Judy (ed.). Simon & Schuster Macmillan, New York, NY 389 p.
ISBN 0-02-864983-4   Q141.S294

S&TA   Science and Technology Almanac. 2002. Lauerman, Lynn. Greenwood Press, Westport, CT 513 p.
ISBN 1-57356-328-5   Q173.S395

S&TF SATF   Science and Technology Firsts. 1997. Bruno, Leonard C. Gale, Detroit and New York 636 p.
ISBN 0-7876-0256-6   T15.B684

SciAm   Scientific American. 1845ff. Scientific American, Inc., New York, NY monthly serial
ISSN 0036-8733   T1.S5 T1.S53

Science   Science. 1883ff. American Association for the Advancement of Science (AAAS). Washington, DC weekly periodical
ISSN 0036-8075   Q1.S35

SFLOE   Selections from Lives of the Engineers by Samuel Smiles. 1966. Hughes, Thomas P. (ed.). MIT Press, Cambridge, MA and London. 447 p.
LCCN 66019360   TA139.S66   (See LOTE)

SIBP   Scientists in the Black Perspective. 1974. Young, Herman A. and Young, Barbara H. Lincoln Foundation, Louisville, KY 185 p.
LCCN 74-190654   Q147.Y68

SMOS   Soviet Men of Science. 1963. Turkevich, John D. D. Van Nostrand Co., Inc., New York, NY
        441 p.
        ISBN 0837182468   Q141.T83

SS1660   Scientists Since 1660—A Bibliography of Bibliographies. 1997. Howssan, Leslie, Ashgate
        Publ., Aldershot, England and Brookfield, VT 150 p.
        ISBN 1-859280358   Z7401.H66

S:TLAW   Scientists: The Lives and Works of 150. 1996. Saar, Peggy and Allison, Stephen (ed.). Gale
        Group, Farmington Hills, MI In 5 vol.
        ISBN 0-7876-0959-5   Q141.S3717

TAAE   The African American Encyclopedia. 1993. Williams, Michael W. (ed.). Marshall Cavendish
        Publ., North Baltimore, NY In 8 vol.
        ISBN 1-85435-545-7   E185.A253

TAE   The Ancient Engineers. 1963. DeCamp, L. Sprague. Ballantine Books, Random House, NY 450 p.
        ISBN 0-345-29347-9   TA16.D4

TAOSAT   The Almanac of Science and Technology. 1990. Golub, Richard and Brus, Eric. HBJ Pub-
        lishers, Boston, MA 530 p. 482
        ISBN 0-151050503   Q158.A47

TAT   The Agricultural Tractor. 1954. 1975. Gray, Roy B. American Society of Agricultural Engineers,
        St. Joseph, MI In 2 parts
        LCCN 76381575   S711.G7

TBDOS   The Biographical Dictionary of Scientists. 1994. 2000. Porter, Ray and Ogilvie, Marilyn.
        Oxford University Press, New York, NY In 2 vol.
        ISBN 0-19-521663-6   Q141.B528

TBOF   The Book of Firsts. 1974. 1982. Robinson, Patrick. Branhall House, New York, NY 512 p.
        ISBN 0-517-216795   AG243.R54

TCBE CBE   Cambridge Biographical Encyclopedia. 1998 (2nd ed.). Crystal, David (ed.). Cambridge
        University Press, Cambridge, England 1179 p.
        ISBN 0-521-63099-1   CT103.C26

TCDOS   The Cambridge Dictionary of Scientists. 1996. Millar, David, Ian, John and Ian, Margaret
        (ed.). Cambridge University Press, Cambridge, England 387 p.
        ISBN 0-521-56185X(hb)   Q141.C128

TCPONT   The Complete Patents of Nikola Tesla. 1944. Glenn, Jim (ed.). Barnes and Noble Books,
        New York, NY 535 p.
        ISBN 1566192668   TK257.T45

TechRev   Technology Review: MITs Magazine of Innovation. 1899ff. Massachusetts Inst. of Technol-
        ogy, Cambridge, MA now 10X/yr
        ISSN 0040-1692   T171.M47

TEE   Through Engineering Eyes. 1941. Cullimore, Allan Reginald. Pitman Publishing Corp., New
        York, NY 166 p.
        LCCN 42-005148   T185.C8

TEOUT   The Evolution of Useful Things. 1992. Petroski, Henry. Alfred A. Knopf, Inc., New York,
        NY 288 p.
        ISBN 0-679-41226-3   T212.P465

TFE&T   Twenty-Five Engineers and Inventors. 1976. Susskind, Charles. San Francisco Press, Inc.,
        San Francisco, CA 122 p.
        ISBN 0-911302-29-8   TA139.S95

TFM   To Fathom More: African American Scientists and Inventors. 1996. Jenkins, Edward Sidney. University Press of American, Lanham, MD 399 p.
ISBN 0-7618-0214-2   Q141.J46

TGE:B   The Great Engineers: The Art of British Engineers (1837–1987). 1987. Walker, Derek. Academy, London and St. Martin's Press, New York, NY 288 p.
ISBN 0312011369   TA57.G74

TGH   The Grain Harvesters. 1978. Quick, Graeme R. and Buchele, Wesley F. American Society of Agricultural Engineers, St. Joseph, MI 269 p.
ISBN 0916150135   S695.Q5

TGS   The Great Scientists. 1989. Magill, Frank N. (ed.). Grolier Educational Corporation, Danbury, CT In 12 vol.
ISBN 0-7172-7134X   Q141.G767

TI   Times Index (London). 1951ff. Newspaper Archives Developments, Reading, England
ISBN 0-903713-97-7   CT120.Q16

TIA   Technology in America. 1999. Marcus, Alan I. and Segal, Hoard P. Harcourt Brace & Co., Ft. Worth, TX 400 p.
ISBN 0-15505531-3   T21.M372

TIY   The Ingenious Yankees. 1976. Gies, Joseph and Gies, Frances. Drowell, New York, NY 376 p. 483
ISBN 0690011504   T21.G5

TIY(2)   The Inventive Yankee. 1989. Chesman, Andrea (ed.). Yankee Publishing, Inc., Dublin, NY 223 p.
ISBN 0-89909-172-5   T39.I58

TLOTE   The Lives of the Engineers. 1968. Smiles, Samuel. Augustus M. Kelley Publ., New York, NY In 3 vol.
ISBN 0715342797   TA139.S65

TMA   They Made America. 2004. Evans, Harold with Gail Buckland and David Lefer. Little, Brown, Inc., New York, NY 498 p.
LCCN 2003065954   T39.E83

TNAS   The National Academy of Sciences: The First Hundred Years, 1863–1963. 1978. Cochrane, Rexmond C. (ed.). National Academy of Sciences, Washington, DC 694 p.
ISBN 0-309-02518-4   Q11.N2862C6

TNISAIE   The New Illustrated Science and Invention Encyclopedia: How It Works. 1987. H. S. Stuttman Publ., Westport, CT 3628 p. In 26 vol.
ISBN 087475450X   Q123.H73

TodChem   Today's Chemist at Work. 1992–2004. Chemical & Engineering News Magazine Group, American Chemical Society, Washington, DC monthly
ISSN 1062-094   QD1.T59

T100   The 100 Most Influential People. 1992. 1998. Hart, Michael H. Carol Publishing Group, Citadel Press, Secaucus, NJ 356 p.
ISBN 0806513500 (pb)   CT105.H32

TOIAD   Timetable of Inventions and Discoveries. 1986. Desmond, Kevin. M. Evans & Co., Inc., New York, NY (unpaged)
ISBN 0-87131-520-3   T15.D44

TP   Tuxedo Park. 2002. Conant, Jennet. Simon & Schuster, New York, NY 330 p.
ISBN 0-684-87287-0   QC16.L647C66

T7-AP   The 7-Ability Plan. 1989. Skromme, Arnold. The Self-Confidence Press, Moline, IL 146 p.
ISBN 0-962-3508-0-X   LB1590.5S57

TS100   The Scientific 100. 1996. Simmons, John. Citadel Press, Secaucus, NJ 504 p.
ISBN 0-8065-1749-2   Q148.S55

TSOE   The Story of Engineering. 1960. Finch, James Kip. Anchor Books, Doubleday & Co., Inc., Garden City, NY 528 p
LCCC No. 60-10669   TA15.F57

TTOS   Timetables™ of Science. 1988. Hellemans, Alexander and Bunch, Bryan. Simon & Schuster, New York, NY 660 p.
ISBN 0-671-62130-0   Q125.H557

TTOT   Timetables™ of Technology. 1993. Bunch, Bryon and Hellemans, Alexander. Simon & Schuster, New York, NY 490 p.
ISBN 0-671-76918-9   T15.B73

TSWABOF   The Second World Almanac Book of Inventions. 1986. Giscard d'Estaing, Varlerie-Anne. World Almanac Publ., New York, NY 352 p.
ISBN 0-345-33730-1   TI5.G57

TWABOI   The World Almanac Book of Inventions. 1985. Giscard d'Estaing, Valerie-Anne. World Almanac Publ., New York, NY 362 p. 484
ISBN 0-345-32661-X   T15.G57

UEOIC   Ullmann's Encyclopedia of Industrial Chemistry. 1986. 1996. 2003. Gerhartz, Wolfgang (ed.). VCH Verlagsgesallschaft mbH, Weinheim, Germany In 30 vol (2003)
ISBN 3-527-20150-5   TP9.U57

VNSE   Van Nostrand's Scientific Encyclopedia. 1995 (8th ed.). Considine, Douglas M. Van Nostrand Reinhold Co., New York, NY In 2 vol.
ISBN 0-442-01864-9   Q121.V3

WA   The World Almanac and Book of Facts. 2000 and earlier. Pharos Books, New York, NY followed by Premedia Reference, Inc., Mahwah, NJ 1024 p. (for year 2000)
ISBN 0-88687-847-0   ISSN 0084-1382 AY71

WBE   The World Book Encyclopedia. 1990. 1999. World Book, Inc., Chicago, IL In 22 vol.
ISBN 0-7166-0100-1   LCCC No. 99-62063   AE5.W55

WBEOS   World Book Encyclopedia of Science. 1990. Zeleny, Robert O. (ed.). A supplement to WBE. World Book, Inc., Chicago, IL Also vol. 8, Men and Women of Science, 128 p.
ISBN 0-7166-3226-8   LCCC No. 90-70521

WI   Women Inventors. 1997. Casey, Susan. Chicago Review Press, Chicago, IL 142 p.
ISBN 1-55652-317-3   T39.C38

WIS(1)   Women in Science: Antiquity through Nineteenth Century. 1986. Ogilvie, Marilyn Bailey. MIT Press, Cambridge, MA & London 254 p.
ISBN 0-262-15031-X   Q141.O34

WIS(2)   Women in Science—Macmillan Profiles. 2001. Macmillan, New York, NY and Gale Group, Farmington Hills, MI 421 p.
ISBN 0-02-865502-8   Q141.W6795

WOI   World of Invention. 1999. McGrath, Kimberley A. Gale, Detroit, MI and London 1043 p.
ISBN 0-7876-2759-3   T15.W82

WOM   World of Mathematics. 2001. Narins, Brigham (ed.). Gale Group, Farmington Hills, MI In 2 vol.
ISBN 0-7876-3652-5   QA5.W67

WPI   The Washington Post Index. 1971ff. Research Publications, Woodbridge, CT, recently by ProQuest Information and Learning Co., Ann Arbor, MI annual.
ISSN 1041-1543 (recent)   A121.W33

WWAAA   Who's Who Among African Americans. 1999. Henderson, Ashyia N. and Phelps, Shirelle. Gale Group, Detroit, MI 1626 p.
ISBN 0-7876-2757-7   E185.96.W52

WWIA   Who's Who in America. 1899ff. Marquis Who's Who, Chicago, IL In 2 vol. in 2000.
ISBN 0-8379-0139-1 (1976)   E663.W56

WWIE   Who's Who in Engineering. 1977ff. American Association of Engineering Societies, Washington, DC preceded by Engineers of Distinction then published by Engineers Joint Council, New York, NY from 1970–1978, also included under this designation
ISBN 0-87615-017-2   TA139.E37

WWIE(I)   Who's Who in Engineering. 1922–1964 (9th ed.). Downs, Winfield Scott (ed.). 1931 and Dodge, Edward N. (ed.). 1964. Lewis Historical Publishing Co., New York and Palm Beach, FL
LCCN 2204132   TA139.W4 485

WWOAW   Who's Who of American Wolmen. 2000 (22nd ed.). Marquis Who's Who, Chicago, IL 1705 p.
ISBN 0-8379-0426-9   E176.W647

WWW   Who Was Who (1897–1993)—Accumulated Index (1897–1990). 1996. Black, Adam and Black, Charles (ed.). St. Martin's Press, New York, NY and A & C Black, Ltd., London In 11 vol.
ISBN 0-7136-3457-X   DA28.W65

WWW?   Who Was When? 1940. 1950. 1976. De Ford, Miriam Allen (ed.). The H. W. Wilson Co., New York, NY c. 200 p.
ISBN 0-8242-0532-4   CT103.D4

WWWIA   Who Was Who in America—1607–2000. 2000. Marquis Who's Who, Chicago, IL. St. Martin's Press, New York, NY and A & C Black, Ltd., London In 15 vol.
ISBN 0-8379-0232-0 (set)
ISSN 0146-8081   E176.W64

WWWIAH-S&T   Who Was Who in American History—Science & Technology. 1976. Marquis Who's Who, Chicago, IL 688 p.
LCCN 76005763   Q149.U5W5

WWWIS   World Who's Who in Science—From Antiquity to Present. 1968. DeBus, Allan G. (ed.). Marquis Who's Who, Chicago, IL 1855 p. LCD 68-56149   Q141.W7

Note: LCCC No. is Library of Congress Card Control Number, which was changed to LCCN now called the Library of Congress Catalogue Number. One of the Cs is sometimes used to represent computer catalogue. ISBN, International Standard Book Number (originated in Great Britain in 1965 and adopted internationally in 1969) and ISSN, International Standard Serial Number are now more consistently used to identify books.